URL	Upper Reject Limit
u	Count of nonconformities per unit
u_0	Standard or reference value; central line
$\bar{u}$	Average of nonconformities per unit
v, df	Degrees of freedom
w	Weight
W	Standardized range value
X_i	Observed value
$\bar{X}(\mu)$	Sample average or average (population mean)
$\bar{X}_o$	Standard or reference value; central line
$\bar{\bar{X}}$	Average of averages or grand average
X^*	Percent tolerance precontrol value
y	Response variable
Z	[illegible] d
$\bar{Z}$	[illegible]
α	Producer… [illegible]
β	Consumer's risk; Type II error; Weibull slope
χ	Failure rate
μ	See $\bar{X}$
Σ	"Sum of"
σ	See s
τ	Target
θ	Mean life; mean time to failure

QUALITY IMPROVEMENT

(formerly entitled Quality Control)

Ninth Edition

Dale H. Besterfield, Ph.D., P.E.

Professor Emeritus
College of Engineering
Southern Illinois University

PEARSON

Boston Columbus Indianapolis New York San Francisco Upper Saddle River
Amsterdam Cape Town Dubai London Madrid Milan Munich Paris Montreal Toronto
Delhi Mexico City São Paulo Sydney Hong Kong Seoul Singapore Taipei Tokyo

Editorial Director: Vernon R. Anthony
Acquisitions Editor: David Ploskonka
Editorial Assistant: Nancy Kesterson
Director of Marketing: David Gesell
Executive Marketing Manager: Derril Trakalo
Senior Marketing Coordinator: Alicia Wozniak
Marketing Assistant: Les Roberts
Senior Managing Editor: JoEllen Gohr
Associate Managing Editor: Alexandrina Benedicto Wolf
Production Project Manager: Maren L. Miller
Production Manager: Laura Messerly
Art Director: Jayne Conte
Cover Designer: Suzanne Duda
Cover Art: Anna Velichkovsky/Fotolia
Media Director: Karen Bretz
Full-Service Project Management: Sandeep Rawat/Aptara®, Inc.
Composition: Aptara®, Inc.
Printer/Binder: Edwards Brothers Malloy
Cover Printer: Lehigh-Phoenix Color/Hagerstown
Text Font: Minion

Credits and acknowledgments for material borrowed from other sources and reproduced, with permission, in this textbook appear on the appropriate page within the text.

Library of Congress Cataloging-in-Publication Data

Besterfield, Dale H.
Quality improvement / Dale H. Besterfield.—9th ed.
p. cm.
Includes bibliographical references and index.
ISBN-13: 978-0-13-262441-1
ISBN-10: 0-13-262441-9
1. Quality control. I. Title.
TS156.B47 2013
658.5'62—dc23

10 9 8 7 6 5 4 3 2

PEARSON

ISBN 10: 0-13-262441-9
ISBN 13: 978-0-13-262441-1

PREFACE

This book was formerly titled *Quality Control* and is now titled *Quality Improvement* to reflect the concept of improvement rather than control. In 1997 the American Society for Quality Control changed its name to the American Society for Quality as an indication of this change in philosophy. The text has been reworked to focus on the quantitative aspects of improvement. Chapters entitled "Lean Enterprise," "Six Sigma," "Experimental Design," and "Taguchi's Quality Engineering" have been added. A fundamental, yet comprehensive, coverage of statistical quality improvement concepts is provided. A practical state-of-the-art approach is stressed throughout. Sufficient theory is presented to ensure that the reader has a sound understanding of the basic principles of quality control. The use of probability and statistical techniques is reduced to simple mathematics or is developed in the form of tables and charts.

The book has served the instructional needs of technology and engineering students in technical institutes, community colleges, and universities. It has also been used by undergraduate and graduate business students. There is enough material for two courses. Professional organizations and industrial corporations have found the book an excellent training and instruction manual for production, quality, inspection, marketing, purchasing, and design personnel.

Quality Improvement begins with an introductory chapter covering quality improvement tools, which is followed by chapters "Lean Enterprise" and "Six Sigma." Readers will find that the first three chapters can be covered and comprehended before beginning to work with the statistical chapters. Subsequent chapters discuss qualitative aspects of statistical process control, fundamentals of statistics, control charts for variables, additional statistical process control (SPC) techniques for variables, fundamentals of probability, and control charts for attributes. The final group of chapters describes acceptance sampling, reliability, management and planning tools, experimental design, and Taguchi's quality engineering.

STUDENT DATA FILES

Microsoft® Excel data files are available for student use on the Companion Website at www.pearsonhighered.com/besterfield.

DOWNLOAD INSTRUCTOR RESOURCES FROM THE INSTRUCTOR RESOURCE CENTER

An online Instructor's Manual is available for this book. To access supplementary materials online, instructors need to request an instructor access code. Go to www.pearsonhighered.com/irc to register for an instructor access code. Within 48 hours of registering, you will receive a confirming e-mail including an instructor access code. Once you have received your code, locate your text in the online catalog and click on the Instructor Resources button on the left side of the catalog product page. Select a supplement, and a login page will appear. Once you have logged in, you can access instructor material for all Prentice Hall textbooks. If you have any difficulties accessing the site or downloading a supplement, please contact Customer Service at http://247pearsoned.custhelp.com/.

I am indebted to the publishers and authors who have given permission to reproduce their charts, graphs, and tables. I thank the following for reviewing the manuscript: Hans Chapman, Morehead State University; Thomas C. DeCanio, New York Institute of Technology; Dr. Steve Dusseau, Indiana Institute of Technology; Susan Ely, Ivy Tech Community College; Dan Fields, Eastern Michigan University; Dana Johnson, Michigan Technological University; Ali E. Kashef, The University of Northern Iowa; Dale Schrimshaw, Northeastern State University; Carrie Steinlicht, Ph.D., South Dakota State University; Amer Zaza, Alabama A&M University; and Donna Zimmerman, Ivy Tech Community College. I am also grateful to the Chinese and Spanish translators who converted the English text into those languages. Professors, practitioners, and students throughout the world have been most helpful in pointing out the need for further clarification and additional material in this ninth edition.

Dale H. Besterfield

CONTENTS

CHAPTER
ONE

INTRODUCTION TO QUALITY IMPROVEMENT

OBJECTIVES

Upon completion of this chapter, the reader is expected to

- be able to define quality, quality control, quality improvement, statistical quality control, quality assurance, and process;
- be able to describe FMEA, QFD, ISO9000, ISO14000, Benchmarking, TPM, Quality by Design, Products Liability, and IT.

INTRODUCTION

Definitions

When the term *quality* is used, we usually think of an excellent product or service that fulfills or exceeds our expectations. These expectations are based on the intended use and the selling price. For example, a customer expects a different performance from a plain steel washer than from a chrome-plated steel washer because they are different grades. When a product surpasses our expectations, we consider that quality. Thus, it is somewhat of an intangible based on perception.

Quality can be quantified as follows:

$$Q = \frac{P}{E}$$

where Q = quality
P = performance
E = expectations

If Q is greater than 1.0, then the customer has a good feeling about the product or service. Of course, the determination of P and E will most likely be based on perception, with the organization determining performance and the customer determining expectations. Customer expectations are continually becoming more demanding.

The American Society for Quality (ASQ) defines *quality* as a subjective term for which each person or sector has its own definition. In technical usage, *quality* can have two meanings: the characteristics of a product or service that bear on its ability to satisfy stated or implied needs, or a product or service that is free of deficiencies.[1]

A more definitive definition of quality is given in ISO 9000. It is defined there as the degree to which a set of inherent characteristics fulfills requirements. *Degree* means that quality can be used with adjectives such as *poor, good,* and *excellent. Inherent* is defined as existing in something, especially as a permanent characteristic. *Characteristics* can be quantitative or qualitative. *Requirement* is a need or expectation that is stated; generally implied by the organization, its customers, and other interested parties; or obligatory.

Quality control is the use of techniques and activities to achieve and sustain the quality of a product or service. *Quality improvement* is the use of tools and techniques to continually make the product or service better and better.

Statistical quality control (SQC) is the collection, analysis, and interpretation of data for use in quality activities. *Quality assurance* is all the planned or systematic actions necessary to provide adequate confidence that a product or service will satisfy given requirements for quality. It involves making sure that quality is what it should be. This includes a continuing evaluation of adequacy and effectiveness with a view to having timely corrective measures and feedback initiated where necessary.

A *process* is a set of interrelated activities that uses specific inputs to produce specific outputs. The output of one process is usually the input to another. Process refers to both business and production activities. *Customer* refers to both internal and external customers, and *supplier* refers to both internal and external suppliers.

QUALITY IMPROVEMENT TOOLS

Quality improvement is not the responsibility of any one person or functional area; it is everyone's job. It includes the equipment operator, the keyboard operator, the purchasing agent, the design engineer, and the president of the company. There are many improvement tools to assist the organization and individuals to improve their product or service. Those

[1]Dave Nelson and Susan E. Daniels, "Quality Glossary," *Quality Progress* (June 2007): 39–59.

provided in this book are check sheets, Pareto diagram, cause and effect diagram, process map, run chart, statistics, control charts, probability, gauge repeatability and reproducibility (GR&R), acceptance sampling, reliability, management and planning tools, experimental design, and Taguchi's quality engineering. A brief description of those tools not covered follows.

Failure Mode and Effect Analysis (FMEA)

FMEA is an analytical technique (a paper test) that combines the technology and experience of people in identifying foreseeable failure modes of a product, service, or process, and planning for its elimination. In other words, FMEA can be explained as a group of activities intended to

- recognize and evaluate the potential failure of a product, service, or process and its effects;
- identify actions that could eliminate or reduce the chance of the potential failure occurring;
- document the process.

FMEA is a before-the-event action requiring a team effort to alleviate most easily and inexpensively changes in design and production. There are two types of FMEA: Design FMEA and Process FMEA.

Quality Function Deployment (QFD)

QFD is a system that identifies and sets the priorities for product, service, and process improvement opportunities that lead to increased customer satisfaction. It ensures the accurate deployment of the "voice of the customer" throughout the organization, from product planning to field service. The multifunctional team approach to QFD improves those processes necessary to provide goods and services that meet or exceed customer expectations.

The QFD process answers the following questions:

1. What do customers want?
2. Are all wants equally important?
3. Will delivering perceived needs yield a competitive advantage?
4. How can we change the product, service, or process?
5. How does an engineering decision affect customer perception?
6. How does an engineering change affect other technical descriptors?
7. What is the relationship to parts deployment, process planning, and production planning?

QFD reduces start-up costs, reduces engineering design changes, and, most important, leads to increased customer satisfaction.

ISO 9000

ISO stands for International Organization for Standards. The 9000 series is a standardized Quality Management System (QMS) that has been approved by over 100 countries. It consists of three standards: (1) ISO 9000, which covers fundamentals and vocabulary; (2) ISO 9001, which is the requirements; and (3) ISO 9004, which provides guidance for performance improvement. The latest revision of ISO 9000 occurred in the year 2008, hence the designation ISO 9001:8000.

The five clauses of QMS are continual improvement; management responsibility; resource management; product/service realization; and measurement, analysis, and improvement. These five clauses are related to customer requirements and customer satisfaction.

ISO 14000

ISO 14000 is the international standard for an environmental management system (EMS). It provides organizations with the EMS elements that can be integrated into other management systems to help achieve environmental and economic goals. The standard describes the requirements for registration and/or self-declaration of the organization's EMS. Demonstration of successful implementation of the system can be used to assure other parties that an appropriate EMS is in place. ISO 14000 was written to be applicable to all types and sizes of organizations and to accommodate diverse geographic, cultural, and social conditions. The requirements are based on the process and not on the product or service. It does, however, require commitment to the organization's EMS policy, applicable regulations, and continual improvement.

The basic approach to EMS begins with the environmental policy, which is followed by planning, implementation, and operation; checking and corrective action; and management review. There is a logical sequence of events to achieve continual improvement. Many of the requirements may be developed concurrently or revisited at any time. The overall aim is to support environmental protection and prevention of pollution in balance with socioeconomic needs.

Benchmarking

It is the search for industry's best practices that leads to superior performance. Benchmarking is a relatively new way of doing business that was developed by Xerox in 1979. The idea is to find another company that is doing a particular process better than your company and then, using that information, improve your process. For example, suppose a small company takes 15 hours to complete a payroll for 75 people, whereas the local bank takes 10 hours to complete one for 80 people. Because both processes are similar, the small company should find out why the bank is more efficient in its payroll process.

Benchmarking requires constant testing of internal processes against industry's best practices. It promotes teamwork by directing attention to business practices as well as production to remain competitive. The technique is unarguable—if another company can do a particular process or practice better, why can't our company? Benchmarking allows a company to establish realistic and credible goals.

Total Productive Maintenance (TPM)

TPM is a technique that utilizes the entire workforce to obtain the optimum use of equipment. There is a continuous search to improve maintenance activities. Emphasis is placed on an interaction between operators and maintenance to maximize uptime. The technical skills in TPM are daily equipment checking, machine inspection, fine-tuning machinery, lubrication, troubleshooting, and repair.

Quality by Design

Quality by design is the practice of using a multidisciplinary team to conduct product or service conceptions, design, and production planning at one time. It is also known as simultaneous engineering or parallel engineering. The team is composed of specialists from design engineering, marketing, purchasing, quality, manufacturing engineering, finance, and the customer. Suppliers of process equipment, purchased parts, and services are included on the team at appropriate times.

In the past, the major functions would complete their task—"throw it over the wall" to the next department in the sequence—and not be concerned with any internal customer problems that may arise. Quality by design requires the major functions to be going on at the same time. This system provides for immediate feedback and prevents quality and productivity problems from occurring.

The major benefits are faster product development, shorter time to market, better quality, less work-in-process, fewer engineering change orders, and increased productivity. Design for Manufacturing and Assembly (DFMA) is an integral part of the process.

Products Liability

Consumers are initiating lawsuits in record numbers as a result of injury, death, and property damage from faulty product or service design or faulty workmanship. The number of liability lawsuits has skyrocketed since 1965. Jury verdicts in favor of the injured party have continued to rise in recent years. The size of the judgment or settlement has also increased significantly, which has caused an increase in product liability insurance costs. Although the larger organizations have been able to absorb the judgment or settlement cost and pass the cost on to the consumer, smaller organizations have occasionally been forced into bankruptcy. Although injured consumers must be compensated, it is also necessary to maintain viable manufacturing entities.

Reasons for injuries fall generally into three areas: the behavior or knowledge of a user, the environment where the product is used, and whether the factory has designed and constructed the product carefully using safety analysis and quality control. The safety and quality of products has been steadily improving. Organizations have met the challenge admirably—for instance, using safety glass where previously glass shards caused many severe injuries, placing safety guards around lawn mower blades to prevent lacerations and amputations, redesigning hot water vaporizers to reduce the risk of burns to children, and removing sharp edges on car dashboards to minimize secondary collision injuries.

Resources are limited; therefore, the perfect product or service is, in many cases, an unattainable goal. In the long term, customers pay for the cost of regulations and lawsuits. It is appropriate to mention the old cliché, "An ounce of prevention is worth a pound of cure." An adequate prevention program can substantially reduce the risk of damaging litigation.

Information Technology (IT)

IT is a tool like the other tools presented in this textbook. And, like the other tools, it helps the organization achieve its goals. Over the past few decades, computers and quality management practices have evolved together and have supported each other. This interdependence will continue in the near future.

Information technology is defined as computer technology (either hardware or software) for processing and storing information, as well as communications technology for transmitting information.[2] There are three levels of information technology:

> *Data* are alphanumeric and can be moved about without regard to meaning.
>
> *Information* is the meaningful arrangement of data that creates patterns and activates meanings in a person's mind. It exists at the point of human perception.
>
> *Knowledge* is the value-added content of human thought, derived from perception and intelligent manipulation of information. Therefore, it is the basis for intelligent action.[3]

Organizations need to become proficient in converting information to knowledge. According to Alan Greenspan, former Chairman of the Federal Reserve, "Our economy is benefiting from structural gains in productivity that have been driven by a remarkable wave of technological innovation. What differentiates this period from other periods in our history is the extraordinary role played by information and communication technologies."[4]

---

[2]E. Wainright Martin, Carol V. Brown, Daniel W. DeHayes, and Jeffrey A. Hoffer, *Managing Information Technology,* 4th ed. (Upper Saddle River, NJ: Prentice-Hall, 2001).

[3]Kurt Albrecht, "Information: The Next Quality Revolution," *Quality Digest* (June 1999): 30–32.

[4]The Associated Press, "Information Technology Raises Productivity, Greenspan Says," *St. Louis Post-Dispatch,* June 14, 2000: p. C2.

COMPUTER PROGRAM

The student version of Microsoft's EXCEL has the capability of performing some of the calculations under the Formulas/More Functions/Statistical or Formulas/Math & TrigTabs. In addition, there are EXCEL program files on the website that will solve many of the exercises. Information on these files is given in the appropriate chapter. However, it is a good idea to solve most exercises either manually or by calculator in order to better understand the concept.

The website address is www.pearsonhighered.com/besterfield.

Bill Gates has observed: "The computer is just a tool to help in solving identified problems. It isn't, as people sometimes seem to expect, a magical panacea. The first rule of any technology used in a business is that automation applied to an efficient operation will magnify the efficiency. The second is that automation applied to an inefficient operation will magnify the inefficiency."[5]

EXERCISES

1. Visit one or more of the following organizations. Determine how they define quality and how it is controlled.
   a. Large bank
   b. Health-care facility
   c. University academic department
   d. University nonacademic department
   e. Large department store
   f. Grade school
   g. Manufacturing facility
   h. Large grocery store
2. Determine the quality improvement tools used by one of the organizations listed in Exercise 1.

[5]Bill Gates. *The Road Ahead* (New York: Viking Penguin, 1995).

CHAPTER TWO

LEAN ENTERPRISE

OBJECTIVES

Upon completion of this chapter, the reader is expected to

- know the definitions of a value-added activity and a non-value-added activity;
- know the types of waste and their categories;
- be able to describe the lean fundamentals;
- understand how to implement lean;
- be able to list at least five benefits of lean;
- be able to construct a value stream map.

INTRODUCTION

A lean enterprise is one that emphasizes the prevention of waste throughout the organization. Waste is defined as any extra time, labor, capital, space, facilities, or material that does not add value to the product or service for the customer. In contrast, a value-added activity is one that must transform the product or service; have a customer who is willing to pay for it with money, time, or other resources; and be produced correctly the first time. This concept embodies a set of principles, tools, and application methodologies that enable organizations to achieve dramatic competitive advantages in development, cost, quality, and delivery performance.

HISTORIAL REVIEW

Most lean concepts were practiced at Ford Motor Company in the 1920s. Frank and Lillian Gilbreth's concept of motion efficiency and Fredrick Taylor's scientific management principles were well known at that time. Shigeo Shingo, the legendary Japanese industrial engineering consultant, cites Taylor's work as his inspiration. Henry Ford's books described lean concepts including just-in-time (JIT) inventory control, authority to stop the production line, time and motion study, standardized work methods, and waste elimination.[1] Eiji Toyota visited the big three Detroit automakers in 1950 to learn about U.S. automobile production. He was amazed at the output per day. At that time, Toyota was producing 40 automobiles a day and the quality of the Datson automobile was terrible. With the eventual assistance of Taichi Ohno and Shigeo Shingo, a system was developed to reduce or eliminate non-value-added activities, for which the customer was not willing to pay. This system became known as the Toyota Production System (TPS). It was recently reintroduced and popularized as Lean Enterprise.[2]

LEAN FUNDAMENTALS

The basics of lean are types of waste, categories of waste, workplace organization, concept of flow, inventory control, visual management, Kaizen, and value stream.

Types of Waste

In order to eliminate or reduce waste, we need to look at the different types. The first type was previously discussed and is non-value added and <u>unnecessary</u> for the system to function. Administration, inspection, and reports are examples of activities that do not add value to the product or service.

The second type is described as non-value added, but <u>necessary</u> for the system to function. For example, to sell oil overseas, it must be transported by an oceangoing vessel, which does not add value to the product, or similarly, the travel cost of a lean consultant going to Uganda would not be value added. However, both the transportation and travel costs are necessary for the activity to occur.

A third type of waste is due to unevenness or variation in quality, cost, or delivery. An excellent example of this concept is given by Taguchi's loss function. As the quality characteristic varies from the target value, the cost increases. The concept is described in Chapter 14.

Another example is illustrated by a work center that delivers an average of 50 units per hour, but has a range of 35 to 65 units per hour.

[1] Levinson, William A., "Lean Manufacturing: Made in the USA," *Quality Digest* (February 2002): 64.

[2] Aluka, George l., "Create a Lean, Mean Machine," *Quality Progress* (April 2003): 29–35.

The fourth type of waste is caused by overstressing people, equipment, or systems. Continually using equipment without proper maintenance will lead to a breakdown, or continually requiring overtime will lead to poorer performance and burnout.

Elimination of waste leads to improved quality, cycle time, cost, and most important, customer satisfaction.

Categories of Waste

Most practitioners agree that there are seven categories of waste. They are described below:

1. **Overproduction.** Producing more, earlier, or faster than required by the next process is waste. It causes inventory, manpower, and transportation to accommodate the excess.
2. **Waiting.** Any idle time or delay that occurs when an operation waits for materials, information, equipment, and so forth, is waste. Idle resources are an obvious form of waste.
3. **Transportation.** Any movement of material around the plant is not value added and, therefore, waste. The more you move material the greater the opportunity for damage and the need for more space.
4. **Defects.** Products or services that do not conform to customer expectations are waste. They incur customer dissatisfaction and frequently reprocessing. Quality is built in at the source.
5. **Inventory.** Any inventory in the value stream is not value added and, therefore, waste. Inventory requires space and hides other wastes.
6. **Motion.** Any movement of a person's body that does not add value is waste. Walking requires time to retrieve materials that are not available in the immediate area.
7. **Extra Processing.** Any extra processing that does not add value to the product or service is waste. An example is the removal of a gate in a molded part.

Many of these categories and definitions may seem extreme; however, when viewed from the customer's viewpoint, they are reasonable forms of waste.

Workplace Organization

In order to establish an effective product or service flow, the workplace must be organized using the 5S's, which are sort, straighten, shine, standardize, and sustain.

1. **Sort.** Divide all the items into three categories: keep those items that are necessary for the activity to function and group by frequency of use; return those items that belong to another customer or location; and move all other items to a staging area and red tag for disposal with appropriate identifiers.
2. **Straighten.** Items that are left are arranged to reduce or eliminate wasted motion.
3. **Shine.** Clean your workplace to eliminate dirt, dust, fluids, and any other debris. Good housekeeping provides an environment to locate equipment problems, to improve productivity, and to reduce accidents.
4. **Standardize.** Documentation is developed to ensure that all parties using the workplace are performing the process in the same manner.
5. **Sustain.** Gains made in the first four S's are maintained by charts, checklists, and audits.

One of the best places to begin your lean journey is with the 5S's at a work center or department. Note that many organizations add a sixth S, which is safety.

Concept of Flow

From the time the first action begins until the product or service reaches the end user, the flow should be continuous with minimum variation. It never stops for an equipment breakdown, delays, inventory, nor any other waste. In order for this utopian situation to exist, there must be one-piece flow, which is one unit at a time rather than a number of units at one time (Batch Processing). One-piece flow

reduces the time between the order and the delivery;

prevents wait times and production delays;

reduces the labor and space to store and move batches;

reveals any defects or problems early in the process;

reduces damage that occurs during batch processing;

provides flexibility to produce a specific product;

reveals non-value activities.

One-piece flow forces employees to concentrate on the process rather than the non-value activities of transportation, inventory, delay, and other forms of waste.[3]

Equipment needs are different for batch processing. Machines are simpler, smaller, and may be slower. Automatic and semiautomatic equipment work well provided they do not reduce uptime or increase changeover time. The blending of automation and human labor requires a high level of operator skill. Equipment needs to be agile and flexible.

The equipment needs to be able to make changeovers quickly. The story of the Single Minute Die Change system follows. One day, Ohno said to Shingo that the setup time on a punch press needed to be changed from four hours to two hours. Shingo said OK, but a short time later Ohno changed his mind and stated that two hours was not good enough—it had to be 10 minutes—and Shingo said OK. It took a while but Shingo was able to reduce the setup time to less than 10 minutes. Today, one-minute setup times are the normal for one-piece flow.

Another element of one-piece flow is cell technology. Machines of different types and performing different

[3]Macinnes, Richard L., *The Lean Enterprise Memory Jogger* (Salem, NH: GOAL/QPC, 2002).

operations are placed in a tight sequence to permit not only one-piece flow, but provide flexible deployment of human resources. With proper cell design, people find work more challenging and interesting.[4]

Production must be leveled by volume and mix so that employees, equipment, and the system are not overstressed.[5]

Inventory Control

Just in time (JIT) is a well-known element of inventory control. It means that the right material arrives at the workplace when it is needed—not too soon and not too late—in the amount required. One-piece flow facilitates this concept.

Products and services are pulled through the system based on actions by the internal or external customer. An example of a pull system is provided by the replenishment of an item in a hardware store. Every item in the store has a particular location or bin that contains space for a specific number of items and has an identifying tag. When the space is empty or almost empty, it is a signal to replenish the item. The signal can also be generated by a computerized system tied to the bar code at the checkout register. In a lean enterprise, a pull signal or *kanban* is used to initiate the replenishment action. Kanbans can be a card, container, empty space, bar code or some other trigger. The kanban contains information about the product and the amount required.

A metric used for inventory control is called *takt* time, which is the German word for "beat." It is the pace of production based on the rate of customer demand. For example, if customer demand for auto insurance is 55 per day and the available processing time is 15 hours, then the takt time is 12.0 minutes $((11 \times 60)/ 55)$. Note that level production is assumed.

A computerized system controls the entire logistics from raw material product to consumer purchase.

Visual Management

The saying "a picture is worth a thousand words" is a well-known saying and one that a lean enterprise utilizes to the fullest. Visual displays are used throughout the facility to inform people about customers, projects, performance, goals, and so forth. In addition, signals are used to alert people to a problem at a particular location.

Kaizen

Kaizen is a Japanese word for the philosophy that defines management's role in continuous encouraging and implementing small improvements involving everyone. It is the process of continuous improvement in small increments that make the process more efficient, effective, under control, and adaptable. Improvements are usually accomplished at little or no expense, without sophisticated techniques or expensive equipment. Kaizen focuses on simplification by breaking down complex processes into their subprocesses and then improving them.

Kaizen relies heavily on a culture that encourages suggestions by operators who continually and incrementally improve their job or process. An example of a Kaizen-type improvement would be the change in color of a welding booth from black to white. This change results in a small improvement in weld quality and a substantial improvement in operator satisfaction. The "Define, Measure, Analyze, Improve, and Control" (DMAIC) problem-solving method described in Chapter 3 can be used to help implement Kaizen concepts as can any problem-solving method.

Fleetwood, a manufacturer of recreational vehicles, has successfully implemented Kaizen. In 1998, there was a 65% overall reduction in work in process and a 22% reduction in cycle times.[6] Copeland Corporation, a manufacturer of air conditioning and refrigeration reciprocating compressors, began adopting Kaizen and lean enterprise in one of its plants in the early 1990s. Since then, productivity has doubled, and there has been a 33% reduction in floor space. In addition, time per unit has been reduced by 35%.

A Kaizen Blitz is a highly focused, action-oriented, 3–5 day improvement workshop where a team takes immediate action to improve a specific process. The end goal is to standardize and implement the improved process without a significant increase in capital expense. A successful blitz uses "Plan, Do, Study, Act" (PDSA) or any problem-solving techniques. It requires a value stream map (VSM) and basic training, which should be spread out over the blitz time period on a need-to-know frequency. It is important to start implementing as soon as practical, rather than wait to have the perfect improvement strategy.

Managers, operators, lawyers, regulators, technicians, and end users are assembled in one room for five days to reduce the time it takes to obtain a coal mining permit. The targeted task is meticulously mapped using colored sticky notes to identify junctures where paperwork must be filed, decisions made, sign-offs obtained, and so forth. Then brainstorming occurs to improve the process.

Value Stream

The term *value stream* refers to the specific activities required to design, order, produce, and deliver a product or service to consumers. It includes the flow of materials from suppliers to customers; the transformation of the raw materials to a completed product or service; and the required information. There may be more than one value stream operating within an organization. Ideally, the value stream would include only value-added activities and the necessary associated information. However, waste in some form is usually present

[4]Bodek, Norman, *Lean Manufacturing 2002* (Detroit, MI: Society of Manufacturing Engineers, 2002). 444

[5]Jones, Daniel T., "Heijunka: Leveling Production." *Manufacturing Engineering* (August 2006): 29–36.

[6]Franco, Vanessa R., and Robert Green, "Kaizen at Fleetwood," *Quality Digest* (March 2000: 24–28).

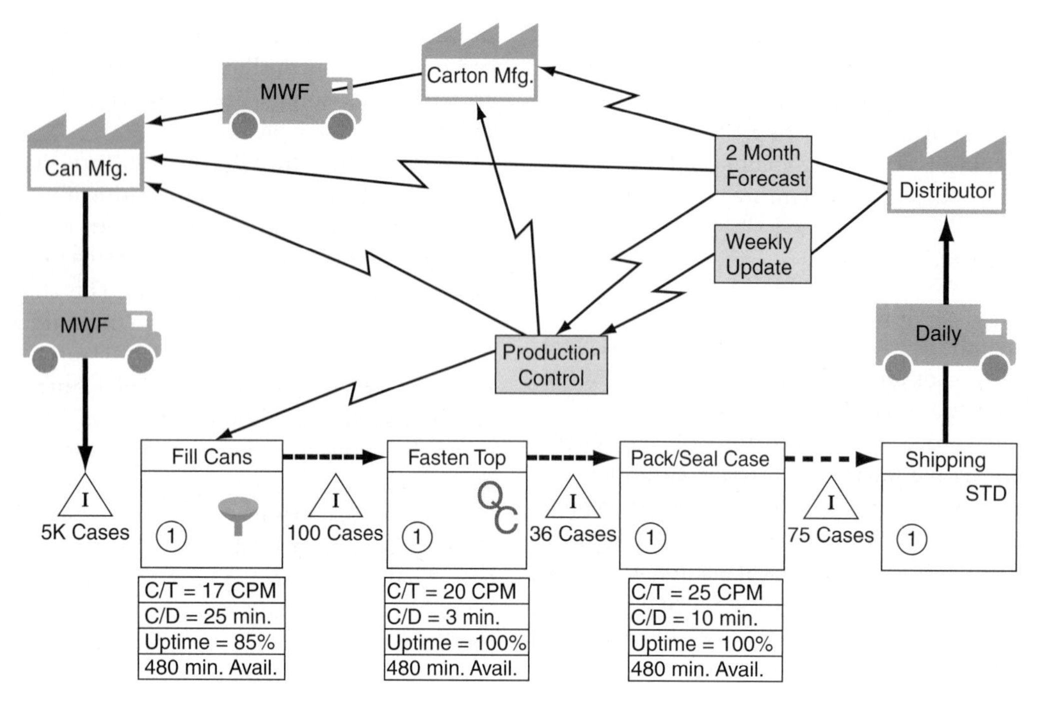

FIGURE 2-1 Sample Value Stream Map

Adapted from Cathy Kingery, Editor, *The Lean Enterprise Memory Jogger*. GOAL/QPC 2002

in the system. The perfect value stream is one where all the operations are

- capable of meeting the quality requirements of the customer;
- available with no unplanned downtime;
- sufficiently efficient to eliminate unnecessary use of energy and materials;
- able to meet customer demand.[7]

The value stream concept can be used for continuous processes such as paper making, petrochemical processes, and sugar refining.

VALUE STREAM MAP

A value stream map (VSM) is used to graphically describe the sequence and movement of the activities associated with the value stream. The first step in the improvement process is to develop the map for the current state. Next, a map for the ideal state is developed with only value-added activities. The difference between the two maps provides valuable information for the improvement team. A VSM is shown in Figure 2-1 and a description of the icons is shown in Figure 2-2.

If the overall management objective is to maximize throughput, then the Theory of Constraints (TOC) can effectively use the VSM. TOC states that at any one time there is usually one operation that limits the throughput of the process system. Improvement of this constraint improves the efficiency of the system and another operation becomes the constraint.

IMPLEMENTING LEAN

There is no one best way to implement lean. Given below is a suggested approach, which can be modified by the organization.

1. With the approval of senior management, establish a cross-function team with a team leader.
2. Train the cross-function team in the lean fundamentals.
3. Construct a VSM for the current state and the ideal state.
4. Analyze the maps to determine the best place for a successful pilot project.
5. Train the people in lean fundamentals and simple tools, such as cause and effect, check sheets, and so forth.
6. Apply 5S and Kaizen techniques.
7. Use the Kaizen Blitz to develop continuous, stable one-piece flow as well as inventory control with the pull system.
8. Expand your activities to include all departments as well as suppliers and end users.
9. Standardize the improvements.

To insure success, a senior manager should be involved or lead the process.

[7]SiewMun Ha, "Continuous Processing Can Be Lean," *Manufacturing Engineering* (June 2007): 103–109.

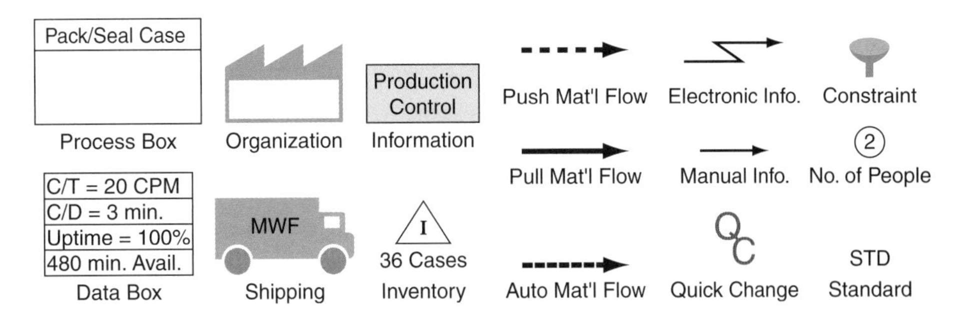

FIGURE 2-2 Value Stream Map Icon Designations

BENEFITS TO LEAN ENTERPRISE

The Tyco Flow Control plant in Chennai, India, is a good example of the benefits that can be achieved with lean enterprise. Some of the results were as follows: on-time delivery improved by 94%; lead time reduced from 150 days to 56 days; material movement reduced by 68%; machining capacity increased 200%; cycle time balanced; and incoming inspection reduced from 16 days to one day.[8]

The success of Toyota is ample evidence that lean enterprise is an effective improvement system for manufacturing by improving quality, eliminating waste, reducing lead time, and reducing total costs. Other types of organizations such as construction, education, and health care can also benefit from lean enterprise.

Veridan Homes of Madison, WI, reported the following lean enterprise results: drafting time reduced by more than one hour; inspection reduced by 50%; defects reduced by more than 50%; and cycle time reduced from 32 to 15 days. Tracer Industries Canada Ltd. reported that total design time was reduced from 20 days to 1.2 days.[9]

At the University of Scranton, marketing students were able to reduce the admission office's response to inquiries from 13 days to less than one day, and eliminate faculty involvement. New student application processing time was reduced from 88 days to less than one day. The students were not able to map the process, because the system was completely out of control.[10]

Heartland Regional Medical Center in Marion, IL, used lean enterprise to improve quality, safety, and cost, as well as survive an accreditation audit.[11]

ADDITIONAL COMMENTS

This chapter would not be complete without mentioning the Toyota accounting system. It is somewhat like a small store where performance is measured by the difference between the income that goes in one pocket and the cost that goes out the other pocket. Accounting tracks what goes in the plant and the product that goes out—it is not concerned with the internal cost as a measure of performance. In the Toyota Production System (TPS), the work provides all the information needed to control the operations.[12]

EXERCISES

1. Describe the difference between waste and a value-added activity and give examples for different types of organizations.
2. Give examples of the four types of waste and the seven categories of waste for different types of organizations.
3. Use the 5S's to organize a workplace.
4. Describe the difference between Kaizen and a Kaizen Blitz.
5. Describe the purpose of the value stream map.
6. Working as a team of three or more people, determine how the lean fundamentals of flow and inventory

[8]Malhotra, Iqbal S., "Moving and Controlling the Flow of Quality," *Quality Progress* (December 2006): 67–69.

[9]Sowards, Dennis, "Lean Construction," *Quality Digest* (November 2007): 32–36.

[10]Tischter, Len, "Bringing Lean to the Office," *Quality Progress* (July 2007): 32–37.

[11]Collins, Kevin F., and Senthil K. Muthusamy, "Applying the Toyota Production System to a Healthcare Organization: A Case Study on a Rural Community Healthcare Provider," *Quality Management Journal* (Vol. 14, No. 4, 2007): 41–52.

[12]Johnson, Thomas H., "Manage a Living System, Not a Ledger," *Manufacturing Engineering* (December 2006): 73–80.

control might improve one or more of the organizations listed below:

a. Large bank
b. Health-care facility
c. University academic department
d. University nonacademic department
e. Large department store
f. Grade or high school
g. Manufacturing facility
h. Large grocery store

7. Veterans have to wait six months for a disability claim to be processed. Who would you select as members of a Kaizen Blitz team?
8. Working as a team of three or more people, construct a VSM for a process of one or more of the organizations listed in Exercise 6.

CHAPTER THREE

SIX SIGMA[1]

OBJECTIVES

Upon completion of this chapter, the reader is expected to

- understand the concept of Six Sigma statistics;
- be able to describe the DMAIC project methodology;
- know the advantages of the methodology.

INTRODUCTION

Six Sigma is both a statistical concept that measures the number of nonconformities in a product or service and a quality improvement methodology. The statistical aspects have been known for decades; however, they have been reemphasized. The essentials of the quality improvement methodology are DMAIC, benefits, and organizational structure, which is not discussed in this book.

HISTORICAL REVIEW

The idea of Six Sigma quality was first conceived by statistical experts at Motorola in the mid-1980s in response to a demanding quality stretch goal by Robert Galvin, CEO. By 1988, there was a staggering increase in the quality level of several products and the organization received the inaugural Malcolm Baldrige National Quality Award, which was primarily due to its TQM initiatives. By 1990 Mikel Harry and his colleagues had refined the methodology. The information was publicized and by the mid-1990s other companies such as General Electric and Allied Signal were obtaining similar quality improvements.

STATISTICAL ASPECTS[2]

Sigma is the Greek word for the symbol, σ, which stands for the population standard deviation. It is the best measure of process variability because the smaller the standard deviation, the less variability in the process. If we can reduce sigma, σ, to the point that the specifications are at ± 6, then 99.9999998% of the product or service will be between specifications, and the nonconformance rate will be 0.002 parts/million (ppm). Another useful measure is the capability index, C_p, which is the ratio of the specification tolerance (upper specification limit minus the lower specification limit) by 6σ. Figure 3-1 graphically illustrates this situation, and Table 3-1 provides information at other specification limits. Chapter 5 provides information on standard deviation and normal curve and Chapter 6 discusses process capability in detail.

According to the Six Sigma philosophy, processes rarely stay centered—the center tends to "shift" above and below the target. Figure 3-2 shows a process that is normally distributed but has shifted within a range of 1.5σ above and 1.5σ below the target. For the diagrammed situation, 99.9996600% of the product or service will be between specifications, and the nonconformance rate will be 3.4 ppm. This off-center situation gives a process capability index (C_{pk}) of 1.5. Table 3-2 shows the percent between specifications, the nonconformance rate, and process capability for different specification limit locations. The magnitude and type of shift is a matter of discovery and should not be assumed ahead of time.

A nonconformance rate of 3.4 ppm is a statistical number for a 1.5σ process shift. Actually, the true non-

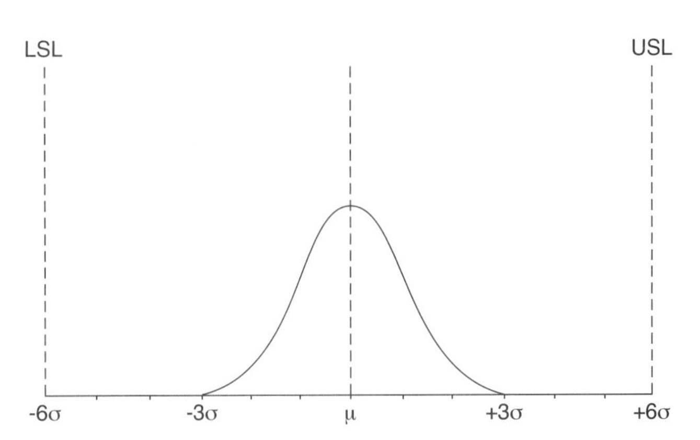

FIGURE 3-1 Nonconformance rate when process is centered.

[1]Six Sigma is a registered trademark of Motorola.

[2]This statistical information is based on parts per million. The original concept, advocated by Smith of Motorola, was based on occurrences per million. He later stated that parts per million would be better, because this is what the customer cares about.

TABLE 3-1 Nonconformance Rate and Process Capability When the Process Is Centered

Specification Limit	Percent Conformance	Nonconformance Rate (ppm)	Process Capability (C_P)
$\pm 1\sigma$	68.7	317,300	0.33
$\pm 2\sigma$	95.45	485,500	0.67
$\pm 3\sigma$	99.73	2,700	1.00
$\pm 4\sigma$	99.9937	63	1.33
$\pm 5\sigma$	99.999943	0.57	1.67
$\pm 6\sigma$	99.9999998	0.002	2.00

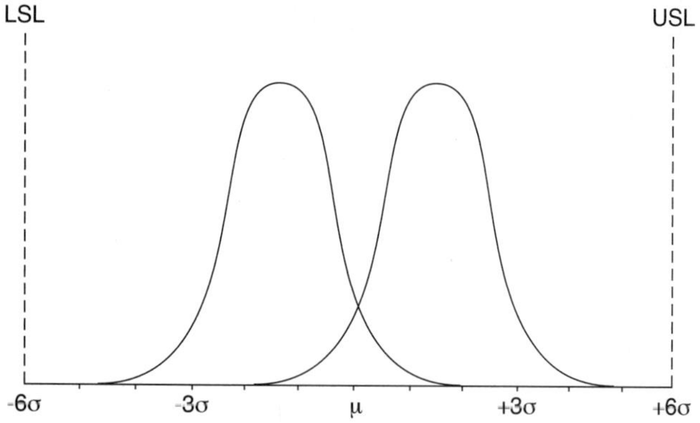

LSL USL

-6σ -3σ μ +3σ +6σ

FIGURE 3-2 Nonconformance rate when process is off-center $\pm 1.5\sigma$.

conformance rate is much closer to 0.002 ppm or 2 parts per billion (ppb). The rational for this statement is as follows:[3]

- First, it is assumed that the process shift is 1.5σ when it may be much less. This shift was originally envisioned around 1990. We are able to control and measure the process much better.
- Second, it is assumed that the process shift is always at 1.5σ, which will not occur.
- Third, the monitoring process of an X bar and R chart with $n = 4$ would most likely catch and correct an out-of-control condition within, say, 2 h. Thus, in a 40-h week, the process would be running at a 3.4 ppm nonconformance rate for only 5% of the time.
- If, in fact, the process does drift around, an attached controller would be able to keep the process centered most of the time.
- A nonconformance rate of 3.4 ppm can be achieved with a 0.5σ process shift at 5.0 sigma and a 1.0σ process shift at 5.5 sigma, as well as the 1.5σ process shift at 6.0 sigma. Thus, keeping the processes centered is another technique to achieve 3.4 ppm.

It is important to understand that this information is based on the individual part and not on the entire product

TABLE 3-2 Nonconformance Rate and Process Capability When the Process Is Off-Center $\pm 1.5\sigma$

Specification Limit	Percent Conformance	Nonconformance Rate (ppm)	Process Capability (C_{PK})
$\pm 1\sigma$	30.23	697,700	−0.167
$\pm 2\sigma$	69.13	308,700	0.167
$\pm 3\sigma$	93.32	66,810	0.500
$\pm 4\sigma$	99.3790	6,210	0.834
$\pm 5\sigma$	99.97670	2,330	1.167
$\pm 6\sigma$	99.9996600	3.4	1.500

[3]Ronald D. Snee, "Why Should Statisticians Pay Attention to Six Sigma," *Quality Progress,* (September 1999): 100–103.

or service. For example, if there are 20 parts or steps in a product or service and each is at 6.0 sigma with the 1.5σ shift, then 68 ppm of the product will be nonconforming [$(1 - 0.9999966^{20}) = 0.000068$ or 68 ppm], which is 0.0068%. If at 4.0 sigma with the 1.5σ shift, then 117,100 ppm of the product or service will be nonconforming, which is 11.781%.

Achieving Six Sigma specifications will not be easy and should only be attempted for critical quality characteristics when the economics of the situation dictate.

IMPROVEMENT METHODOLOGY

DMAIC are the letters for the acronym that stands for the five phases of the improvement methodology, which is Define, Measure, Analyze, Improve, and Control. Its purpose is to improve cycle time, quality, and cost. Although, it is not a new concept, no other improvement methodology uses and sequences improvement tools as effectively as Six Sigma.[3] Each of the phases requires a progress report to senior management. Motorola developed the MAIC model. The define step was added by General Electric practitioners.

Define

This phase consists of a project charter, process map, and the voice of the customer. www.pearsonhighered.com/besterfield

Project Charter The charter documents the problem statement, project management, and progress toward goals.

An example of a well-written *problem statement* is:

As a result of a customer satisfaction survey, a sample of 150 billing errors showed that 18 had errors that required one hour to correct.

The above statement describes the current state. In the measure phase, we have the opportunity to describe a goal such as "Reduce billing errors by 75%."

Identifying problems for improvement is not difficult, as there are many more problems than can be analyzed. The quality council or work group must prioritize problems using the following selection criteria:

1. Is the problem important and not superficial, and why?
2. Will problem solution contribute to the attainment of goals?
3. Can the problem be defined clearly using objective measures?

In selecting an initial improvement opportunity, the quality council or work group should find a problem that, if solved, gives the maximum benefit for the minimum amount of effort. An affinity diagram and Pareto analysis may help in this step.

Project Management is vested in the quality council or other authority, which will authorize the project, the available resources, and guidelines for internal operation. If the problem relates primarily to a natural work group, then they will constitute the basic team. In addition, the team should include the process owner (if not included as part of the work group) and a representative from the upstream and downstream processes. If the problem is of a multifunctional nature, as most are, then the team should be selected and tasked to improve a particular process or series of processes.

Goals and Progress are a necessary component of the charter. The Six Sigma philosophy emphasizes a financial benefit from the project. However, goals can also be stated in terms of improved quality, worker safety and satisfaction, internal and external customer satisfaction, and environmental impact. Progress should be expressed in a time line or milestones.

Process Map A process map helps the team understand the process, which refers to the business and production activities of an organization. Figure 3-3 shows the Supplier, Input, Process, Output, Customer (SIPOC) process model. In addition to production processes, business processes such as purchasing, engineering, accounting, and marketing are areas where nonconformance can represent an opportunity for substantial improvement.

Inputs may be materials, money, information, data, etc. *Outputs* may be information, data, products, service, etc. Suppliers and customers can be both internal and external. The output of one process also can be the input to another process. Outputs usually require performance measures. They are designed to achieve certain desirable *outcomes* such as customer satisfaction. *Feedback* is provided in order to improve the process.

The *process* is the interaction of some combination of people, materials, equipment, method, measurement, and the environment to produce an outcome such as a product, a service, or an input to another process. In addition to having measurable input and output, a process must have value-added activities and repeatability. It must be effective, efficient, under control, and adaptable. In addition, it must adhere to certain *conditions* imposed by policies and constraints or regulations. Examples of such conditions may include constraints related to union-based job descriptions of employees, state and federal regulations related to storage of environmental waste, or bio-ethical policies related to patient care.

Process definition begins with defining the internal and/or external customers. The customer defines the purpose of the organization and every process within it. Because the organization exists to serve the customer, process improvements must be defined in terms of increased customer satisfaction as a result of higher quality products and services.

All processes have at least one owner. In some cases, the owner is obvious, because there is only one person performing the activity. However, frequently the process will cross multiple organizational boundaries, and supporting subprocesses will be owned by individuals within each of the

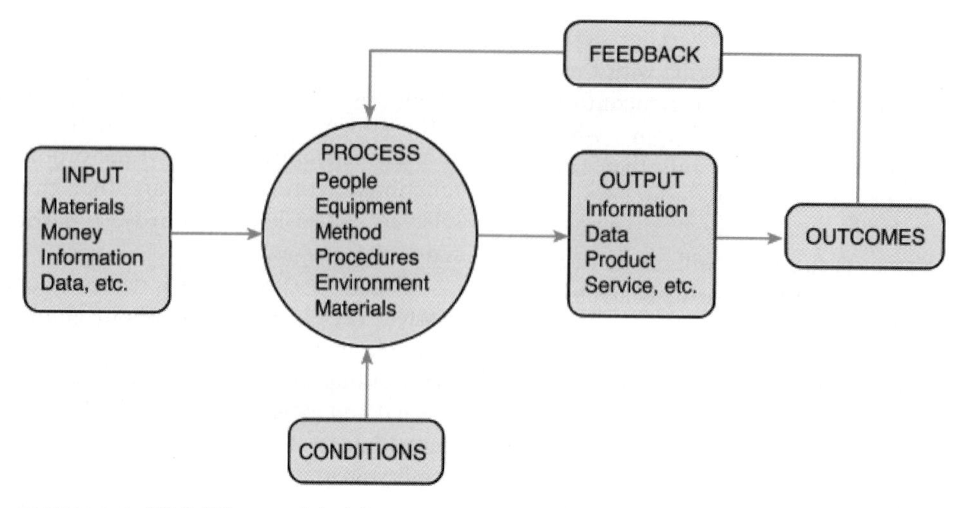

FIGURE 3-3 SIPOC Process Model.

organizations. Thus, ownership should be part of the process improvement initiatives.

Processes are inherent to all functional areas. There are processes for customer satisfaction, leadership, design, suppliers, maintenance, finance, human resources, product realization, marketing, service, and so forth.

Voice of the Customer The voice of the customer provides information that leads to those problems that have the greatest potential for improvement and have the greatest need for solution. Problems can be identified from a variety of inputs, such as the following:

- Analysis of repetitive external alarm signals, such as field failures, complaints, returns, and others
- Analysis of repetitive internal alarm signals (for example, scrap, rework, sorting, and the 100% test)
- Proposals from key insiders (managers, supervisors, professionals, and union stewards)
- Proposals from suggestion schemes
- Field study of users' needs.
- Data on performance of competitors (from users and from laboratory tests)
- Comments of key people outside the organization (customers, suppliers, journalists, and critics)
- Findings and comments of government regulators and independent laboratories
- Customer and employee surveys and focus groups
- Brainstorming by work groups

The Critical to Quality tree is an applicable tool.

Measure

The objective of the measure phase is to understand the process, validate the data accuracy, and determine the process capability. This information is used to review the define phase, establish a baseline, and obtain a better knowledge of the process.

Understand the Process A value stream map provides information concerning the waste in the process. This technique is a beginning point to understand the logistics of the process. A diagram or map translates complex work into an easily understood graphic description. This activity is an eye-opening experience for the team, because it is rare that all members of the team understand the entire process.

Next, the target performance measures are defined for inputs and outputs. Measurement is fundamental to meaningful process improvement. If something cannot be measured, it cannot be improved. There is an old saying that what gets measured gets done. The team will determine if the measurements needed to understand and improve the process are presently being used and if new ones are needed, the team will

- establish performance measures with respect to customer requirements;
- determine data needed to manage the process;
- establish regular feedback with customers and suppliers;
- establish measures for quality, cost, waste, and timeliness.

Once the target performance measures are established, the team can collect all available data and information. If these data are not enough, then additional new information is obtained. Gathering data (1) helps confirm that a problem exists, (2) enables the team to work with facts, (3) makes it possible to establish measurement criteria for baseline, and (4) enables the team to measure the effectiveness of an implemented solution. It is important to collect only needed data and to get the right data for the problem. The team should develop a plan that includes input from internal and external customers and ensures the plan answers the following questions:

1. What do we want to learn about the problem or operation?
2. What are the uses for the data?

3. How many data are needed?
4. What conclusions will we be able to draw from the collected data?
5. What action will we be able to take as a result of the conclusions?

Data can be collected by a number of different methods, such as check sheets, computers with application software, data-collection devices like handheld gauges, or an online system.

The team will systematically review the procedures currently being used. Common items of data and information include

- customer information, such as complaints and surveys;
- design information, such as specifications, drawings, function, bills of materials, costs, design reviews, field data, service, and maintainability;
- process information, such as routing, equipment, operators, raw material, and component parts and supplies;
- statistical information, such as average, median, range, standard deviation, skewness, kurtosis, and frequency distribution;
- quality information, such as Pareto diagrams, cause-and-effect diagrams, check sheets, scatter diagrams, control charts, histograms, process capability, acceptance sampling, Taguchi's loss function, run charts, life testing, inspection steps, and operator and equipment matrix analysis;
- supplier information, such as process variation, on-time delivery, and technical competency;
- data mining—an analytical process using the computer to explore large amounts of data in search of consistent patterns.

With this information, a data collection plan can be developed. It will decide what, who, where, when, why, and how data will be collected. Once the plan is completed, data can be collected using check sheets, Pareto analysis, histogram, run chart, control charts, flowchart, and Taguchi's loss function.

Validate the Data Accuracy All devices used to collect data must be calibrated by established procedures, which may require verification by an independent laboratory. Measurements must be accurate, which means on target, and precise, which means very little variation. One of the most common tools to evaluate a measuring system is called GR&R, which stands for gauge repeatability and reproducibility and is covered in Chapter 7.

Determine the Process Capability Process capability is a statistical measure that compares process variation to the specifications. In order to have a reliable measure the variation must be stable over time as measured by a control chart. It is covered in Chapter 6.

Analyze

This phase consists of process analysis, cause investigation, and charter updating. The objective is to pinpoint and verify causes affecting the problem.

Process Analysis Perform a detailed review of the value stream map to calculate takt time, identify non-value-added activities, and bottlenecks. Review data collected in the measure phase.

Cause Investigation This activity begins with identifying all the potential causes. It requires experience, brainstorming, and a thorough knowledge of the process. The cause-and-effect diagram is particularly effective in this phase. Other tools are why why, tree diagram, and interrelationship diagram. One word of caution: the object is to seek causes, not solutions. Therefore, only possible causes, no matter how trivial, should be listed. Where data is not readily available, many organizations are using simulation modeling to identify possible causes.

The list of potential causes can be narrowed by multivoting, Pareto analysis, and stratification. A review of the problem statement can also reduce the number of potential causes.

The most likely or root cause(s) must be verified, because a mistake here can lead to the unnecessary waste of time and money by investigating possible solutions to the wrong cause. Some verification techniques are given below:

1. Examine the most likely cause against the problem statement.
2. Recheck all data that support the most likely cause.
3. Use statistical techniques such as scatter diagram, hypothesis testing, ANOVA, experimental design, Taguchi's quality engineering, and other advanced techniques to determine the critical factors and their levels.
4. Calculations that show improvement from elimination of non-value-added activities.

Once the root cause is determined, the next phase can begin.

Charter Review Results of the analysis phase may require changes to the charter phase—in particular, the problem statement, team membership, schedule, resources needed, and goals.

Improve

The phase selects the optimal solution(s), tests a pilot, and implements the solution. Its objective is to develop an improved process that will meet goals.

Optimal Solution Once all the information is available, the project team begins its search for possible solutions. More than one solution is frequently required to remedy a situation. Sometimes the solutions are quite evident from a cursory analysis of the data. In this phase, creativity plays the major role, and brainstorming is the principal technique. Brainstorming on possible solutions requires not only a knowledge of the problem but also innovation and creativity.

There are three types of creativity: (1) create new processes, (2) combine different processes, or (3) modify the existing process. The first type is innovation in its highest form, such as the invention of the transistor. Combining two or more processes is a synthesis activity to create a better process. It is a unique combination of what already exists. This type of creativity relies heavily on benchmarking. Modification involves altering a process that already exists so that it does a better job. It succeeds when managers utilize the experience, education, and energy of empowered work groups or project teams. There is not a distinct line between the three types—they overlap.[4]

Creativity is the unique quality that separates mankind from the rest of the animal kingdom. Most of the problems that cause inefficiency and ineffectiveness in organizations are simple problems. There is a vast pool of creative potential available to solve these problems. Quality is greatly improved because of the finding and fixing of a large number of problems, and morale is greatly increased because it is enormously satisfying to be allowed to create.[5]

Once possible solutions have been determined, evaluation or testing of the solutions comes next. As mentioned, more than one solution can contribute to the situation. Evaluation and/or testing determine which of the possible solutions have the greatest potential for success and the advantages and disadvantages of these solutions. Criteria for judging the possible solutions include such things as cost, feasibility, effect, resistance to change, consequences, and training. Solutions also may be categorized as short range and long range. Statistical techniques such as design of experiments, histogram, prioritization matrix, run charts and control charts will facilitate the decision. One of the features of control charts is the ability to evaluate possible solutions. Whether the idea is good, poor, or has no effect is evident from the chart. The value stream map should be revised to determine the process performance after recommended changes have been made. At a minimum, the solution must prevent recurrence.

Pilot Testing Prior to full scale implementation, it is a good idea to run a pilot. This activity will frequently require approval, because it will disrupt normal production. Participants will need to be trained. Results will need to be evaluated to verify that goals have been met.

Implementation This step has the objective of preparing the implementation plan, obtaining approval, and implementing the process improvements.

Although the project team usually has some authority to institute remedial action, more often than not the approval of the quality council or other appropriate authority is required. If such approval is needed, a written and/or oral report is given.

The contents of the implementation plan report must fully describe,

- Why will it be done?
- How will it be done?
- When will it be done?
- Who will do it?
- Where will it be done?

Answers to these questions will designate required actions, assign responsibility, and establish implementation milestones. The length of the report is determined by the complexity of the change. Simple changes may require only an oral report, whereas other changes require a detailed, written report.

After approval by the quality council, it is desirable to obtain the advice and consent of departments, functional areas, teams, and individuals that may be affected by the change. A presentation to these groups will help gain support from those involved in the process and provide an opportunity for feedback with improvement suggestions.

The final element of the implementation plan is the monitoring activity that answers the following:

- What information will be monitored or observed, and what resources are required?
- Who will be responsible for taking the measurements?
- Where will the measurements be taken?
- How will the measurements be taken?
- When will the measurements be taken?

Measurement tools such as run charts, control charts, Pareto diagrams, histograms, check sheets, and questionnaires are used to monitor and evaluate the process change.

Pylipow provides a combination map to help formulate an action plan to help measure the results of an improvement. The map, shown in Table 3-3 provides the dimensions of what is being inspected, the type of data, timing of data collection, by whom, how the results will be recorded, the necessary action that needs to be taken based on the results, and who is to take the action.

Control

The final phase consists of evaluating the process, standardizing the procedures, and conclusion. It has the objective of monitoring and evaluating the change by tracking and studying the effectiveness of the improvement efforts through data collection and review.

Evaluating the Process This step has the objective of monitoring and evaluating the change by tracking and

[4]Paul Mallette, "Improving Through Creativity," *Quality Digest* (May 1993): 81–85.

[5]George Box, "When Murphy Speaks—Listen," *Quality Progress* (October 1989): 79–84.

TABLE 3-3 Combination Map of Dimensions for Process Control

What's Inspected	Type of Data	Timing	By Whom?	Type of Record	Action	By Whom?
Process Variable Continuous	Variable	During run: on-line	Device	Electronic control chart	Process improved	Automated equipment
Process Variable Sample				Paper control chart		
		During run: off-line	Process Operator	Electronic trend chart	Process adjusted	Operator
Product Sample	Attribute			Paper trend chart	Lot sorted	
		After lot: complete	Inspector	Electronic list	Sample repaired or discarded	Inspector or mechanic
100% of product				Paper list		

Reproduced, with permission, from Peter E. Pylipow, "Understanding the Hierarchy of Process Control: Using a Combination Map to Formulate an Action Plan," *Quality Progress* (October 2000): 63–66.

studying the effectiveness of the improvement efforts through data collection and review of progress. It is vital to institutionalize meaningful change and ensure ongoing measurement and evaluation efforts to achieve continuous improvement.

The team should meet periodically during this step to evaluate the results to see that the problem has been solved or if fine-tuning is required. In addition, the team will want to see if any unforeseen problems have developed as a result of the change. If the team is not satisfied, then some of the phases will need to be repeated. Run charts, control charts, histogram, process capability, and combination map are applicable tools.

Standardize the Procedures Once the team is satisfied with the change, it must be institutionalized by positive control of the process, process certification, and operator certification. Positrol (positive control) assures that important variables are kept under control. It specifies the what, who, how, where, and when of the process and is an updating of the monitoring activity. Standardizing the solution prevents "backsliding."

In addition, the quality peripherals—the system, environment, and supervision must be certified. A checklist provides the means to initially evaluate the peripherals and periodically audit them to ensure the process will meet or exceed customer requirements for the product or service. The *system* includes shutdown authority, preventative maintenance, visible and/or audible alarm signals, foolproof inspection, and self-inspection. The *environment* includes water/air purity, dust/chemical control, temperature/humidity control, electrostatic discharge, and storage/inventory control. *Supervision* involves coach—not boss, clear instructions, encouraging suggestions, and feedback of results.

Finally, operators must be certified to know what to do and how to do it for a particular process. Also needed is cross-training in other jobs within the process to ensure next-customer knowledge and job rotation. Total product knowledge is also desirable. Operator certification is an ongoing process that must occur periodically.

A graphical storybook is provided at bulletin boards throughout the organization.

Final Actions The success of the project should be celebrated by the participants in order to provide satisfaction for the team. If for some reason, the process owner has not been a member of the team, then the responsibility for the improved process is returned. Lessons learned during the project and suggestions for future activities are reported to the appropriate authority.

ADDITIONAL COMMENTS

The information given above describes the basic methodology. For a new process or a radically redesigned one, the control step is replaced with a verify "V" step. Also Harry and Schroeder, methodology leaders, have a version that has eight steps, where there is a recognize step at the beginning and standardize and integrate steps at the end. If there are multiple facilities producing the same product or service, then a replicate "R" step is added, which would be similar to the standardize and integration steps.

The Six Sigma methodology works because it gives bottom line results; trains leaders to complete successful projects; reduces process variation to improve quality and generate satisfied customers; and utilizes a sound statistical approach. But, the greatest contribution comes from a dedicated, involved, and enthusiastic CEO.

Regardless of how successful initial improvement efforts are, the improvement process continues. Everyone in the organization is involved in a systematic, long-term endeavor to constantly improve quality by developing processes that are customer oriented, flexible, and responsive. Continuous improvement requires organizations and individuals to not only be satisfied with doing a good job or process but also striving to improve that job or process. It is accomplished by incorporating process measurement and team problem solving in all work activities.

A key activity is to conduct regularly scheduled reviews of progress by the quality council and/or work group. Management must establish the systems to identify areas for future improvement and to track performance with respect to internal and external customers. They must also track changing customer requirements. Lessons learned in problem solving, communications, and group dynamics, as well as technical know-how, must be transferred to appropriate activities within the organization.

Although the DMAIC method is no guarantee of success, experience has indicated that an orderly approach will yield the highest probability of success.

EXERCISES

1. If there are 8 parts in a product and they are each at 6 sigma with a 1.5 shift, determine the ppm for the product.
2. Working as an individual or in a team of three or more people, use DMAIC to improve a process in one or more of the organizations listed below:
   a. Large bank
   b. Health-care facility
   c. University academic department
   d. University nonacademic department
   e. Large department store
   f. Grade school
   g. Manufacturing facility
   h. Large grocery store
3. List the benefits of Six Sigma.

CHAPTER
FOUR

STATISTICAL PROCESS CONTROL (SPC)

OBJECTIVES

Upon completion of this chapter, the reader is expected to

- be able to construct a Pareto diagram;
- be able to construct a cause-and-effect diagram;
- explain how to construct a check sheet;
- be able to construct a process flow diagram.

INTRODUCTION

SPC usually comprises the following tools: Pareto diagram, cause-and-effect diagram, check sheet, process flow diagram, scatter diagram, histogram, control charts, and run chart. The first four are discussed in detail, the last four in summary form with reference to the appropriate chapters, because they (except for the run chart) are based on statistics, which are discussed in Chapter 5.

PARETO DIAGRAM

Alfredo Pareto (1848–1923) conducted extensive studies of the distribution of wealth in Europe. He found that there were a few people with a lot of money and many people with little money. This unequal distribution of wealth became an integral part of economic theory. Joseph Juran recognized this concept as a universal that could be applied to many fields. He coined the phrases *vital few* and *useful many.*[1]

A Pareto diagram is a graph that ranks data classifications in descending order from left to right, as shown in Figure 4-1. In this case, the data classifications are types of field failures. Other possible data classifications are problems, causes, types of nonconformities, and so forth. The vital few are on the left, and the useful many are on the right. It is sometimes necessary to combine some of the useful many into one classification called *other* and labeled O in the figure. When the *other* category is used, it is always on

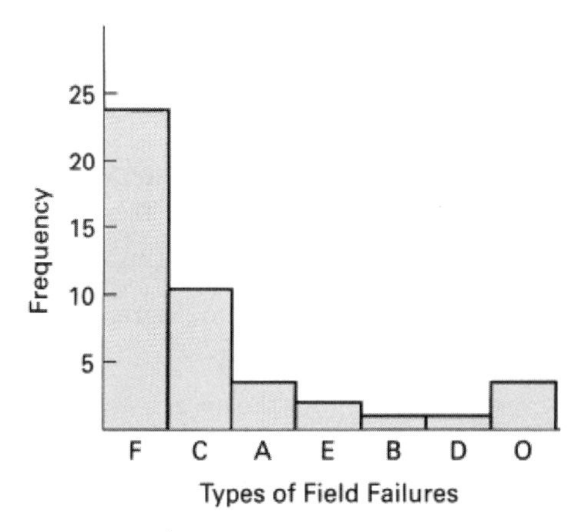

FIGURE 4-1 Pareto diagram.

the far right. The vertical scale is dollars, frequency, or percent. Pareto diagrams can be distinguished from histograms (to be discussed) by the fact that the horizontal scale of a Pareto is categorical, whereas the scale for the histogram is numerical.

Sometimes a Pareto diagram has a cumulative line, as shown in Figure 4-2. This line represents the sum of the data as they are added together from left to right. Two scales are used: The one on the left is either frequency or dollars, and the one on the right is percent.

Pareto diagrams are used to identify the most important problems. Usually, 80% of the total results from 20% of the items. This fact is shown in Figure 4-2, where the F and C types of field failures account for almost 80% of the total. Actually, the most important items could be identified by listing the items in descending order. However, the graph has the advantage of providing a visual impact of those vital few characteristics that need attention. Resources are then directed to take the necessary corrective action. Examples of the vital few are given below:

- A few customers account for the majority of sales.
- A few products, processes, or quality characteristics account for the bulk of the scrap or rework cost.
- A few nonconformities account for the majority of customer complaints.

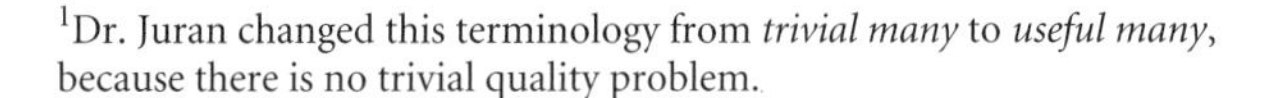

[1]Dr. Juran changed this terminology from *trivial many* to *useful many*, because there is no trivial quality problem.

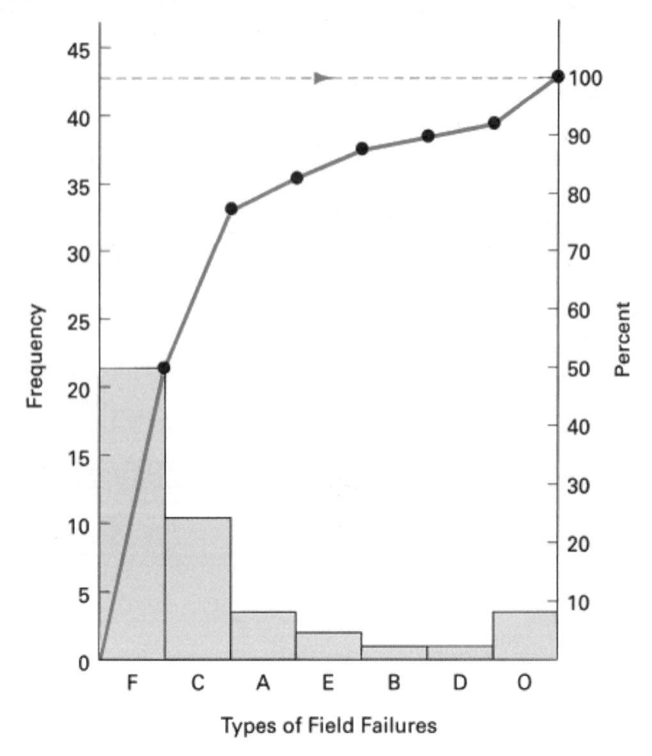

FIGURE 4-2 Cumulative line of Pareto diagram.

- A few vendors account for the majority of rejected parts.
- A few problems account for the bulk of the process downtime.
- A few products account for the majority of the profit.
- A few items account for the bulk of the inventory cost.

Construction of a Pareto diagram is very simple. There are six steps:

1. Determine the method of classifying the data: by problem, cause, type of nonconformity, etc.
2. Decide if dollars (best), weighted frequency, or frequency is to be used to rank the characteristics.
3. Collect data for an appropriate time interval.
4. Summarize the data and rank-order categories from largest to smallest.
5. Compute the cumulative percentage if it is to be used.
6. Construct the diagram and find the vital few.

The cumulative percentage scale, when used, must match with the dollar or frequency scale such that 100% is at the same height as the total dollars or frequency. See the arrow in Figure 4-2.

It is noted that a quality improvement of the vital few, say 50%, is a much greater return on investment than a 50% improvement of the useful many. Also, experience has shown that it is easier to make a 50% improvement in the vital few.

The use of a Pareto diagram is a never-ending process. For example, let's assume that F is the target for correction in the improvement program. A project team is assigned to investigate and make improvements. The next time a Pareto analysis is made, another field failure, say C, becomes the target for correction, and the improvement process continues until field failures become an insignificant quality problem.

The Pareto diagram is a powerful quality improvement tool. It is applicable to problem identification and the measurement of progress.

CAUSE-AND-EFFECT DIAGRAM

A cause-and-effect (C&E) diagram is a picture composed of lines and symbols designed to represent a meaningful relationship between an effect and its causes. It was developed by Kaoru Ishikawa in 1943 and is sometimes referred to as an Ishikawa diagram.

C&E diagrams are used to investigate either a "bad" effect and to take action to correct the causes or a "good" effect and to learn those causes responsible. For every effect, there are likely to be numerous causes. Figure 4-3 illustrates a C&E diagram with the effect on the right and causes on the left. The effect is the quality characteristic that needs improvement. Causes are usually broken down into the major causes of work methods, materials, measurement, people, and the environment. Management and maintenance are also sometimes used for the major cause. Each major cause is further subdivided into numerous minor causes. For example, under Work Methods, we might have training, knowledge, ability, physical characteristics, and so forth. C&E diagrams (frequently called "fish-bone diagrams" because of their shape) are the means of picturing all these major and minor causes.

The first step in the construction of a C&E diagram is for the project team to identify the effect or quality problem. It is placed on the right side of a large piece of paper by the team leader. Next, the major causes are identified and placed on the diagram.

Determining all the minor causes requires brainstorming by the project team. Brainstorming is an idea-generating technique that is well suited to the C&E diagram. It uses the creative thinking capacity of the team.

Attention to a few essentials will provide a more accurate and usable result:

1. Participation by every member of the team is facilitated by each member taking a turn, suggesting one idea at a time. If a member cannot think of a minor cause, he or she passes for that round. Another idea may occur at a later round. By following this procedure, one or two individuals do not dominate the brainstorming session.
2. Quantity of ideas, rather than quality, is encouraged. One person's idea will trigger someone else's idea, and a chain reaction occurs. Frequently, a trivial or "dumb" idea will lead to the best solution.
3. Criticism of an idea is not allowed. There should be a freewheeling exchange of information that liberates the

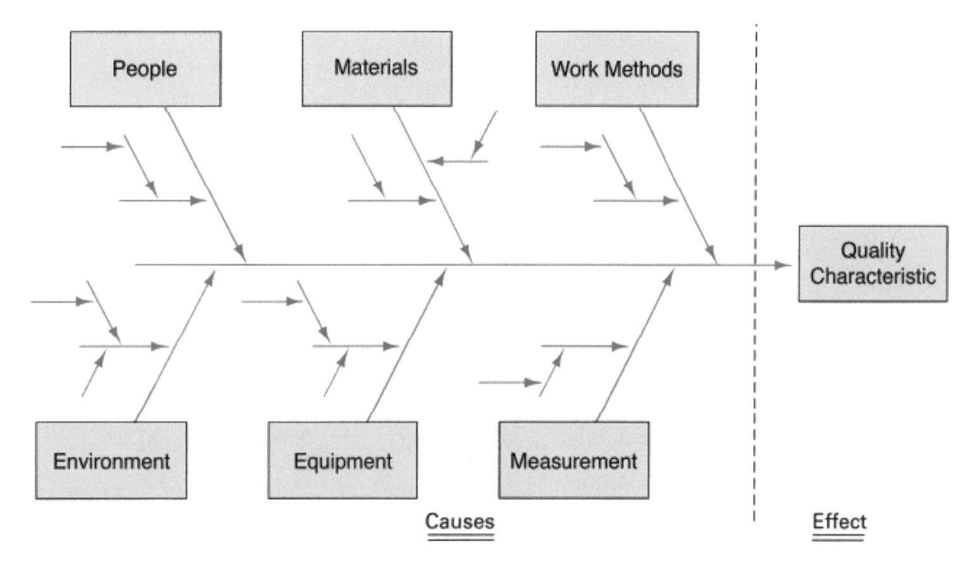

FIGURE 4-3 Cause-and-effect diagram.

imagination. All ideas are placed on the diagram. Evaluation of ideas occurs at a later time.

4. Visibility of the diagram is a primary factor of participation. In order to have space for all the minor causes, a 2-ft by 3-ft piece of paper is recommended. It should be taped to a wall for maximum visibility.
5. Create a solution-oriented atmosphere and not a gripe session. Focus on solving a problem rather than discussing how it began. The team leader should ask questions using the why, what, where, when, who, and how techniques.
6. Let the ideas incubate for a period of time (at least overnight), and then have another brainstorming session. Provide team members with a copy of the ideas after the first session. When no more ideas are generated, the brainstorming activity is terminated.

Once the C&E diagram is complete, it must be evaluated to determine the most likely causes. This activity is accomplished in a separate session. The procedure is to have each person vote on the minor causes. Team members may vote on more than one cause, and they do not need to vote on a cause they presented. Those causes with the most votes are circled, and the four or five most likely causes of the effect are determined.

Solutions are developed to correct the causes and improve the process. Criteria for judging the possible solutions include cost, feasibility, resistance to change, consequences, training, and so forth. Once the solutions have been agreed by the team, testing and implementation follow.

Diagrams are posted in key locations to stimulate continued reference as similar or new problems arise. The diagrams are revised as solutions are found and improvements are made.

The cause-and-effect diagram has nearly unlimited application in research, manufacturing, marketing, office operations, and so forth. One of its strongest assets is the participation and contribution of everyone involved in the brainstorming process. The diagrams are useful in

1. *Analyzing* actual conditions for the purpose of product or service quality improvement, more efficient use of resources, and reduced costs.
2. *Elimination* of conditions causing nonconforming product or service and customer complaints.
3. *Standardization* of existing and proposed operations.
4. *Education and training* of personnel in decision making and corrective-action activities.

The previous paragraphs have described the *cause enumeration* type of C&E diagram, which is the most common type. There are two other types of C&E diagrams that are similar to the cause enumeration type. They are the dispersion analysis and process analysis types. The only difference among the three methods is the organization and arrangement.

The *dispersion analysis* type of C&E diagram looks just like the cause enumeration type when both are complete. The difference is in the approach to constructing it. For this type, each major branch is filled in completely before starting work on any of the other branches. Also, the objective is to analyze the causes of dispersion or variability.

The *process analysis* type of C&E diagram is the third type, and it does look different from the other two. In order to construct this diagram, it is necessary to write each step of the production process. Steps in the production process such as load, cut, bore, c'sink, chamfer, and unload become the *major causes,* as shown in Figure 4-4. Minor causes are then connected to the major ones. This C&E diagram is for elements within an operation. Other possibilities are operations within a process, an assembly process, a continuous chemical process, and so forth. The advantage of this type of C&E diagram is the ease of construction and its simplicity, because it follows the production sequence.

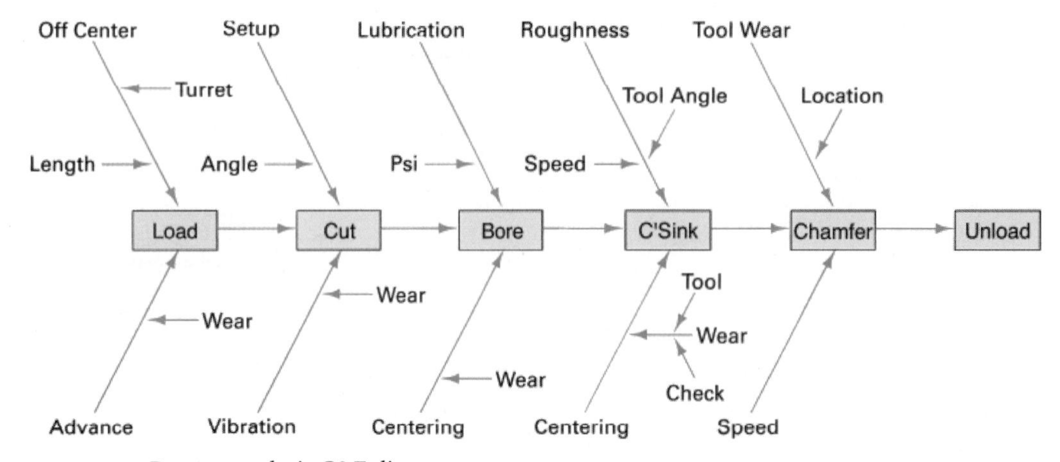

FIGURE 4-4 Process analysis C&E diagram.

CHECK SHEETS

The main purpose of check sheets is to ensure that data are collected carefully and accurately by operating personnel for process control and problem solving. Data should be presented in such a form that it can be quickly and easily used and analyzed. The form of the check sheet is individualized for each situation and is designed by the project team. Figure 4-5 shows a check sheet for paint nonconformities for bicycles. Figure 4-6 shows a maintenance check sheet for the swimming pool of a major motel chain. Checks are made on a daily and weekly basis, and some checks, such as temperature, are measured. This type of check sheet ensures that a check or test is made.

Figure 4-7 shows a check sheet for temperature. The scale on the left represents the midpoint and boundaries for each temperature range. Data collection for this type of check sheet is frequently accomplished by placing an × in the appropriate square. In this case, the time has been recorded in order to provide additional information for problem solving.

Whenever possible, check sheets are also designed to show location. For example, the check sheet for bicycle paint nonconformities could have shown an outline of a bicycle with small ×'s indicating the location of the nonconformities.

Figure 4-8 shows a check sheet for a 9-cavity plastic mold. This check sheet clearly shows that there are quality

CHECK SHEET

Product: Bicycle—32 **Date:** Jan. 21
Stage: Final Inspection **ID:** Paint
Number Inspected: 2217 **Inspector/Operator:** Jane Doe

Nonconformity Type	Check	Total
Blister	卌 卌 卌 卌 |	21
Light Spray	卌 卌 卌 卌 卌 卌 卌 |||	38
Drips	卌 卌 卌 卌 ||	22
Overspray	卌 卌 |	11
Splatter	卌 |||	8
Runs	卌 卌 卌 卌 卌 卌 卌 卌 卌 ||	47
Others	卌 卌 ||	12
	Total	159
Number Nonconforming	卌 |||	113

FIGURE 4-5 Check sheet for paint nonconformities.

D = Daily A = As Needed

Hot Tub		Mon.	Tues.	Wed.	Th.	Fri.	Sat.	Sun.
Chemical Test (Add if Needed) ph/chlorine	(D)	7.4						
Temperature	(D)	81°						
Add Water (If Needed)	(D)							
Clean Deck Around Hot Tub	(D)	✓						
Pool								
Chemical Test (Add if Needed)	(D)	7.6						
Add Water (If Needed)	(D)	300 gals.						
Check Temperature	(D)	78°						
Vacuum Pool (If Needed)	(A)							
Filter Backwash (20 lb.)	(A)	✓						
Lint Filter	(D)	✓						
Sweep and Hose Off Deck	(D)	✓						
General Cleaning								
Vacuum Carpets	(D)	✓						
Vacuum and Sweep Building B	(D)	✓						
Clean Tables	(D)	✓						
Sweep and Mop Wooden Deck	(D)	✓						
Clean Outside Deck, Bring in Chairs	(D)	✓						
Take Out Trash	(D)	✓						
Empty Building B Trash Cans	(D)	✓						
Wash Windows	(D)	✓						
Bathrooms								
Scrub Sinks, Toilets, and Showers	(D)	✓						
Sweep and Mop Floors	(D)	✓						
Empty Trash and Check Lockers	(D)	✓						
Cover Hot Tub (At End of the Night)	(D)	✓						
Check Pool Filters—Be Sure It's On	(D)	✓						

List any and all deviations from this work schedule on the reverse side, date it, and initial it.

FIGURE 4-6 Check sheet for swimming pool.

problems at the upper corners of the mold. What additional information would the reader suggest?

Creativity plays a major role in the design of a check sheet. It should be user-friendly and, whenever possible, include information on time and location.

PROCESS FLOW DIAGRAM

For many products and services, it may be useful to construct a flow diagram, which is also called a process map. It is a schematic diagram that shows the flow of the product or service as it moves through the various processing stations or operations. The diagram makes it easy to visualize the entire system, identify potential trouble spots, and locate control activities.

Standardized symbols are used by industrial engineers; however, they are not necessary for problem solving. Figure 4-9 shows a flow diagram for the order entry activity of a made-to-order company. Enhancements to the diagram are the addition of time to complete an operation and the number of people performing an operation. The diagram shows who is the next customer in the process, thereby increasing the understanding of the process.

Flow diagrams are best constructed by a team, because it is rare for one individual to understand the entire process.

TEMPERATURE CHECK SHEET

385	387.4 382.5						
380	382.4 377.5						
375	377.4 372.5	10.0					
370	372.4 367.5						
365	367.4 362.5	7.0	7.5	9.0			
360	362.4 357.5	8.0	8.5				
355	357.4 352.5	9.5					

FIGURE 4-7 Check sheet for temperature with time recorded.

XXXX XX	X	XXXX X
	XX	
	X	X

FIGURE 4-8 Check sheet for plastic mold nonconformities.

Improvements to the process can be accomplished by eliminating steps, combining steps, or making frequently occurring steps more efficient.

SCATTER DIAGRAM

A scatter diagram is a graphical presentation of the relationship between two variables. One variable, which is usually the controllable one, is placed on the x axis and the other, or dependent variable, is placed on the y axis as shown in Figure 4-10. The plotted points are ordered pairs (x, y) of the variables. The subject is discussed in detail in Chapter 5.

HISTOGRAM

Histograms are discussed in Chapter 5. They describe the variation in the process as illustrated by Figure 4-11. The histogram graphically shows the process capability and, if desired, the relationship to the specifications and the nominal. It also suggests the shape of the population and indicates if there are any gaps in the data.

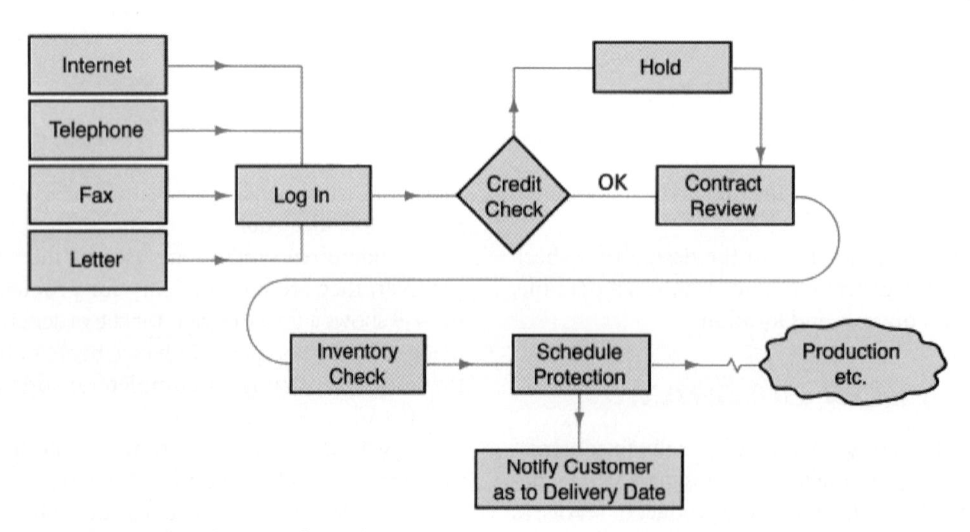

FIGURE 4-9 Flow diagram for order entry.

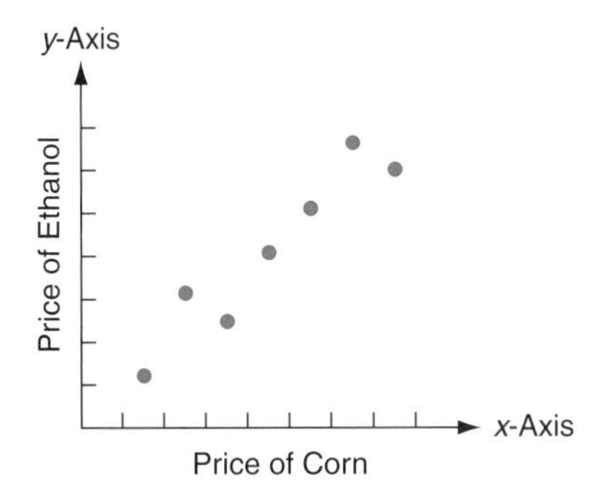

FIGURE 4-10 Scatter diagram for price of corn and price of ethanol.

CONTROL CHARTS

Control charts are discussed in Chapters 6, 7, and 9. A control chart, illustrating quality improvement, is shown in Figure 4-12. Control charts are an outstanding technique for problem solving and the resulting quality improvement.

Quality improvement occurs in two situations. When a control chart is first introduced, the process usually is unstable. As assignable causes for out-of-control conditions are identified and corrective action taken, the process becomes stable, with a resulting quality improvement.

The second situation concerns the testing or evaluation of ideas. Control charts are excellent decision makers because the pattern of the plotted points will determine if the idea is a good one, a poor one, or has no effect on the process. If the idea is a good one, the pattern of plotted points of the $\overline{X}$ chart will converge on the central line, $\overline{X}_0$. In other words, the pattern will get closer to perfection, which is the central line. For the R chart and the attribute charts, the pattern will tend toward zero, which is perfection. These improvement patterns are illustrated in Figure 4-12. If the idea is a poor one, an opposite pattern will occur. If the pattern of plotted points does not change, then the idea has no effect on the process.

Although the control charts are excellent for problem solving by improving the quality, they have limitations when

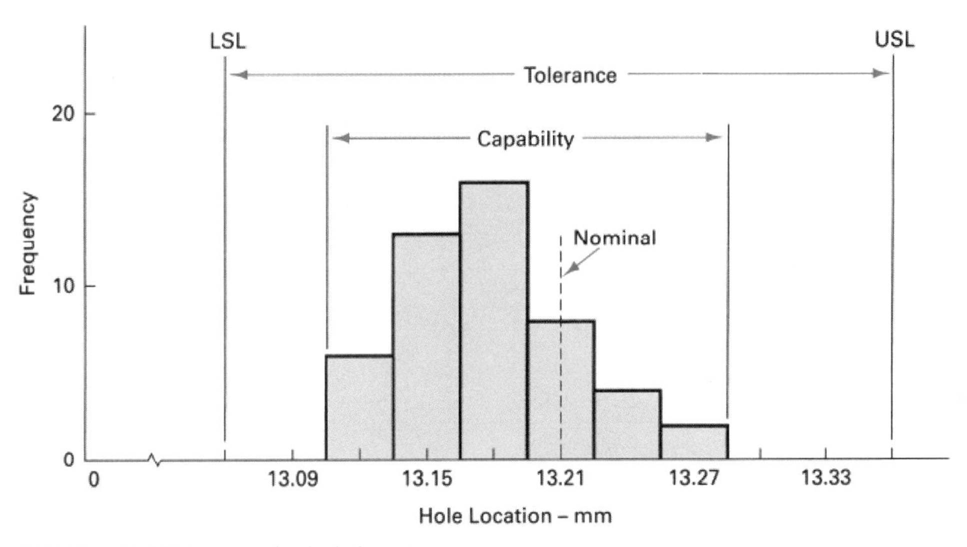

FIGURE 4-11 Histogram for hole location.

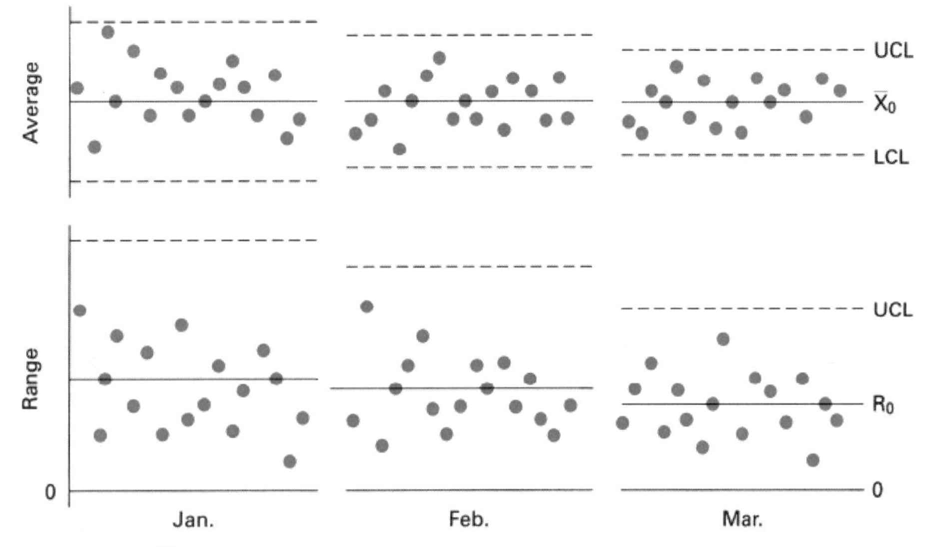

FIGURE 4-12 $\overline{X}$ and R charts, showing quality improvement.

used to monitor or maintain a process. The precontrol technique is much better at monitoring.

RUN CHART

A run chart is a graphical presentation of data collected from a process running over time. It is similar to a control chart, except it does not have control limits—it just shows the process variation. If the control limits are removed from Figure 4-12, the result is a run chart.

COMPUTER PROGRAM

The EXCEL program file on the website will solve for the Pareto diagram. Its file name is *Pareto.*

EXERCISES

1. Construct a Pareto diagram for replacement parts for an electric stove. Data for a six-month period are: oven door, 193; timer, 53; front burners, 460; rear burners, 290; burner control, 135; drawer rollers, 46; other, 84; and oven regulators, 265.
2. A project team is studying the downtime cost of a soft drink bottling line. Data analysis in thousands of dollars for a three-month period are: back-pressure regulator, 30; adjust feed worm, 15; jam copper head, 6; lost cooling, 52; valve replacement, 8; and other, 5. Construct a Pareto diagram.
3. Approximately two-thirds of all automobile accidents are due to improper driving. Construct a Pareto diagram without the cumulative line for the data: improper turn, 3.6%; driving too fast for conditions, 28.1%; following too closely, 8.1%; right-of-way violations, 30.1%; driving left of center, 3.3%; improper overtaking, 3.2%; and other, 23.6%.
4. A major DVD-of-the-month club collected data on the reasons for returned shipments during a quarter. Results are: wrong selection, 50,000; refused, 195,000; wrong address, 68,000; order canceled, 5000; and other, 15,000. Construct a Pareto diagram.
5. Paint nonconformities for a one-month period for a riding lawn mower manufacturer are: blister, 212; light spray, 582; drips, 227; overspray, 109; splatter, 141; bad paint, 126; runs, 434; and other, 50. Construct a Pareto diagram.
6. Construct a Pareto diagram for the analysis of internal failures for the following data:

Type of Cost	Dollars (in Thousands)
Purchasing—rejects	205
Design—scrap	120
Operations—rework	355
Purchasing—rework	25
All other	65

7. Construct a Pareto diagram for the analysis of the external failure costs for a wireless telephone manufacturer using the following data:

Type of Cost	Dollars (in Thousands)
Customer complaints	20
Returned goods	30
Retrofit costs	50
Warranty claims	90
Liability costs	10
Penalties	5
Customer goodwill	25

8. A building construction company needs a Pareto diagram for the analysis of the following design department cost of poor quality:

Element	Dollars (in Thousands)
Progress reviews	5
Support activities	3
Qualification tests	2
Corrective action	15
Rework	50
Scrap	25
Liaison	2

9. Form a project team of six or seven people, elect a leader, and construct a cause-and-effect diagram for bad coffee from a 22-cup appliance used in the office.
10. Form a project team of six or seven people, elect a leader, and construct a C&E diagram for
    a. Dispersion analysis type for a quality characteristic.
    b. Process analysis type for a sequence of office activities on an insurance form.
    c. Process analysis type for a sequence of production activities on a lathe: load 25 mm dia.—80 mm long rod, rough turn 12 mm dia.—40 mm long, UNF thread—12 mm dia., thread relief, finish turn 25 mm dia.—20 mm long, cut off, and unload.
11. Design a check sheet for the maintenance of a piece of equipment such as a gas furnace, laboratory scale, or computer.
12. Construct a flow diagram for the manufacture of a product or the providing of a service.
13. Using the EXCEL program file , solve Exercises 1, 2, 3, 4, 5, 6, 7, and 8.

CHAPTER
FIVE

FUNDAMENTALS OF STATISTICS

OBJECTIVES

Upon completion of this chapter, the reader is expected to

- know the difference between a variable and an attribute;
- perform mathematical calculations to the correct number of significant figures;
- construct histograms for simple and complex data and know the parts of a histogram;
- calculate and effectively use the measures of central tendency, dispersion, and interrelationship;
- understand the concept of a universe and a sample;
- understand the concept of a normal curve and the relationship to the mean and standard deviation;
- calculate the percent of items below a value, above a value, or between two values for data that are normally distributed. Calculate the process center given the percent of items below a value;
- calculate the various tests of normality;
- construct and perform the necessary calculations for a scatter diagram.

INTRODUCTION

Definition of Statistics

The word *statistics* has two generally accepted meanings:

1. A collection of quantitative data pertaining to any subject or group, especially when the data are systematically gathered and collated. Examples of this meaning are blood pressure statistics, statistics of a football game, employment statistics, and accident statistics, to name a few.
2. The science that deals with the collection, tabulation, analysis, interpretation, and presentation of quantitative data.

It is noted that the second meaning is broader than the first, because it, too, is concerned with collection of data. The use of statistics in quality deals with the second and broader meaning and involves the divisions of collecting, tabulating, analyzing, interpreting, and presenting the quantitative data. Each division is dependent on the accuracy and completeness of the preceding one. Data may be collected by a technician measuring the tensile strength of a plastic part or by a market researcher determining consumer color preferences. It may be tabulated by simple paper-and-pencil techniques or by the use of a computer. Analysis may involve a cursory visual examination or exhaustive calculations. The final results are interpreted and presented to assist in making decisions concerning quality.

There are two phases of statistics:

1. *Descriptive* or *deductive statistics,* which endeavor to describe and analyze a subject or group.
2. *Inductive statistics,* which endeavor to determine from a limited amount of data (sample) an important conclusion about a much larger amount of data (population). Because these conclusions or inferences cannot be stated with absolute certainty, the language of *probability* is often used.

This chapter covers the statistical fundamentals necessary to understand the subsequent quality control techniques. Fundamentals of probability are discussed in Chapter 8. An understanding of statistics is vital for an understanding of quality and, for that matter, many other disciplines.

Collecting the Data

Data may be collected by direct observation or indirectly through written or verbal questions. The latter technique is used extensively by market research personnel and public opinion pollsters. Data that are collected for quality purposes are obtained by direct observation and are classified as either variables or attributes. *Variables* are those quality characteristics that are measurable, such as a weight measured in grams. *Attributes,* on the other hand, are those quality characteristics that are classified as either conforming or not conforming to specifications, such as a "go/no go gage."

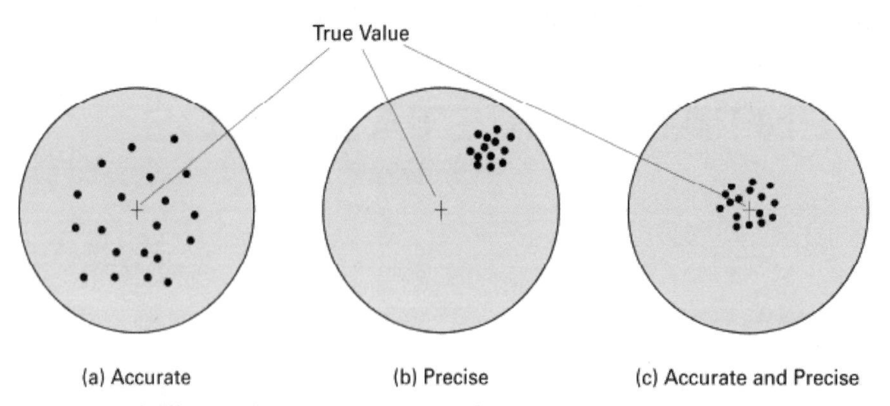

FIGURE 5-1 Difference between Accuracy and Precision

A variable that is capable of any degree of subdivision is referred to as *continuous*. The weight of a gray iron casting, which can be measured as 11 kg, 11.33 kg, or 11.3398 kg (25 lb), depending on the accuracy of the measuring instrument, is an example of a continuous variable. Measurements such as meters (feet), liters (gallons), and pascals (pounds per square inch) are examples of continuous data. Variables that exhibit gaps are called *discrete*. The number of nonconforming rivets in a travel trailer can be any whole number, such as 0, 3, 5, 10, 96, . . . ; however, there cannot be, say, 4.65 nonconforming rivets in a particular trailer. In general, continuous data are measurable, whereas discrete data are countable.

Sometimes it is convenient for verbal or nonnumerical data to assume the nature of a variable. For example, the quality of the surface finish of a piece of furniture can be classified as poor, average, or good. The poor, average, or good classification can be replaced by the numerical values of 1, 2, or 3, respectively. In a similar manner, educational institutions assign to the letter grades of A, B, C, D, and F the numerical values of 4, 3, 2, 1, and 0, respectively, and use those discrete numerical values as discrete variables for computational purposes.

Although many quality characteristics are stated in terms of variables, there are many characteristics that must be stated as attributes. Frequently, those characteristics that are judged by visual observation are classified as attributes. The wire on an electric motor is either attached to the terminal or it is not; the words on this page are correctly spelled or they are incorrectly spelled; the switch is on or it is off; and the answer is right or it is wrong. The examples given in the previous sentence show conformance to a particular specification or nonconformance to that specification.

It is sometimes desirable for variables to be classified as attributes. Factory personnel are frequently interested in knowing if the product they are producing conforms to the specifications. For example, the numerical values for the weight of a package of sugar may not be as important as the information that the weight is within the prescribed limits. Therefore, the data, which are collected on the weight of the package of sugar, are reported as conforming or not conforming to specifications.

In collecting data, the number of figures is a function of the intended use of the data. For example, in collecting data on the life of light bulbs, it is acceptable to record 995.6 h; however, recording a value of 995.632 h is too accurate and unnecessary. Similarly, if a keyway specification has a lower limit of 9.52 mm (0.375 in.) and an upper limit of 9.58 mm (0.377 in.), data would be collected to the nearest 0.001 mm and rounded to the nearest 0.01 mm. In general, the more number of figures to the right of the decimal point, the more sophisticated is the measuring instrument.

Measuring instruments may not give a true reading because of problems due to accuracy and precision. Figure 5-1(a) shows an accurate series of repeated measurements because their average is close to the true value, which is at the center. In Figure 5-1(b) the repeated measurements in the series are precise (very close together), but are not close to the true value. Figure 5-1(c) shows the series of repeated measurements tightly grouped around the true value, and these measurements are both accurate and precise.

Many failures occur because a team tried to solve a problem without an adequate measurement system. Thus, the first step is to improve that variation.[1]

The rounding of data requires that certain conventions be followed. In rounding the numbers 0.9530, 0.9531, 0.9532, 0.9533, and 0.9534 to the nearest thousandth, the answer is 0.953, because all the numbers are closer to 0.953 than they are to 0.954. And in rounding the numbers 0.9535, 0.9536, 0.9537, 0.9538, and 0.9539, the answer is 0.954, because all the numbers are closer to 0.954 than to 0.953. In other words, if the last digit is 5 or greater, the number is rounded up.[2]

Based on this rounding rule, a rounded number is an approximation of the exact number. Thus, the rounded number 6.23 lies between 6.225 and 6.235 and is expressed as

$$6.225 \leq 6.23 < 6.235$$

The precision is 0.010, which is the difference between 6.225 and 6.235. An associated term is *greatest possible error* (g.p.e.), which is one-half the precision or $0.010 \div 2 = 0.005$.

[1]Stefan Steiner and Jock MacKay, "Statistical Engineering: A Case Study," *Quality Progress* (June 2006): 33–39.

[2]This rounding rule is the simplest to use. Another one decides whether to round up or down based on the previous digit. Thus, a number that ends in *x*5 will be rounded up if *x* is odd and down if *x* is even. For example, rounding to two decimal places, the number 6.415 would equal or approximate 6.42 and the number 3.285 would equal 3.28.

Sometimes precision and g.p.e. are not adequate to describe error. For example, the numbers 8765.4 and 3.2 have the same precision (0.10) and g.p.e. (0.05); however, the relative error (r.e.) is much different. It equals the g.p.e. of the number divided by the number. Thus,

$$\text{r.e. of } 8765.4 = 0.05 \div 8765.4 = 0.000006$$
$$\text{r.e. of } 3.2 = 0.05 \div 3.2 = 0.02$$

The following examples will help to clarify these concepts.

Rounded Number with Boundaries	Precision	g.p.e.	r.e.
$5.645 \le 5.65 < 5.655$	0.01	0.005	0.0009
$431.5 \le 432 < 432.5$	1.0	0.5	0.001

In working with numerical data, *significant figures* are very important. The significant figures of a number are the digits exclusive of any leading zeros needed to locate the decimal point. For example, the number 3.69 has three significant figures; 36.900 has five significant figures; 2700 has four significant figures; 22.0365 has six significant figures; and 0.00270 has three significant figures. Trailing zeros are counted as being significant, whereas leading zeros are not. This rule leads to some difficulty when working with whole numbers, because the number 300 can have one, two, or three significant figures. This difficulty can be eliminated by using scientific notation. Therefore, 3×10^2 has one significant figure, 3.0×10^2 has two significant figures, and 3.00×10^2 has three significant figures. Numbers with leading zeros can be written as 2.70×10^{-3} for 0.00270. Numbers that are associated with counting have an unlimited number of significant figures, and the counting number 65 can be written as 65 or 65.000. . . .

The following examples illustrate the number 600 with three, two, and one significant figures, respectively:

Rounded Number with Boundaries	Precision	g.p.e.	r.e.
$599.5 \le 6.00 \times 10^2 < 600.5$	1	0.5	0.0008
$595 \le 6.0 \times 10^2 < 605$	10	5	0.008
$550 \le 6 \times 10^2 < 650$	100	50	0.08

When performing the mathematical operations of multiplication, division, and exponentiation, the answer has the same number of significant figures as the number with the fewest significant figures. The following examples will help to clarify this rule.

$\sqrt{81.9} = 9.05$

$6.59 \times 2.3 = 15$

$32.65 \div 24 = 1.4$ (24 is not a counting number)

$32.65 \div 24 = 1.360$ (24 is a counting number with a value of 24.00. . .)

When performing the mathematical operations of addition and subtraction, the final answer can have no more significant figures after the decimal point than the number with the fewest significant figures after the decimal point. In cases involving numbers without decimal points, the final answer has no more significant figures than the number with the fewest significant figures. Examples to clarify this rule are as follows:

$38.26 - 6 = 32$ (6 is not a counting number)

$38.26 - 6 = 32.26$ (6 is a counting number)

$38.26 - 6.1 = 32.2$ (answer is rounded from 32.16)

$8.1 \times 10^3 - 1232 = 6.9 \times 10^3$ (fewest significant figures are two)

$8.100 \times 10^3 - 1232 = 6868$ (fewest significant figures are four)

Following these rules will avoid discrepancies in answers among quality personnel; however, some judgment may sometimes be required. When a series of calculations is made, significant figure and rounding determinations can be evaluated at the end of the calculations. In any case, the final answer can be no more accurate than the incoming data. For example, if the incoming data is to two decimal places (x.xx), then the final answer must be to two decimal places (x.xx).

Describing the Data

In industry, business, health care, education, and government, the mass of data that has been collected is voluminous. Even one item, such as the number of daily billing errors of a large organization, can represent such a mass of data that it can be more confusing than helpful. Consider the data shown in Table 5-1. Clearly these data, in this form, are difficult to use and are not effective in describing the data's characteristics. Some means of summarizing the data are needed to show what value or values the data tend to cluster about and how the data are dispersed or spread out. Two techniques are available to accomplish this summarization of data—graphical and analytical.

The graphical technique is a plot or picture of a *frequency distribution,* which is a summarization of how the data points (observations) occur within each subdivision of observed values or groups of observed values. Analytical techniques summarize data by computing a *measure of*

TABLE 5-1 Number of Daily Billing Errors

0	1	3	0	1	0	1	0
1	5	4	1	2	1	2	0
1	0	2	0	0	2	0	1
2	1	1	1	2	1	1	
0	4	1	3	1	1	1	
1	3	4	0	0	0	0	
1	3	0	1	2	2	3	

central tendency and a *measure of dispersion*. Sometimes both the graphical and analytical techniques are used.

These techniques will be described in the subsequent sections of this chapter.

FREQUENCY DISTRIBUTION

Ungrouped Data

Ungrouped data comprise a listing of the observed values, whereas grouped data represent a lumping together of the observed values. The data can be discrete, as they are in this section, or continuous, as they are in the next section.

Because unorganized data are virtually meaningless, a method of processing the data is necessary. Table 5-1 will be used to illustrate the concept. An analyst reviewing the information as given in this table would have difficulty comprehending the meaning of the data. A much better understanding can be obtained by tallying the frequency of each value, as shown in Table 5-2.

The first step is to establish an *array,* which is an arrangement of raw numerical data in ascending or descending order of magnitude. An array of ascending order from 0 to 5 is shown in the first column of Table 5-2. The next step is to tabulate the frequency of each value by placing a tally mark under the tabulation column and in the appropriate row. Start with the numbers 0, 1, 1, 2, . . . of Table 5-1 and continue placing tally marks until all the data have been tabulated. The last column of Table 5-2 is the numerical value for the number of tallies and is called the *frequency*.

Analysis of Table 5-2 shows that one can visualize the distribution of the data. If the "Tabulation" column is eliminated, the resulting table is classified as a *frequency distribution,* which is an arrangement of data to show the frequency of values in each category.

The frequency distribution is a useful method of visualizing data and is a basic statistical concept. To think of a set of numbers as having some type of distribution is fundamental for solving quality problems. There are different types of frequency distributions, and the type of distribution can indicate the problem-solving approach.

TABLE 5-2 Tally of Number of Daily Billing Errors

Number Nonconforming	Tabulation	Frequency
0																	15
1																						20
2										8
3							5
4					3
5			1

Frequency distributions are presented in graphical form when greater visual clarity is desired. There are a number of different ways to present the frequency distribution.

A *histogram* consists of a set of rectangles that represent the frequency in each category. It represents graphically the frequencies of the observed values. Figure 5-2(a) is a histogram for the data in Table 5-2. Because this is a discrete variable, a vertical line in place of a rectangle would have been theoretically correct (see Figure 5-5). However, the rectangle is commonly used.

Another type of graphic representation is the relative frequency distribution. Relative, in this sense, means the proportion or fraction of the total. Relative frequency is calculated by dividing the frequency for each data value (in this case, number nonconforming) by the total, which is the sum of the frequencies. These calculations are shown in the third column of Table 5-3. Graphical representation is shown in Figure 5-2(b). Relative frequency has the advantage of a reference. For example, the proportion of 15 nonconforming units is 0.29. Some practitioners prefer to use percents for the vertical scale rather than fractions.

Cumulative frequency is calculated by adding the frequency of each data value to the sum of the frequencies for the previous data values. As shown in the fourth column of Table 5-3, the cumulative frequency for 0 nonconforming units is 15; for 1 nonconforming unit, 15 + 20 = 35; for 2 nonconforming units, 35 + 8; and so on. Cumulative frequency is the number of data points equal to or less than a data value.

TABLE 5-3 Different Frequency Distributions of Data Given in Table 5-1

Number Nonconforming	Frequency	Relative Frequency	Cumulative Frequency	Relative Cumulative Frequency
0	15	$15 \div 52 = 0.29$	15	$15 \div 52 = 0.29$
1	20	$20 \div 52 = 0.38$	$15 + 20 = 35$	$35 \div 52 = 0.67$
2	8	$8 \div 52 = 0.15$	$35 + 8 = 43$	$43 \div 52 = 0.83$
3	5	$5 \div 52 = 0.10$	$43 + 5 = 48$	$48 \div 52 = 0.92$
4	3	$3 \div 52 = 0.06$	$48 + 3 = 51$	$51 \div 52 = 0.98$
5	1	$1 \div 52 = 0.02$	$51 + 1 = 52$	$52 \div 52 = 1.00$
Total	52	1.00		

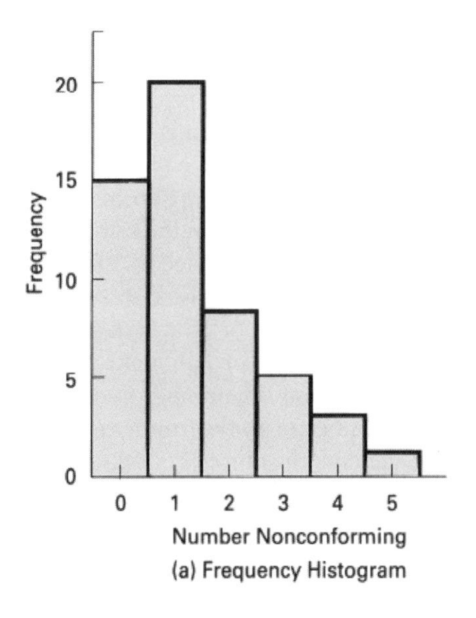

(a) Frequency Histogram

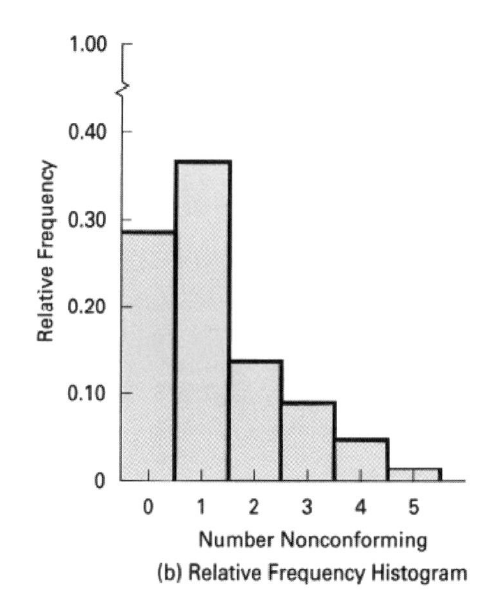

(b) Relative Frequency Histogram

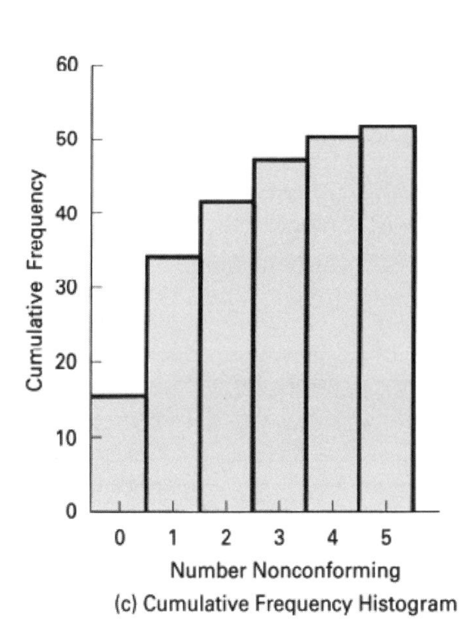

(c) Cumulative Frequency Histogram

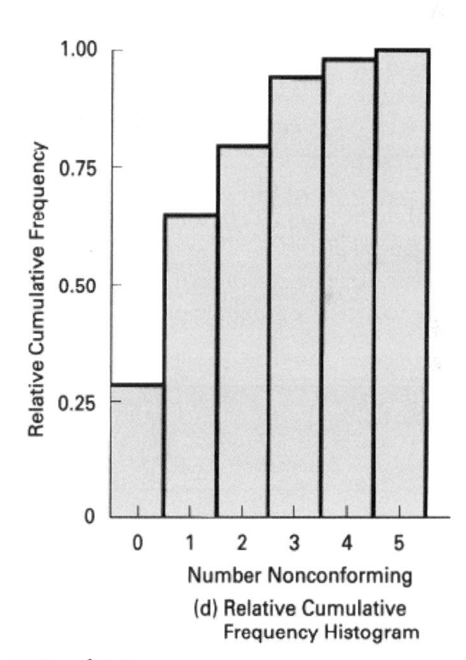

(d) Relative Cumulative Frequency Histogram

FIGURE 5-2 Graphic Representation of Data in Tables 5-2 and 5-3

For example, this value for 2 or less nonconforming units is 43. Graphic representation is shown in Figure 5-2(c).

Relative cumulative frequency is calculated by dividing the cumulative frequency for each data value by the total. These calculations are shown in the fifth column of Table 5-3, and the graphical representation is shown in Figure 5-2(d). The graph shows that the proportion of the billing errors that have 2 or fewer nonconforming units is 0.83 or 83%.

The preceding example was limited to a discrete variable with six values. Although this example is sufficient for a basic introduction to the frequency distribution concept, it does not provide a thorough knowledge of the subject. Most data are continuous rather than discrete and require grouping.

Grouped Data

The construction of a frequency distribution for grouped data is more complicated because there is usually a larger number of categories. An example problem using a continuous variable illustrates the concept.

1. **Collect data and construct a tally sheet.** Data collected on the weights of 110 steel shafts are shown in Table 5-4. The first step is to make a tally of the values, as shown in Table 5-5. In order to be more efficient, the weights are coded from 2.500 kg, which is a technique used to simplify data. Therefore, a weight with a value of 31 is equivalent to 2.531 kg (2.500 + 0.031). Analysis of Table 5-5 shows that more infor-

TABLE 5-4 Steel Shaft Weight (kg)

2.559	2.556	2.566	2.546	2.561
2.570	2.546	2.565	2.543	2.538
2.560	2.560	2.545	2.551	2.568
2.546	2.555	2.551	2.554	2.574
2.568	2.572	2.550	2.556	2.551
2.561	2.560	2.564	2.567	2.560
2.551	2.562	2.542	2.549	2.561
2.556	2.550	2.561	2.558	2.556
2.559	2.557	2.532	2.575	2.551
2.550	2.559	2.565	2.552	2.560
2.534	2.547	2.569	2.559	2.549
2.544	2.550	2.552	2.536	2.570
2.564	2.553	2.558	2.538	2.564
2.552	2.543	2.562	2.571	2.553
2.539	2.569	2.552	2.536	2.537
2.532	2.552	2.575(h)	2.545	2.551
2.547	2.537	2.547	2.533	2.538
2.571	2.545	2.545	2.556	2.543
2.551	2.569	2.559	2.534	2.561
2.567	2.572	2.558	2.542	2.574
2.570	2.542	2.552	2.551	2.553
2.546	2.531(l)	2.563	2.554	2.544

mation is conveyed to the analyst than from the data of Table 5-4; however, the general picture is still somewhat blurred.

In this problem there are 45 categories, which are too many and must be reduced by grouping into cells.[3] A *cell* is a grouping within specified boundaries of observed values along the *abscissa* (horizontal axis) of the histogram. The grouping of data by cells simplifies the presentation of the distribution; however, some of the detail is lost. When the number of cells is large, the true picture of the distribution is distorted by cells having an insufficient number of items or none at all. Or, when the number of cells is small, too many items are concentrated in a few cells and the true picture of the distribution is also distorted.

The number of cells or groups in a frequency distribution is largely a matter of judgment by the analyst. This judgment is based on the number of observations and can require trial and error to determine the optimum number of cells. In general, the number of cells should be between 5 and 20. Broad guidelines are as follows: Use 5 to 9 cells when the number of observations is less than 100; use 8 to 17 cells when the number of observations is between 100 and 500; and use 15 to 20 cells when the number of observations is greater than 500. To provide flexibility, the numbers of cells in the guidelines are overlapping. It is emphasized that these guidelines are not rigid and can be adjusted when necessary to present an acceptable frequency distribution.

2. **Determine the range.** It is the difference between the highest observed value and the lowest observed value, as shown by the formula

$$R = X_h - X_l$$

where R = range

X_h = highest number

X_l = lowest number

TABLE 5-5 Tally Sheet Shaft Weight (Coded from 2.500 kg)

Weight	Tabulation	Weight	Tabulation	Weight	Tabulation
31	ǀ	46	ǀǀǀǀ	61	卌
32	ǀǀ	47	ǀǀǀ	62	ǀǀ
33	ǀ	48		63	ǀ
34	ǀǀ	49	ǀǀ	64	ǀǀǀ
35		50	ǀǀǀǀ	65	ǀǀ
36	ǀǀ	51	卌ǀǀǀ	66	ǀ
37	ǀǀ	52	卌ǀ	67	ǀǀ
38	ǀǀǀ	53	ǀǀǀ	68	ǀǀ
39	ǀ	54	ǀǀ	69	ǀǀǀ
40		55	ǀ	70	ǀǀǀ
41		56	卌	71	ǀǀ
42	ǀǀǀ	57	ǀ	72	ǀǀ
43	ǀǀǀ	58	ǀǀǀ	73	
44	ǀǀ	59	卌	74	ǀǀ
45	ǀǀǀǀ	60	卌	75	ǀǀ

[3]The word *class* is sometimes used in place of *cell*.

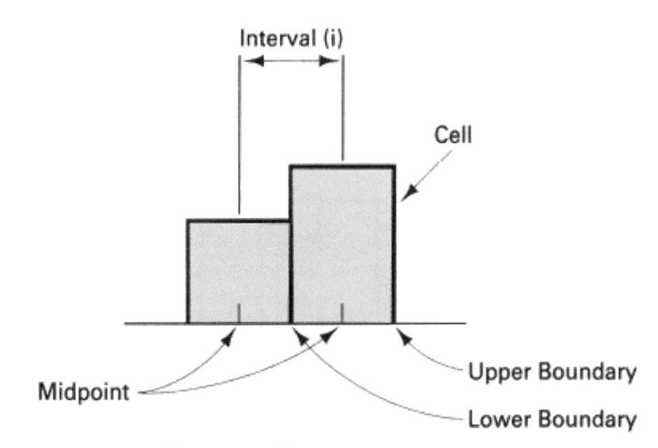

FIGURE 5-3 Cell Nomenclature

From Table 5-4 or Table 5-5, the highest number is 2.575 and the lowest number is 2.531. Thus,

$$\begin{aligned} R &= X_h - X_l \\ &= 2.575 - 2.531 \\ &= 0.044 \end{aligned}$$

3. **Determine the cell interval.** The *cell interval* is the distance between adjacent cell midpoints as shown in Figure 5-3. Whenever possible, an odd interval such as 0.001, 0.07, 0.5, or 3 is recommended, so that the midpoint values will be to the same number of decimal places as the data values. The simplest technique is to use Sturgis' rule, which is

$$i = \frac{R}{1 + 3.322 \log n}$$

For the example problem, the answer is

$$i = \frac{R}{1 + 3.322 \log n} = \frac{0.044}{1 + 3.322(2.041)} = 0.0057$$

and the closest odd interval for the data is 0.005.

Another technique uses trial and error. The cell interval (i) and the number of cells (h) are interrelated by the formula, $h = R/i$. Because h and i are both unknown, a trial-and-error approach is used to find the interval that will meet the guidelines.

$$\text{Assume that } i = 0.003; \text{ then } h = \frac{R}{i} = \frac{0.044}{0.003} = 15$$

$$\text{Assume that } i = 0.005; \text{ then } h = \frac{R}{i} = \frac{0.044}{0.005} = 9$$

$$\text{Assume that } i = 0.007; \text{ then } h = \frac{R}{i} = \frac{0.044}{0.007} = 6$$

A cell interval of 0.005 with 9 cells will give the best presentation of the data based on the guidelines for the number of cells given in step 1.

Both techniques give similar answers.

4. **Determine the cell midpoints.** The lowest cell midpoint must be located to include the lowest data value in its cell. The simplest technique is to select the lowest data point (2.531) as the midpoint value for the first cell. A better technique is to use the formula

$$MP_l = X_l + \frac{i}{2} \text{ (Do not round answer)}$$

where MP_l = midpoint for lowest cell

For the example problem, the answer is

$$MP_l = X_l + \frac{i}{2} = 2.531 + \frac{0.005}{2} = 2.533$$

The answer cannot be rounded using this formula. Because the interval is 0.005, there are 5 data values in each cell; therefore, a midpoint value of 2.533 can be used for the first cell. This value will have the lowest data value (2.531) in the first cell, which will have data values of 2.531, 2.532, 2.533, 2.534, and 2.535.

Midpoint selection is a matter of judgment, and in this case a midpoint of 2.533 was selected and the number of cells is 9. Selection of any other midpoint, although not incorrect, would have given 10 cells in the frequency distribution. Selection of different midpoint values will produce different frequency distributions—for this example 5 are possible. The midpoints for the other 8 cells are obtained by adding the cell interval to the previous midpoint: 2.533 + 0.005 = 2.538, 2.538 + 0.005 = 2.543, 2.543 + 0.005 = 2.548, . . . , and 2.568 + 0.005 = 2.573. These midpoints are shown in Table 5-6.

The midpoint value is the most representative value within a cell, provided that the number of observations in a cell is large and the difference in boundaries is not too great. Even if this condition is not met, the number of observations above and below the midpoint of a cell will frequently be equal. And even if the number of observations above and below a cell midpoint is unbalanced in one direction, it will probably be offset by an unbalance in the opposite direction of another cell. Midpoint values should be to the same degree of accuracy as the original observations.

5. **Determine the cell boundaries.** *Cell boundaries* are the extreme or limit values of a cell, referred to as the

TABLE 5-6 Frequency Distribution of Steel Shaft Weight (kg)

Cell Boundaries	Cell Midpoint	Frequency
2.531–2.535	2.533	6
2.536–2.540	2.538	8
2.541–2.545	2.543	12
2.546–2.550	2.548	13
2.551–2.555	2.553	20
2.556–2.560	2.558	19
2.561–2.565	2.563	13
2.566–2.570	2.568	11
2.571–2.575	2.573	8
Total		110

upper boundary and the lower boundary. All the observations that fall between the upper and lower boundaries are classified into that particular cell. Boundaries are established so there is no question as to the location of an observation. Therefore, the boundary values are an extra decimal place or significant figure in accuracy than the observed values. Because the interval is odd, there will be

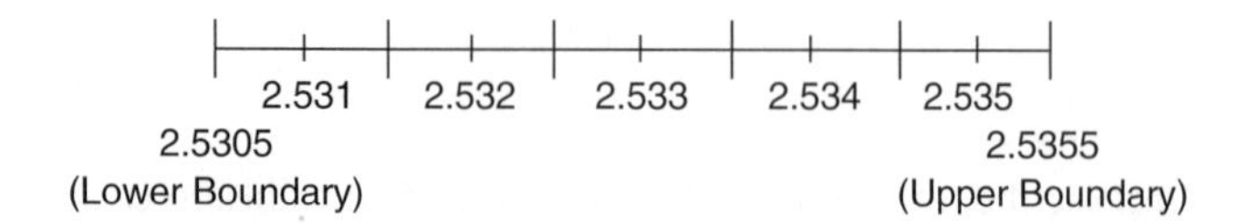

an equal number of data values on each side of the midpoint. For the first cell with a midpoint of 2.533 and an interval of 0.005, there will be two values on each side. Therefore, that cell will contain the values 2.531, 2.532, 2.533, 2.534, and 2.535. To prevent any gaps, the true boundaries are extended about halfway to the next number, which gives values of 2.5305 and 2.5355. The following number line illustrates this principle:

Some analysts prefer to leave the boundaries at the same number of decimal places as the data. No difficulty is encountered with this practice as long as the cell interval is odd and it is understood that the true boundaries are extended halfway to the next number. This practice is followed in this book and in EXCEL. It is shown in Table 5-6. Therefore, the lower boundary for the first cell is 2.531.

Once the boundaries are established for one cell, the boundaries for the other cells are obtained by successive additions of the cell interval. Therefore, the lower boundaries are $2.531 + 0.005 = 2.536$, $2.536 + 0.005 = 2.541, \ldots,$ $2.566 + 0.005 = 2.571$. The upper boundaries are obtained in a similar manner and are shown in the first column of Table 5-6.

6. **Post the cell frequency.** The amount of numbers in each cell is posted to the frequency column of Table 5-6. An analysis of Table 5-5 shows that for the lowest cell there are one 2.531, two 2.532, one 2.533, two 2.534, and zero 2.535. Therefore, there is a total of six values in the lowest cell, and the cell with a midpoint of 2.533 has a frequency of 6. The amounts are determined for the other cells in a similar manner.

The completed frequency distribution is shown in Table 5-6. This frequency distribution gives a better conception of the central value and how the data are dispersed about that value than the unorganized data or a tally sheet. The histogram is shown in Figure 5-4.

Information on the construction of the relative frequency, cumulative frequency, and relative cumulative frequency histograms for grouped data is the same as for ungrouped data, but with one exception. With the two cumulative frequency histograms, the true upper cell boundary is the value labeled on the abscissa. Construction of these histograms for the example problem is left to the reader as an exercise.

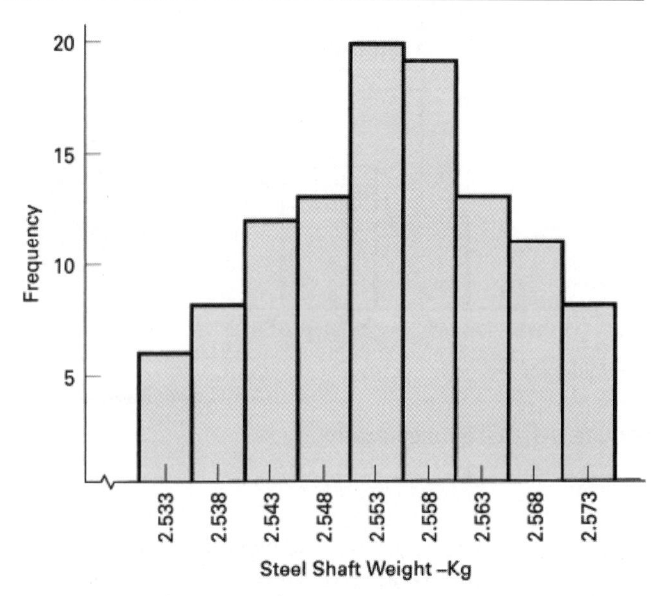

FIGURE 5-4 Histogram of Data in Table 5-6

The histogram describes the variation in the process. It is used to

1. Solve problems.
2. Determine the process capability.
3. Compare with specifications.
4. Suggest the shape of the population.
5. Indicate discrepancies in data such as gaps.

Other Types of Frequency Distribution Graphs

The bar graph can also represent frequency distributions, as shown in Figure 5-5(a) using the data of Table 5-1. As mentioned previously, the bar graph is theoretically correct for discrete data but is not commonly used.

The *polygon* or *frequency polygon* is another graphical way of presenting frequency distributions and is illustrated in Figure 5-5(b) using the data of Table 5-6. It is constructed by placing a dot over each cell midpoint at the height indicated for each frequency. The curve is extended at each end in order for the figure to be enclosed. Because the histogram shows the area in each cell, it is considered to present a better graphical picture than the polygon and is the one most commonly used.

The graph that is used to present the frequency of all values less than the upper cell boundary of a given cell is called a *cumulative frequency*, or *ogive*. Figure 5-5(c) shows a cumulative frequency distribution curve for the data in Table 5-6. The cumulative value for each cell is plotted on the graph and joined by a straight line. The true upper cell boundary is labeled on the abscissa except for the first cell, which also has the true lower boundary.

Characteristics of Frequency Distribution Graphs

The graphs of Figure 5-6 use smooth curves rather than the rectangular shapes associated with the histogram. A smooth curve represents a population frequency distribution, whereas the histogram represents a sample frequency distribution. The difference between a population and a sample is discussed in a later section of this chapter.

Frequency distribution curves have certain identifiable characteristics. One characteristic of the distribution concerns the symmetry or lack of symmetry of the data. Are the data equally distributed on each side of the central value, or are the data skewed to the right or to the left? Another characteristic concerns the number of modes or peaks to the data. There can be one mode, two modes (bimodal), or multiple modes. A final characteristic concerns the "peakedness" of the data. When the curve is quite peaked, it is referred to as *leptokurtic;* and when it is flatter, it is referred to as *platykurtic.*

Frequency distributions can give sufficient information about a quality control problem to provide a basis for decision making without further analysis. Distributions can also be compared in regard to location, spread, and shape as illustrated in Figure 5-7.

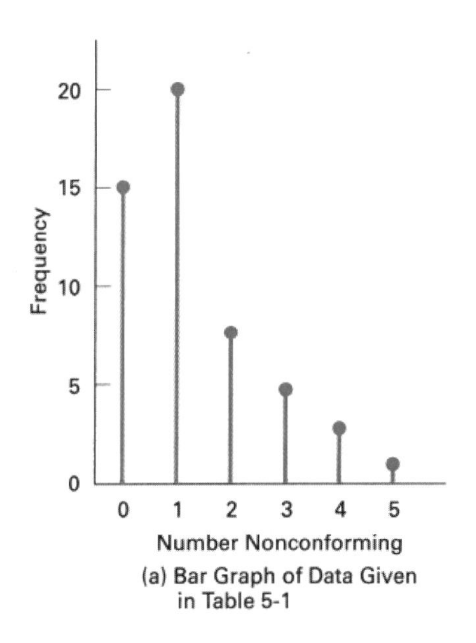

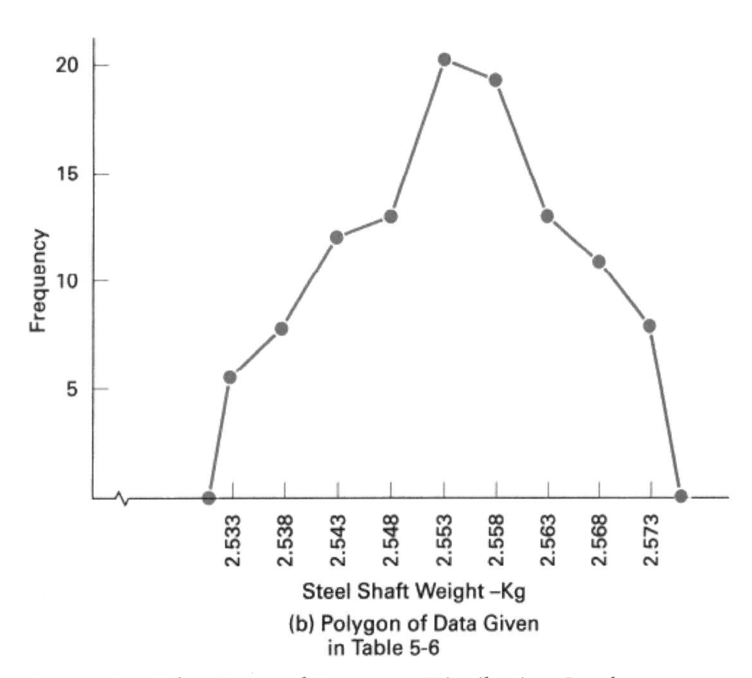

FIGURE 5-5 Other Types of Frequency Distribution Graphs

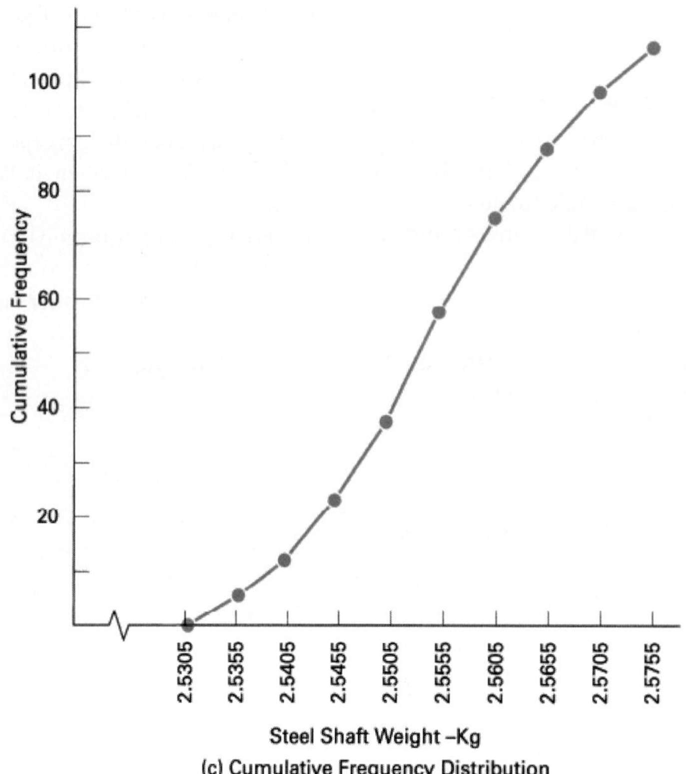

FIGURE 5-5 *(continued)*

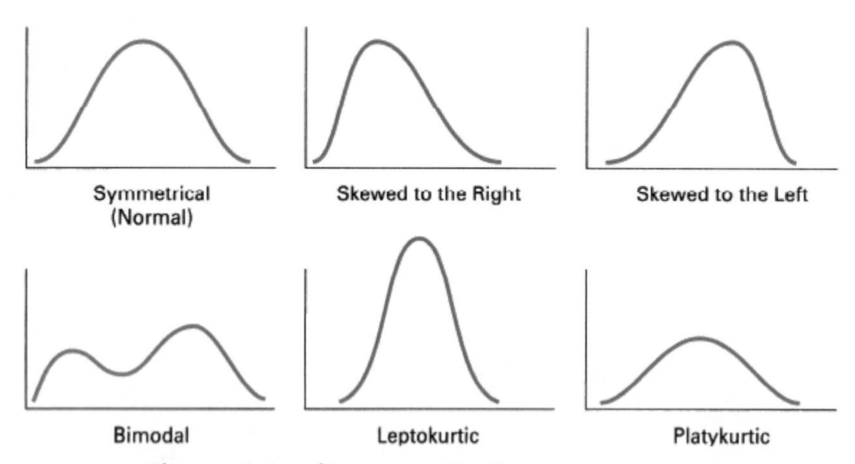

FIGURE 5-6 Characteristics of Frequency Distributions

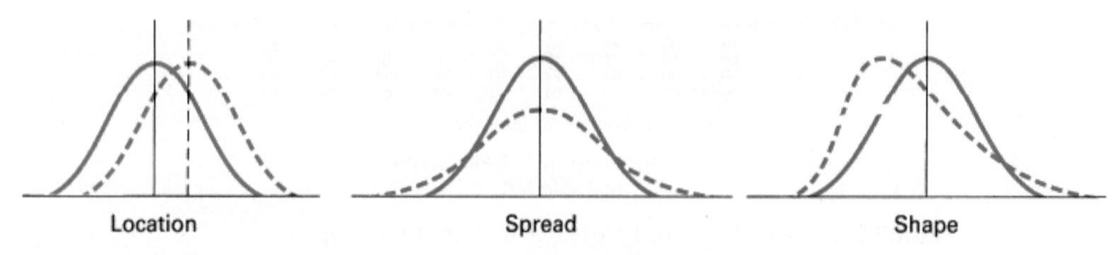

FIGURE 5-7 Differences Due to Location, Spread, and Shape

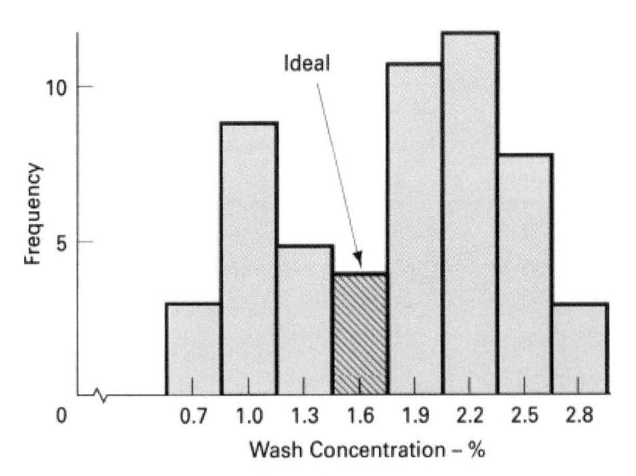

FIGURE 5-8 Histogram of Wash Concentration

Analysis of Histograms

Analysis of a histogram can provide information concerning specifications, the shape of the population frequency distribution, and a particular quality problem. Figure 5-8 shows a histogram for the percentage of wash concentration in a steel tube cleaning operation prior to painting. The ideal concentration is between 1.45% and 1.74%, as shown by the crosshatched rectangle. Concentrations less than 1.45% produce poor quality; concentrations greater than 1.75%, while producing more than adequate quality, are costly and therefore reduce productivity. No complex statistics are needed to show that corrective measures are needed to bring the spread of the distribution closer to the ideal value of 1.6%.

By adding specifications to the graph, additional problem-solving information is created. Because the spread of the distribution is a good indication of the process capability, the graph will show the degree of capability.

Final Comments

Another type of distribution that is similar to the histogram is the Pareto diagram. The reader is referred to Chapter 4 for a discussion of this type of distribution. A Pareto analysis is a very effective technique for determining the location of major quality problems. The differences between a Pareto diagram and a frequency distribution are twofold. Categories are used for the abscissa rather than data values, and the categories are in descending order from the highest frequency to the lowest one, rather than in numerical order.

One limitation of a frequency distribution is the fact that it does not show the order in which the data were produced. In other words, initial data could all be located on one side and later data on the other side. When this situation occurs, the interpretation of the frequency distribution will be different. A time series plot or run chart, discussed in Chapter 6, shows the order in which the data were produced and can aid in the analysis. An additional aid is an analysis of the data source or sources.[4]

MEASURES OF CENTRAL TENDENCY

A frequency distribution is sufficient for many quality problems. However, for a broad range of problems, a graphical technique is either undesirable or needs the additional information provided by analytical techniques. Analytical methods of describing a collection of data have the advantage of occupying less space than a graph. They also have the advantage of allowing for comparisons between collections of data. And, they also allow for additional calculations and inferences. There are two principal analytical methods of describing a collection of data—measures of central tendency and measures of dispersion. The latter measure is described in the next section, while this section covers measures of central tendency.

A *measure of central tendency* of a distribution is a numerical value that describes the central position of the data or how the data tend to build up in the center. There are three measures in common use: (1) the average, (2) the median, and (3) the mode.

Average

The *average* is the sum of the observations divided by the number of observations. It is the most common measure of central tendency. There are three different techniques available for calculating the average: (1) ungrouped data, (2) grouped data, and (3) weighted average.

1. **Ungrouped data.** This technique is used when the data are unorganized. The average is represented by the notation $\overline{X}$, which is read as "X bar" and is given by the formula

$$\overline{X} = \frac{\sum_{i=1}^{n}}{n} = \frac{X_1 + X_2 + \cdots + X_n}{n}$$

where $\overline{X}$ = average

n = number of observed values

$X_1, X_2, \ldots, X_n$ = observed value identified by the subscript $1, 2, \ldots, n$ or general subscript i

Σ = symbol meaning "sum of"

The first expression is a simplified method of writing the formula, in which $\sum_{i=1}^{n} X_i$ is read as "summation from 1 to n of X sub i" and means to add together the values of the observations.

[4]Matthew Barrows, "Use Distribution Analysis to Understand Your Data Source," *Quality Progress* (December 2005): 50–56.

Example Problem 5-1

A technician checks the resistance value of 5 coils and records the values in ohms (Ω): $X_1 = 3.35$, $X_2 = 3.37$, $X_3 = 3.28$, $X_4 = 3.34$, and $X_5 = 3.30$. Determine the average.

$$\overline{X} = \frac{\sum_{i=1}^{n} X_i}{n}$$

$$= \frac{3.35 + 3.37 + 3.28 + 3.34 + 3.30}{5}$$

$$= 3.33\Omega$$

Most electronic hand calculators have the capability of automatically calculating the average after the data are entered.

2. **Grouped data.** When the data have been grouped into a frequency distribution, the following technique is applicable. The formula for the average of grouped data is

$$\overline{X} = \frac{\sum_{i=1}^{n} f_i X_i}{n} = \frac{f_1X_1 + f_2X_2 + \cdots + f_hX_h}{f_1 + f_2 + \cdots + f_h}$$

where n = sum of the frequencies

f_i = frequency in a cell or frequency of an observed value

X_i = cell midpoint or an observed value

h = number of cells or number of observed values

The formula is applicable when the grouping is by cells with more than one observed value per cell, as illustrated by the steel shaft problem (Table 5-6). It is also applicable when each observed value, X_i, has its own frequency, f_i, as illustrated by the billing-error problem (Table 5-1). In this situation, h is the number of observed values.

In other words, if the frequency distribution has been grouped into cells, X_i is the cell midpoint and f_i is the number of observations in that cell. If the frequency distribution has been grouped by individual observed values, X_i is the observed value and f_i is the number of times that observed value occurs in the data. This practice holds for both discrete and continuous variables.

Each cell midpoint is used as the representative value of that cell. The midpoint is multiplied by its cell frequency, the products are summed, and they are divided by the total number of observations. In the example problem that follows, the first three columns are those of a typical frequency distribution. The fourth column is derived from the product of the second column (midpoint) and third column (frequency) and is labeled "f_iX_i."

Example Problem 5-2

Given the frequency distribution of the life of 320 automotive tires in 1000 km (621.37 mi) as shown in Table 5-7, determine the average.

TABLE 5-7 Frequency Distributions of the Life of 320 Tires in 1000 km

Boundaries	Midpoint X_i	Frequency f_i	Computation f_iX_i
23.6–26.5	25.0	4	100
26.6–29.5	28.0	36	1,008
29.6–32.5	31.0	51	1,581
32.6–35.5	34.0	63	2,142
35.6–38.5	37.0	58	2,146
38.6–41.5	40.0	52	2,080
41.6–44.5	43.0	34	1,462
44.6–47.5	46.0	16	736
47.6–50.5	49.0	6	294
Total		$n = 320$	$\Sigma f_iX_i = 11{,}549$

$$\overline{X} = \frac{\sum_{i=1}^{h} f_iX_i}{n}$$

$$= \frac{11{,}549}{320}$$

$$= 36.1 \text{ (which is in 1000 km)}$$

Therefore, $\overline{X} = 36.1 \times 10^3$ km.

When comparing an average calculated from this technique with one calculated using the ungrouped technique, there can be a slight difference. This difference is caused by the observations in each cell being unevenly distributed in the cell. In actual practice the difference will not be of sufficient magnitude to affect the accuracy of the problem.

3. **Weighted average.** When a number of averages are combined with different frequencies, a *weighted average* is computed. The formula for the weighted average is given by

$$\overline{X}_w = \frac{\sum_{i=1}^{n} w_i\overline{X}_i}{\sum_{i=1}^{n} w_i}$$

where $\overline{X}_w$ = weighted average

w_i = weight of the ith average

Example Problem 5-3

Tensile tests on aluminum alloy rods are conducted at three different times, which results in three different average values in megapascals (MPa). On the first occasion, 5 tests are conducted with an average of 207 MPa (30,000 psi); on the second occasion, 6 tests, with an average of 203 MPa; and on the last occasion,

3 tests, with an average of 206 MPa. Determine the weighted average.

$$\bar{X}_w = \frac{\sum_{i=1}^{n} w_i X_i}{\sum_{i=1}^{n} w_i}$$

$$= \frac{(5)(207) + (6)(203) + (3)(206)}{5 + 6 + 3}$$

$$= 205 \text{ Mpa}$$

The weighted average technique is a special case of the grouped data technique in which the data are not organized into a frequency distribution. In Example Problem 5-3, the weights are whole numbers. Another method of solving the same problem is to use proportions. Thus,

$$w_1 = \frac{5}{5 + 6 + 3} = 0.36$$

$$w_2 = \frac{6}{5 + 6 + 3} = 0.43$$

$$w_3 = \frac{3}{5 + 6 + 3} = 0.21$$

and the sum of the weights equals 1.00. The latter technique will be necessary when the weights are given in percent or the decimal equivalent.

Unless otherwise noted, $\bar{X}$ stands for the average of observed values, $\bar{X}_x$. The same equation is used to find

$\bar{X}_x$ or $\bar{\bar{X}}$—average of averages
$\bar{R}$—average of ranges
$\bar{c}$—average of count of nonconformities
$\bar{s}$—average of sample standard deviations, etc.

A bar on top of any variable indicates that it is an average.

Median

Another measure of central tendency is the *median*, which is defined as the value that divides a series of ordered observations so that the number of items above it is equal to the number below it.

1. **Ungrouped technique.** Two situations are possible in determining the median of a series of ungrouped data—when the number in the series is odd and when the number in the series is even. When the number in the series is odd, the median is the midpoint of the values. Thus, the ordered set of numbers 3, 4, 5, 6, 8, 8, and 10 has a median of 6, and the ordered set of numbers 22, 24, 24, 24, and 30 has a median of 24. When the number in the series is even, the median is the average of the two middle numbers. Thus, the ordered set of numbers 3, 4, 5, 6, 8, and 8 has a median that is the average of 5 and 6, which is $(5 + 6)/2 = 5.5$. If both middle numbers are the same, as in the ordered set of numbers 22, 24, 24, 24, 30, and 30, it is still computed as the average of the two middle numbers, because $(24 + 24)/2 = 24$. The reader is cautioned to be sure the numbers are ordered before computing the median.

2. **Grouped technique.** When data are grouped into a frequency distribution, the median is obtained by finding the cell that has the middle number and then interpolating within the cell. The interpolation formula for computing the median is given by

$$\text{Md} = L_m + \left(\frac{\frac{n}{2} - cf_m}{f_m} \right) i$$

where Md = median
L_m = lower boundary of the cell with the median
n = total number of observations
cf_m = cumulative frequency of all cells below L_m
f_m = frequency of median cell
i = cell interval

To illustrate the use of the formula, data from Table 5-7 will be used. By counting up from the lowest cell (midpoint 25.0), the halfway point ($320/2 = 160$) is reached in the cell with a midpoint value of 37.0 and a lower limit of 35.6. The cumulative frequency (cf_m) is 154, the cell interval is 3, and the frequency of the median cell is 58.

$$\text{Md} = L_m + \left(\frac{\frac{n}{2} - cf_m}{f_m} \right) i$$

$$= 35.6 + \left(\frac{\frac{320}{2} - 154}{58} \right) 3$$

$$= 35.9 \text{ (which is in 1000 km)}$$

If the counting is begun at the top of the distribution, the cumulative frequency is counted to the cell upper limit and the interpolated quantity is subtracted from the upper limit. However, it is more common to start counting at the bottom of the distribution.

The median of grouped data is not used too frequently.

Mode

The *mode* (Mo) of a set of numbers is the value that occurs with the greatest frequency. It is possible for the mode to be nonexistent in a series of numbers or to have more than one value. To illustrate, the series of numbers 3, 3, 4, 5, 5, 5, and 7 has a mode of 5; the series of numbers 22, 23, 25, 30, 32, and 36 does not have a mode; and the series of numbers 105, 105, 105, 107, 108, 109, 109, 109, 110, and 112 has two modes, 105 and 109. A series of numbers is referred to as *unimodal* if it has one mode, *bimodal* if it has two modes, and *multimodal* if there are more than two modes.

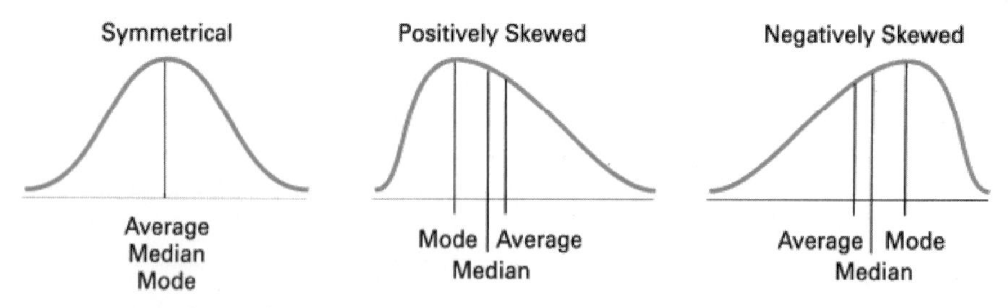

FIGURE 5-9 Relationship Among Average, Median, and Mode

When data are grouped into a frequency distribution, the midpoint of the cell with the highest frequency is the mode, because this point represents the highest point (greatest frequency) of the histogram. It is possible to obtain a better estimate of the mode by interpolating in a manner similar to that used for the median. However, this is not necessary, because the mode is employed primarily as an inspection method for determining the central tendency, and greater accuracy than the cell midpoint is not required.

Relationship Among the Measures of Central Tendency

Differences among the three measures of central tendency are shown in the smooth polygons of Figure 5-9. When the distribution is symmetrical, the values for the average, median, and mode are identical; when the distribution is skewed, the values are different.

The average is the most commonly used measure of central tendency. It is used when the distribution is symmetrical or not appreciably skewed to the right or left; when additional statistics, such as measures of dispersion, control charts, and so on, are to be computed based on the average; and when a stable value is needed for inductive statistics.

The median becomes an effective measure of the central tendency when the distribution is positively (to the right) or negatively (to the left) skewed. It is used when an exact midpoint of a distribution is desired. When a distribution has extreme values, the average will be adversely affected but the median will remain unchanged. Thus, in a series of numbers such as 12, 13, 14, 15, 16, the median and average are identical and equal to 14. However, if the first value is changed to a 2, the median remains at 14, but the average becomes 12. A control chart based on the median is user-friendly and excellent for monitoring quality.

The mode is used when a quick and approximate measure of the central tendency is desired. Thus, the mode of a histogram is easily found by visual examination. In addition, the mode is used to describe the most typical value of a distribution, such as the modal age of a particular group.

Other measures of central tendency are the geometric mean, harmonic mean, and quadratic mean. These measures are not used in quality.

MEASURES OF DISPERSION

Introduction

In the preceding section, techniques for describing the central tendency of data were discussed. A second tool of statistics is composed of the *measures of dispersion,* which describe how the data are spread out or scattered on each side of the central value. Measures of dispersion and measures of central tendency are both needed to describe a collection of data. To illustrate, the employees of the plating and the assembly departments of a factory have identical average weekly wages of \$625.36; however, the plating department has a high of \$630.72 and a low of \$619.43, while the assembly department has a high of \$680.79 and a low of \$573.54. The data for the assembly department are spread out or dispersed farther from the average than are those of the plating department.

Measures of dispersion discussed in this section are range, standard deviation, and variance. Other measures, such as mean deviation and quartile deviation, are not used in quality.

Range

The *range* of a series of numbers is the difference between the largest and smallest values or observations. Symbolically, it is given by the formula

$$R = X_h = X_l$$

where R = range
X_h = highest observation in a series
X_l = lowest observation in a series

Example Problem 5-4

If the highest weekly wage in the assembly department is \$680.79 and the lowest weekly wage is \$573.54, determine the range.

$$\begin{aligned} R &= X_h - X_l \\ &= \$680.79 - \$573.54 \\ &= \$107.25 \end{aligned}$$

The range is the simplest and easiest to calculate of the measures of dispersion. A related measure, which is occasionally used, is the *midrange,* which is the range divided by $2(R/2)$.

Standard Deviation

The *standard deviation* is a numerical value in the units of the observed values that measures the spreading tendency of the data. A large standard deviation shows greater variability of the data than does a small standard deviation. In symbolic terms it is given by the formula

$$s = \sqrt{\frac{\sum_{i=1}^{n}(X_i - \overline{X})^2}{n-1}}$$

where s = sample standard deviation
X_i = observed value
$\overline{X}$ = average
n = number of observed values

Table 5-8 will be used to explain the standard deviation concept. The first column (X_i) gives six observed values in kilograms, and from these values the average, $\overline{X} = 3.0$, is obtained. The second column $(X_i - \overline{X})$ is the deviation of the individual observed values from the average. If we sum the deviations, the answer will be 0, which is always the case, but it will not lead to a measure of dispersion. However, if the deviations are squared, they will all be positive and their sum will be greater than 0. Calculations are shown in the third column, $(X_i - \overline{X})^2$ with a resultant sum of 0.08, which will vary depending on the observed values. The average of the squared deviations can be found by dividing by n; however, for theoretical reasons we divide by $n - 1$.[5] Thus,

$$\frac{\Sigma(X_i - \overline{X})^2}{n-1} = \frac{0.08}{6-1} = 0.016 \text{ kg}^2$$

which gives an answer that has the units squared. This result is not acceptable as a measure of the dispersion but is valuable as a measure of variability for advanced statistics. It is called the *variance* and is given the symbol s^2. If we take the square root, the answer will be in the same units as the observed values. The calculation is

$$s = \sqrt{\frac{\Sigma(X_i - \overline{X})^2}{n-1}} = \sqrt{\frac{0.08}{6-1}} = 0.13 \text{ kg}$$

This formula is for explanation rather than for the purpose of calculation. Because the form of the data can be either grouped or ungrouped, there are different computing techniques.

1. **Ungrouped technique.** The formula used in the definition of standard deviation can be used for ungrouped data. However, an alternative formula is more convenient for computation purposes:

$$s = \sqrt{\frac{n\sum_{i=1}^{n}X_i^2 - \left(\sum_{i=1}^{n}X_i\right)^2}{n(n-1)}}$$

Example Problem 5-5

Determine the standard deviation of the moisture content of a roll of kraft paper. The results of six readings across the paper web are 6.7, 6.0, 6.4, 6.4, 5.9, and 5.8%.

$$s = \sqrt{\frac{n\sum_{i=1}^{n}X_i^2 - \left(\sum_{i=1}^{n}X_i\right)^2}{n(n-1)}}$$

$$= \sqrt{\frac{6(231.26) - (37.2)^2}{6(6-1)}}$$

$$= 0.35\%$$

After entry of the data, many hand calculators compute the standard deviation on command.

2. **Grouped technique.** When the data have been grouped into a frequency distribution, the following technique is applicable. The formula for the standard deviation of grouped data is

$$s = \sqrt{\frac{n\sum_{i=1}^{h}(f_iX_i^2) - \left(\sum_{i=1}^{h}f_iX_i\right)^2}{n(n-1)}}$$

where the symbols f_i, X_i, n, and h have the same meaning as given for the average of grouped data.

To use this technique, two additional columns are added to the frequency distribution. These additional columns are labeled "f_iX_i" and $f_iX_2^i$, as shown in Table 5-9. It will be recalled that the "f_iX_i" column is needed for the average computations; therefore, only one additional column is required to compute the standard deviation. The technique is shown by Example Problem 5-6.

Do not round ΣfX or ΣfX^2, as this action will affect accuracy. Most hand calculators have the capability to enter grouped data and calculate s on command.

Unless otherwise noted, s stands for s_x, the sample standard deviation of observed values. The same formula is used to find

$s_{\overline{X}}$—sample standard deviation of averages

s_p—sample standard deviation of proportions

TABLE 5-8 Standard Deviation Analysis

X_i	$X_i - \overline{X}$	$(X_i - \overline{X})^2$
3.2	+0.2	0.04
2.9	−0.1	0.01
3.0	0.0	0.00
2.9	−0.1	0.01
3.1	+0.1	0.01
2.9	−0.1	0.01
$\overline{X} = 3.0$	$\Sigma = 0$	$\Sigma = 0.08$

[5]We use $n - 1$ because one degree of freedom is lost due to the use of the sample statistic, $\overline{X}$ rather than the population parameter, μ.

TABLE 5-9 Passenger Car Speeds (in km/h) During a 15-Minute Interval on I-57 at Location 236

Boundaries	Midpoint X_i	Frequency f_i	Computations f_iX_i	$f_iX_i^2$
72.6–81.5	77.0	5	385	29,645
81.6–90.5	86.0	19	1,634	140,524
90.6–99.5	95.0	31	2,945	279,775
99.6–108.5	104.0	27	2,808	292,032
108.6–117.5	113.0	14	1,582	178,766
Total		$n = 96$	$\Sigma fX = 9{,}354$	$\Sigma fX^2 = 920{,}742$

s_R—sample standard deviation of ranges

s_s—sample standard deviation of standard deviations, etc.

Example Problem 5-6

Given the frequency distribution of Table 5-9 for passenger car speeds during a 15-minute interval on I-57, determine the average and standard deviation.

$$\bar{X} = \frac{\sum_{i=1}^{h} f_i X_i}{n} = \frac{9{,}354}{96} = 97.4 \text{ km/h}$$

$$s = \sqrt{\frac{n\sum_{i=1}^{h}(f_i X_i^2) - \left(\sum_{i=1}^{h} f_i X_i\right)^2}{n(n-1)}} = \sqrt{\frac{96(920{,}742) - (9{,}354)^2}{96(96-1)}} = 9.9 \text{ km/h}$$

The standard deviation is a reference value that measures the dispersion in the data. It is best viewed as an index that is defined by the formula. The smaller the value of the standard deviation, the better the quality, because the distribution is more closely compacted around the central value. Also, the standard deviation helps to define populations.

This method is only valid when the process is stable. The best methods use $s = \bar{s}/c_4$ or $s = \bar{R}/d_2$, which are discussed in Chapter 6.[6]

Relationship Between the Measures of Dispersion

The range is a very common measure of the dispersion; it is used in one of the principal control charts. The primary advantages of the range are in providing knowledge of the total spread of the data and in its simplicity. It is also valuable when the amount of data is too small or too scattered to justify the calculation of a more precise measure of dispersion. The range is not a function of a measure of central tendency. As the number of observations increases, the accuracy of the range decreases, because it becomes easier for extremely high or low readings to occur. It is suggested that the use of the range be limited to a maximum of 10 observations.

The standard deviation is used when a more precise measure is desired. Figure 5-10 shows two distributions with the same average, $\bar{X}$ and range, R; however, the distribution on the bottom is much better. The sample standard deviation is much smaller on the bottom distribution, indicating that the data are more compact around the central value, $\bar{X}$. As the sample standard deviation gets smaller, the quality gets better. It is also the most common measure of the dispersion and is used when subsequent statistics are to be calculated. When the data have an extreme value for the high or the low, the standard deviation is more desirable than the range.

OTHER MEASURES

There are three other measures that are frequently used to analyze a collection of data: skewness, kurtosis, and coefficient of variation.

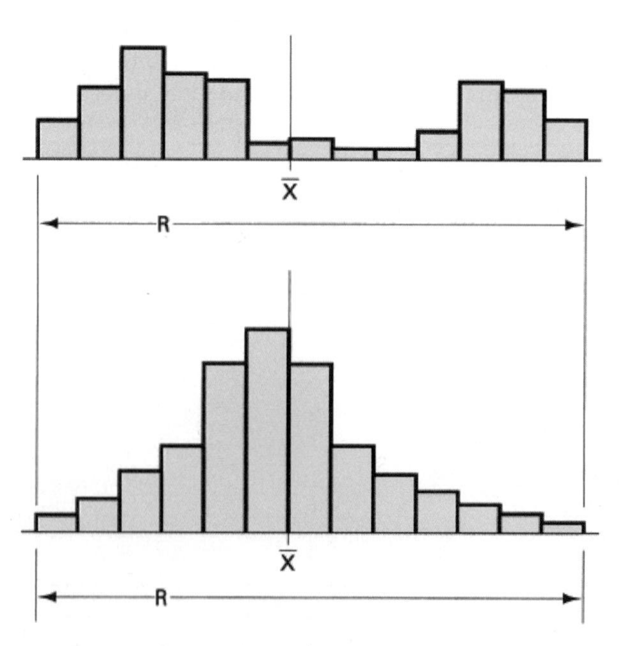

FIGURE 5-10 Comparison of Two Distributions with Equal Average and Range

[6]Thomas Pyzdek, "How Do I Compute σ? Let Me Count the Ways," *Quality Progress* (May 1998): 24–25.

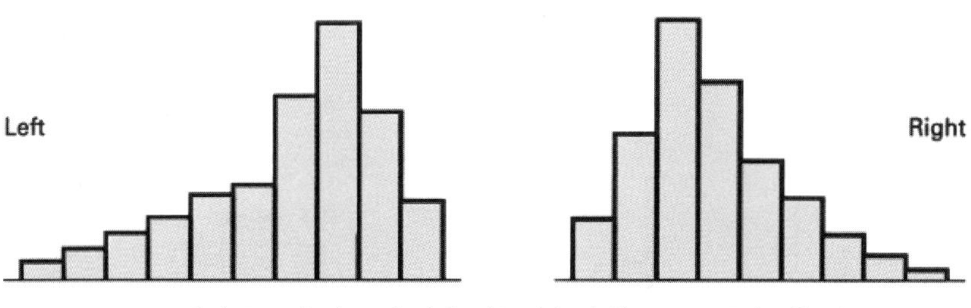

FIGURE 5-11 Left (Negative) and Right (Positive) Skewness Distributions

Skewness

As indicated previously, *skewness* is a lack of symmetry of the data. The formula is given by[7]

$$a_3 = \frac{\sum_{i=1}^{h} f_i(X_i - \overline{X})^3/n}{s^3}$$

where a_3 represents skewness.

Skewness is a number whose size tells us the extent of the departure from symmetry. If the value of a_3 is 0, the data are symmetrical; if it is greater than 0 (positive), the data are skewed to the right, which means that the long tail is to the right; and if it is less than 0 (negative), the data are skewed to the left, which means that the long tail is to the left. See Figure 5-11 for a graphical representation of skewness. Values of +1 or −1 imply a strongly unsymmetrical distribution.

Example Problem 5-7

Determine the skewness of the frequency distribution of Table 5-10. The average and sample standard deviation are calculated and are 7.0 and 2.30, respectively.

$$a_3 = \frac{\sum_{i=1}^{h} f_i(X_i - \overline{X})^3/n}{s^3} = \frac{-648/124}{2.30^3} = -0.43$$

The skewness value of −0.43 tells us that the data are skewed to the left. Visual examination of the X and f columns or a histogram would have indicated the same information.

In order to determine skewness, the value of n must be large—say, at least 100. Also, the distribution must be unimodal. The skewness value provides information concerning the shape of the population distribution. For example, a normal distribution has a skewness value of zero, $a_3 = 0$.

Kurtosis

As indicated previously, *kurtosis* is the peakedness of the data. The formula is given by[8]

$$a_4 = \frac{\sum_{i=1}^{h} f_i(X_i - \overline{X}^4)/n}{s^4}$$

where a_4 represents kurtosis.

Kurtosis is a dimensionless value that is used as a measure of the height of the peak in a distribution. Figure 5-12 shows a *leptokurtic* (more peaked) distribution and a *platykurtic* (flatter) distribution. Between these two distributions is one referred to as *mesokurtic*, which is the normal distribution.

Example Problem 5-8

Determine the kurtosis of the frequency distribution of Table 5-10, which has $\overline{X} = 7.0$ and $s = 2.30$.

$$a_4 = \frac{\sum_{i=1}^{h} f_i(X_i - \overline{X})^4/n}{s^4} = \frac{9720/124}{2.30^4} = 2.80$$

TABLE 5-10 Data for Skewness and Kurtosis Example Problems

X_i	f_i	$X_i - \overline{X}$	$f_i(X_i - \overline{X})^3$	$f_i(X_i - \overline{X})^4$
1	4	(1 − 7) = −6	$4(-6)^3 = -864$	$4(-6)^4 = 5184$
4	24	(4 − 7) = −3	$24(-3)^3 = -648$	$24(-3)^4 = 1944$
7	64	(7 − 7) = 0	$64(0)^3 = 0$	$64(0)^4 = 0$
10	32	(10 − 7) = +3	$32(+3)^3 = +864$	$32(+3)^4 = 2592$
	Σ = 124		Σ = −648	Σ = 9720

[7]This formula is an approximation that is good enough for most purposes.

[8]This formula is an approximation that is good enough for most purposes.

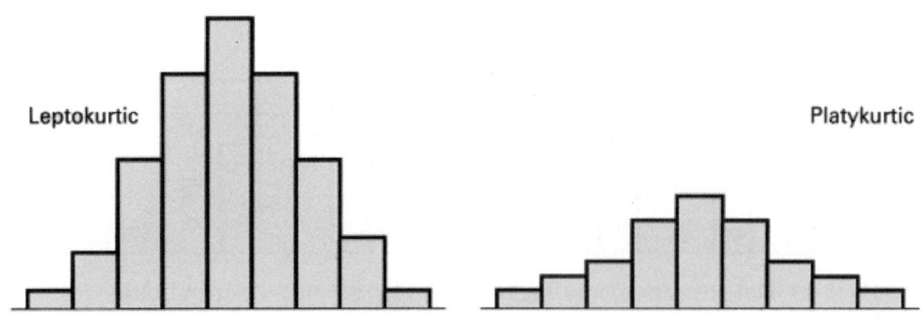

FIGURE 5-12 Leptokurtic and platykurtic distributions.

The kurtosis value of 2.80 does not provide any information by itself—it must be compared to another distribution. Use of the kurtosis value is the same as skewness—large sample size, n, and unimodal distribution. It provides information concerning the shape of the population distribution. For example, a normal distribution, mesokurtic, has a kurtosis value of 3, $a_4 = 3$. If $a_4 > 3$, then the height of the distribution is more peaked than normal, leptokurtic; and if $a_4 < 3$, the height of the distribution is less peaked than normal, platykurtic. Some software packages such as EXCEL normalize the data to 0 by subtracting 3 from the answer.

The concepts of skewness and kurtosis are useful in that they provide some information about the shape of the distribution. Calculations are best made by a computer program.

Coefficient of Variation[9]

Coefficient of variation is a measure of how much variation exists in relation to the mean. The standard deviation alone is not particularly useful without a context. For example, a standard deviation of 15 kg would be very good with data at a mean of 2600 kg, but very poor with data at a mean of 105 kg. The coefficient of variation (CV) provides a reference. The formula is given by

$$\text{CV} = \frac{s(100\%)}{\overline{X}}$$

Note that the units for s and $\overline{X}$ cancel; therefore, the answer will be in percent.

Example Problem 5-9

Compare the coefficient of variation for a standard deviation of 15 kg and means of 2600 kg and 105 kg.

$$\text{CV} = \frac{s(100\%)}{\overline{X}} = \frac{15(100\%)}{2600} = 0.58\%$$

$$\text{CV} = \frac{s(100\%)}{\overline{X}} = \frac{15(100\%)}{105} = 14.3\%$$

The smaller the value, the smaller is the amount of variation relative to the mean.

[9]Michael J. Cleary, "Beyond Deviation," *Quality Progress* (August 2004): 30, 70.

CONCEPT OF A POPULATION AND A SAMPLE

At this point, it is desirable to examine the concept of a population and a sample. In order to construct a frequency distribution of the weights of steel shafts, a small portion, or *sample,* is selected to represent all the steel shafts. Similarly, the data collected concerning the passenger car speeds represented only a small portion of all the passenger cars. The *population* is the whole collection of measurements, and in the preceding examples, the populations were all the steel shafts and all the passenger cars. When averages, standard deviations, and other measures are computed from samples, they are referred to as *statistics.* Because the composition of samples will fluctuate, the computed statistics will be larger or smaller than their true population values, or *parameters.* Parameters are considered to be fixed reference (standard) values, or the best estimate of these values available at a particular time.

The population may have a finite number of items, such as a day's production of steel shafts. It may be infinite or almost infinite, such as the number of rivets in a year's production of jet airplanes. The population may be defined differently depending on the particular situation. Thus, a study of a product could involve the population of an hour's production, a week's production, 5000 pieces, and so on.

Because it is rarely possible to measure all of the population, a sample is selected. Sampling is necessary when it may be impossible to measure the entire population; when the expense to observe all the data is prohibitive; when the required inspection destroys the product; or when a test of the entire population may be too dangerous, as would be the case with a new medical drug. Actually, an analysis of the entire population may not be as accurate as sampling. It has been shown that 100% manual inspection is not as accurate as sampling when the percent nonconforming is very small. This is probably due to the fact that boredom and fatigue cause inspectors to prejudge each inspected item as being acceptable.

When designating a population, the corresponding Greek letter is used. Thus, the sample average has the symbol $\overline{X}$, and the population mean has the symbol μ (mu).

Note that the word *average* changes to *mean* when used for the population. The symbol $\overline{X}_0$ is the standard or

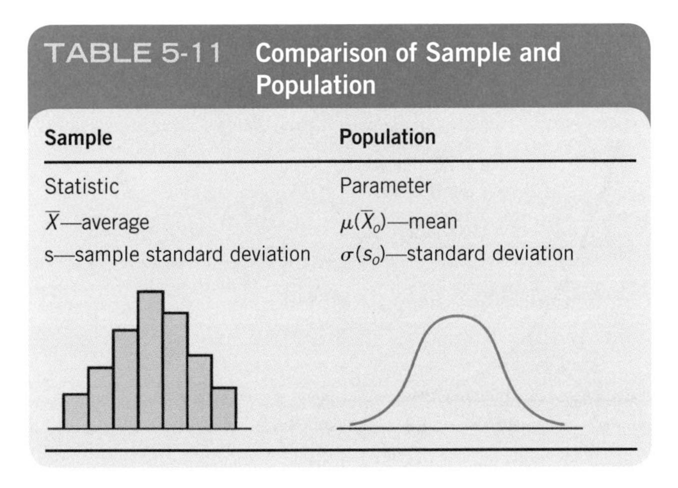

TABLE 5-11 Comparison of Sample and Population

Sample	Population
Statistic	Parameter
$\overline{X}$—average	$\mu(\overline{X}_o)$—mean
s—sample standard deviation	$\sigma(s_o)$—standard deviation

reference value. Mathematical concepts are based on μ, which is the true value—$\overline{X}_0$ represents a practical equivalent in order to use the concepts. The sample standard deviation has the symbol s, and the population standard deviation the symbol σ (sigma). The symbol s_0 is the standard or reference value and has the same relationship to σ that $\overline{X}_0$ has to μ. The true population value may never be known; therefore, the symbols $\hat{\mu}$ and $\hat{\sigma}$ are sometimes used to indicate "estimate of." A comparison of sample and population is given in Table 5-11. A sample frequency distribution is represented by a histogram, whereas a population frequency distribution is represented by a smooth curve. Additional comparisons will be given as they occur.

The primary objective in selecting a sample is to learn something about the population that will aid in making some type of decision. The sample selected must be of such a nature that it tends to resemble or represent the population. How successfully the sample represents the population is a function of the size of the sample, chance, sampling method, and whether the conditions change or not.

Table 5-12 shows the results of an experiment that illustrates the relationship between samples and the population.

TABLE 5-12 Results of Eight Samples of Blue and Green Spheres from a Known Population

Sample Number	Sample Size	Number of Blue Spheres	Number of Green Spheres	Percentage of Green Spheres
1	10	9	1	10
2	10	8	2	20
3	10	5	5	50
4	10	9	1	10
5	10	7	3	30
6	10	10	0	0
7	10	8	2	20
8	10	9	1	10
Total	80	65	15	18.8

A container holds 800 blue and 200 green spheres 5 mm (approximately 3/16 in.) in diameter. The 1000 spheres are considered to be the population, with 20% being green. Samples of size 10 are selected and posted to the table and then replaced in the container. The table illustrates the differences between the sample results and what should be expected from the known population. Only in samples 2 and 7 are the sample statistics equal to the population parameter. There definitely is a chance factor that determines the composition of the sample. When the eight individual samples are combined into one large one, the percentage of green spheres is 18.8, which is close to the population value of 20%.

Although inferences are made about the population from samples, it is equally true that a knowledge of the population provides information for analysis of the sample. Thus, it is possible to determine whether a sample came from a particular population. This concept is necessary to understand control chart theory. A more detailed discussion is delayed until Chapter 6.

THE NORMAL CURVE

Description

Although there are as many different populations as there are conditions, they can be described by a few general types. One type of population that is quite common is called the *normal curve* or *Gaussian distribution*. The normal curve is a symmetrical, unimodal, bell-shaped distribution with the mean, median, and mode having the same value.

A population curve or distribution is developed from a frequency histogram. As the sample size of a histogram gets larger and larger, the cell interval gets smaller and smaller. When the sample size is quite large and the cell interval is very small, the histogram will take on the appearance of a smooth polygon or a curve representing the population. A curve of the normal population of 1000 observations of the resistance in ohms of an electrical device with population mean, μ, of 90 Ω and population standard deviation, σ, of 2 Ω is shown in Figure 5-13. The interval between dotted lines is equal to one standard deviation, σ.

Much of the variation in nature and in industry follows the frequency distribution of the normal curves. Thus, the variations in the weight of elephants, the speed of antelopes, and the height of human beings will follow a normal curve. Also, the variations found in industry, such as the weight of gray iron castings, the life of 60-W light bulbs, and the dimensions of a steel piston ring, will be expected to follow the normal curve. When considering the heights of human beings, we can expect a small percentage of them to be extremely tall and a small percentage to be extremely short, with the majority of human heights clustering about the average value. The normal curve is such a good description of the variations that occur in most quality characteristics in industry that it is the basis for many techniques.

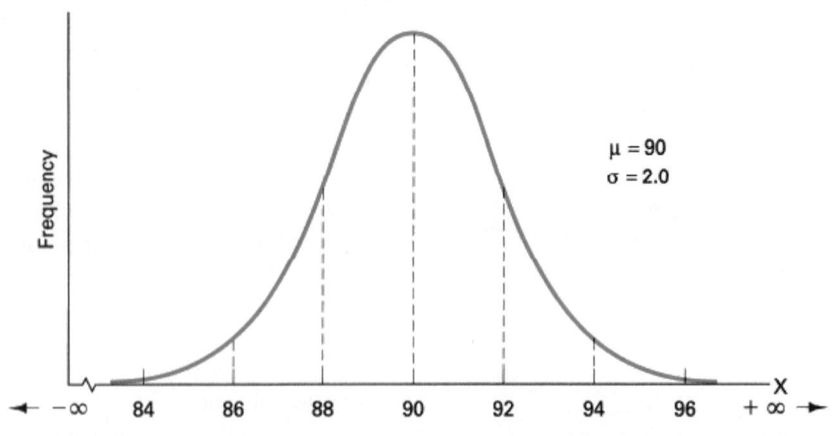

FIGURE 5-13 Normal Distribution for Resistance of an Electrical Device with $\mu = 90\ \Omega$ and $\sigma = 2.0\ \Omega$

All normal distributions of continuous variables can be converted to the standardized normal distribution (see Figure 5-14) by using the *standardized normal value, Z.* For example, consider the value of 92 Ω in Figure 5-14, which is one standard deviation above the mean $[\mu + 1\sigma = 90 + 1(2) = 92]$. Conversion to the Z value is

$$Z = \frac{X_i - \mu}{\sigma} = \frac{92 - 90}{2} = +1$$

which is also 1 σ above μ on the Z scale of Figure 5-14.

The formula for the standardized normal curve is

$$f(Z) = \frac{1}{\sqrt{2\pi}} e^{-Z^2} 2 = 0.3989 e^{-Z^2} 2$$

where $\pi = 3.14159$

$e = 2.71828$

$$Z = \frac{X_i - \mu}{\sigma}$$

A table is provided in the Appendix (Table A); therefore, it is not necessary to use the formula. Figure 5-14 shows the standardized curve with its mean of 0 and standard deviation of 1. It is noted that the curve is asymptotic at $Z = -3$ and $Z = +3$.

The area under the curve is equal to 1.0000 or 100% and therefore can easily be used for probability calculations. Because the area under the curve between various points is a very useful statistic, a normal area table is provided as Table A in the Appendix.

The normal distribution can be referred to as a normal probability distribution. Although it is the most important population distribution, there are a number of other ones for continuous variables. There are also a number of probability distributions for discrete variables. These distributions are discussed in Chapter 8.

Relationship to the Mean and Standard Deviation

As seen by the formula for the standardized normal curve, there is a definite relationship among the mean, the standard deviation, and the normal curve. Figure 5-15 shows three normal curves with different mean values; it is noted that the only change is in the location. Figure 5-16 shows three normal curves with the same mean but different standard deviations. The figure illustrates the principle that the larger the standard deviation, the flatter the curve (data are widely dispersed), and the smaller the standard deviation, the more

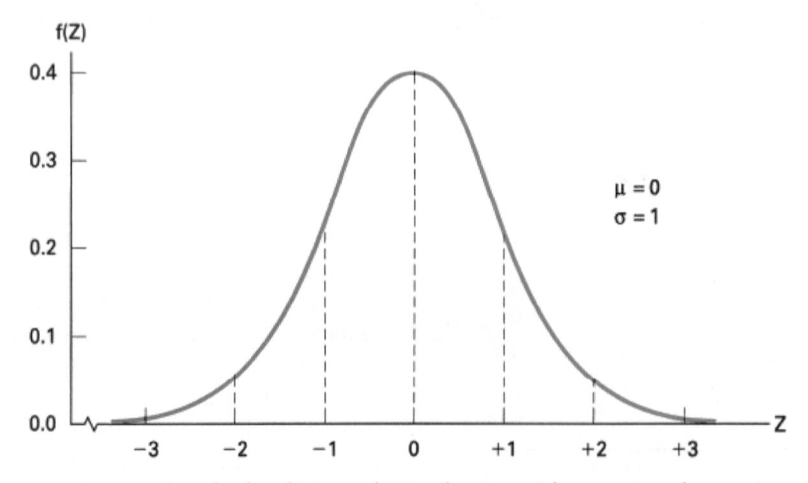

FIGURE 5-14 Standardized Normal Distribution with $\sigma = 0$ and $\sigma = 1$

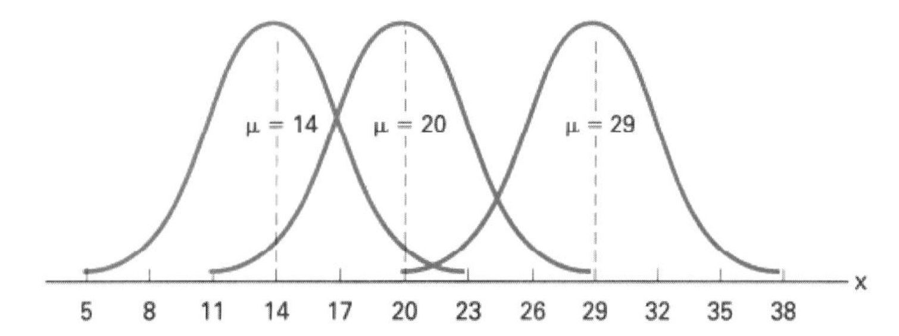

FIGURE 5-15 Normal Curve with Different Means but Identical Standard Deviations

peaked the curve (data are narrowly dispersed). If the standard deviation is 0, all values are identical to the mean and there is no curve.

The normal distribution is fully defined by the population mean and population standard deviation. Also, as seen by Figures 5-15 and 5-16, these two parameters are independent. In other words, a change in one has no effect on the other.

A relationship exists between the standard deviation and the area under the normal curve as shown in Figure 5-17. The figure shows that in a normal distribution, 68.26% of the items are included between the limits of $\mu + 1\ \sigma$ and $\mu - 1$, 95.46% of the items are included between the limits $\mu + 2$ and $\mu - 2\ \sigma$, and 99.73% of the items are included between $\mu + 3\ \sigma$ and $\mu - 3\ \sigma$. One hundred percent of the items are included between the limits $+\infty$ and $-\infty$. These percentages hold true regardless of the shape of the normal curve. The fact that 99.73% of the items are included between $\pm 3\sigma$ is the basis for control charts that are discussed in Chapter 6.

Applications

The percentage of items included between any two values can be determined by calculus. However, this is not necessary, because the areas under the curve for various Z values are given in Table A in the Appendix. Table A, "Areas Under the Normal Curve," is a left-reading table,[10] which means that the given areas are for that portion of the curve from $-\infty$ to a particular value, X_i.

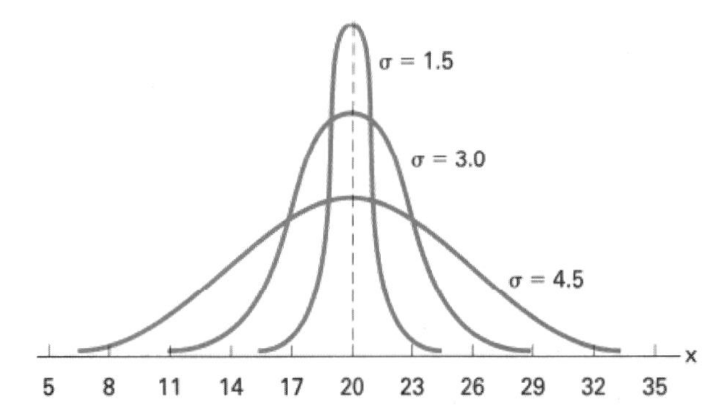

FIGURE 5-16 Normal Curve with Different Standard Deviations but Identical Means

[10] In some texts the table for the areas under the normal curve is arranged in a different manner.

The first step is to determine the Z value using the formula

$$Z = \frac{X_i - \mu}{\sigma}$$

where Z = standard normal value
X_i = individual value
μ = mean
σ = population standard deviation

Next, using the calculated Z value, the area under the curve to the left of X_i is found in Table A. Thus, if a calculated Z value is -1.76, the value for the area is 0.0392. Because the total area under the curve is 1.0000, the 0.0392 value for the area can be changed to a percent of the items under the curve by moving the decimal point two places to the right. Therefore, 3.92% of the items are less than the particular X_i value.

Assuming that the data are normally distributed, it is possible to find the percent of the items in the data that are less than a particular value, greater than a particular value, or between two values. When the values are upper and/or lower specifications, a powerful statistical tool is available. The following example problems illustrate the technique.

Example Problem 5-10

The mean value of the weight of a particular brand of cereal for the past year is 0.297 kg (10.5 oz) with a standard deviation of 0.024 kg. Assuming normal distribution, find the percent of the data that falls below the lower specification limit of 0.274 kg. (*Note:* Because the mean and standard deviation were determined from

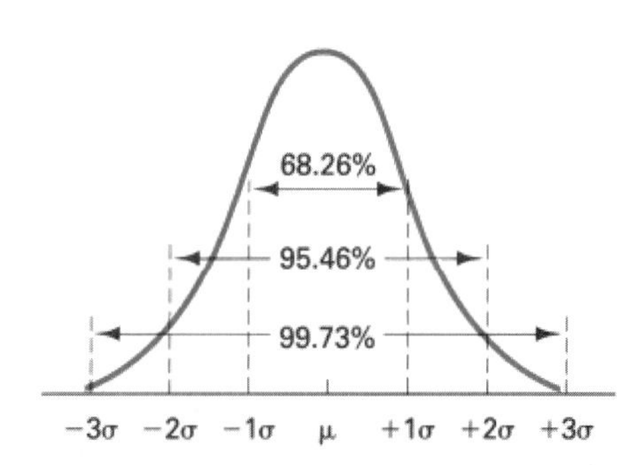

FIGURE 5-17 Percent of Items Included between Certain Values of the Standard Deviation

a large number of tests during the year, they are considered to be valid estimates of the population values.)

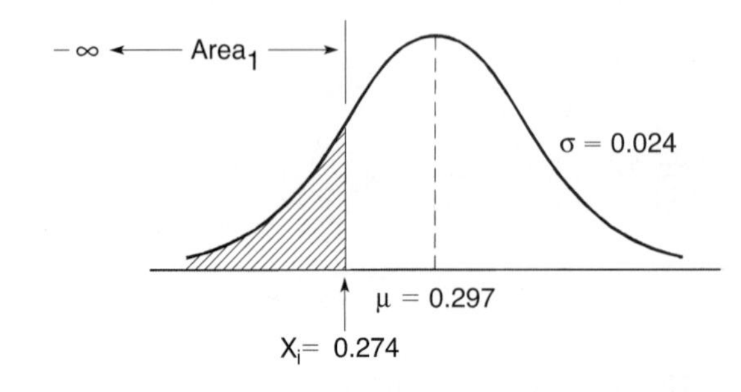

$$Z = \frac{X_i - \mu}{\sigma}$$
$$= \frac{0.274 - 0.297}{0.024}$$
$$= -0.96$$

From Table A it is found that for $Z = -0.96$,

$$\text{Area}_1 = 0.1685 \text{ or } 16.85\%$$

Thus, 16.85% of the data are less than 0.274 kg.

Example Problem 5-11

Using the data from Example Problem 5-10, determine the percentage of the data that fall above 0.347 kg.

Because Table A is a left-reading table, the solution to this problem requires the use of the relationship: $\text{Area}_1 + \text{Area}_2 = \text{Area}_T = 1.0000$. Therefore, Area_2 is determined and subtracted from 1.0000 to obtain Area_1.

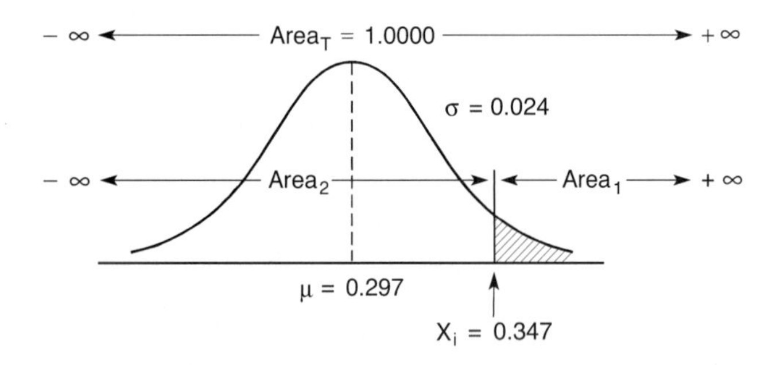

$$Z_2 = \frac{X_i - \mu}{\sigma}$$
$$= \frac{0.347 - 0.297}{0.024}$$
$$= +2.08$$

From Table A it is found that for $Z_2 = +2.08$,

$$\text{Area}_2 = 0.9812$$
$$\text{Area}_1 = \text{Area}_T - \text{Area}_2$$
$$= 1.0000 - 0.9812$$
$$= 0.0188 \text{ or } 1.88\%$$

Thus, 1.88% of the data are above 0.347 kg.

Example Problem 5-12

A large number of tests of line voltage to home residences show a mean of 118.5 V and a population standard deviation of 1.20 V. Determine the percentage of data between 116 and 120 V.

Because Table A is a left-reading table, the solution requires that the area to the left of 116 V be subtracted from the area to the left of 120 V. The graph and calculations show the technique.

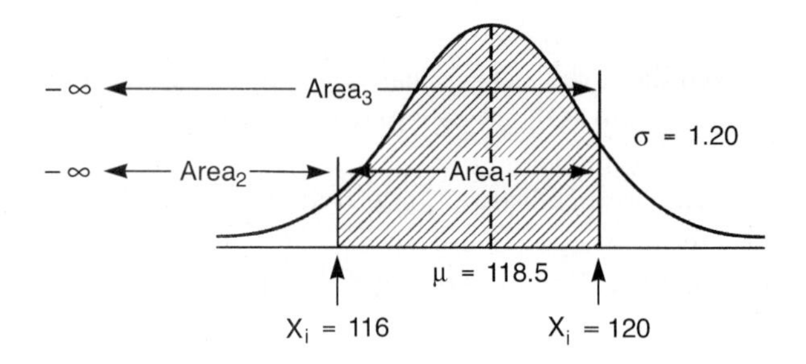

$$Z_2 = \frac{X_i - \mu}{\sigma} = \frac{116 - 118.5}{1.20} = -2.08 \qquad Z_3 = \frac{X_i - \mu}{\sigma} = \frac{120 - 118.5}{1.20} = +1.25$$

From Table A it is found that for $Z_2 = -2.08$, $\text{Area}_2 = 0.0188$; for $Z_3 = +1.25$, $\text{Area}_3 = 0.8944$.

$$\text{Area}_1 = \text{Area}_3 - \text{Area}_2$$
$$= 0.8944 - 0.0188$$
$$= 0.8756 \text{ or } 87.56\%$$

Thus, 87.56% of the data are between 116 and 120 V.

Example Problem 5-13

If it is desired to have 12.1% of the line voltage below 115 V, how should the mean voltage be adjusted? The dispersion is $\sigma = 1.20$ V.

The solution to this type of problem is the reverse of that for the other problems. First, 12.1%, or 0.1210, is found in the body of Table A. This gives a Z value, and using the formula for Z, we can solve for the mean voltage. From Table A with $\text{Area}_1 = 0.1210$, the Z value of -1.17 is obtained.

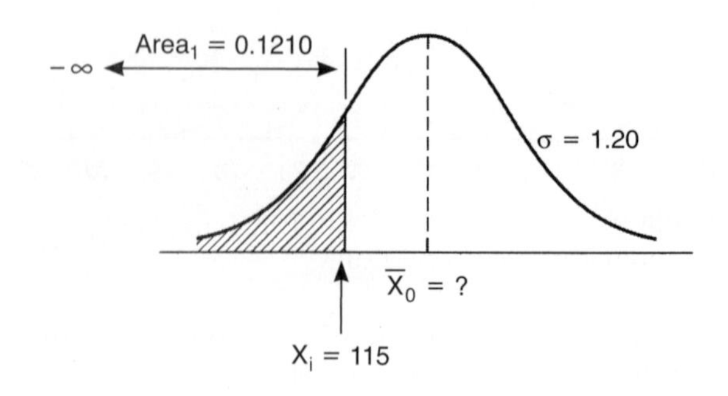

$$Z_1 = \frac{X_i - \overline{X}_0}{\sigma}$$

$$-1.17 = \frac{115 - \overline{X}_0}{1.20}$$

$$\overline{X}_0 = 116.4\text{ V}$$

Thus, the mean voltage should be centered at 116.4 V for 12.1% of the values to be less than 115 V.

Note that $\overline{X}_0$ has been substituted in the equation. The concept of the normal curve is based on the values of μ and σ; however, $\overline{X}_0$ and s_0 can substitute provided there is some evidence that the distribution is normal. Example Problem 5-13 illustrates the independence of μ and σ. A small change in the centering of the process does not affect the dispersion.

TESTS FOR NORMALITY

Because of the importance of the normal distribution, it is frequently necessary to determine if the data are normal. In using these techniques the reader is cautioned that none are 100% certain. The techniques of histogram, skewness and kurtosis, probability plots, and chi-square test are also applicable with some modification to other population distributions.

Histogram

Visual examination of a histogram developed from a large amount of data will give an indication of the underlying population distribution. If a histogram is unimodal, is symmetrical, and tapers off at the tails, normality is a definite possibility and may be sufficient information in many practical situations. The histogram of Figure 5-4 of steel shaft weight is unimodal, tapers off at the tails, and is somewhat symmetrical except for the upper tail. If a sorting operation had discarded shafts with weights above 2.575, this would explain the upper tail cutoff.

The larger the sample size, the better the judgment of normality. A minimum sample size of 50 is recommended.

Skewness and Kurtosis

Skewness and kurtosis measurements are another test of normality. From the steel shaft data of Table 5-6, we find that $a_3 = -0.11$ and $a_4 = 2.19$. These values indicate that the data are moderately skewed to the left but are close to the normal value of 0 and that the data are not as peaked as the normal distribution, which would have an a_4 value of 3.0.

These measurements tend to give the same information as the histogram. As with the histogram, the larger the sample size, the better is the judgment of normality. A minimum sample size of 100 is recommended.

Probability Plots

Another test of normality is to plot the data on normal probability paper. This type of paper is shown in Figure 5-18. Different probability papers are used for different distributions. To illustrate the procedure, we will again use the steel

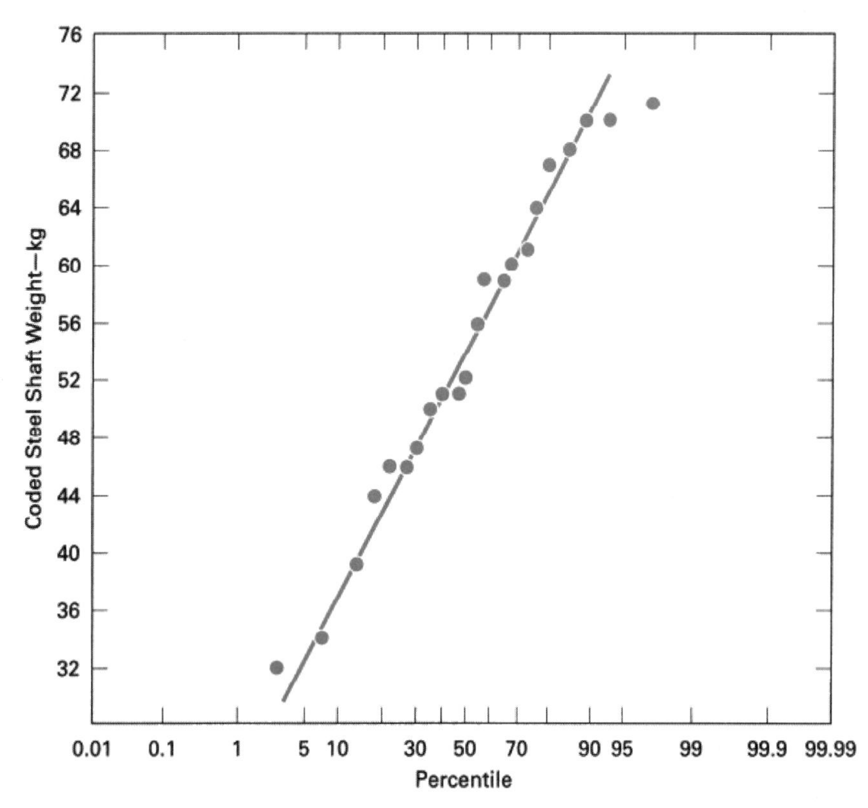

FIGURE 5-18 Probability Plots of Data from Table 5-13

TABLE 5-13 Data on Steel Shaft Weight for Probability Plotting

Observation X_i	Rank i	Plotting Position	Observation X_i	Rank i	Plotting Position
32	1	2.3	56	12	52.3
34	2	6.8	59	13	56.8
39	3	11.4	59	14	61.4
44	4	15.9	60	15	65.9
46	5	20.5	61	16	70.5
46	6	25.0	64	17	75.0
47	7	29.5	67	18	79.5
50	8	34.1	68	19	84.1
51	9	38.6	70	20	88.6
51	10	43.2	70	21	93.2
52	11	47.7	71	22	97.7

shaft data in its coded form. The step-by-step procedure follows.

1. **Order the data.** The data from the first column of Table 5-4 are used to illustrate the concept. Each observation is recorded as shown in Table 5-13, from the smallest to the largest. Duplicate observations are recorded as shown by the value 46.

2. **Rank the observations.** Starting at 1 for the lowest observation, 2 for the next lowest observation, and so on, rank the observations. The ranks are shown in column 2 of Table 5-13.

3. **Calculate the plotting position.** This step is accomplished using the formula

$$PP = \frac{100(i - 0.5)}{n}$$

where i = rank

PP = plotting position in percent

n = sample size

The first plotting position is $100(1 - 0.5)/22$, which is 2.3%; the others are calculated similarly and posted to Table 5-13.

4. **Label the data scale.** The coded values range from 32 to 71, so the vertical scale is labeled appropriately and is shown in Figure 5-18. The horizontal scale represents the normal curve and is preprinted on the paper.

5. **Plot the points.** The plotting position and the observation are plotted on the normal probability paper.

6. **Attempt to fit by eye a "best" line.** A clear plastic straightedge will be most helpful in making this judgment. When fitting this line, greater weight should be given to the center values than to the extreme ones.

7. **Determine normality.** This decision is one of judgment as to how close the points are to the straight line. If we disregard the extreme points at each end of the line, we can reasonably assume that the data are normally distributed.

If normality appears reasonable, additional information can be obtained from the graph. The mean is located at the 50th percentile, which gives a value of approximately 55. Standard deviation is two-fifths the difference between the 90th percentile and the 10th percentile, which would be approximately $14[(2/5)(72 - 38)]$. We can also use the graph to determine the percent of data below, above, or between data values. For example, the percent less than 48 is approximately 31%. Even though the example problem used 22 data points with good results, a minimum sample size of 30 is recommended.

The eyeballing technique is a judgment decision, and for the same data, different people will have different slopes to their straight line. An analytical technique using the Weibull distribution effectively overcomes this limitation. It is similar to normal probability plots and can use spreadsheet software such as EXCEL.[11] The Weibull distribution is discussed in Chapter 11, where normality is determined if the slope parameter, β, is approximately equal to 3.4.

Chi-Square Goodness of Fit

The chi-square (χ^2) test is another technique of determining whether the sample data fit a normal distribution or some other distribution.

This test uses the equation

$$\chi^2 = \sum_{i=1}^{k} \frac{(O_i - E_i)^2}{E_i}$$

where χ^2 = chi-squared

O_i = observed value in a cell

E_i = expected value for a cell

The expected value is determined from the normal distribution or from any distribution. After χ^2 is determined, it is compared to the χ^2 distribution to determine whether the observed data is from the expected distribution. An example is given in the EXCEL program file chi-squared. Although the χ^2 test is the best method of determining normality, it does require a minimum sample size of 125.

It is important for the analyst to understand that none of these techniques proves that the data are normally distributed. We can only conclude that there is no evidence that the data cannot be treated as if they were normally distributed.

SCATTER DIAGRAM

The simplest way to determine if a cause-and-effect relationship exists between two variables is to plot a scatter diagram. Figure 5-19 shows the relationship between automotive speed and gas mileage. The figure shows that as speed increases, gas mileage decreases. Automotive speed is plotted on the x axis and is the independent variable. The independent variable is usually controllable. Gas mileage is on

[11]EXCEL has a normal probability plot available under regression analysis.

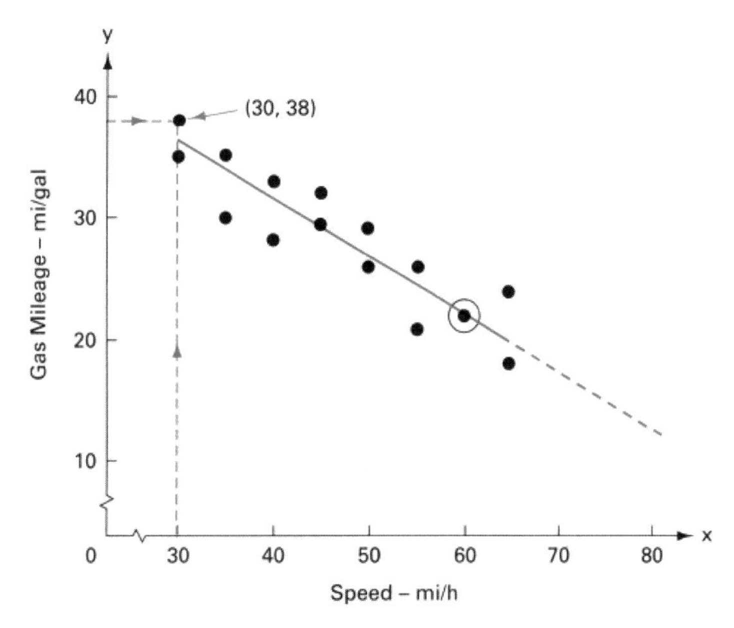

FIGURE 5-19 Scatter Diagram

the y axis and is the dependent, or response, variable. Other examples of relationship are

Cutting speed and tool life
Moisture content and thread elongation
Temperature and lipstick hardness
Striking pressure and electrical current
Temperature and percent foam in soft drinks
Yield and concentration
Breakdowns and equipment age

There are a few simple steps in constructing a scatter diagram. Data are collected as ordered pairs (x, y). The automotive speed (cause) is controlled and the gas mileage (effect) is measured. Table 5-14 shows resulting x, y paired data.

The horizontal and vertical scales are constructed with the higher values on the right for the x axis and on the top for the y axis. After the scales are labeled, the data are plotted. Using dotted lines, the technique of plotting sample number 1 (30, 38) is illustrated in Figure 5-19. The x value is 30, and the y value is 38. Sample numbers 2 through 16 are plotted, and the scatter diagram is complete. If two points are identical, concentric circles can be used, as illustrated at 60 mi/h.

Once the scatter diagram is complete, the relationship or correlation between the two variables can be evaluated. Figure 5-20 shows different patterns and their interpretation. In (a) we have a positive correlation between the two variables because as x increases, y increases. In (b) there is a negative correlation between the two variables because as x increases, y decreases. In (c) there is no correlation, and this pattern is sometimes referred to as a shotgun pattern.

The patterns described in (a), (b), and (c) are easy to understand; however, those described in (d), (e), and (f) are more difficult. In (d) there may or may not be a relationship between the two variables. There appears to be a negative relationship between x and y, but it is not too strong. Further statistical analysis is needed to evaluate this pattern. In (e) the data have been stratified to represent different causes for the same effect. Some examples are gas mileage with the wind versus against the wind, two different suppliers of material, and two different machines. One cause is plotted with a small solid circle, and the other cause is plotted with a solid triangle. When the data are separated, we see that there is a strong correlation. In (f) there is a curvilinear relationship rather than a linear one.

When all the plotted points fall on a straight line, there is perfect correlation. Because of variations in the experiment and measurement error, this perfect situation will rarely, if ever, occur.

TABLE 5-14 Data on Automotive Speed Versus Gas Mileage

Sample Number	Speed (mi/h)	Mileage (mi/gal)	Sample Number	Speed (mi/h)	Mileage (mi/gal)
1	30	38	9	50	26
2	30	35	10	50	29
3	35	35	11	55	32
4	35	30	12	55	21
5	40	33	13	60	22
6	40	28	14	60	22
7	45	32	15	65	18
8	45	29	16	65	24

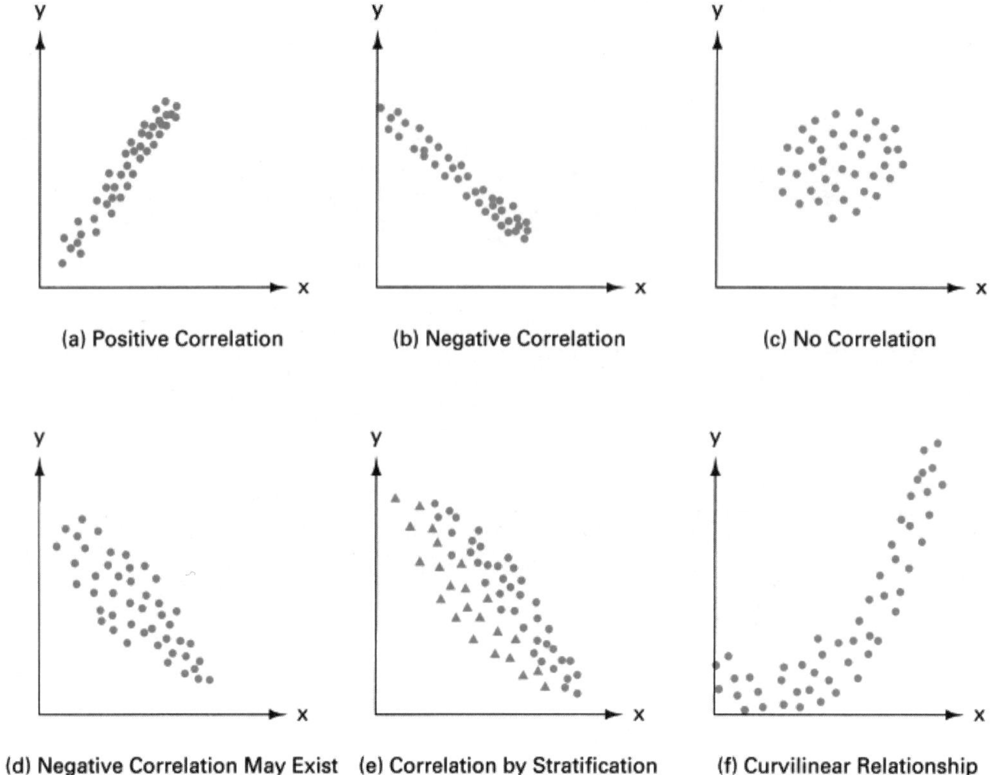

FIGURE 5-20 Scatter Diagram Patterns

It is sometimes desirable to fit a straight line to the data in order to write a prediction equation. For example, we may wish to estimate the gas mileage at 43 mi/h. A line can be placed on the scatter diagram by sight or mathematically using least-squares analysis. In either approach, the idea is to make the deviation of the points on each side of the line approximately equal. Where the line is extended beyond the data, a dashed line is used because there are no data in that area.

In order to fit a straight line to the data mathematically, we need to determine its slope, m, and its intercept with the y axis, a. These formulas are given by

$$m = \frac{\Sigma xy - [(\Sigma x)(\Sigma y)/n]}{\Sigma x^2 - [(\Sigma x)^2/n]}$$

$$a = \Sigma \tfrac{y}{n} - m(\Sigma \tfrac{x}{n})$$

$$y = a + mx$$

Another useful statistic is the *coefficient of correlation,* which describes the goodness of fit of the linear model. It is a dimensionless number, r, that lies between +1 and −1. The + and − signs tell whether there is a negative correlation, Figure 5-20(a), or a positive correlation, Figure 5-20(b). The closer the value is to 1.00, the better is the fit, with a value of 1 meaning that all points fall on the line. Its formula is given by

$$r = \frac{\Sigma xy - [(\Sigma x)(\Sigma y)/n]}{(\Sigma x^2 - [(\Sigma x)^2/n])(\Sigma y^2 - [(\Sigma y)^2/n])}$$

Example Problem 5-14

Using the data about gas mileage versus speed, determine the straight line and the coefficient of correlation. Table 5-15 extends Table 5-14 to calculate Σx^2, Σy^2, and Σxy. Also determine the mileage at 57 mi/h.

TABLE 5-15 Extension of Data in Table 5-14

Sample Number	Speed (mi/h) x	Mileage (mi/gal) y	X^2	Y^2	XY
1	30	38	900	1,444	1,140
2	30	35	900	1,225	1,050
3	35	35	1,225	1,225	1,225
4	35	30	1,225	900	1,050
5	40	33	1,600	1,089	1,320
6	40	28	1,600	784	1,120
7	45	32	2,025	1,024	1,440
8	45	29	2,025	841	1,305
9	50	26	2,500	676	1,300
10	50	29	2,500	841	1,450
11	55	32	3,025	1,024	1,760
12	55	21	3,025	441	1,155
13	60	22	3,600	484	1,320
14	60	22	3,600	484	1,320
15	65	18	4,225	324	1,170
16	65	24	4,225	576	1,560
SUM	760	454	38,200	13,382	20,685

$$m = \frac{\Sigma xy - [(\Sigma x)(\Sigma y)/n]}{\Sigma x^2 - [(\Sigma x)^2/n]}$$

$$= \frac{20.685 - [(760)(454)/16]}{38.200 - [(760)^2/16]} = -0.42$$

$$a = \Sigma\frac{y}{n} = -m\left(\Sigma\frac{x}{n}\right)$$

$$= \frac{454}{16} - (-0.42)\left(\frac{760}{16}\right) = 48.4$$

Thus, $y = 48.4 + (-0.42)x$.

For 57 mi/h, the gas mileage will be $y = 48.4 + (-0.42)(57) = 24.5$.

$$r = \frac{\Sigma xy - [(\Sigma x)(\Sigma y)/n]}{(\Sigma x^2 - [(\Sigma x)^2/n])(\Sigma y^2 - [(\Sigma y^2)/n])}$$

$$= \frac{20{,}685 - (760)(454)/16}{38{,}200 - 760^2/16\ 13{,}382 - 454^2/16} = -0.86$$

This correlation of –0.86 is good, but not great.

COMPUTER PROGRAM

Microsoft's EXCEL has the capability of performing calculations under the Formulas/More Functions/Statistical tabs. These calculations are for average, median, mode, standard deviation, frequency, kurtosis, skewness, normal distribution, Z test, Weibell distribution, and correlation. The histogram with descriptive statistics is an add-in under Data Analysis on the Tools menu. The EXCEL program files on the website will solve chi-squared, and the scatter diagram. The file names are *chi-squared*, and *scatter diagram.*

EXERCISES

1. Round the following numbers to two decimal places.
   a. 0.862
   b. 0.625
   c. 0.149
   d. 0.475
2. Find the g.p.e. of the following numbers.
   a. 8.24
   b. 522
   c. 6.3×10^2
   d. 0.02
3. Find the relative error of the numbers in Exercise 2.
4. Perform the operation indicated and leave the answer in the correct number of significant figures.
   a. (34.6)(8.20)
   b. (0.035)(635)
   c. 3.8735/6.1
   d. 5.362/6 (6 is a counting number)
   e. 5.362/6 (6 is not a counting number)
5. Perform the operation indicated and leave the answer in the correct number of significant figures.
   a. 64.3 + 2.05
   b. 381.0 − 1.95
   c. 8.652 − 4 (4 is not a counting number)
   d. 8.652 − 4 (4 is a counting number)
   e. $6.4 \times 10^2 + 24.32$
6. In his last 70 games, a professional basketball player made the following scores:

10	17	9	17	18	20	16
7	17	19	13	15	14	13
12	13	15	14	13	10	14
11	15	14	11	15	15	16
9	18	15	12	14	13	14
13	14	16	15	16	15	15
14	15	15	16	13	12	16
10	16	14	13	16	14	15
6	15	13	16	15	16	16
12	14	16	15	16	13	15

   a. Make a tally sheet in ascending order.
   b. Using the data above, construct a histogram.
7. A company that fills bottles of shampoo tries to maintain a specific weight of the product. The table gives the weight of 110 bottles that were checked at random intervals. Make a tally of these weights and construct a frequency histogram. (Weight is in kilograms.)

6.00	5.98	6.01	6.01	5.97	5.99	5.98	6.01	5.99	5.98	5.96
5.98	5.99	5.99	6.03	5.99	6.01	5.98	5.99	5.97	6.01	5.98
5.97	6.01	6.00	5.96	6.00	5.97	5.95	5.99	5.99	6.01	6.00
6.01	6.03	6.01	5.99	5.99	6.02	6.00	5.98	6.01	5.98	5.99
6.00	5.98	6.05	6.00	6.00	5.98	5.99	6.00	5.97	6.00	6.00
6.00	5.98	6.00	5.94	5.99	6.02	6.00	5.98	6.02	6.01	6.00
5.97	6.01	6.04	6.02	6.01	5.97	5.99	6.02	5.99	6.02	5.99
6.02	5.99	6.01	5.98	5.99	6.00	6.02	5.99	6.02	5.95	6.02
5.96	5.99	6.00	6.00	6.01	5.99	5.96	6.01	6.00	6.01	5.98
6.00	5.99	5.98	5.99	6.03	5.99	6.02	5.98	6.02	6.02	5.97

8. Listed next are 125 readings obtained in a hospital by a motion-and-time study analyst who took 5 readings each day for 25 days. Construct a tally sheet. Prepare a table showing cell midpoints, cell boundaries, and observed frequencies. Plot a frequency histogram.

Day	Duration of Operation Time (min)				
1	1.90	1.93	1.95	2.05	2.20
2	1.76	1.81	1.81	1.83	2.01
3	1.80	1.87	1.95	1.97	2.07
4	1.77	1.83	1.87	1.90	1.93
5	1.93	1.95	2.03	2.05	2.14
6	1.76	1.88	1.95	1.97	2.00
7	1.87	2.00	2.00	2.03	2.10
8	1.91	1.92	1.94	1.97	2.05
9	1.90	1.91	1.95	2.01	2.05
10	1.79	1.91	1.93	1.94	2.10
11	1.90	1.97	2.00	2.06	2.28
12	1.80	1.82	1.89	1.91	1.99
13	1.75	1.83	1.92	1.95	2.04
14	1.87	1.90	1.98	2.00	2.08
15	1.90	1.95	1.95	1.97	2.03
16	1.82	1.99	2.01	2.06	2.06
17	1.90	1.95	1.95	2.00	2.10
18	1.81	1.90	1.94	1.97	1.99
19	1.87	1.89	1.98	2.01	2.15
20	1.72	1.78	1.96	2.00	2.05
21	1.87	1.89	1.91	1.91	2.00
22	1.76	1.80	1.91	2.06	2.12
23	1.95	1.96	1.97	2.00	2.00
24	1.92	1.94	1.97	1.99	2.00
25	1.85	1.90	1.90	1.92	1.92

9. The relative strengths of 150 silver solder welds are tested, and the results are given in the following table. Tally these figures and arrange them in a frequency distribution. Determine the cell interval and the approximate number of cells. Make a table showing cell midpoints, cell boundaries, and observed frequencies. Plot a frequency histogram.

1.5	1.2	3.1	1.3	0.7	1.3
0.1	2.9	1.0	1.3	2.6	1.7
0.3	0.7	2.4	1.5	0.7	2.1
3.5	1.1	0.7	0.5	1.6	1.4
1.7	3.2	3.0	1.7	2.8	2.2
1.8	2.3	3.3	3.1	3.3	2.9
2.2	1.2	1.3	1.4	2.3	2.5
3.1	2.1	3.5	1.4	2.8	2.8
1.5	1.9	2.0	3.0	0.9	3.1
1.9	1.7	1.5	3.0	2.6	1.0
2.9	1.8	1.4	1.4	3.3	2.4
1.8	2.1	1.6	0.9	2.1	1.5
0.9	2.9	2.5	1.6	1.2	2.4
3.4	1.3	1.7	2.6	1.1	0.8
1.0	1.5	2.2	3.0	2.0	1.8
2.9	2.5	2.0	3.0	1.5	1.3
2.2	1.0	1.7	3.1	2.7	2.3
0.6	2.0	1.4	3.3	2.2	2.9
1.6	2.3	3.3	2.0	1.6	2.7
1.9	2.1	3.4	1.5	0.8	2.2
1.8	2.4	1.2	3.7	1.3	2.1
2.9	3.0	2.1	1.8	1.1	1.4
2.8	1.8	1.8	2.4	2.3	2.2
2.1	1.2	1.4	1.6	2.4	2.1
2.0	1.1	3.8	1.3	1.3	1.0

10. Using the data of Exercise 6, construct:
    a. A relative frequency histogram
    b. A cumulative frequency histogram
    c. A relative cumulative frequency histogram
11. Using the data of Exercise 7, construct:
    a. A relative frequency histogram
    b. A cumulative frequency histogram
    c. A relative cumulative frequency histogram
12. Using the data of Exercise 8, construct:
    a. A relative frequency histogram
    b. A cumulative frequency histogram
    c. A relative cumulative frequency histogram
13. Using the data of Exercise 9, construct:
    a. A relative frequency histogram
    b. A cumulative frequency histogram
    c. A relative cumulative frequency histogram
14. Construct a bar graph of the data in:
    a. Exercise 6
    b. Exercise 7
15. Using the data of Exercise 8, construct:
    a. A polygon
    b. An ogive
16. Using the data of Exercise 9, construct:
    a. A polygon
    b. An ogive
17. An electrician testing the incoming line voltage for a residential house obtains 5 readings: 115, 113, 121, 115, 116. What is the average?
18. An employee makes 8 trips to load a trailer. If the trip distances in meters are 25.6, 24.8, 22.6, 21.3, 19.6, 18.5, 16.2, and 15.5, what is the average?
19. Tests of noise ratings at prescribed locations throughout a large stamping mill are given in the following frequency distribution. Noise is measured in decibels. Determine the average.

Cell Midpoint	Frequency
148	2
139	3
130	8
121	11
112	27
103	35
94	43
85	33
76	20
67	12
58	6
49	4
40	2

20. The weight of 65 castings in kilograms is distributed as follows:

Cell Midpoint	Frequency
3.5	6
3.8	9
4.1	18
4.4	14
4.7	13
5.0	5

Determine the average.

21. Destructive tests on the life of an electronic component were conducted on 2 different occasions. On the first occasion, 3 tests had a mean of 3320 h; on the second occasion, 2 tests had a mean of 3180 h. What is the weighted average?

22. The average height of 24 students in Section 1 of a course in quality control is 1.75 m; the average height of 18 students in Section 2 of quality control is 1.79 m; and the average height of 29 students in Section 3 of quality control is 1.68 m. What is the average height of the students in the 3 sections of quality control?

23. Determine the median of the following numbers.
    a. 22, 11, 15, 8, 18
    b. 35, 28, 33, 38, 43, 36

24. Determine the median for the following:
    a. The frequency distribution of Exercise 8
    b. The frequency distribution of Exercise 9
    c. The frequency distribution of Exercise 19
    d. The frequency distribution of Exercise 20
    e. The frequency distribution of Exercise 30
    f. The frequency distribution of Exercise 32

25. Given the following series of numbers, determine the mode.
    a. 50, 45, 55, 55, 45, 50, 55, 45, 55
    b. 89, 87, 88, 83, 86, 82, 84
    c. 11, 17, 14, 12, 12, 14, 14, 15, 17, 17

26. Determine the modal cell of the data in:
    a. Exercise 6
    b. Exercise 7
    c. Exercise 8
    d. Exercise 9
    e. Exercise 19
    f. Exercise 20

27. Determine the range for each set of numbers.
    a. 16, 25, 18, 17, 16, 21, 14
    b. 45, 39, 42, 42, 43
    c. The data in Exercise 6
    d. The data in Exercise 7

28. Frequency tests of a brass rod 145 cm long give values of 1200, 1190, 1205, 1185, and 1200 vibrations per second. What is the sample standard deviation?

29. Four readings of the thickness of the paper in this textbook are 0.076 mm, 0.082 mm, 0.073 mm, and 0.077 mm. Determine the sample standard deviation.

30. The frequency distribution given here shows the percent of organic sulfur in Illinois No. 5 coal. Determine the sample standard deviation.

Cell Midpoint (%)	Frequency (Number of Samples)
0.5	1
0.8	16
1.1	12
1.4	10
1.7	12
2.0	18
2.3	16
2.6	3

31. Determine the sample standard deviation for the following:
    a. The data of Exercise 9
    b. The data of Exercise 19

32. Determine the average and sample standard deviation for the frequency distribution of the number of inspections per day as follows:

Cell Midpoint	Frequency
1000	6
1300	13
1600	22
1900	17
2200	11
2500	8

33. Using the data of Exercise 19, construct:
    a. A polygon
    b. An ogive

34. Using the data of Exercise 20, construct:
    a. A polygon
    b. An ogive
35. Using the data of Exercise 30, construct:
    a. A polygon
    b. An ogive
36. Using the data of Exercise 32, construct:
    a. A polygon
    b. An ogive
37. Using the data of Exercise 19, construct:
    a. A histogram
    b. A relative frequency histogram
    c. A cumulative frequency histogram
    d. A relative cumulative frequency histogram
38. Using the data of Exercise 20, construct:
    a. A histogram
    b. A relative frequency histogram
    c. A cumulative frequency histogram
    d. A relative cumulative frequency histogram
39. Using the data of Exercise 30, construct:
    a. A histogram
    b. A relative frequency histogram
    c. A cumulative frequency histogram
    d. A relative cumulative frequency histogram
40. Using the data of Exercise 32, construct:
    a. A histogram
    b. A relative frequency histogram
    c. A cumulative frequency histogram
    d. A relative cumulative frequency histogram
41. Determine the skewness, kurtosis, and coefficient of variation of:
    a. Exercise 6
    b. Exercise 7
    c. Exercise 8
    d. Exercise 9
    e. Exercise 20
    f. Exercise 32
42. If the maximum allowable noise is 134.5 decibel (db), what percent of the data of Exercise 19 is above that value?
43. Evaluate the histogram of Exercise 20, where the specifications are 4.25, 6, and 0.60 kg.
44. A utility company will not use coal with a sulfur content of more than 2.25%. Based on the histogram of Exercise 30, what percent of the coal is in that category?
45. The population mean of a company's racing bicycles is 9.07 kg (20.0 lb) with a population standard deviation of 0.40 kg. If the distribution is approximately normal, determine (a) the percentage of bicycles less than 8.30 kg, (b) the percentage of bicycles greater than 10.00 kg, and (c) the percentage of bicycles between 8.00 and 10.10 kg.
46. If the mean time to clean a motel room is 16.0 min and the standard deviation is 1.5 min, what percentage of the rooms will take less than 13.0 min to complete? What percentage of the rooms will take more than 20.0 min to complete? What percentage of the rooms will take between 13.0 and 20.5 min to complete? The data are normally distributed.
47. A cold cereal manufacturer wants 1.5% of the product to be below the weight specification of 0.567 kg (1.25 lb). If the data are normally distributed and the standard deviation of the cereal filling machine is 0.018 kg, what mean weight is required?
48. In the precision grinding of a complicated part, it is more economical to rework the part than to scrap it. Therefore, it is decided to establish the rework percentage at 12.5%. Assuming normal distribution of the data, a standard deviation of 0.01 mm, and an upper specification limit of 25.38 mm (0.99 in.), determine the process center.
49. Using the information of Exercise 41, what is your judgment concerning the normality of the distribution in each of the following?
    a. Exercise 6
    b. Exercise 7
    c. Exercise 8
    d. Exercise 9
    e. Exercise 20
    f. Exercise 32
50. Using normal probability paper, determine (judgment) the normality of the distribution of the following.
    a. Second column of Table 5-4
    b. First three columns of Exercise 7
    c. Second column of Exercise 8
51. By means of a scatter diagram, determine if a relationship exists between product temperatures and percent foam for a soft drink. Data are as follows:

Day	Product Temperature (°F)	Foam (%)	Day	Product Temperature (°F)	Foam (%)
1	36	15	11	44	32
2	38	19	12	42	33
3	37	21	13	38	20
4	44	30	14	41	27
5	46	36	15	45	35
6	39	20	16	49	38
7	41	25	17	50	40
8	47	36	18	48	42
9	39	22	19	46	40
10	40	23	20	41	30

52. By means of a *scatter diagram*, determine whether there is a relationship between hours of machine use and millimeters off the target. Data for 20 (x, y) pairs with hours of machine use as the x variable are (30, 1.10), (31, 1.21), (32, 1.00), (33, 1.21), (34, 1.25), (35, 1.23), (36, 1.24), (37, 1.28), (38, 1.30), (39, 1.30), (40, 1.38), (41, 1.35), (42, 1.38), (43, 1.38), (44, 1.40), (45, 1.42), (46, 1.45), (47, 1.45), (48, 1.50), and (49, 1.58). Draw a line for the data using eyesight only and estimate the number of millimeters off the target at 55 h.
53. Data on gas pressure (kg/cm^2) and its volume (liters) are as follows: (0.5, 1.62), (1.5, 0.75), (2.0, 0.62), (3.0, 0.46), (2.5, 0.52), (1.0, 1.00), (0.8, 1.35), (1.2, 0.89), (2.8, 0.48), (3.2, 0.43), (1.8, 0.71), and (0.3, 1.80). Construct a scatter diagram. Determine the coefficient of correlation, the equation of the line, and the value at 2.7 kg/cm.
54. The following data (tensile strength, hardness) are for tensile strength (100 psi) and hardness (Rockwell E) of die-cast aluminum. Construct a scatter diagram and determine the relationship: (293, 53), (349, 70), (368, 40), (301, 55), (340, 78), (308, 64), (354, 71), (313, 53), (322, 82), (334, 67), (377, 70), (247, 56), (348, 86), (298, 60), (287, 72), (292, 51), (345, 88), (380, 95), (257, 51), (258, 75).
55. Data on the amount of water applied in inches and the yield of alfalfa in tons per acre are as follows:

Water	12	18	24	30	36	42	48	60
Yield	5.3	5.7	6.3	7.2	8.2	8.7	8.4	8.2

Prepare a scatter diagram and analyze the results. What is the coefficient of correlation?

56. Using the EXCEL software, determine the descriptive statistics and histogram using the data of the following:
    a. Exercise 6
    b. Exercise 7
    c. Exercise 8
    d. Exercise 9
    e. Exercise 19
    f. Exercise 20
    g. Exercise 30
    h. Exercise 32
57. Using the EXCEL program file with the file name *Weibull*, determine the normality of Exercises 50(a), 50(b), and 50(c).
58. Using the EXCEL program file with the file name *chi-squared*, determine the normality of the distribution of the following:
    a. Exercise 19
    b. Exercise 20
    c. Exercise 30
    d. Exercise 32
59. Using the EXCEL program file with the file name scatter diagram, obtain data from production or the laboratory and determine the graph.

CHAPTER
SIX

CONTROL CHARTS FOR VARIABLES[1]

OBJECTIVES

Upon completion of this chapter, the reader is expected to:

- know the three categories of variation and their sources;
- understand the concept of the control chart method;
- know the purposes of variable control charts;
- know how to select the quality characteristics, the rational subgroup, and the method of taking samples;
- calculate the central value, trial control limits, and the revised control limits for an $\overline{X}$ and R chart;
- compare an R chart with an s chart;
- explain what is meant by a process in control and the advantages that accrue;
- explain what is meant by a process out of control and the various out-of-control patterns;
- know the difference between individual measurements and averages; and between control limits and specifications;
- know the different situations between the process spread and specifications and what can be done to correct an undesirable situation;
- be able to calculate process capability;
- be able to identify the different types of variable control charts and the reasons for their use.

INTRODUCTION

Variation

One of the truisms of manufacturing is that no two objects are ever made exactly alike. In fact, the variation concept is a law of nature, in that no two natural items in any category are exactly the same. The variation may be quite large and easily noticeable, such as the height of human beings, or the variation may be very small, such as the weight of fiber-tipped pens or the shapes of snowflakes. When variations are very small, it may appear that items are identical; however, precision instruments will show differences. If two items appear to have the same measurement, it is due to the limits of our measuring instruments. As measuring instruments have become more refined, variation has continued to exist—only the increment of variation has changed. The ability to measure variation is necessary before it can be controlled.

There are three categories of variations in piece part production.

1. **Within-piece variation.** This type of variation is illustrated by the surface roughness of a piece, wherein one portion of the surface is rougher than another portion, or the printing on one end of the page is better than on the other end.
2. **Piece-to-piece variation.** This type of variation occurs among pieces produced at the same time. Thus, the light intensity of four consecutive light bulbs produced from the same machine will be different.
3. **Time-to-time variation.** This type of variation is illustrated by the difference in a product or service produced at different times of the day. Thus, a service given early in the morning will be different from one given later in the day; or, as a cutting tool wears, the cutting characteristics will change.

Categories of variation for other types of processes, such as a continuous chemical process or an income tax audit, will not be exactly the same; however, the concept will be similar.

Variation is present in every process because of a combination of the equipment, materials, environment, and operator. The first source of variation is the *equipment.* This source includes tool wear, machine vibration, workholding device positioning, and hydraulic and electrical fluctuations. When all these variations are put together, there is a certain capability or precision within which the equipment operates. Even supposedly identical machines have different capabilities, and this fact becomes a very important consideration when scheduling the manufacture of critical parts.

[1]The information in this chapter is based on ANSI/ASQC B1-B3.

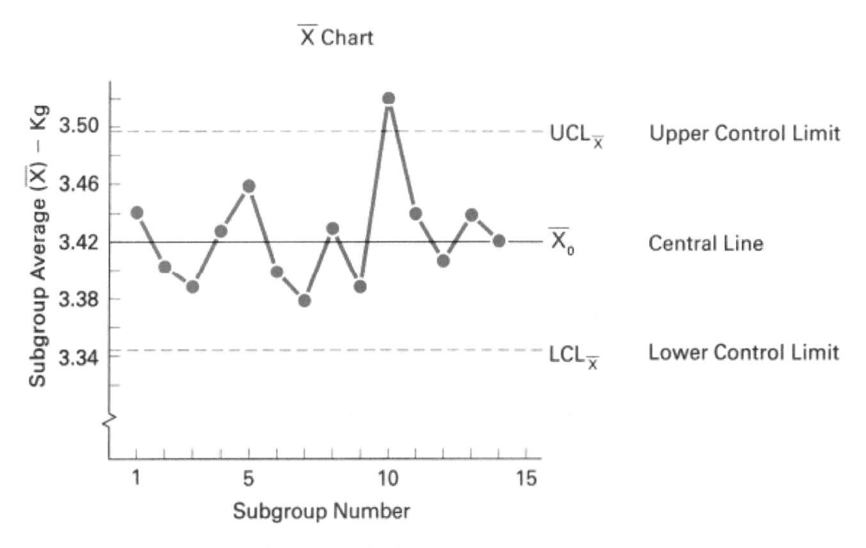

FIGURE 6-1 Example of a Control Chart

The second source of variation is the *material.* Because variation occurs in the finished product, it must also occur in the raw material (which was someone else's finished product). Such quality characteristics as tensile strength, ductility, thickness, porosity, and moisture content can be expected to contribute to the overall variation in the final product.

A third source of variation is the *environment.* Temperature, light, radiation, electrostatic discharge, particle size, pressure, and humidity can all contribute to variation in the product. In order to control this source, products are sometimes manufactured in "white rooms." Experiments are conducted in outer space to learn more about the effect of the environment on product variation.

A fourth source is the *operator.* This source of variation includes the method by which the operator performs the operation. The operator's physical and emotional well-being also contributes to the variation. A cut finger, a twisted ankle, a personal problem, or a headache can cause an operator's quality performance to vary. An operator's lack of understanding of equipment and material variations for want of training may lead to frequent machine adjustments, thereby compounding the variability. As our equipment has become more automated, the operator's effect on variation has lessened.

These four sources together account for the true variation. There is also a reported variation, which is due to the *inspection* activity. Faulty inspection equipment, incorrect application of a quality standard, or too heavy a pressure on a micrometer can be the cause of the incorrect reporting of variation. In general, variation due to inspection should be one-tenth of the four other sources of variations. It should be noted that three of these sources are present in the inspection activity: an inspector, inspection equipment, and the environment.

As long as these sources of variation fluctuate in a natural or expected manner, a stable pattern of many *chance causes* (random causes) of variation develops. Chance causes of variation are inevitable. Because they are numerous and individually of relatively small importance, they are difficult to detect or identify. Those causes of variation that are large in magnitude, and therefore readily identified, are classified as *assignable causes.*[2] When only chance causes are present in a process, the process is considered to be in a state of statistical control. It is stable and predictable. However, when an assignable cause of variation is also present, the variation will be excessive, and the process is classified as out of control or beyond the expected natural variation.

The Control Chart Method

In order to indicate when observed variations in quality are greater than could be left to chance, the control chart method of analysis and presentation of data is used. The control chart method for variables is a means of visualizing the variations that occur in the central tendency and dispersion of a set of observations. It is a graphical record of the quality of a particular characteristic. It shows whether or not the process is in a stable state.

An example of a control chart is given in Figure 6-1. This particular chart is referred to as an $\overline{X}$ chart (average) and is used to record the variation in the average value of samples. Another chart, such as the *R* chart (range), would have also served for explanation purposes. The horizontal axis is labeled "Subgroup Number," which identifies a particular sample consisting of a fixed number of observations. These subgroups are in order, with the first one inspected being 1 and the last one inspected being 14. The vertical axis of the graph is the variable, which in this particular case is weight measured in kilograms.

Each small solid circle represents the average value within a subgroup. Thus, subgroup number 5 consists of, say, four observations, 3.46, 3.49, 3.45, and 3.44, and their

[2]W. Edwards Deming uses the words *common* and *special* for *chance* and *assignable.*

average is 3.46 kg. This value is the one posted on the chart for subgroup number 5. Averages are usually used on control charts rather than individual observations because average values will indicate a change in variation much faster.[3] Also, with two or more observations in a sample, a measure of the dispersion can be obtained for a particular subgroup.

The solid line in the center of the chart can have three different interpretations, depending on the available data. First, and most commonly, it can be the average of the plotted points, which in the case of an $\overline{X}$ chart is the average of the averages or "X-double bar," $\overline{\overline{X}}$. Second, it can be a standard or reference value, $\overline{X}_0$, based on representative prior data, an economic value based on production costs or service needs, or an aimed-at value based on specifications. Third, it can be the population mean, μ, if that value is known.

The two dashed outer lines are the upper and lower control limits. These limits are established to assist in judging the significance of the variation in the quality of the product or service. Control limits are frequently confused with *specification limits,* which are the permissible limits of a quality characteristic of each *individual* unit of a product. However, *control limits* are used to evaluate the variations in quality from subgroup to subgroup. Therefore, for the $\overline{X}$ chart, the control limits are a function of the subgroup averages. A frequency distribution of the subgroup averages can be determined with its corresponding average and standard deviation. The control limits are usually established at ± 3 standard deviations from the central line. One recalls, from the discussion of the normal curve, that the number of items between $+3\sigma$ and -3σ equals 99.73%. Therefore, it is expected that over 9973 times out of 10,000, the subgroup values will fall between the upper and lower limits, and when this occurs, the process is considered to be in control. When a subgroup value falls outside the limits, the process is considered to be out of control and an assignable cause for the variation is present. There are other types of out-of-control conditions, and these are discussed later in the chapter. Subgroup number 10 in Figure 6-1 is beyond the upper control limit; therefore, there has been a change in the stable nature of the process at that time, causing the out-of-control point. It is also true that the out-of-control condition could be due to a chance cause, and this fact can occur 27 times out of 10,000. The average run length (ARL) before a chance cause occurs at 3.70 units ($10{,}000 \div 27$).

In practice, control charts are posted at work centers to control a particular quality characteristic. Usually, an $\overline{X}$ chart for the central tendency and an R chart for the dispersion are used together. An example of this dual charting is illustrated in Figure 6-2, which shows a method of charting and reporting inspection results for rubber durometers. At work center number 365-2 at 8:30 A.M., the operator selects four items for testing, and records the observations of 55, 52, 51, and 53 in the rows marked X_1, X_2, X_3, and X_4, respectively. A subgroup average value of 52.8 is obtained by summing the observation and dividing by 4, and the range value of 4 is obtained by subtracting the low value, 51, from the high value, 55. The operator places a small solid circle at 52.8 on the $\overline{X}$ chart and a small solid circle at 4 on the R chart and then proceeds with other duties.

The frequency with which the operator inspects a product at a particular work center is determined by the quality of the product. When the process is in control and no difficulties are being encountered, fewer inspections may be required; conversely, when the process is out of control or during startup, more inspections may be needed. The inspection frequency at a work center can also be determined by the amount of time that must be spent on non-inspection activities. In the example problem, the inspection frequency appears to be every 60 or 65 minutes.

At 9:30 A.M. the operator performs the activities for subgroup 2 in the same manner as for subgroup 1. It is noted that the range value of 7 falls right on the upper control limit. Whether to consider this in control or out of control would be a matter of organization policy. It is suggested that it be classified as in control and a cursory examination for an assignable cause be conducted by the operator. A plotted point that falls on the control limit will be a rare occurrence.

The inspection results for subgroup 2 show that the third observation, X_3, has a value of 57, which exceeds the upper control limit. The reader is cautioned to remember the earlier discussion on control limits and specifications. In other words, the 57 value is an individual observation and does not relate to the control limits. Therefore, the fact that an individual observation is greater than or less than a control limit is meaningless.

Subgroup 4 has an average value of 44, which is less than the lower control limit of 45. Therefore, subgroup 4 is out of control, and the operator will report this fact to the departmental supervisor. The operator and supervisor will then look for an assignable cause and, if possible, take corrective action. Whatever corrective action is taken will be noted by the operator on the $\overline{X}$ and R chart or on a separate form. The control chart indicates when and where trouble has occurred; the identification and elimination of the difficulty is a production problem. Ideally, the control chart should be maintained by the operator, provided time is available and proper training has been given. When the operator cannot maintain the chart, then it is maintained by quality control.

A control chart is a statistical tool that distinguishes between natural and unnatural variation as shown in Figure 6-3. Unnatural variation is the result of assignable causes. It usually, but not always, requires corrective action by people close to the process, such as operators, technicians, clerks, maintenance workers, and first-line supervisors.

Natural variation is the result of chance causes. It requires management intervention to achieve quality improvement. In this regard, between 80% and 85% of quality problems are

[3]For a proof of this statement, see J. M. Juran, ed., *Quality Control Handbook,* 4th ed. (New York: McGraw-Hill, 1988), sec. 24, p. 10.

$\overline{X}$ AND R CHART

Work Center Number 365–2
Quality Characteristic Durometer Date 3/6

Time	8 30 AM	9 30 AM	10 40 AM	11 50 AM	1 30 PM									
Subgroup	1	2	3	4	5	6	7	8	9	10	11	12	13	14
X_1	55	51	48	45	53									
X_2	52	52	49	43	50									
X_3	51	57	50	45	48									
X_4	53	50	49	43	50									
Sum	211	210	196	176	201									
$\overline{X}$	52.8	52.5	49	44	50.2									
R	4	7	2	2	5									

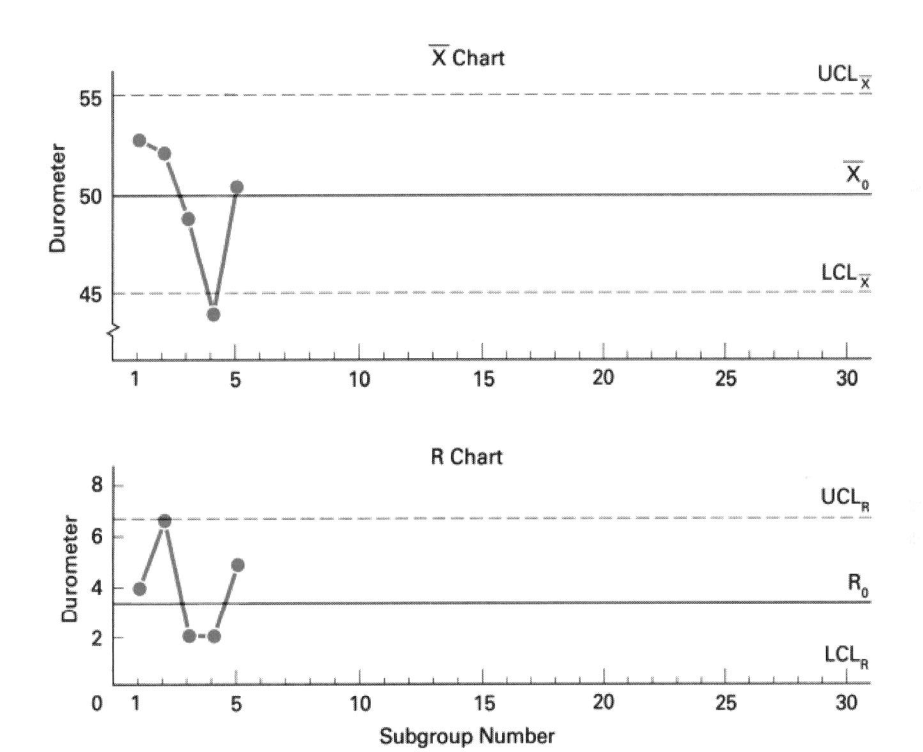

FIGURE 6-2 Example of a Method of Reporting Inspection Results

due to management or the system, and 15% to 20% are due to operations.

The control chart is used to keep a continuing record of a particular quality characteristic. It is a picture of the process over time. When the chart is completed, it is replaced by a fresh chart, and the completed chart is stored in an office file. The chart is used to improve the process quality, to determine the process capability, to help determine effective specifications, to determine when to leave the process alone and when to make adjustments, and to investigate causes of unacceptable or marginal quality.[4]

[4]Wallace Davis III, "Using Corrective Action to Make Matters Worse," *Quality Progress* (October 2000): 56–61.

Objectives of Variable Control Charts

Variable control charts provide information:

1. **For quality improvement.** Having a variable control chart merely because it indicates that there is a quality control program is missing the point. A variable control chart is an excellent technique for achieving quality improvement.

2. **To determine the process capability.** The true process capability can be achieved only after substantial quality improvement has been achieved. During the quality improvement cycle, the control chart will indicate that no further improvement is possible without a large dollar expenditure. At that point the true process capability is obtained.

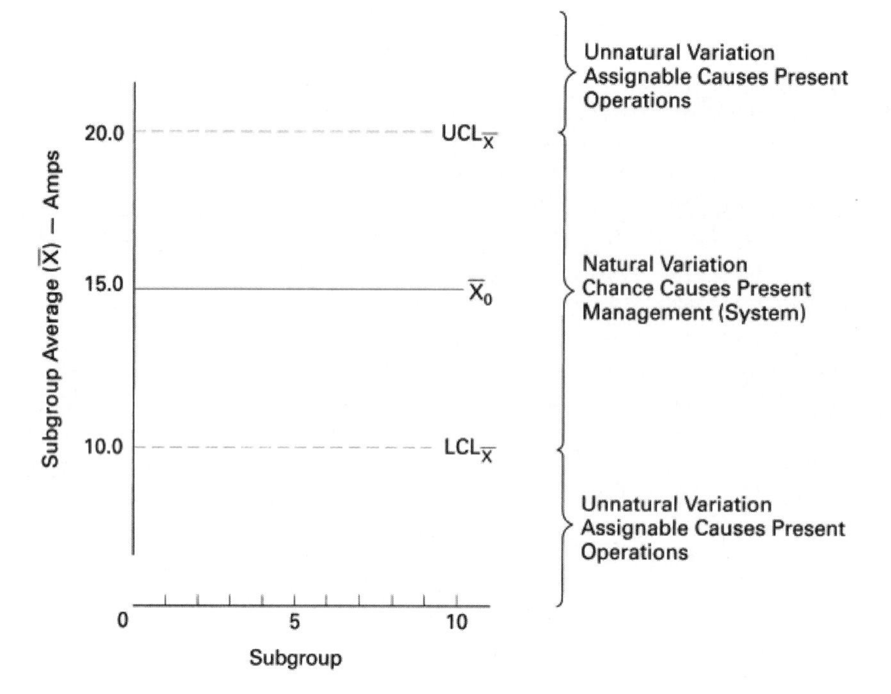

FIGURE 6-3 Natural and Unnatural Causes of Variation

3. **For decisions in regard to product specifications.** Once the true process capability is obtained, effective specifications can be determined. If the process capability is ± 0.003, then specifications of ± 0.004 are realistically obtainable by operating personnel.
4. **For current decisions in regard to the production process.** First a decision is needed to judge whether control exists. If not, the control chart is used to attain control. Once control is obtained, the control chart is used to maintain control. Thus, the control chart is used to decide when a natural pattern of variation occurs and the process should be left alone, and when an unnatural pattern of variation is occurring, which requires action to find and eliminate the assignable causes.

   In this regard, operating personnel are giving a quality performance as long as the plotted points are within the control limits. If this performance is not satisfactory, the solution is the responsibility of the system rather than the operator.
5. **For current decisions in regard to recently produced items.** Thus, the control chart is used as one source of information to help decide whether an item or items should be released to the next phase of the sequence or some alternative disposition made, such as sorting and repairing.

These purposes are frequently dependent on each other. For example, quality improvement is needed prior to determining the true process capability, which is needed prior to determining effective specifications. Control charts for variables should be established to achieve a particular purpose. Their use should be discontinued when the purpose has been achieved or their use continued with inspection substantially reduced.

CONTROL CHART TECHNIQUES

Introduction

In order to establish a pair of control charts for the average ($\overline{X}$) and the range (R), it is desirable to follow a set procedure. The steps in this procedure are as follows:

1. Select the quality characteristic.
2. Choose the rational subgroup.
3. Collect the data.
4. Determine the trial central line and control limits.
5. Establish the revised central line and control limits.
6. Achieve the objective.

The procedure presented in this section relates to an $\overline{X}$ and R chart. Information on an s chart is also presented.

Select the Quality Characteristic

The variable that is chosen for an $\overline{X}$ and R chart must be a quality characteristic that is measurable and can be expressed in numbers. Quality characteristics that can be expressed in terms of the seven basic units—length, mass, time, electrical current, temperature, substance, or luminous intensity—are appropriate, as well as any of the derived units, such as power, velocity, force, energy, density, and pressure.

Those quality characteristics affecting the performance of the product or service will normally be given first attention. These may be a function of the raw materials, component parts, subassemblies, or finished parts. In other words, give high priority to the selection of those characteristics that are giving difficulty in terms of production problems and/or cost. An excellent opportunity for cost savings is frequently selected where spoilage and rework costs are high. A Pareto analysis is also useful to establish priorities. Another possibility occurs where destructive testing is used to inspect a product.

In any manufacturing plant there are a large number of variables that make up a product. It is therefore impossible to place $\overline{X}$ and R charts on all variables, and a judicious selection of those quality characteristics is required. Because all variables can be treated as attributes, an attribute control chart (see Chapter 8) can also be used to achieve quality improvement.

Choose the Rational Subgroup

As previously mentioned, the data that are plotted on the control chart consist of groups of items that are called rational subgroups. It is important to understand that data collected in a random manner do *not* qualify as rational. A rational subgroup is one in which the variation within the group is due only to chance causes. This within-subgroup variation is used to determine the control limits. Variation between subgroups is used to evaluate long-term stability. There are two schemes for selecting the subgroup samples:

1. The first scheme is to select the subgroup samples from product or service produced at one instant of time or as close to that instant as possible. Four consecutive parts from a machine or four parts from a tray of recently produced parts would be an example of this subgrouping technique. The next subgroup sample would be similar but for product or service produced at a later time—say, 1 h later. This scheme is called the instant-time method.

2. The second scheme is to select product or service produced over a period of time so that it is representative of all the product or service. For example, an inspector makes a visit to a circuit-breaker assembling process once every hour. The subgroup sample of, say, four is randomly selected from all the circuit breakers produced in the previous hour. On his next visit, the subgroup is selected from the product produced between visits, and so forth. This scheme is called the period-of-time method.

In comparing the two schemes, the instant-time method will have a minimum variation *within* a subgroup and a maximum variation *among* subgroups. The period-of-time method will have a maximum variation *within* a subgroup and a minimum variation *among* subgroups. Some numerical values may help to illustrate this difference. Thus, for the instant-time method, subgroup average values ($\overline{X}$'s) could be from, say, 26 to 34 with subgroup range values (R's) from 0 to 4; whereas for the period-of-time method, the subgroup average values ($\overline{X}$'s) would vary from 28 to 32 with the subgroup range values (R's) from 0 to 8.

The instant-time method is more commonly used because it provides a particular time reference for determining assignable causes. It also provides a more sensitive measure of changes in the process average. Because all the values are close together, the variation will most likely be due to chance causes and thereby meet the rational subgroup criteria.

The advantage of the period-of-time method is that it provides better overall results, and, therefore, quality reports will present a more accurate picture of the quality. It is also true that because of process limitations this method may be the only practical method of obtaining the subgroup samples. Assignable causes of variation *may* be present in the subgroup, which will make it difficult to ensure that a rational subgroup is present.

In rare situations, it may be desirable to use both subgrouping methods. When this occurs, two charts with different control limits are required.

Regardless of the scheme used to obtain the subgroup, the lots from which the subgroups are chosen must be homogeneous. By *homogeneous* is meant that the pieces in the lot are as alike as possible—same machine, same operator, same mold cavity, and so on. Similarly, a fixed quantity of material, such as that produced by one tool until it wears out and is replaced or resharpened, should be a homogeneous lot. Homogeneous lots can also be designated by equal time intervals, because this technique is easy to organize and administer. No matter how the lots are designated, the items in any one subgroup should have been produced under essentially the same conditions.

Decisions on the size of the sample or subgroup require a certain amount of empirical judgment; however, some helpful guidelines can be given:

1. As the subgroup size increases, the control limits become closer to the central value, which makes the control chart more sensitive to small variations in the process average.
2. As the subgroup size increases, the inspection cost per subgroup increases. Does the increased cost of larger subgroups justify the greater sensitivity?
3. When destructive testing is used and the item is expensive, a small subgroup size of 2 or 3 is necessary, because it will minimize the destruction of expensive product or service.
4. Because of the ease of computation, a sample size of 5 is quite common in industry; however, when inexpensive electronic hand calculators are used, this reason is no longer valid.
5. From a statistical basis, a distribution of subgroup averages, $\overline{X}$'s, are nearly normal for subgroups of 4 or more, even when the samples are taken from a nonnormal population. Proof of this statement is given later in the chapter.
6. When the subgroup size exceeds 10, the s chart should be used instead of the R chart for the control of the dispersion.

There is no rule for the frequency of taking subgroups, but the frequency should be often enough to detect process changes. The inconveniences of the factory or office layout and the cost of taking subgroups must be balanced with the value of the data obtained. In general, it is best to sample

TABLE 6-1 Sample Sizes

Lot Size	Sample Size
91–150	10
151–280	15
281–400	20
401–500	25
501–1,200	35
1,201–3,200	50
3,201–10,000	75
10,001–35,000	100
35,001–150,000	150

Source: ANSI/ASQ Z1.9, Normal Inspection, Level II.

quite often at the beginning and to reduce the sampling frequency when the data permit.

The use of Table 6-1, which was obtained from ANSI/ASQ Z1.9, can be a valuable aid in making judgments on the amount of sampling required. If a process is expected to produce 4000 pieces per day, then 75 total inspections are suggested. Therefore, with a subgroup size of four, 19 subgroups would be a good starting point.

The precontrol rule (see Chapter 7) for the frequency of sampling could also be used. It is based on how often the process is adjusted. If the process is adjusted every hour, then sampling should occur every 10 min; if the process is adjusted every 2 h, then sampling should occur every 20 min; if the process is adjusted every 3 h, then sampling should occur every 30 min; and so forth.

The frequency of taking a subgroup is expressed in terms of the percent of items produced or in terms of a time interval. In summary, the selection of the rational subgroup is made in such a manner that only chance causes are present in the subgroup.

Collect the Data

The next step is to collect the data. This step can be accomplished using the type of form shown in Figure 6-2, wherein the data are recorded in a vertical fashion. By recording the measurements one below the other, the summing operation for each subgroup is somewhat easier. An alternative method of recording the data is shown in Table 6-2, wherein the data are recorded in a horizontal fashion. The particular method makes no difference when an electronic hand calculator is available. For illustrative purposes, the latter method will be used.

TABLE 6-2 Data on the Depth of the Keyway (mm)[a]

Subgroup Number	Date	Time	MEASUREMENTS X_1	X_2	X_3	X_4	Average $\overline{X}$	Range R	Comment
1	12/26	8:50	35	40	32	37	6.36	0.08	
2		11:30	46	37	36	41	6.40	0.10	
3		1:45	34	40	34	36	6.36	0.06	
4		3:45	69	64	68	59	6.65	0.10	New, temporary
5		4:20	38	34	44	40	6.39	0.10	operator
6	12/27	8:35	42	41	43	34	6.40	0.09	
7		9:00	44	41	41	46	6.43	0.05	
8		9:40	33	41	38	36	6.37	0.08	
9		1:30	48	44	47	45	6.46	0.04	
10		2:50	47	43	36	42	6.42	0.11	
11	12/28	8:30	38	41	39	38	6.39	0.03	
12		1:35	37	37	41	37	6.38	0.04	
13		2:25	40	38	47	35	6.40	0.12	
14		2:35	38	39	45	42	6.41	0.07	
15		3:55	50	42	43	45	6.45	0.08	
16	12/29	8:25	33	35	29	39	6.34	0.10	
17		9:25	41	40	29	34	6.36	0.12	
18		11:00	38	44	28	58	6.42	0.30	Damaged oil line
19		2:35	35	41	37	38	6.38	0.06	
20		3:15	56	55	45	48	6.51	0.11	Bad material
21	12/30	9:35	38	40	45	37	6.40	0.08	
22		10:20	39	42	35	40	6.39	0.07	
23		11:35	42	39	39	36	6.39	0.06	
24		2:00	43	36	35	38	6.38	0.08	
25		4:25	39	38	43	44	6.41	0.06	
Sum							160.25	2.19	

[a]For simplicity in recording, the individual measurements are coded from 6.00 mm.

Assuming that the quality characteristic and the plan for the rational subgroup have been selected, a technician can be assigned the task of collecting the data as part of his or her normal duties. The first-line supervisor and the operator should be informed of the technician's activities; however, no charts or data are posted at the work center at this time.

Because of difficulty in the assembly of a gear hub to a shaft using a key and keyway, the project team recommends using an $\overline{X}$ and R chart. The quality characteristic is the shaft keyway depth of 6.35 mm (0.250 in.). Using a rational subgroup of 4, a technician obtains 5 subgroups per day for 5 days. The samples are measured, the subgroup average $(\overline{X})$ and range R are calculated, and the results are recorded on the form. Additional recorded information includes the date, time, and any comments pertaining to the process. For simplicity, individual measurements are coded from 6.00 mm. Thus, the first measurement of 6.35 is recorded as 35.

It is necessary to collect a minimum of 25 subgroups of data. Fewer subgroups would not provide a sufficient amount of data for the accurate computation of the central line and control limits, and more subgroups would delay the introduction of the control chart. When subgroups are obtained slowly, it may be desirable to obtain preliminary conclusions with fewer subgroups.

The data are plotted in Figure 6-4, and this chart is called a *run chart*. It does not have control limits but can be used to analyze the data, especially in the development stage of a product or prior to a state of statistical control. The data points are plotted by order of production, as shown in the figure. Plotting the data points is a very effective way of finding out about the process. This should be done as the first step in data analysis.

Statistical limits are needed to determine if the process is stable.

Determine the Trial Central Line and Control Limits

The central lines for the $\overline{X}$ and R charts are obtained using the formulas

$$\overline{\overline{X}} = \frac{\sum_{i=1}^{g} \overline{X}_i}{g} \quad \text{and} \quad \overline{R} = \frac{\sum_{i=1}^{g} R_i}{g}$$

where $\overline{\overline{X}}$ = average of the subgroup averages (read "X double bar")
$\overline{X}_i$ = average of the ith subgroup
g = number of subgroups
$\overline{R}$ = average of the subgroup ranges
R_i = range of the ith subgroup

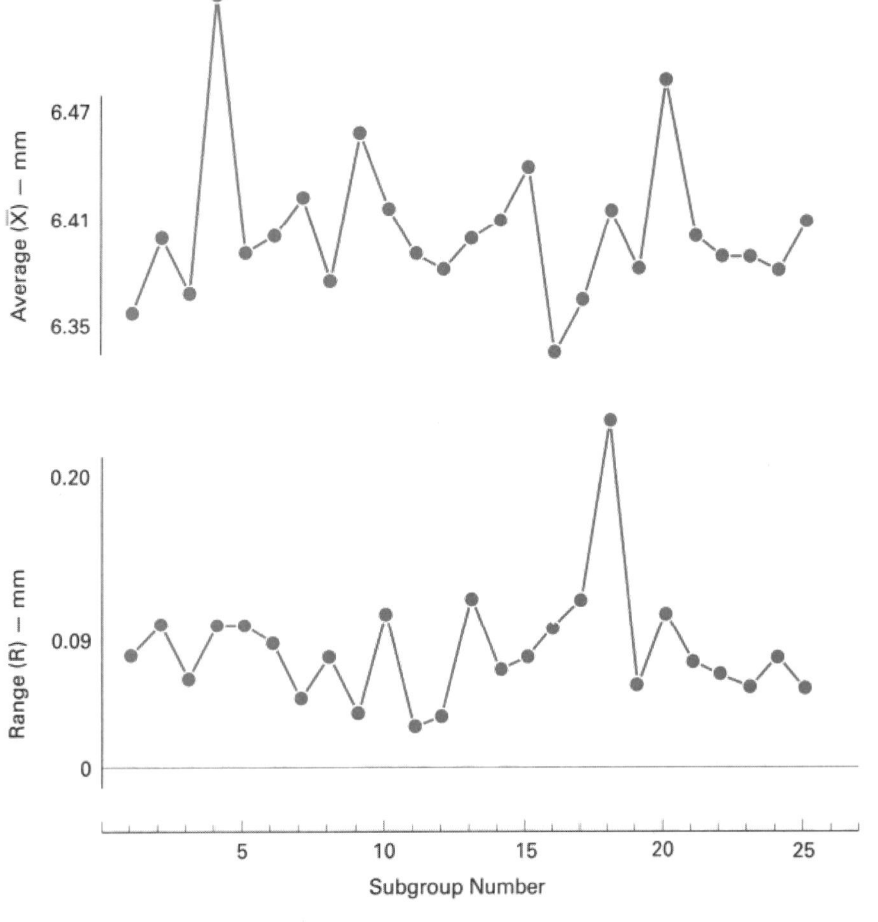

FIGURE 6-4 Run Chart for Data of Table 6-2

Trial control limits for the charts are established at ±3 standard deviations from the central value, as shown by the formulas

$$\mathrm{UCL}_{\bar{X}} = \bar{\bar{X}} + 3\sigma_{\bar{X}} \qquad \mathrm{UCL}_R = \bar{R} + 3\sigma_R$$
$$\mathrm{LCL}_{\bar{X}} = \bar{\bar{X}} - 3\sigma_{\bar{X}} \qquad \mathrm{LCL}_R = \bar{R} - 3\sigma_R$$

where UCL = upper control limit
LCL = lower control limit
$\sigma_{\bar{X}}$ = population standard deviation of the subgroup averages ($\bar{X}$'s)
σ_R = population standard deviation of the range

In practice, the calculations are simplified by using the product of the range ($\bar{R}$) and a factor (A_2) to replace the 3 standard deviations ($A_2\bar{R} = 3\sigma_{\bar{X}}$)[5] in the formulas for the $\bar{X}$ chart. For the R chart, the range $\bar{R}$ is used to estimate the standard deviation of the range (σ_R).[6] Therefore, the derived formulas are

$$\mathrm{UCL}_{\bar{X}} = \bar{\bar{X}} + A_2\bar{R} \qquad \mathrm{UCL}_R = D_4\bar{R}$$
$$\mathrm{LCL}_{\bar{X}} = \bar{\bar{X}} - A_2\bar{R} \qquad \mathrm{LCL}_R = D_3\bar{R}$$

where A_2, D_3, and D_4 are factors that vary with the subgroup size and are found in Table B of the Appendix. For the $\bar{X}$ chart, the upper and lower control limits are symmetrical about the central line. Theoretically, the control limits for an R chart should also be symmetrical about the central line. But, for this situation to occur, with subgroup sizes of 6 or less, the lower control limit would need to have a negative value. Because a negative range is impossible, the lower control limit is located at 0 by assigning to D_3 the value of 0 for subgroups of 6 or less.

When the subgroup size is 7 or more, the lower control limit is greater than 0 and symmetrical about the central line. However, when the R chart is posted at the work center, it may be more practical to keep the lower control limit at 0. This practice eliminates the difficulty of explaining to the operator that points below the lower control limit on the R chart are the result of exceptionally good performance rather than poor performance. However, quality personnel should keep their own charts with the lower control limit in its proper location, and any out-of-control low points should be investigated to determine the reason for the exceptionally good performance. Because subgroup sizes of 7 or more are uncommon, the situation occurs infrequently.

[5]The derivation of $3\sigma_{\bar{X}} = A_2\bar{R}$ is based on the substitution of $\sigma_{\bar{X}} = \sigma/\sqrt{n}$ and an estimate of $\sigma = R/d_2$, where d_2 is a factor for the subgroup size.

$$3\sigma_x = \frac{3\sigma}{\sqrt{n}} = \frac{3}{d_2\sqrt{n}}\bar{R}; \quad \text{therefore, } A_2 = \frac{3}{d_2\sqrt{n}}$$

[6]The derivation of the simplified formula is based on the substitution of $d_3\sigma = \sigma R$ and $\sigma = \bar{R}/d_2$, which gives

$$\left(1 + \frac{3d_3}{d_2}\right)\bar{R} \quad \text{and} \quad \left(1 - \frac{3d_3}{d_2}\right)\bar{R}$$

for the control limits. Thus, D_4 and D_3 are set equal to the coefficients of $\bar{R}$.

Example Problem 6-1

In order to illustrate the calculations necessary to obtain the trial control limits and the central line, the data in Table 6-2 concerning the depth of the shaft keyway will be used. From that table, $\Sigma\bar{X} = 160.25$, $\Sigma R = 2.19$, and $g = 25$; thus, the central lines are

$$\bar{\bar{X}} = \frac{\sum_{i=1}^{g} \bar{X}_i}{g} = \frac{160.25}{25} = 6.41 \text{ mm} \qquad \bar{R} = \frac{\sum_{i=1}^{g} R_i}{g} = \frac{2.19}{25} = 0.0876 \text{ mm}$$

From Table B in the Appendix, the values for the factors for a subgroup size (n) of 4 are $A_2 = 0.729$, $D_3 = 0$, and $D_4 = 2.282$. Trial control limits for the $\bar{X}$ chart are

$$\mathrm{UCL}_{\bar{X}} = \bar{\bar{X}} + A_2\bar{R} = 6.41 + (0.729)(0.0876) = 6.47 \text{ mm}$$

$$\mathrm{LCL}_{\bar{X}} = \bar{\bar{X}} - A_2\bar{R} = 6.41 - (0.729)(0.0876) = 6.35 \text{ mm}$$

Trial control limits for the R chart are

$$\mathrm{UCL}_R = D_4\bar{R} = (2.282)(0.0876) = 0.20 \text{ mm} \qquad \mathrm{LCL}_R = D_3\bar{R} = (0)(0.0876) = 0 \text{ mm}$$

Figure 6-5 shows the central lines and the trial control limits for the $\bar{X}$ and R charts for the preliminary data.

Establish the Revised Central Line and Control Limits

The first step is to post the preliminary data to the chart along with the control limits and central lines. This has been accomplished and is shown in Figure 6-5.

The next step is to adopt standard values for the central lines or, more appropriately stated, the best estimate of the standard values with the available data. If an analysis of the preliminary data shows good control, then $\bar{\bar{X}}$ and $\bar{R}$ can be considered as representative of the process and these become the standard values, $\bar{X}_0$ and R_0. Good control can be briefly described as that which has no out-of-control points, no long runs on either side of the central line, and no unusual patterns of variation. More information concerning in control and out of control is provided later in the chapter.

Most processes are not in control when first analyzed. An analysis of Figure 6-5 shows that there are out-of-control points on the $\bar{X}$ chart at subgroups 4, 16, and 20 and an out-of-control point on the R chart at subgroup 18. It also

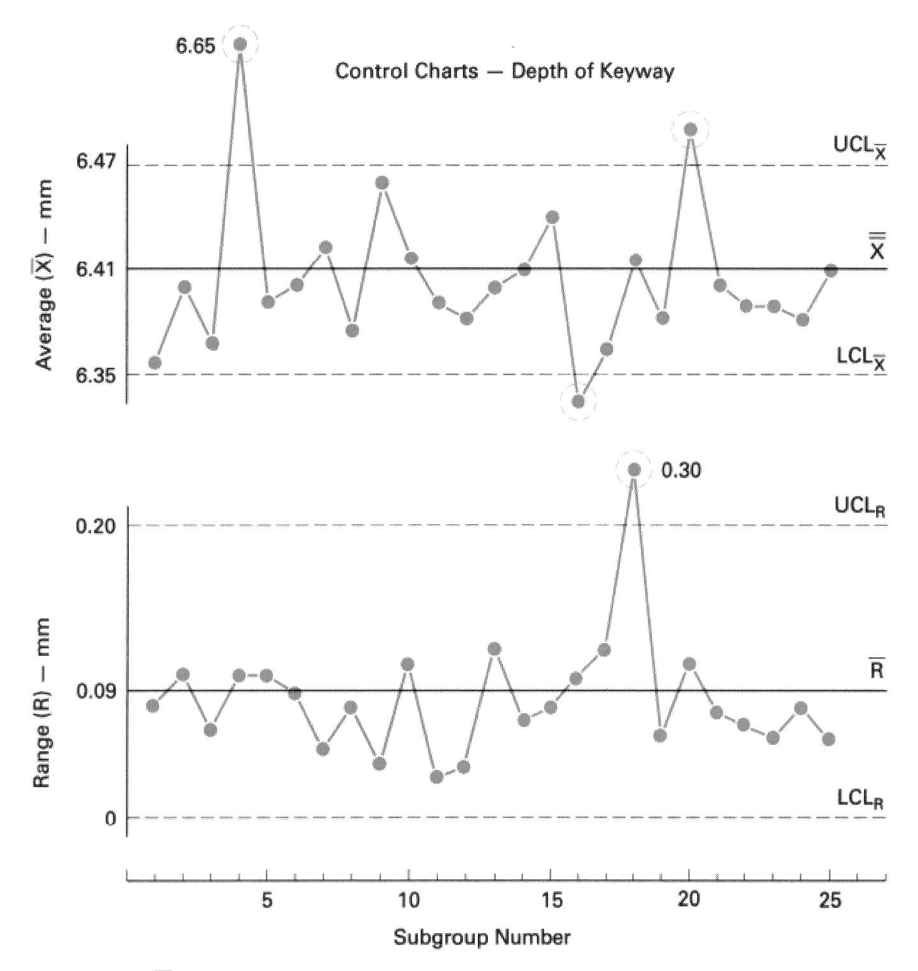

FIGURE 6-5 $\bar{X}$ and R Chart for Preliminary Data with Trial Control Limits

appears that there are a large number of points below the central line, which is no doubt due to the influence of the high points.

The R chart is analyzed first to determine if it is stable. Because the out-of-control point at subgroup 18 on the R chart has an assignable cause (damaged oil line), it can be discarded from the data. The remaining plotted points indicate a stable process.

The $\bar{X}$ chart can now be analyzed. Subgroups 4 and 20 had an assignable cause, but the out-of-control condition for subgroup 16 did not. It is assumed that subgroup 16's out-of-control state is due to a chance cause and is part of the natural variation.

Subgroups 4 and 20 for the $\bar{X}$ chart and subgroup 18 for the R chart are not part of the natural variation and are discarded from the data, and new $\bar{\bar{X}}$ and $\bar{R}$ values are computed with the remaining data. The calculations are simplified by using the following formulas:

$$\bar{\bar{X}}_{\text{new}} = \frac{\Sigma\bar{X} - \bar{X}_d}{g - g_d} \qquad \bar{R}_{\text{new}} = \frac{\Sigma R - R_d}{g - g_d}$$

where $\bar{X}_d$ = discarded subgroup averages
g_d = number of discarded subgroups
R_d = discarded subgroup ranges

There are two techniques used to discard data. If either the $\bar{X}$ or the R value of a subgroup is out of control and has an assignable cause, both are discarded, or only the out-of-control value of a subgroup is discarded. In this book the latter technique is followed; thus, when an $\bar{X}$ value is discarded, its corresponding R value is not discarded and vice versa. A knowledge of the process may indicate which technique is most appropriate at any given time.

Example Problem 6-2

Calculations for a new $\bar{\bar{X}}$ are based on discarding the $\bar{X}$ values of 6.65 and 6.51 for subgroups 4 and 20, respectively. Calculations for a new $\bar{R}$ are based on discarding the R value of 0.30 for subgroup 18.

$$\bar{\bar{X}}_{\text{new}} = \frac{\Sigma\bar{X} - \bar{X}_d}{g - g_d} = \frac{160.25 - 6.65 - 6.51}{25 - 2} = 6.40 \text{ mm}$$

$$\bar{R}_{\text{new}} = \frac{\Sigma R - R_d}{g - g_d} = \frac{2.19 - 0.30}{25 - 1} = 0.079 \text{ mm}$$

These new values of $\bar{\bar{X}}$ and $\bar{R}$ are used to establish the standard values of $\bar{X}_0$, R_0, and σ_0. Thus,

$$\bar{X}_0 = \bar{\bar{X}}_{\text{new}} \qquad R_0 = \bar{R}_{\text{new}} \qquad \sigma_0 = \frac{R_0}{d_2}$$

where d_2 = a factor from Table B in the Appendix for estimating σ_0 from R_0. The standard or reference values can be considered to be the best estimate with the data available. As more data become available, better estimates or more confidence in the existing standard values are obtained. *Our objective is to obtain the best estimate of these population standard values.*

Using the standard values, the central lines and the 3σ control limits for actual operations are obtained using the formulas

$$\text{UCL}_{\overline{X}} = \overline{X}_0 + A\sigma_0 \qquad \text{LCL}_{\overline{X}} = \overline{X}_0 - A\sigma_0$$
$$\text{UCL}_R = D_2\sigma_0 \qquad \text{LCL}_R = D_1\sigma_0$$

where A, D_1, and D_2 are factors from Table B for obtaining the $3\sigma_0$ control limits from $\overline{X}_0$ and σ_0.

Example Problem 6-3

From Table B in the Appendix and for a subgroup size of 4, the factors are $A = 1.500$, $d_2 = 2.059$, $D_1 = 0$, and $D_2 = 4.698$. Calculations to determine $\overline{X}_0$, R_0, and σ_0 using the data previously given are

$$\overline{X}_0 = \overline{\overline{X}}_{new} = 6.40 \text{ mm}$$
$$R_0 = \overline{R}_{new} = 0.079\,(0.08 \text{ for the chart})$$
$$\sigma_0 = \frac{R_0}{d_2} = \frac{0.079}{2.059} = 0.038 \text{ mm}$$

Thus, the control limits are

$$\text{UCL}_{\overline{X}} = \overline{X}_0 + A\sigma_0 = 6.40 + (1.500)(0.038) = 6.46 \text{ mm}$$
$$\text{LCL}_{\overline{X}} = \overline{X}_0 - A\sigma_0 = 6.40 - (1.500)(0.038) = 6.34 \text{ mm}$$
$$\text{UCL}_R = D_2\sigma_0 = (4.698)(0.038) = 0.18 \text{ mm} \qquad \text{LCL}_R = D_1\sigma_0 = (0)(0.038) = 0 \text{ mm}$$

The central lines and control limits are drawn on the $\overline{X}$ and R charts for the next period and are shown in Figure 6-6. For illustrative purposes, the trial control limits and the revised control limits are shown on the same chart. The limits for both the $\overline{X}$ and R charts became narrower, as was expected. No change occurred in LCL_R because the subgroup size is 6 or less. Figure 6-6 also illustrates a simpler charting technique in that lines are not drawn between the points.

The preliminary data for the initial 25 subgroups are not plotted with the revised control limits. These revised control limits are for reporting the results for future subgroups. To make effective use of the control chart during production, it should be displayed in a conspicuous place where it can be seen by operators and supervisors.

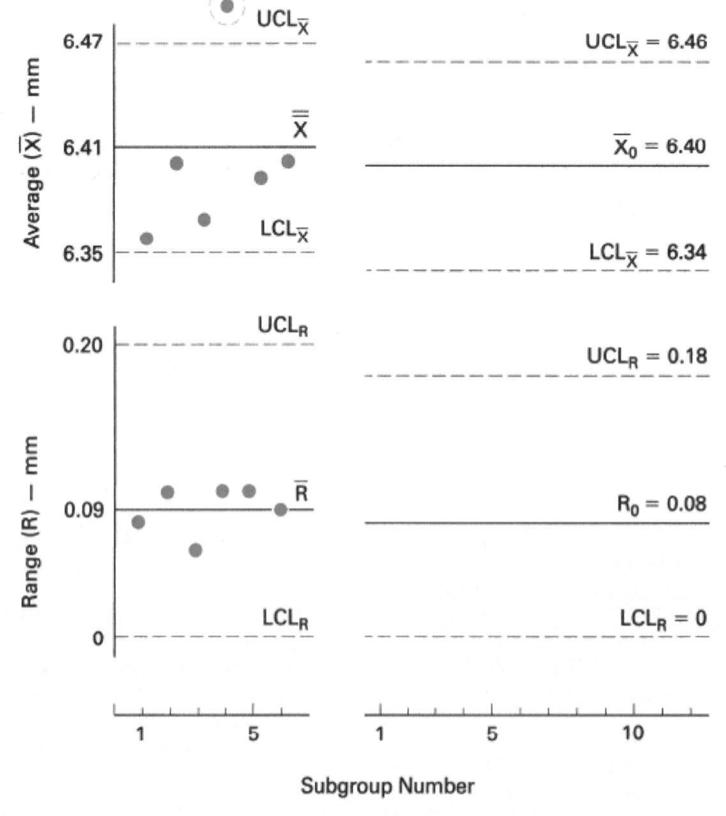

FIGURE 6-6 Trial Control Limits and Revised Control Limits for $\overline{X}$ and R Charts

Before proceeding to the action step, some final comments are appropriate. First, many analysts eliminate this step in the procedure because it appears to be somewhat redundant. However, by discarding out-of-control points with assignable causes, the central line and control limits are more representative of the process. This step may be too complicated for operating personnel. Its elimination would not affect the next step.

Second, the formulas for the control limits are mathematically equal. Thus, for the upper control limit, $\overline{X}_0 + A\sigma_0 = \overline{\overline{X}}_{new} + A_2\overline{R}_{new}$. Similar equivalences are true for the lower control limit and both control limits for the R chart.

Third, the parameter σ_0 is now available to obtain the initial estimate of the process capability, which is $6\sigma_0$. The true process capability is obtained in the next step (Achieve the Objective). Also, as mentioned in Chapter 5, a better estimate of the standard deviation is obtained from $\sigma_0 = R_0/d_2$ and the best estimate is obtained from $\sigma_0 = s_0/c_4$ from the $\overline{X}$, s control charts in the sample standard deviation control chart section.

Fourth, the central line, $\overline{X}_0$, for the $\overline{X}$ chart is frequently based on the specifications. In such a case, the procedure is used only to obtain R_0 and σ_0. If in our example problem the nominal value of the characteristic is 6.38 mm, then $\overline{X}_0$ is set to that value and the upper and lower control limits are

$$\begin{aligned} UCL_{\overline{X}} &= \overline{X}_0 + A\sigma_0 & LCL_{\overline{X}} &= \overline{X}_0 - A\sigma_0 \\ &= 6.38 + (1.500)(0.038) & &= 6.38 - (1.500)(0.038) \\ &= 6.44 & &= 6.32 \end{aligned}$$

The central line and control limits for the R chart do not change. This modification can be taken only if the process is adjustable. If the process is not adjustable, then the original calculations must be used.

Fifth, it follows that adjustments to the process should be made while taking data. It is not necessary to run nonconforming material while collecting data, because we are primarily interested in obtaining R_0, which is not affected by the process setting. The independence of μ and σ provide the rationale for this concept.

Sixth, the process determines the central line and control limits. They are not established by design, manufacturing, marketing, or any other department, except for $\overline{X}_0$ when the process is adjustable.

Finally, when population values are known (μ and σ), the central lines and control limits may be calculated immediately, saving time and work. Thus $\overline{X}_0 = \mu, \sigma_0 = \sigma$, and $R_0 = d_2\sigma$, and the limits are obtained using the appropriate formulas. This situation would be extremely rare.

Achieve the Objective

When control charts are first introduced at a work center, an improvement in the process performance usually occurs. This initial improvement is especially noticeable when the process is dependent on the skill of the operator. Posting a quality control chart appears to be a psychological signal to the operator to improve performance. Most workers want to produce a quality product or service; therefore, when management shows an interest in the quality, the operator responds.

Figure 6-7 illustrates the initial improvement that occurred after the introduction of the $\overline{X}$ and R charts in January. Owing to space limitations, only a representative number of subgroups for each month is shown in the figure. During January, the subgroup averages had less variation and tended to be centered at a slightly higher point. A reduction in the range variation also occurred.

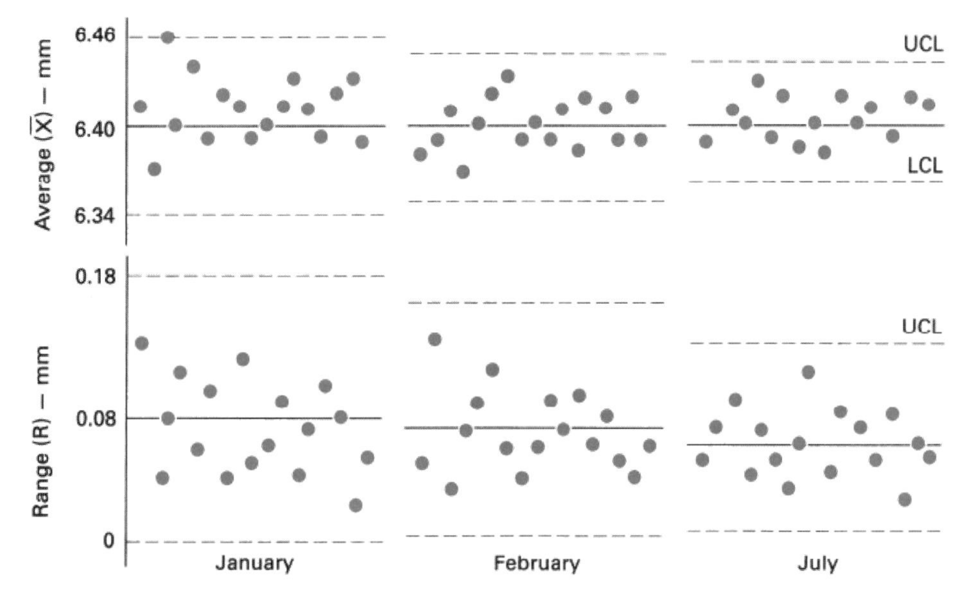

FIGURE 6-7 Continuing Use of Control Charts, Showing Improved Quality

Not all the improved performance in January was the result of operator effort. The first-line supervisor initiated a program of tool wear control, which was a contributing factor.

At the end of January, new central lines and control limits were calculated using the data from subgroups obtained during the month. It is a good idea, especially when a chart is being initiated, to calculate standard values periodically to see if any changes have occurred. This reevaluation can be done for every 25 or more subgroups, and the results compared to the previous values.[7]

New control limits were established for the $\overline{X}$ and R charts and central line for the R chart for the month of February. The central line for the $\overline{X}$ chart was not changed because it is the nominal value. During the ensuing months the maintenance department replaced a pair of worn gears; purchasing changed the material supplier; and tooling modified a workholding device. All these improvements were the result of investigations that tracked down the causes for out-of-control conditions or were ideas developed by a project team. The generation of ideas by many different personnel is the most essential ingredient for continuous quality improvement. Ideas by the operator, first-line supervisor, quality assurance, maintenance, manufacturing engineering, and industrial engineering should be evaluated. This evaluation or testing of an idea requires 25 or more subgroups. The control chart will tell if the idea is good, poor, or has no effect on the process. Quality improvement occurs when the plotted points of the $\overline{X}$ chart converge on the central line, or when the plotted points of the R chart trend downward, or when both actions occur. If a poor idea is tested, then the reverse occurs. Of course, if the idea is neutral, it will have no affect on the plotted point pattern.

In order to speed up the testing of ideas, the taking of subgroups can be compressed in time as long as the data represent the process by accounting for any hourly or day-to-day fluctuations. Only one idea should be tested at a time; otherwise, the results will be confounded.

At the end of June, the periodic evaluation of the past performance showed the need to revise the central lines and the control limits. The performance for the month of July and subsequent months showed a natural pattern of variation and no quality improvement. At this point no further quality improvement is possible without a substantial investment in new equipment or equipment modification.

W. Edwards Deming has stated "that if he were a banker, he would not lend any money to a company unless statistical methods were used to prove that the money was necessary." This is precisely what the control chart can achieve, provided that all personnel use the chart as a method of quality improvement rather than just a maintenance function.

When the objective for initiating the charts has been achieved, its use should be discontinued or the frequency of inspection be substantially reduced to a monitoring action by the operator. Efforts should then be directed toward the improvement of some other quality characteristic. If a project team was involved, it should be congratulated for its performance and disbanded.

The Sample Standard Deviation Control Chart

Although the $\overline{X}$ and R charts are the most common charts for variables, some organizations prefer the sample standard deviation, s, as the measure of the subgroup dispersion. In comparing an R chart with an s chart, an R chart is easier to compute and easier to explain. On the other hand, the subgroup sample standard deviation for the s chart is calculated using all the data rather than just the high and the low value, as done for the R chart. An s chart is therefore more accurate than an R chart. When subgroup sizes are less than 10, both charts will graphically portray the same variation;[8] however, as subgroup sizes increase to 10 or more, extreme values have an undue influence on the R chart. Therefore, at larger subgroup sizes, the s chart must be used.

The steps necessary to obtain the $\overline{X}$ and s trial control and revised control limits are the same as those used for the $\overline{X}$ and R chart except for different formulas. In order to illustrate the method, the same data will be used. They are reproduced in Table 6-3 with the addition of an s column and the elimination of the R column. The appropriate formulas used in the computation of the trial control limits are

$$\bar{s} = \frac{\sum_{i=1}^{g} s_i}{g} \qquad \overline{\overline{X}} = \frac{\sum_{i=1}^{g} \overline{X}_i}{g}$$

$$UCL_{\overline{X}} = \overline{\overline{X}} + A_3\bar{s} \qquad UCL_s = B_4\bar{s}$$

$$LCL_{\overline{X}} = \overline{\overline{X}} - A_3\bar{s} \qquad LCL_s = B_3\bar{s}$$

where s_i = sample standard deviation of the subgroup values

$\bar{s}$ = average of the subgroup sample standard deviations

A_3, B_3, B_4 = factors found in Table B of the Appendix for obtaining the 3σ control limits for $\overline{X}$ and s charts from $\bar{s}$

Formulas for the computation of the revised control limits using the standard values of $\overline{X}_0$ and σ_0 are

$$\overline{X}_0 = \overline{\overline{X}}_{new} = \frac{\Sigma\overline{X} - \overline{X}_d}{g - g_d}$$

$$s_0 = \bar{s}_{new} = \frac{\Sigma s - s_d}{g - g_d} \qquad \sigma_0 = \frac{s_0}{c_4}$$

$$UCL_{\overline{X}} = \overline{X}_0 + A\sigma_0 \qquad UCL_s = B_6\sigma_0$$

$$LCL_{\overline{X}} = \overline{X}_0 - A\sigma_0 \qquad LCL_s = B_5\sigma_0$$

[7] These values are usually compared without the use of formal tests. An exact evaluation can be obtained by mathematically comparing the central lines to see if they are from the same population.

[8] A proof of this statement can be observed by comparing the R chart of Figure 6-5 with the s chart of Figure 6-8.

where s_d = sample standard deviation of the discarded subgroup
c_4 = factor found in Table B for computing σ_0 from $\bar{s}$
A, B_5, B_6 = factors found in Table B for computing 3σ process control limits for $\bar{X}$ and s charts

The first step is to determine the standard deviation for each subgroup from the preliminary data. For subgroup 1, with values of 6.35, 6.40, 6.32, and 6.37, the standard deviation is

$$s = \sqrt{\frac{n\sum_{i=1}^{n} X_i^2 - \left(\sum_{i=1}^{n} X_i\right)^2}{n(n-1)}}$$

$$= \sqrt{\frac{4(6.35^2 + 6.40^2 + 6.32^2 + 6.37^2) - (6.35 + 6.40 + 6.32 + 6.37)^2}{4(4-1)}}$$

$$= 0.034 \text{ mm}$$

The standard deviation for subgroup 1 is posted to the s column, as shown in Table 6-3, and the process is repeated for the remaining 24 subgroups. Continuation of the X and s charts is accomplished in the same manner as the $\bar{X}$ and R charts.

Example Problem 6-4

Using the data of Table 6-3, determine the revised central line and control limits for $\bar{X}$ and s charts. The first step is to obtain $\bar{s}$ and $\bar{\bar{X}}$, which are computed from $\sum s$ and $\sum \bar{X}$, whose values are found in Table 6-3.

$$\bar{s} = \frac{\sum_{i=1}^{g} s_i}{g} = \frac{0.965}{25} = 0.039 \text{ mm} \qquad \bar{\bar{X}} = \frac{\sum_{i=1}^{g} \bar{X}_i}{g} = \frac{160.25}{25} = 6.41 \text{ mm}$$

From Table B the values of the factors—$A_3 = 1.628$, $B_3 = 0$, and $B_4 = 2.266$—are obtained, and the trial control limits are

$$\text{UCL}_{\bar{X}} = \bar{\bar{X}} + A_3\bar{s} = 6.41 + (1.628)(0.039) = 6.47 \text{ mm}$$

$$\text{LCL}_{\bar{X}} = \bar{\bar{X}} - A_3\bar{s} = 6.41 - (1.628)(0.039) = 6.35 \text{ mm}$$

TABLE 6-3 Data on the Depth of the Keyway (mm)[a]

Subgroup Number	Date	Time	MEASUREMENTS X_1	X_2	X_3	X_4	Average $\bar{X}$	Sample Standard Deviation s	Comment
1	12/26	8:50	35	40	32	37	6.36	0.034	
2		11:30	46	37	36	41	6.40	0.045	
3		1:45	34	40	34	36	6.36	0.028	
4		3:45	69	64	68	59	6.65	0.045	New, temporary
5		4:20	38	34	44	40	6.39	0.042	operator
6	12/27	8:35	42	41	43	34	6.40	0.041	
7		9:00	44	41	41	46	6.43	0.024	
8		9:40	33	41	38	36	6.37	0.034	
9		1:30	48	44	47	45	6.46	0.018	
10		2:50	47	43	36	42	6.42	0.045	
11	12/28	8:30	38	41	39	38	6.39	0.014	
12		1:35	37	37	41	37	6.38	0.020	
13		2:25	40	38	47	35	6.40	0.051	
14		2:35	38	39	45	42	6.41	0.032	
15		3:55	50	42	43	45	6.45	0.036	
16	12/29	8:25	33	35	29	39	6.34	0.042	
17		9:25	41	40	29	34	6.36	0.056	
18		11:00	38	44	28	58	6.42	0.125	Damaged oil line
19		2:35	35	41	37	38	6.38	0.025	
20		3:15	56	55	45	48	6.51	0.054	Bad material
21	12/30	9:35	38	40	45	37	6.40	0.036	
22		10:20	39	42	35	40	6.39	0.029	
23		11:35	42	39	39	36	6.39	0.024	
24		2:00	43	36	35	38	6.38	0.036	
25		4:25	39	38	43	44	6.41	0.029	
Sum							160.25	0.965	

[a]For simplicity in recording, the individual measurements are coded from 6.00 mm.

$$\text{UCL}_s = B_4\bar{s} \qquad \text{LCL}_s = B_3\bar{s}$$
$$= (2.266)(0.039) \qquad = (0)(0.039)$$
$$= 0.088 \text{ mm} \qquad = 0 \text{ mm}$$

The next step is to plot the subgroup $\overline{X}$ and s on graph paper with the central lines and control limits. This step is show in Figure 6-8. Subgroups 4 and 20 are out of control on the $\overline{X}$ chart and, because they have assignable causes, they are discarded. Subgroup 18 is out of control on the s chart and, because it has an assignable cause, it is discarded. Computation to obtain the standard values of $\overline{X}_0$, s_0, and σ_0 are as follows:

$$\overline{X}_0 = \overline{\overline{X}}_{new} = \frac{\Sigma\overline{X} - \overline{X}_d}{g - g_d} \qquad s_0 = \bar{s}_{new} = \frac{\Sigma s - s_d}{g - g_d}$$
$$= \frac{160.25 - 6.65 - 6.51}{25 - 2} \qquad = \frac{0.965 - 0.125}{25 - 1}$$
$$= 6.40 \text{ mm} \qquad = 0.035 \text{ mm}$$

$$\sigma_0 = \frac{s_0}{c_4} \qquad \text{from Table B, } c_4 = 0.9213$$
$$= \frac{0.035}{0.9213}$$
$$= 0.038 \text{ mm}$$

The reader should note that the standard deviation, σ_0, is the same as the value obtained from the range in the preceding section. Using the standard values of $\overline{X}_0 = 6.40$ and $\sigma_0 = 0.038$, the revised control limits are computed.

$$\text{UCL}_{\overline{X}} = \overline{X} + A\sigma_0 \qquad \text{LCL}_{\overline{X}} = \overline{X}_0 - A\sigma_0$$
$$= 6.40 + (1.500)(0.038) \qquad = 6.40 - (1.500)(0.038)$$
$$= 6.46 \text{ mm} \qquad = 6.34 \text{ mm}$$

$$\text{UCL}_s = B_6\sigma_0 \qquad \text{LCL}_s = B_5\sigma_0$$
$$= (2.088)(0.038) \qquad = (0)(0.038)$$
$$= 0.079 \text{ mm} \qquad = 0 \text{ mm}$$

STATE OF CONTROL

Process in Control

When the assignable causes have been eliminated from the process to the extent that the points plotted on the control chart remain within the control limits, the process is in a state of control. No higher degree of uniformity can be attained with the existing process. Greater uniformity can, however, be attained through a change in the basic process through quality improvement ideas.

When a process is in control, there occurs a natural pattern of variation, which is illustrated by the control chart in Figure 6-9. This natural pattern of variation has (1) about 34% of the plotted points in an imaginary band between 1 standard deviation on both sides of the central

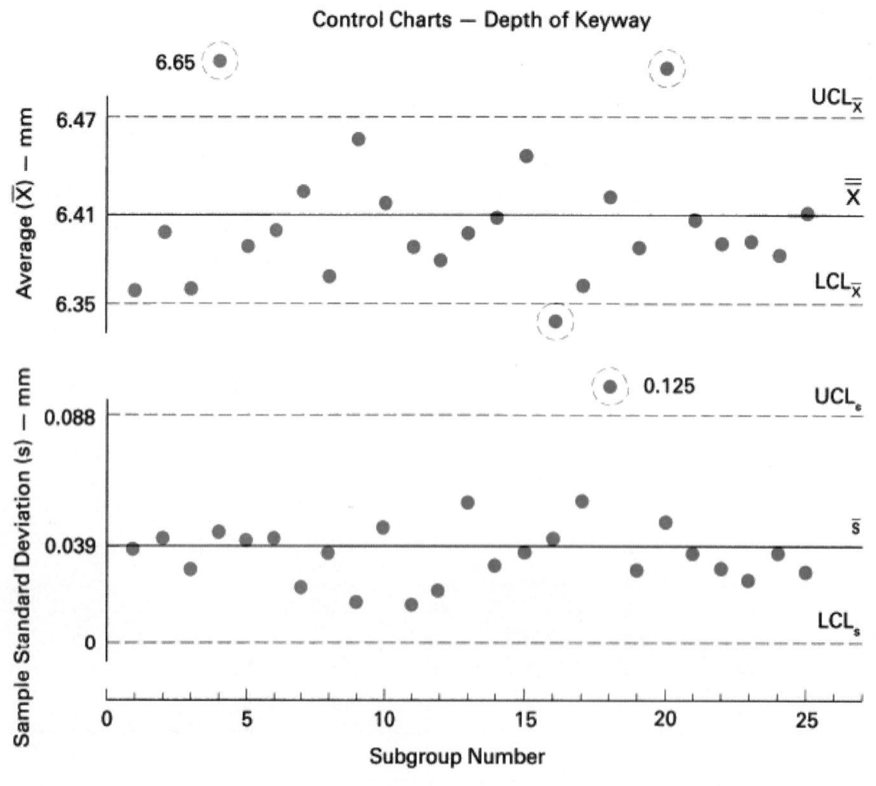

FIGURE 6-8 $\overline{X}$ and s Chart for Preliminary Data with Trial Control Limits

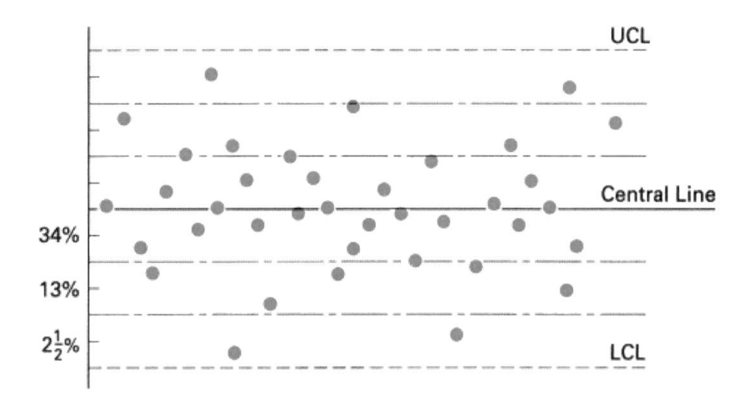

FIGURE 6-9 Natural Pattern of Variation of a Control Chart

line, (2) about 13.5% of the plotted points in an imaginary band between 1 and 2 standard deviations on both sides of the central line, and (3) about 2.5% of the plotted points in an imaginary band between 2 and 3 standard deviations on both sides of the central line. The points are located back and forth across the central line in a random manner, with no points beyond the control limits. The natural pattern of the points or subgroup average values forms its own frequency distribution. If all of the points were stacked up at one end, they would form a normal curve (see Figure 6-11).

Control limits are usually established at 3 standard deviations from the central line. They are used as a basis to judge whether there is evidence of lack of control. The choice of 3σ limits is an economic one with respect to two types of errors that can occur. One error, called Type I by statisticians, occurs when looking for an assignable cause of variation, when in reality a chance cause is present. When the limits are set at 3 standard deviations, a Type I error will occur 0.27% (3 out of 1000) of the time. In other words, when a point is outside the control limits, it is assumed to be due to an assignable cause, even though it would be due to a chance cause 0.27% of the time. We might think of this situation as "guilty until proven innocent." The other type error, called Type II, occurs when assuming that a chance cause of variation is present, when in reality there is an assignable cause. In other words, when a point is inside the control limits, it is assumed to be due to a chance cause, even though it might be due to an assignable cause. We might think of this situation as "innocent until proven guilty." Table 6-4 illustrates the difference between the

TABLE 6-4 Type I and Type II Errors

	PLOTTED POINT IS	
	Outside Control Limits	Inside Control Limits
Assignable cause present	OK	Type II Error
Chance cause present	Type I Error	OK

Type I and Type II errors. If control limits are established at, say, ± 2.5 standard deviations, Type I errors would increase and Type II decrease. Abundant experience since 1930 in all types of industry indicates that 3σ limits provide an economic balance between the costs resulting from the two types of errors. Unless there are strong practical reasons for doing otherwise, the ± 3 standard deviation limits should be used.[9]

When a process is in control, only chance causes of variation are present. Small variations in machine performance, operator performance, and material characteristics are expected and are considered to be part of a stable process.

When a process is in control, certain practical advantages accrue to the producer and consumer.

1. Individual units of the product or service will be more uniform—or, stated another way, there will be less variation and fewer rejections.

2. Because the product or service is more uniform, fewer samples are needed to judge the quality. Therefore, the cost of inspection can be reduced to a minimum. This advantage is extremely important when 100% conformance to specifications is not essential.

3. The process capability or spread of the process is easily attained from 6σ. With a knowledge of process capability, a number of reliable decisions relative to specifications can be made, such as

   a. to decide the product or service specifications or requirements,
   b. to decide the amount of rework or scrap when there is insufficient tolerance, and
   c. to decide whether to produce the product to tight specifications and permit interchangeability of components or to produce the product to loose specifications and use selective matching of components.

4. Trouble can be anticipated before it occurs, thereby speeding up production by avoiding rejections and interruptions.

[9]Elisabeth J. Umble and M. Michael Umble, "Developing Control Charts and Illustrating Type I and Type II Errors," *Quality Management Journal*, Vol. 7, No. 4 (2000): 23–30.

5. The percentage of product that falls within any pair of values may be predicted with the highest degree of assurance. For example, this advantage can be very important when adjusting filling machines to obtain different percentages of items below, between, or above particular values.

6. It permits the consumer to use the producer's data and, therefore, to test only a few subgroups as a check on the producer's records. The $\overline{X}$ and R charts are used as statistical evidence of process control.

7. The operator is performing satisfactorily from a quality viewpoint. Further improvement in the process can be achieved only by changing the input factors: materials, equipment, environment, and operators. These changes require action by management.

When only chance causes of variation are present, the process is stable and predictable over time, as shown in Figure 6-10(a). We know that future variation as shown by the dotted curve will be the same, unless there has been a change in the process due to an assignable cause.

Process Out of Control

The term *out of control* is usually thought of as being undesirable; however, there are situations where this condition is desirable. It is best to think of the term *out of control* as a change in the process due to an assignable cause.

When a point (subgroup value) falls outside its control limits, the process is out of control. This means that an assignable cause of variation is present. Another way of viewing the out-of-control point is to think of the subgroup value as coming from a different population than the one from which the control limits were obtained.

Figure 6-11 shows a frequency distribution of plotted points that are all stacked up at one end for educational purposes to form a normal curve for averages. The data were developed from a large number of subgroups and, therefore, represent the population mean, $\mu = 450$ g, and the population standard deviation for the averages, $\sigma_{\overline{X}} = 8$ g. The frequency distribution for subgroup averages is shown by a dashed line. Future explanations will use the dashed line to represent the frequency distribution of

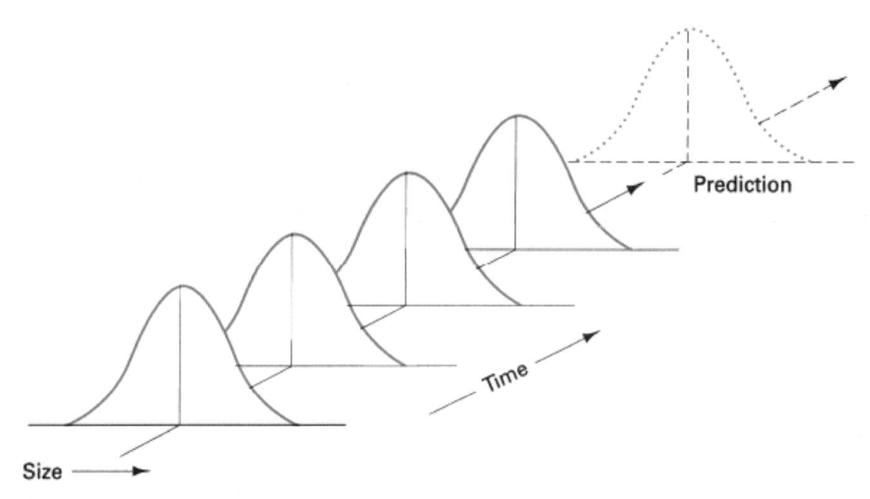

(a) Only Chance Causes of Variation Present

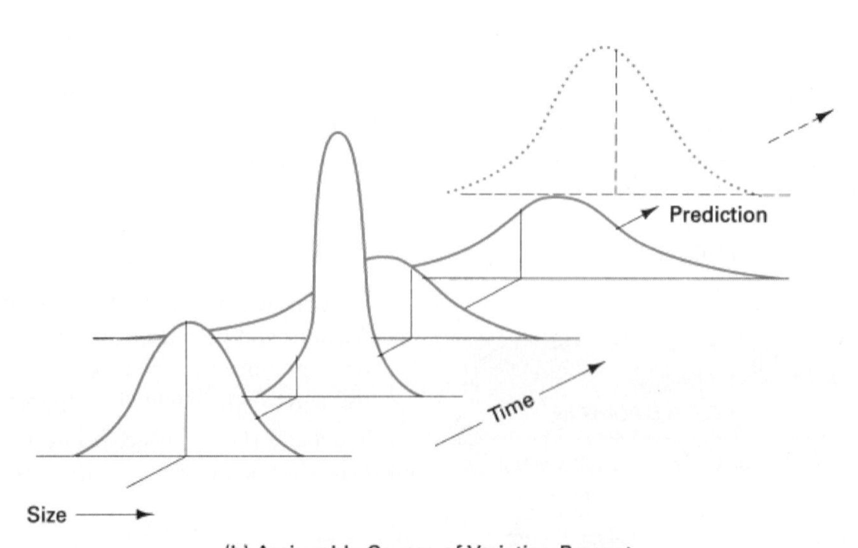

(b) Assignable Causes of Variation Present

FIGURE 6-10 Stable and Unstable Variation

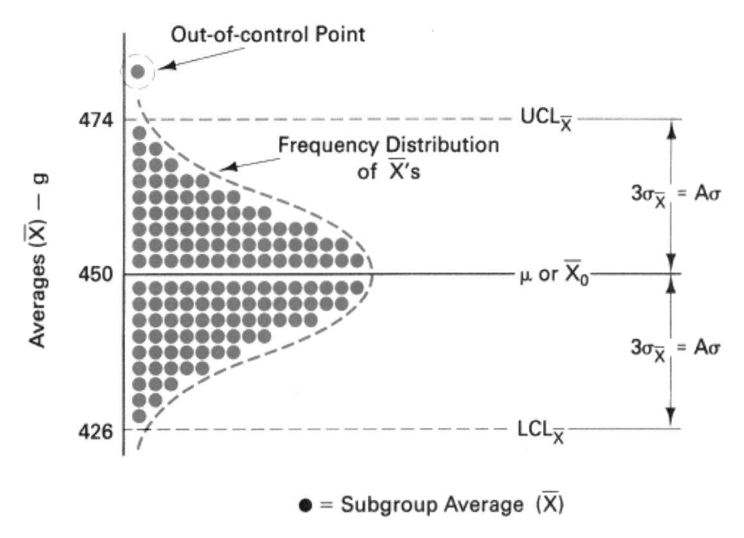

FIGURE 6-11 Frequency Distribution of Subgroup Averages with Control Limits

averages and will use a solid line for the frequency distribution of individual values. The out-of-control point has a value of 483 g. This point is so far away from the 3σ limits (99.73%) that it can only be considered to have come from another population. In other words, the process that produced the subgroup average of 483 g is a different process than the stable process from which the 3σ control limits were developed. Therefore, the process has changed; some assignable cause of variation is present.

Figure 6-10(b) illustrates the effect of assignable causes of variation over time. The unnatural, unstable nature of the variation makes it impossible to predict future variation. The assignable causes must be found and corrected before a natural, stable process can continue.

A process can also be considered out of control even when the points fall inside the 3σ limits. This situation occurs when unnatural runs of variation are present in the process. First, let's divide the control chart into 6 equal standard deviation bands in the same manner as Figure 6-9. For identification purposes the bands are labeled A, B, and C zones, as shown in Figure 6-12.

It is not natural for seven or more consecutive points to be above or below the central line as shown in Figure 6-12(a). Also, when 10 out of 11 points or 12 out of 14 points, etc., are

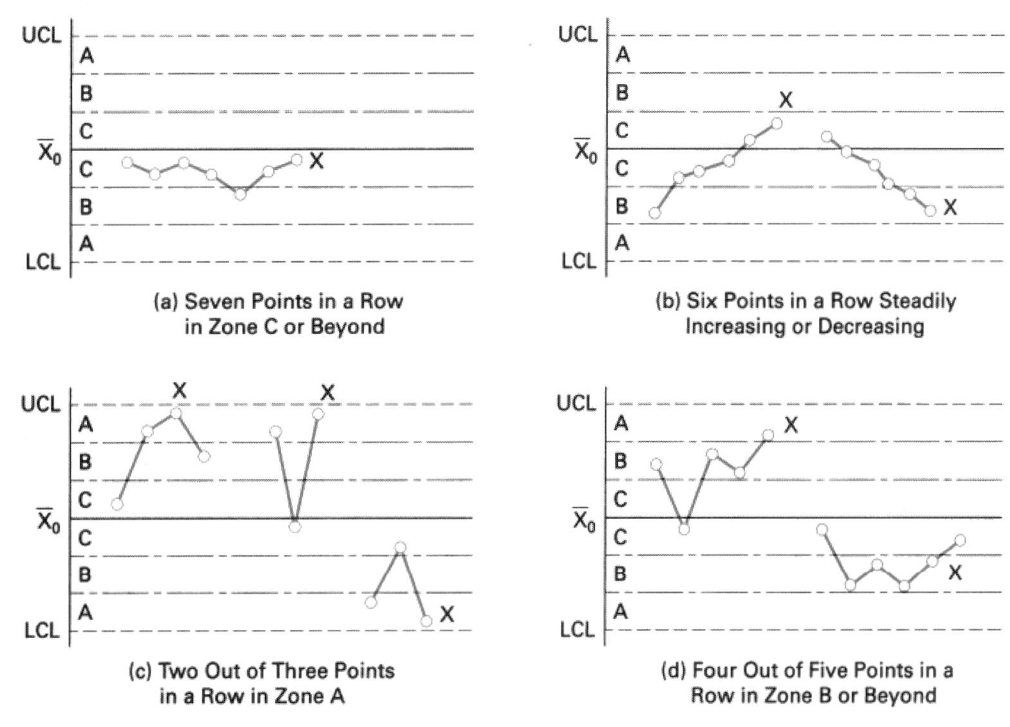

FIGURE 6-12 Some Unnatural Runs—Process Out of Control

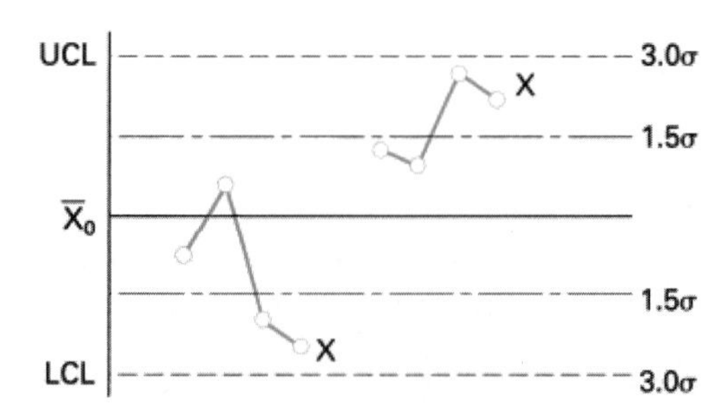

FIGURE 6-13 Simplified Rule for Out-of-Control Process

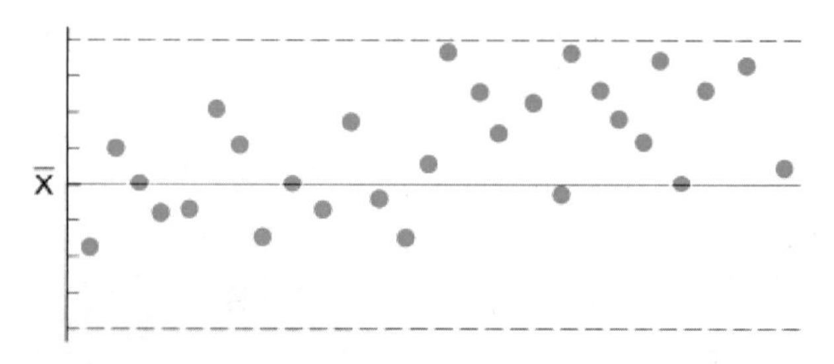

FIGURE 6-14 Out-of-Control Pattern: Change or Jump in Level

located on one side of the central line, it is unnatural. Another unnatural run occurs at (b), where six points in a row are steadily increasing or decreasing. In Figure 6-12(c) we have two out of three points in a row in zone A and at (d) four out of five points in a row in zone B and beyond. There are many statistical possibilities, with the four common ones being shown in the figure. Actually, any significant divergence from the natural pattern as shown in Figure 6-9 would be unnatural and would be classified as an out-of-control condition.

Rather than divide the space into three equal zones of 1 standard deviation, a simplified technique would divide the space into two equal zones of 1.5 standard deviations. The process is out of control when there are two successive points at 1.5 standard deviations or beyond. The simplified rule makes for greater ease of implementation by operators without drastically sacrificing power.[10] It is shown in Figure 6-13 and replaces the information of Figures 6-12(c) and (d).

Analysis of Out-of-Control Condition

When a process is out of control, the assignable cause responsible for the condition must be found. The detective work necessary to locate the cause of the out-of-control condition can be minimized by knowledge of the types of out-of-control patterns and their assignable causes. Types of out-of-control $\overline{X}$ and R patterns are (1) change or jump in level, (2) trend or steady change in level, (3) recurring cycles, (4) two populations, and (5) mistakes.

1. **Change or jump in level.** This type is concerned with a sudden change in level to the $\overline{X}$ chart, to the R chart, or to both charts. Figure 6-14 illustrates the change in level. For an $\overline{X}$ chart, the change in the process average can be due to

   a. An intentional or unintentional change in the process setting
   b. A new or inexperienced operator
   c. A different material
   d. A minor failure of a machine part

Some causes for a sudden change in the process spread or variability as shown on the R chart are

   a. Inexperienced operator
   b. Sudden increase in gear play
   c. Greater variation in incoming material

Sudden changes in level can occur on both the $\overline{X}$ and the R charts. This situation is common during the beginning of control chart activity, prior to the attainment of a state of control. There may be more than one assignable cause, or it may be a cause that could affect both charts, such as an inexperienced operator.

2. **Trend or steady change in level.** Steady changes in control chart level are a very common industrial phenomena. Figure 6-15 illustrates a trend or steady change that is occurring in the upward direction; the trend could have been illustrated in the downward direction. Some causes of steady progressive changes on an $\overline{X}$ chart are

   a. Tool or die wear
   b. Gradual deterioration of equipment
   c. Gradual change in temperature or humidity
   d. Viscosity breakdown in a chemical process
   e. Buildup of chips in a workholding device

A steady change in level or trend on the R chart is not as common as on the $\overline{X}$ chart. It does, however, occur, and some possible causes are

   a. An improvement in operator skill (downward trend)
   b. A decrease in operator skill due to fatigue, boredom, inattention, etc. (upward trend)
   c. A gradual improvement in the homogeneity of incoming material

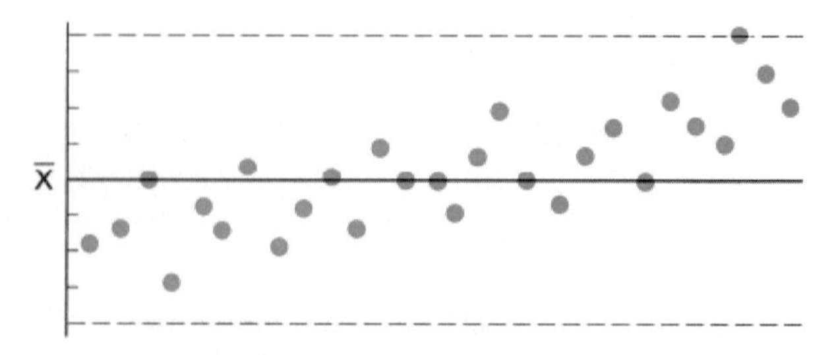

FIGURE 6-15 Out-of-Control Pattern: Trend or Steady Change in Level

[10] For more information, see A. M. Hurwitz and M. Mathur, "A Very Simple Set of Process Control Rules," *Quality Engineering*, Vol. 5, No. 1 (1992–93): 21–29.

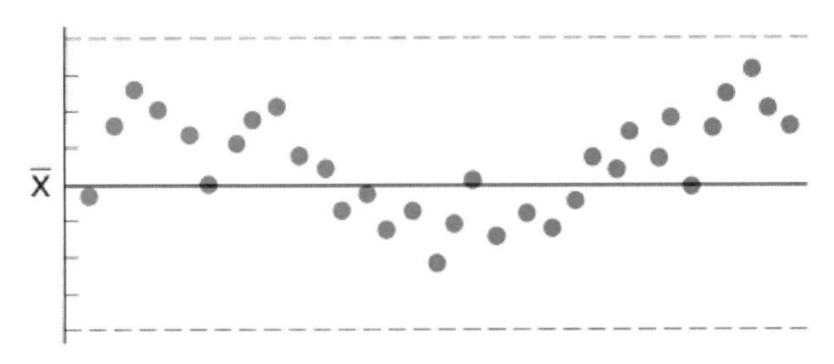

FIGURE 6-16 Out-of-Control Pattern: Recurring Cycles

3. **Recurring cycles.** When the plotted points on an $\overline{X}$ or R chart show a wave or periodic high and low points, it is called a *cycle*. A typical recurring out-of-control pattern is shown in Figure 6-16.

For an $\overline{X}$ chart, some of the causes of recurring cycles are

a. The seasonal effects of incoming material
b. The recurring effects of temperature and humidity (cold morning start-up)
c. Any daily or weekly chemical, mechanical, or psychological event
d. The periodic rotation of operators

Periodic cycles on an R chart are not as common as for an $\overline{X}$ chart. Some affecting the R chart are due to

a. Operator fatigue and rejuvenation resulting from morning, noon, and afternoon breaks
b. Lubrication cycles

The out-of-control pattern of a recurring cycle sometimes goes unreported because of the inspection cycle. Thus, a cyclic pattern of a variation that occurs approximately every 2 h could coincide with the inspection frequency. Therefore, only the low points on the cycle are reported, and there is no evidence that a cyclic event is present.

4. **Two populations (also called mixture).** When there are a large number of points near or outside the control limits, a two-population situation may be present. This type of out-of-control pattern is illustrated in Figure 6-17.

For an $\overline{X}$ chart the out-of-control pattern can be due to

a. Large differences in material quality
b. Two or more machines on the same chart
c. Large differences in test method or equipment

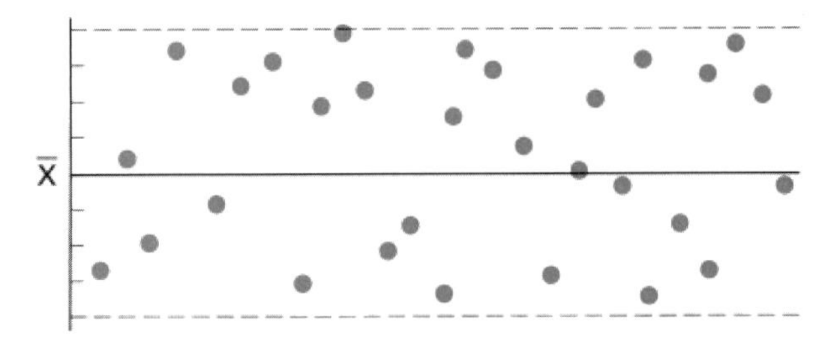

FIGURE 6-17 Out-of-Control Pattern: Two Populations

Some causes for an out-of-control pattern on an R chart are due to

a. Different operators using the same chart
b. Materials from different suppliers

5. **Mistakes.** Mistakes can be very embarrassing to quality assurance. Some causes of out-of-control patterns resulting from mistakes are

a. Measuring equipment out of calibration
b. Errors in calculations
c. Errors in using test equipment
d. Taking samples from different populations

Many of the out-of-control patterns that have been described can also be attributed to inspection error or mistakes.

The causes given for the different types of out-of-control patterns are suggested possibilities and are not meant to be all-inclusive. These causes will give production and quality personnel ideas for the solution of problems. They can be a start toward the development of an assignable cause checklist, which is applicable to their particular organization.

When out-of-control patterns occur in relation to the lower control limit of the R chart, it is the result of outstanding performance. The cause should be determined so that the outstanding performance can continue.

The preceding discussion has used the R chart as the measure of the dispersion. Information on patterns and causes also pertains to an s chart.

In the sixth step of the control chart method, it was stated that 25 subgroups were necessary to test an idea. The information given above on out of control can be used to make a decision with a fewer number of subgroups. For example, a run of six consecutive points in a downward trend on an R chart would indicate that the idea was a good one.

SPECIFICATIONS

Individual Values Compared to Averages

Before discussing specifications and their relationship with control charts, it appears desirable, at this time, to obtain a better understanding of individual values and average values. Figure 6-18 shows a tally of individual values (X's) and a tally of the subgroup averages ($\overline{X}$'s) for the data on keyway depths given in Table 6-2. The four out-of-control subgroups were not used in the two tallys; therefore, there are 84 individual values and 21 averages. It is observed that the averages are grouped much closer to the center than the individual values. When four values are averaged, the effect of an extreme value is minimized because the chance of four extremely high or four extremely low values in one subgroup is slight.

Calculations of the average for both the individual values and for the subgroup averages are the same, $\overline{X} = 38.9$. However, the sample standard deviation of the individual values (s) is 4.16, whereas the sample standard deviation of the subgroup average ($s_{\overline{X}}$) is 2.77.

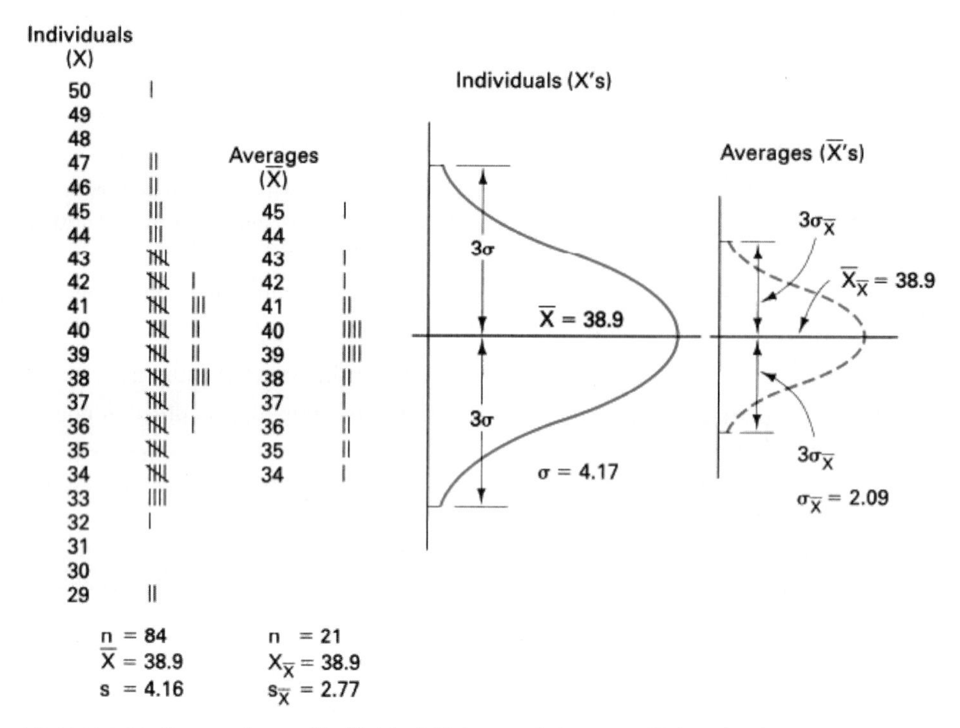

FIGURE 6-18 Comparison of Individual Values and Averages Using the Same Data

If there are a large number of individual values and subgroup averages, the smooth polygons of Figure 6-18 would represent their frequency distributions if the distribution is normal. The curve for the frequency distribution of the averages is a dashed line, whereas the curve for the frequency distribution of individual values is a solid line; this convention will be followed throughout the book. In comparing the two distributions, it is observed that both distributions are normal in shape; in fact, even if the curve for individual values were not quite normal, the curve for averages would be close to a normal shape. The base of the curve for individual values is about twice as large as the base of the curve for averages. When population values are available for the standard deviation of individual values ($\hat{\sigma}$) and for the standard deviation of averages ($\sigma_{\overline{X}}$), there is a definite relationship between them, as given by the formula

$$\sigma_{\overline{X}} = \frac{\sigma}{\sqrt{n}}$$

where $\sigma_{\overline{X}}$ = population standard deviation of subgroup averages $\overline{X}$'s
σ = population standard deviation of individual values (X's)
n = subgroup size

Thus, for a subgroup of size 5, $\sigma_{\overline{X}} = 0.45\sigma$; and for a subgroup of size 4, $\sigma_{\overline{X}} = 0.50\sigma$.

If we assume normality (which may or may not be true), the population standard deviation can be estimated from

$$\hat{\sigma} = \frac{s}{c_4}$$

where $\hat{\sigma}$ is the "estimate" of the population standard deviation and c_4 is approximately equal to 0.996997 for $n = 84$.[11] Thus, $\sigma = s/c_4 = 4.16/0.996997 = 4.17$ and $\sigma_{\overline{X}} = \sigma/\sqrt{n} = 4.17/\sqrt{4} = 2.09$. Note that $s_{\overline{X}}$, which was calculated from sample data, and $\sigma_{\overline{X}}$, which was calculated above, are different. This difference is due to sample variation or the small number of samples, which was only 21, or some combination thereof. The difference would not be caused by a nonnormal population of X's.

Because the height of the curve is a function of the frequency, the curve for individual values is higher. This is easily verified by comparing the tally sheet in Figure 6-18. However, if the curves represent relative or percentage frequency distributions, then the area under the curve must be equal to 100%. Therefore, the percentage frequency distribution curve for averages, with its smaller base, would need to be much higher to enclose the same area as the percentage frequency distribution curve for individual values.

Central Limit Theorem

Now that you are aware of the difference between the frequency distribution of individual values, X's, and the frequency distribution of averages, $\overline{X}$'s, the central limit theorem can be discussed. In simple terms it is:

> If the population from which samples are taken is *not* normal, the distribution of sample averages will tend toward normality provided that the sample size, n, is at least 4. This tendency gets better and better as the sample size gets larger. Furthermore, the standardized

[11] Values of c_4 are given in Table B of the Appendix up to $n = 20$. For values greater than 20, $c_4 = 4(n-1)/4n-3$.

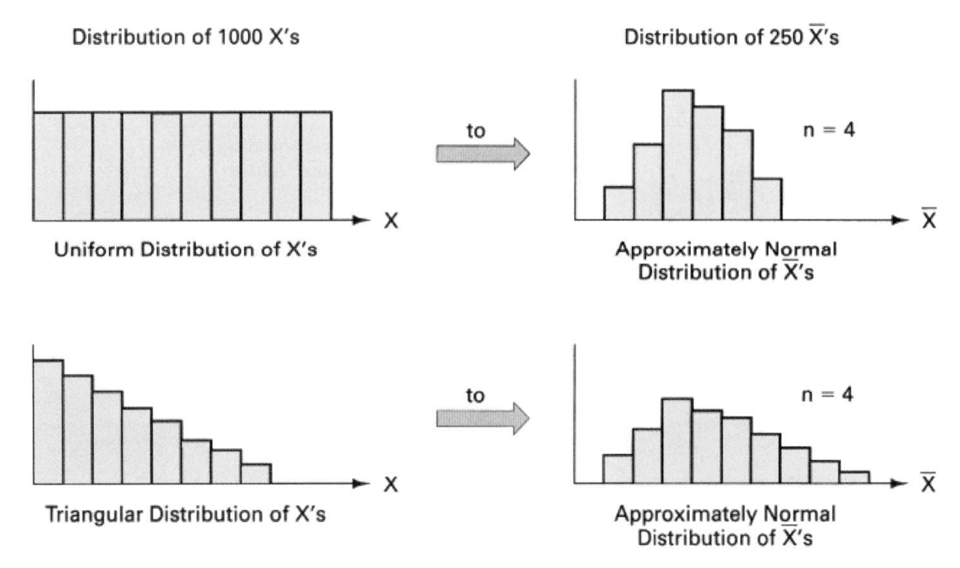

FIGURE 6-19 Illustration of Central Limit Theorem

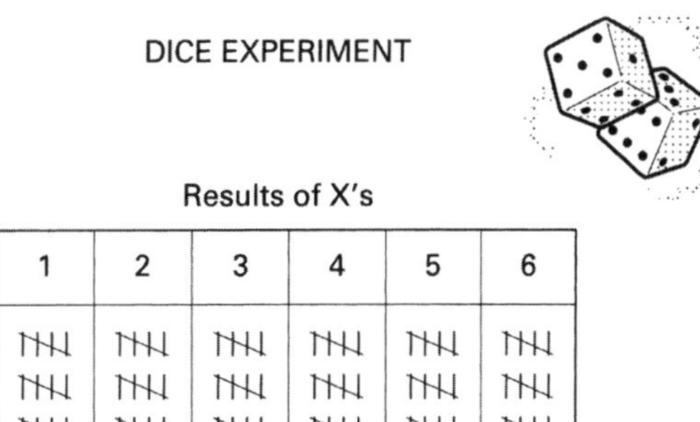

Results of X's

1	2	3	4	5	6
卌 卌 卌 卌	卌 卌 卌 卌	卌 卌 卌 卌	卌 卌 卌 卌	卌 卌 卌 卌	卌 卌 卌 卌

Results of $\overline{X}$'s, n = 2

1.0	1.5	2.0	2.5	3.0	3.5	4.0	4.5	5.0	5.5	6.0
||||	卌 |||	卌 卌 ||	卌 卌 卌 |	卌 卌 卌 卌	卌 卌 卌 卌 ||||	卌 卌 卌 卌	卌 卌 卌 |	卌 卌 ||	卌 |||	||||

FIGURE 6-20 Dice illustration of Central Limit Theorem

normal can be used for the distribution of averages with the modification,

$$Z = \frac{\overline{X} - \mu}{\sigma_{\overline{X}}} = \frac{\overline{X} - \mu}{\sigma\sqrt{n}}$$

This theorem was illustrated by Shewhart[12] for a uniform population distribution and a triangular population distribution of individual values as shown in Figure 6-19. Obviously, the distribution of X's is considerably different than a normal distribution; however, the distribution of $\overline{X}$'s is approximately normal.

The central limit theorem is one of the reasons the $\overline{X}$ chart works, in that we do not need to be concerned if the distribution of X's is not normal, provided that the sample size is 4 or more. Figure 6-20 shows the results of a dice experiment. First is a distribution of individual rolls of a six-sided die; second is a distribution of the average of rolls of two dice. The distribution of the averages $(\overline{X}\text{'s})$ is unimodal, symmetrical, and tapers off at the tails. This experiment provides practical evidence of the validity of the central limit theorem.

[12] W. A. Shewhart, *Economic Control of Quality of Manufactured Product* (Princeton, NJ: Van Nostrand Reinhold, 1931), pp. 180–186.

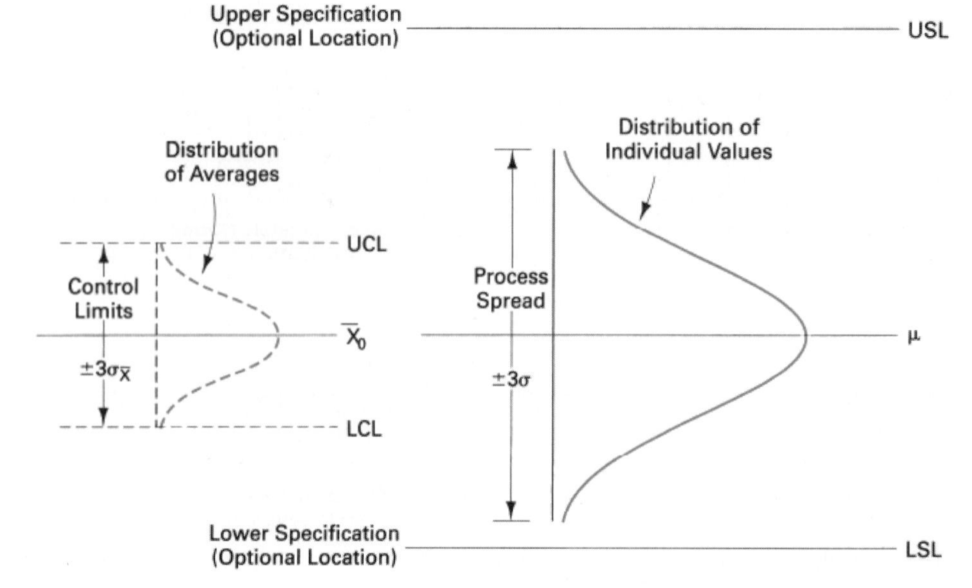

FIGURE 6-21 Relationship of Limits, Specifications, and Distributions

Control Limits and Specifications

Control limits are established as a function of the averages; in other words, control limits are for averages. Specifications, on the other hand, are the permissible variation in the size of the part and are, therefore, for individual values. The specification or tolerance limits are established by design engineers to meet a particular function. Figure 6-21 shows that the location of the specifications is optional and is not related to any of the other features in the figure. The control limits, process spread, distribution of averages, and distribution of individual values are interdependent. They are determined by the process, whereas the specifications have an optional location. Control charts cannot determine whether the process is meeting specifications.

Process Capability and Tolerance

Hereafter the process spread will be referred to as the *process capability* and is equal to 6σ. Also, the difference between specifications is called the *tolerance*. When the tolerance is established by the design engineer without regard to the spread of the process, undesirable situations can result. Three situations are possible: (1) when the process capability is less than the tolerance, (2) when the process capability is equal to the tolerance, and (3) when the process capability is greater than the tolerance.

Case I: 6σ < USL − LSL This situation, where the process capability (6σ) is less than the tolerance (USL – LSL), is the most desirable case. Figure 6-22 illustrates this ideal relationship by showing the distribution of individual values

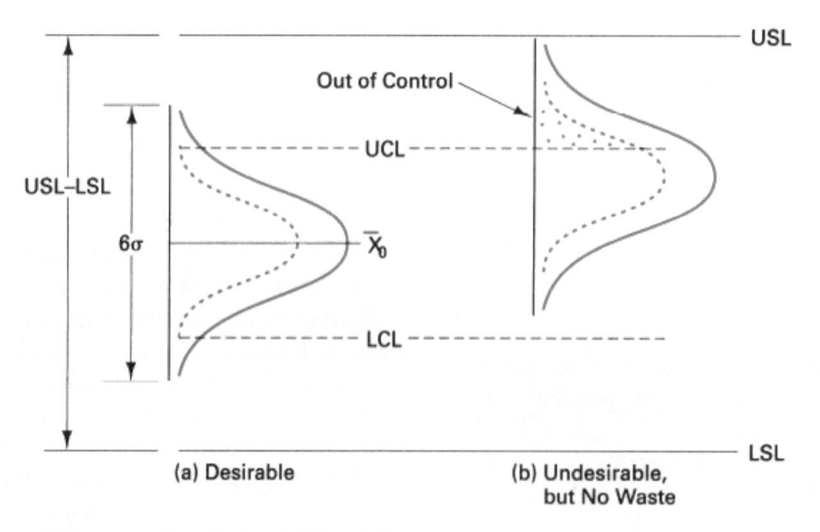

FIGURE 6-22 Case I: 6σ < USL – LSL

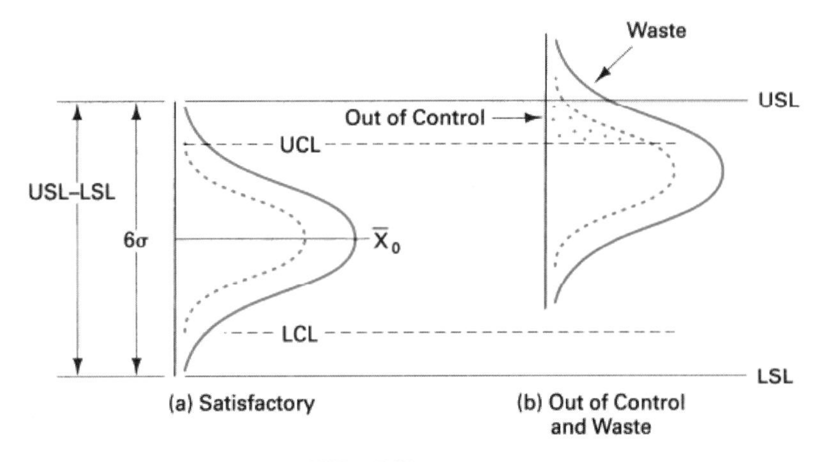

FIGURE 6-23 Case II: $6\sigma = \text{USL} - \text{LSL}$

(X's), the $\overline{X}$ control chart limits, and distribution of averages ($\overline{X}$'s). The process is in control in (a). Because the tolerance is appreciably greater than the process capability, no difficulty is encountered even when there is a substantial shift in the process average, as shown in (b). This shift has resulted in an out-of-control condition as shown by the plotted points. However, no waste is produced, because the distribution of individual values (X's) has not exceeded the upper specification. Corrective action is required to bring the process into control.

Case II: $6\sigma = \text{USL} - \text{LSL}$ Figure 6-23 illustrates this case, where the process capability is equal to the tolerance. The frequency distribution of X's in (a) represents a natural pattern of variation. However, when there is a shift in the process average, as indicated in (b), the individual values (X's) exceed the specifications. As long as the process remains in control, no nonconforming product is produced; however, when the process is out of control as indicated in (b), nonconforming product is produced. Therefore, assignable causes of variation must be corrected as soon as they occur.

Case III: $6\sigma > \text{USL} - \text{LSL}$ When the process capability is greater than the tolerance, an undesirable situation exists. Figure 6-24 illustrates this case. Even though a natural pattern of variation is occurring, as shown by the frequency distribution of X's in (a), some of the individual values are greater than the upper specification and are less than the lower specification. This case presents the unique situation where the process is in control as shown by the control limits and frequency distribution of $\overline{X}$'s, but nonconforming product is produced. In other words, the process is not capable of manufacturing a product that will meet the specifications. When the process changes as shown in (b), the problem is much worse.

When this situation occurs, 100% inspection is necessary to eliminate the nonconforming product.

One solution is to discuss with the design engineer the possibility of increasing the tolerance. This solution may require reliability studies with mating parts to determine if the product can function with an increased tolerance. Selective assembly might also be considered by the engineer.

A second possibility is to change the process dispersion so that a more peaked distribution occurs. To obtain a substantial reduction in the standard deviation might require new material, a more experienced operator, retraining, a new or overhauled machine, or possibly automatic in-process control.

Another solution is to shift the process average so that all of the nonconforming product occurs at one tail of the

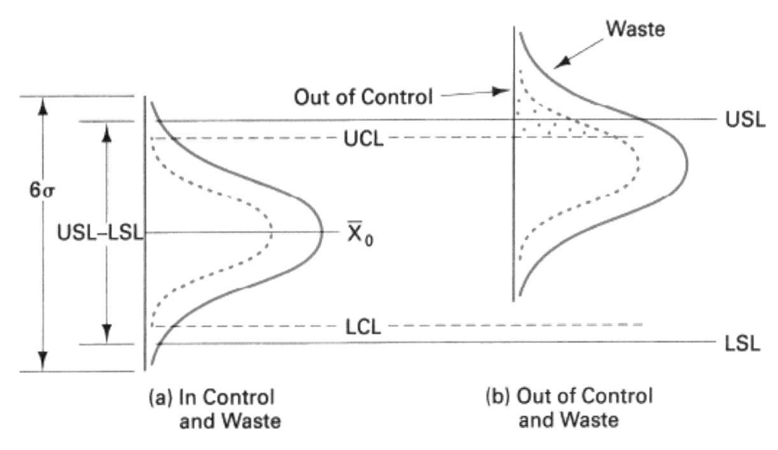

FIGURE 6-24 Case III: $6\sigma > \text{USL} - \text{LSL}$

frequency distribution, as indicated in Figure 6-24(b) . To illustrate this solution, assume that a shaft is being ground to tight specifications. If too much metal is removed, the part is scrapped; if too little is removed, the part must be reworked. By shifting the process average, the amount of scrap is eliminated and the amount of rework is increased. A similar situation exists for an internal member such as a hole or keyway, except that scrap occurs above the upper specification and rework occurs below the lower specification. This type of solution is feasible when the cost of the part is sufficient to economically justify the reworking operation.

Example Problem 6-5

Location pins for workholding devices are ground to a diameter of 12.50 mm (approximately 1/2 in.), with a tolerance of ±0.05 mm. If the process is centered at 12.50 mm (μ) and the dispersion is 0.02 mm (σ), what percent of the product must be scrapped and what percent can be reworked? How can the process center be changed to eliminate the scrap? What is the rework percentage?

The techniques for solving this problem were given in Chapter 4 and are shown below.

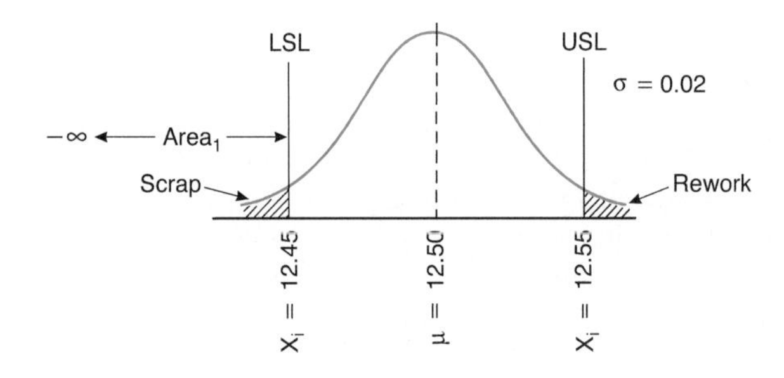

$$USL = \mu + 0.05 = 12.50 + 0.05 = 12.55 \text{ mm}$$

$$LSL = \mu - 0.05 = 12.50 - 0.05 = 12.45 \text{ mm}$$

$$Z = \frac{X_i - \mu}{\sigma} = \frac{12.45 - 12.50}{0.02} = -2.50$$

From Table A of the Appendix, for a Z value of −2.50:

$$\text{Area}_1 = 0.0062 \text{ or } 0.62\% \text{ scrap}$$

Because the process is centered between the specifications and a symmetrical distribution is assumed, the rework percentage will be equal to the scrap percentage of 0.62%. The second part of the problem is solved using the following sketch:

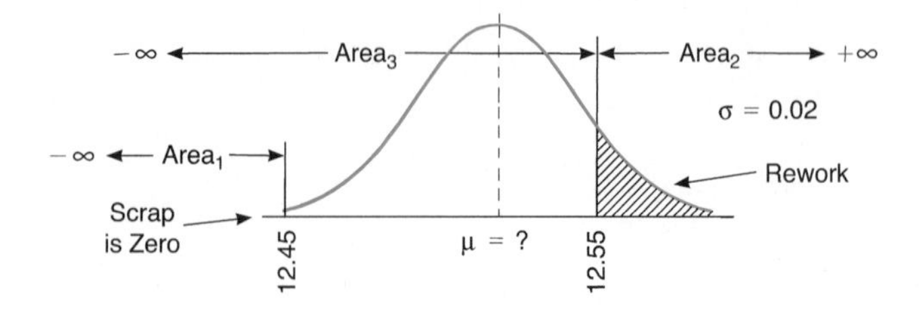

If the amount of scrap is to be zero, then $\text{Area}_1 = 0$. From Table A, the closest value to an Area_1 value of zero is 0.00017, which has a Z value of −3.59. Thus,

$$Z = \frac{X_i - \mu}{\sigma}$$

$$-3.59 = \frac{12.45 - \mu}{0.02}$$

$$\mu = 12.52 \text{ mm}$$

The percentage of rework is obtained by first determining Area_3.

$$Z = \frac{X_i - \mu}{\sigma} = \frac{12.55 - 12.52}{0.02} = +1.50$$

From Table A, $\text{Area}_3 = 0.9332$ and

$$\text{Area}_2 = \text{Area}_T - \text{Area}_3 = 1.0000 - 0.9332 = 0.0668, \text{ or } 6.68\%$$

The amount of rework is 6.68%, which, incidentally, is considerably more than the combined rework and scrap percentage (1.24%) when the process is centered.

The preceding analysis of the process capability and the specifications was made utilizing an upper and a lower specification. Many times there is only one specification, and it may be either the upper or lower. A similar and much simpler analysis would be for a single specification limit.

PROCESS CAPABILITY

The true process capability cannot be determined until the $\overline{X}$ and R charts have achieved the optimal quality improvement without a substantial investment for new equipment or equipment modification. Process capability is equal to $6\sigma_0$ when the process is in statistical control.

In Example Problem 6-1 for the $\overline{X}$ and R charts, the quality improvement process began in January with $\sigma_0 = 0.038$. The process capability is $6\sigma = (6)(0.038) = 0.228$ mm. By July, $\sigma_0 = 0.030$, which gives a process capability of 0.180 mm. This is a 20% improvement in the process capability, which in most situations will be sufficient to solve a quality problem.

It is frequently necessary to obtain the process capability by a quick method rather than by using the $\overline{X}$ and R charts. This method assumes the process is stable or in statistical control, which may or may not be the case. The procedure is as follows:

1. Take 25 subgroups of size 4 for a total of 100 measurements.
2. Calculate the range, R, for each subgroup.
3. Calculate the average range, $\overline{R} = \Sigma R/g = \Sigma R/25$.

4. Calculate the estimate of the population standard deviation,

$$\hat{\sigma} = \frac{\bar{s}}{d_2}$$

where d_2 is obtained from Table B in the Appendix and is 2.059 for $n = 4$.

5. Process capability will equal $6\sigma_0$.

Remember that this technique does not give the true process capability and should be used only if circumstances require its use. Also, more than 25 subgroups can be used to improve accuracy.

Example Problem 6-6

An existing process is not meeting the Rockwell-C specifications. Determine the process capability based on the range values for 25 subgroups of size 4. Data are 7, 5, 5, 3, 2, 4, 5, 9, 4, 5, 4, 7, 5, 7, 3, 4, 4, 5, 6, 4, 7, 7, 5, 5, and 7.

$$\bar{R} = \frac{\Sigma R}{g} = \frac{129}{25} = 5.16$$

$$\sigma_0 = \frac{\bar{R}}{d_2} = \frac{5.16}{2.059} = 2.51$$

$$6\sigma_0 = (6)(2.51) = 15.1$$

The process capability can also be obtained by using the standard deviation. Statistical control of the process is assumed. The procedure is as follows:

1. Take 25 subgroups of size 4 for a total of 100 measurements.
2. Calculate the sample standard deviation, s, for each subgroup.
3. Calculate the average sample standard deviation, $\bar{s} = \Sigma s/g = \Sigma s/25$.
4. Calculate the estimate of the population standard deviation,

$$\hat{\sigma} = \frac{\bar{R}}{c_4}$$

where c_4 is obtained from Table B in the Appendix and is 0.9213 for $n = 4$.

5. Process capability will equal $6\sigma_0$.

More than 25 subgroups will improve the accuracy.

Example Problem 6-7

A new process is started, and the sum of the sample standard deviations for 25 subgroups of size 4 is 105. Determine the process capability,

$$\bar{s} = \frac{\Sigma s}{g} = \frac{105}{25} = 4.2$$

$$\sigma_0 = \frac{\bar{s}}{c_4} = \frac{4.2}{0.9213} = 4.56$$

$$6\sigma_0 = (6)(4.56) = 27.4$$

Either the range or the standard deviation method can be used, although, as previously stated, the standard deviation method is more accurate. A histogram should be constructed to graphically present the process capability. Actually, a minimum of 50 measurements is required for a histogram. Therefore, the histograms made from the same data that were used to calculate the process capability should adequately represent the process at that time.

Process capability and the tolerance are combined to form a *capability index*, defined as

$$C_p = \frac{\text{USL} - \text{LSL}}{6\sigma_0}$$

where C_p = capability index
USL − LSL = upper specification − lower specification, or tolerance
$6\sigma_0$ = process capability

If the capability index is 1.00, we have the Case II situation discussed in the preceding section; if the ratio is greater than 1.00, we have the Case I situation, which is desirable; and if the ratio is less than 1.00, we have the Case III situation, which is undesirable. Figure 6-25 shows these three cases.

Example Problem 6-8

Assume that the specifications are 6.50 and 6.30 in the depth of keyway problem. Determine the capability index before ($\sigma_0 = 0.038$) and after ($\sigma_0 = 0.030$) improvement.

$$C_p = \frac{\text{USL} - \text{LSL}}{6\sigma_0} = \frac{6.50 - 6.30}{6(0.038)} = 0.88$$

$$C_p = \frac{\text{USL} - \text{LSL}}{6\sigma_0} = \frac{6.50 - 6.30}{6(0.030)} = 1.11$$

In Example Problem 6-8, the improvement in quality resulted in a desirable capability index (Case I). The minimum capability index is frequently established at 1.33. Below this value, design engineers may be required to seek approval from manufacturing before the product can be released to production. A capability index of 1.33 is considered by most organizations to be a de facto standard, with even larger values of 2.00 desired, which requires the specifications to be set at $\pm 6\sigma$.

Using the capability index[13] concept, we can measure quality provided the process is centered. The larger the capability index, the better the quality. We should strive to make the capability index as large as possible. This is accomplished by having realistic specifications and continually striving to improve the process capability.

[13] Another measure of the capability is called the *capability ratio*, which is defined as

$$C_r = \frac{6\sigma_0}{\text{USL} - \text{LSL}}$$

The only difference between the two measures is the change in the numerator and denominator. They are used for the same purpose; however, the interpretation is different. The de facto standard for a capability ratio is 0.75, with even smaller values desired. In both cases the de facto standard is established with the tolerance at $8\sigma_0$. To avoid misinterpretation between two parties, they should be sure which process capability measure is being used. In this book the capability index is used.

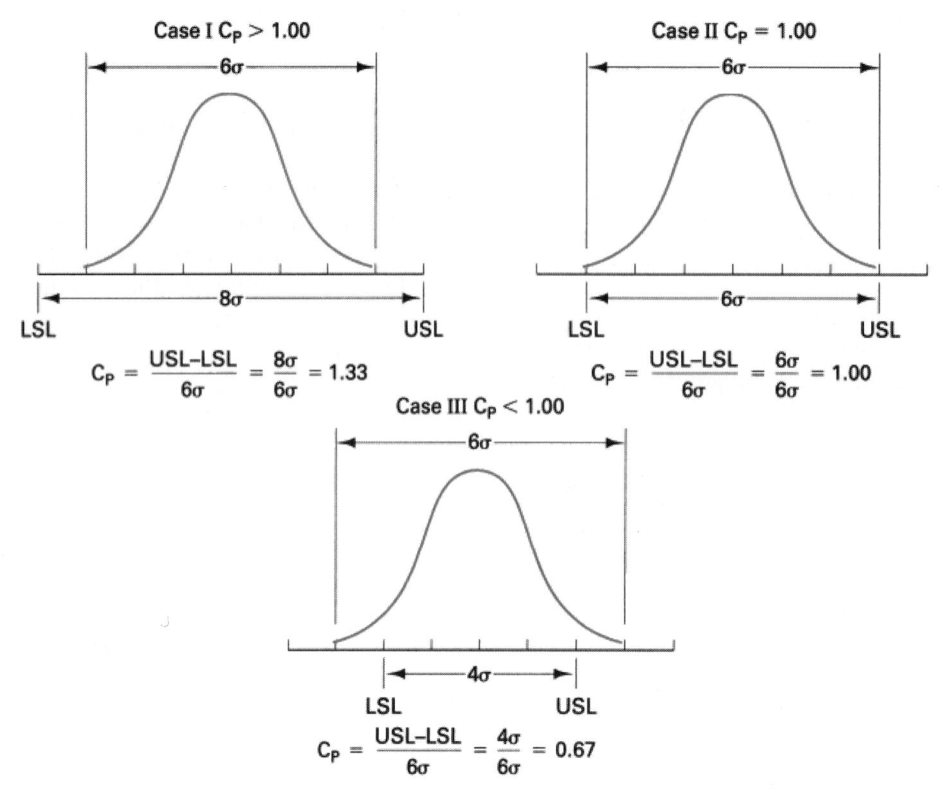

FIGURE 6-25 Capability Index and Three Cases

The capability index does not measure process performance in terms of the nominal or target value. This measure is accomplished using C_{pk}, which is defined as

$$C_{pk} = \frac{\text{Min}\{(\text{USL} - \overline{X}) \text{ or } (\overline{X} - \text{LSL})\}}{3\sigma}$$

Example Problem 6-9

Determine C_{pk} for the previous example problem (USL = 6.50, LSL = 6.30, and $\sigma = 0.030$) when the average is 6.45.

$$C_{pk} = \frac{\text{Min}\{(\text{USL} - \overline{X}) \text{ or } (\overline{X} - \text{LSL})\}}{3\sigma}$$

$$= \frac{\text{Min}\{(6.50 - 6.45) \text{ or } (6.45 - 6.30)\}}{3(0.030)}$$

$$= \frac{0.05}{0.090} = 0.56$$

Find C_{pk} when the average is 6.38.

$$C_{pk} = \frac{\text{Min}\{(\text{USL} - \overline{X}) \text{ or } (\overline{X} - \text{LSL})\}}{3\sigma}$$

$$= \frac{\text{Min}\{(6.50 - 6.38) \text{ or } (6.38 - 6.30)\}}{3(0.030)}$$

$$= \frac{0.08}{0.090} = 0.89$$

Figure 6-26 illustrates C_p and C_{pk} values for a process that is centered and one that is off-center by 1σ for the three cases. Comments concerning C_p and C_{pk} are as follows.

1. The C_p value does not change as the process center changes.
2. $C_p = C_{pk}$ when the process is centered.
3. C_{pk} is always equal to or less than C_p.
4. A C_{pk} value of 1.00 is a de facto standard. It indicates that the process is producing product that conforms to specifications.
5. A C_{pk} value less than 1.00 indicates that the process is producing product that does not conform to specifications.
6. A C_p value less than 1.00 indicates that the process is not capable.
7. A C_{pk} value of 0 indicates the average is equal to one of the specification limits.
8. A negative C_{pk} value indicates that the average is outside the specifications.

OTHER CONTROL CHARTS

The basic control charts for variables were discussed in previous sections. Although most quality control activity for variables is concerned with the $\overline{X}$ and R chart or the $\overline{X}$ and s chart, there are other charts which find application in some situations. These charts are discussed briefly in this section.

Charts for Better Operator Understanding

Because production personnel have difficulty understanding the relationships between averages, individual values, control

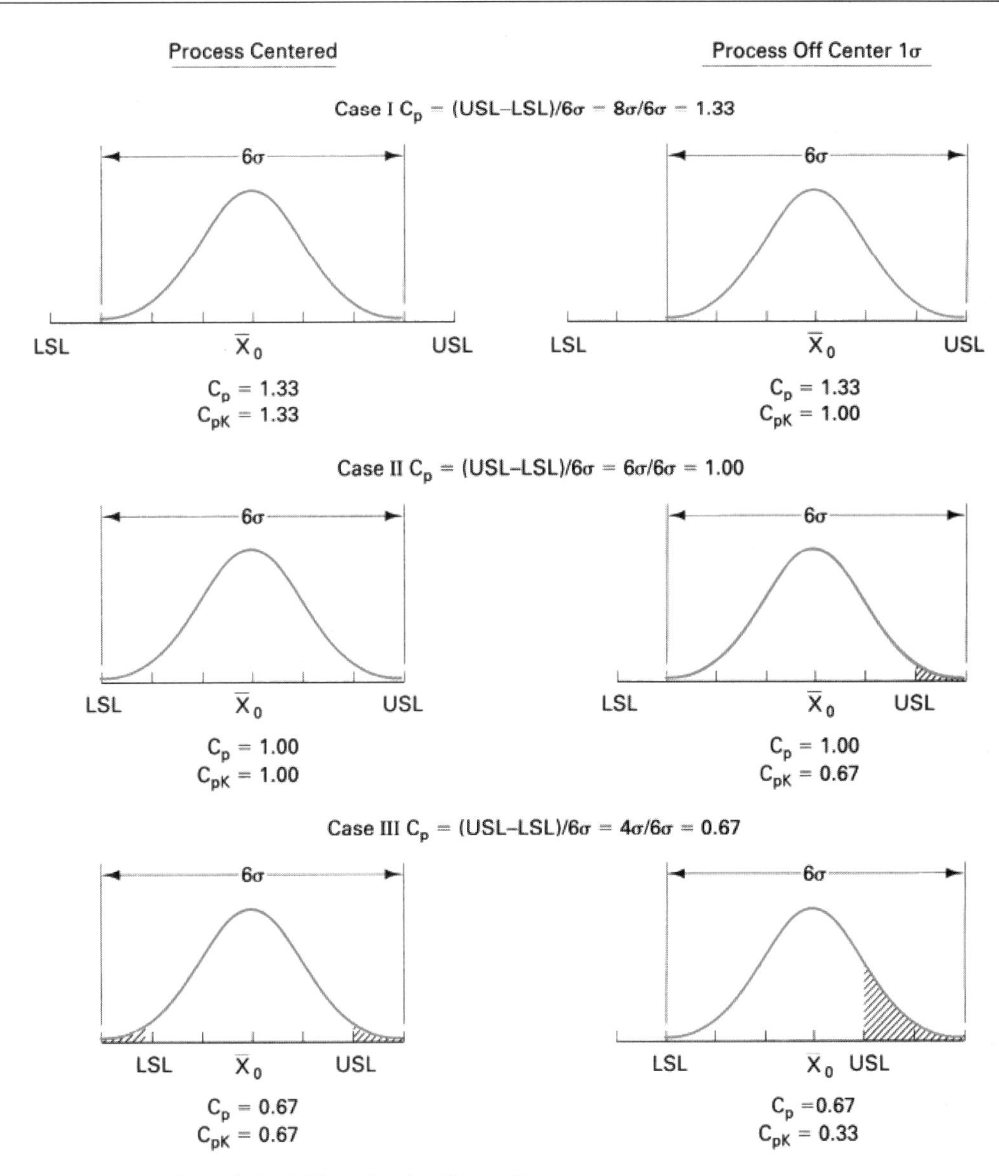

FIGURE 6-26 C_p and C_{pk} Values for the Three Cases

limits, and specifications, various charts have been developed to overcome this difficulty.

1. **Placing individual values on the chart.** This technique plots both the individual values and the subgroup average and is illustrated in Figure 6-27. A small dot represents an individual value and a larger circle represents the subgroup average. In some cases, an individual value and a subgroup average are identical, in which case the small dot is located inside the circle. When two individual values are identical, the two dots are placed side by side. A further refinement of the chart can be made by the addition of upper and lower specification lines; however, this practice is not recommended. In fact, the plotting of individual values is an unnecessary activity that can be overcome by proper operator training.

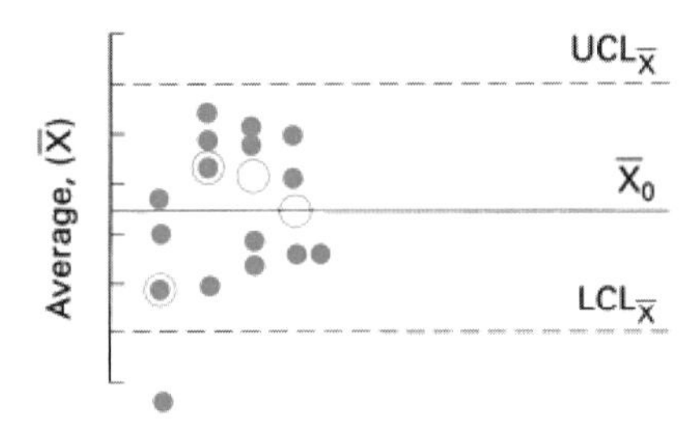

FIGURE 6-27 Chart Showing a Technique for Plotting Individual Values and Subgroup Averages

2. **Chart for subgroup sums.** This technique plots the subgroup sum, $\sum X$, rather than the subgroup average, $\overline{X}$. Because the values on the chart are of a different magnitude than the specifications, there is no chance for confusion. Figure 6-28 shows a subgroup sum chart, which is an $\overline{X}$ chart with the scale magnified by the subgroup size, n. The central line is $n\overline{X}_0$ and the control limits are obtained by the formulas

$$UCL_{\sum X} = n(UCL_{\overline{X}})$$
$$LCL_{\sum X} = n(LCL_{\overline{X}})$$

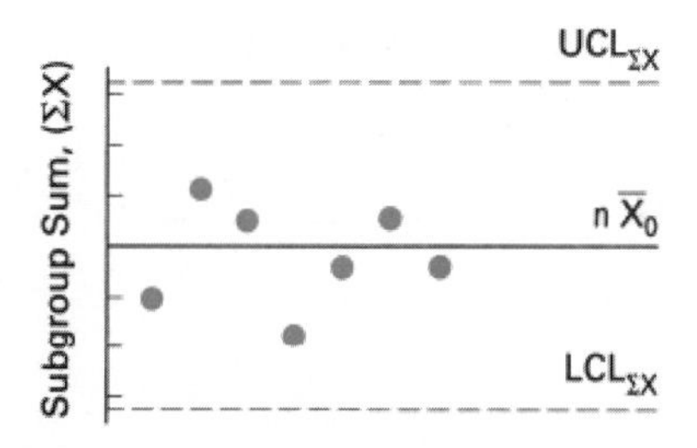

FIGURE 6-28 Subgroup Sum Chart

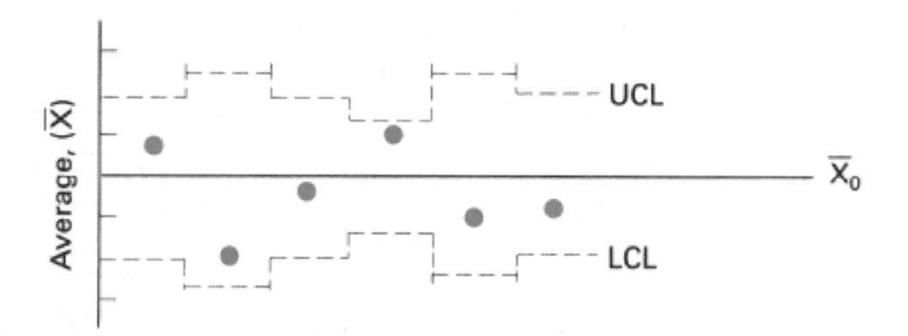

FIGURE 6-29 Chart for Variable Subgroup Size

This chart is mathematically equal to the $\overline{X}$ chart and has the added advantage of simpler calculations. Only addition and subtraction are required.

Chart for Variable Subgroup Size

Every effort should be made to keep the subgroup size constant. Occasionally, however, because of lost material, laboratory tests, production problems, or inspection mistakes, the subgroup size varies. When this situation occurs, the control limits will vary with the subgroup size. As the subgroup size, n, increases, the control limits become narrower; as the subgroup size decreases, the control limits become wider apart (Figure 6-29). This fact is confirmed by an analysis of the control limit factors A_2, D_1, and D_2, which are functions of the subgroup size and which are part of the control limit formulas. Control limits will also vary for the R chart.

One of the difficulties associated with a chart for variable subgroup size is the need to make a number of control limit calculations. A more serious difficulty involves the task of explaining to production people the reason for the different control limits. Therefore, this type of chart should be avoided.

Chart for Trends

When the plotted points of a chart have an upward or downward trend, it can be attributed to an unnatural pattern of variation or to a natural pattern of variation such as tool wear. In other words, as the tool wears, a gradual change in the average is expected and considered to be normal. Figure 6-30 illustrates a chart for a trend that reflects die wear. As the die wears, the measurement gradually increases until it reaches the upper reject limit. The die is then replaced or reworked.

Because the central line is on a slope, its equation must be determined. This is best accomplished using the least-squares method of fitting a line to a set of points. The equation for the trend line, using the slope–intercept form, is

$$\overline{X} = a + bG$$

where $\overline{X}$ = subgroup average and represents the vertical axis

G = subgroup number and represents the horizontal axis

a = point on the vertical axis where the line intercepts the vertical axis

$$a = \frac{(\Sigma\overline{X})(\Sigma G^2) - (\Sigma G)(\Sigma G\overline{X})}{g\Sigma G^2 - (\Sigma G)^2}$$

b = the slope of the line

$$b = \frac{g\Sigma G\overline{X} - (\Sigma G)(\Sigma\overline{X})}{g\Sigma G^2 - (\Sigma G)^2}$$

g = number of subgroups

The coefficients of a and b are obtained by establishing columns for G, $\overline{X}$, $G\overline{X}$, and G^2, as illustrated in Table 6-5; determining their sums; and inserting the sums in the equation.

Once the trend-line equation is known, it can be plotted on the chart by assuming values of G and calculating $\overline{X}$. When two points are plotted, the trend line is drawn between them.

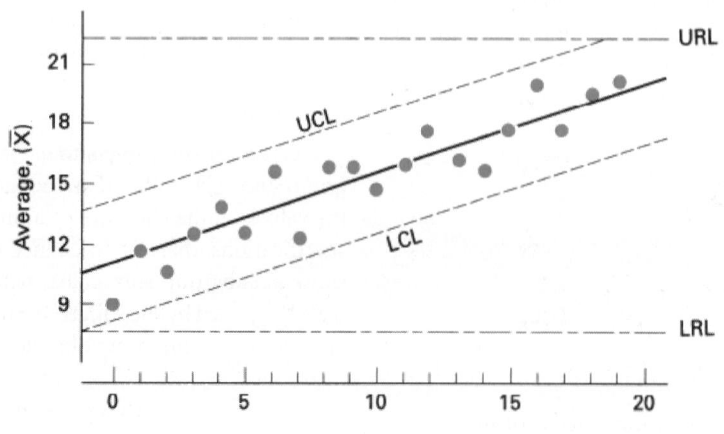

FIGURE 6-30 Chart for Trend

TABLE 6-5 Least-Squares Calculations for Trend Line

Subgroup Number G	Subgroup Average $\overline{X}$	Product of G and $G\overline{X}$ $G\overline{X}$	G^2
1	9	9	1
2	11	22	4
3	10	30	9
.	.	.	.
.	.	.	.
.	.	.	.
.	.	.	.
g			
ΣG	$\Sigma\overline{X}$	$\Sigma G\overline{X}$	ΣG^2

The control limits are drawn on each side of the trend line, a distance (in the perpendicular direction) equal to $A_2\overline{R}$ or $A\sigma_0$.

The R chart will generally have the typical appearance shown in Figure 6-7. However, the dispersion may also be increasing.

Chart for Moving Average and Moving Range

In some situations a chart is used to combine a number of individual values and plot them on a control chart. This type is referred to as a moving-average and moving-range chart and is quite common in the chemical industry, where only one reading is possible at a time. Table 6-6 illustrates the technique. In the development of Table 6-6, no calculations are made until the third period, when the sum of the three values is posted to the three-period moving-sum column $(35 + 26 + 28 = 89)$. The average and range are calculated $(\overline{X} = \frac{89}{3} = 29.6)(R = 35 - 26 = 9)$ and posted to the $\overline{X}$ and R columns. Subsequent calculations are accomplished by adding a new value and dropping the earliest one; therefore, 32 is added and 35 is dropped, making the sum $26 + 28 + 32 = 86$. The average and range calculations are $\overline{X} = \frac{86}{3} = 28.6$ and $R = 32 - 26 = 6$. Once the columns for $\overline{X}$ and R are completed, the charts are developed and used in the same manner as regular $\overline{X}$ and R charts.

The preceding discussion used a time period of 3 h; the time period could have been 2 h, 5 days, 3 shifts, and so on.

In comparing the moving-average and moving-range charts with conventional charts, it is observed that an extreme reading has a greater effect on the former charts. This is true because an extreme value is used a number of times in the calculations, and therefore will detect small changes much faster.

TABLE 6-6 Calculations of Moving Average and Moving Range

Value	Three-Period Moving Sum	$\overline{X}$	R
35	—	—	—
26	—	—	—
28	89	29.6	9
32	86	28.6	6
36	96	32.0	8
.	.	.	.
.	.	.	.
.	.	.	.
.	.	.	.
.	.	.	.
		$\Sigma\overline{X} =$	$\Sigma R =$

Chart for Median and Range

A simplified variable control chart that minimizes calculations is the median and range. The data are collected in the conventional manner and the median, Md, and range, R, of each subgroup are found. When using manual methods, these values are arranged in ascending or descending order. The median of the subgroup medians or grand median, Md_{Md}, and the median of the subgroup range, R_{Md}, are found by counting to the midpoint value. The median control limits are determined from the formulas

$$\text{UCL}_{\text{Md}} = \text{Md}_{\text{Md}} + A_5R_{\text{Md}}$$
$$\text{LCL}_{\text{Md}} = \text{Md}_{\text{Md}} - A_5R_{\text{Md}}$$

where Md_{Md} = grand median (median of the medians); Md_0 can be substituted in the formula

A_5 = factor for determining the 3σ control limits (see Table 6-7)

R_{Md} = median of subgroup ranges

The range control limits are determined from the formulas

$$\text{UCL}_R = D_6R_{\text{Md}}$$
$$\text{LCL}_R = D_5R_{\text{Md}}$$

TABLE 6-7 Factors for Computing 3σ Control Limits for Median and Range Charts from the Median Range

Subgroup Size	A_5	D_5	D_6	D_3
2	2.224	0	3.865	0.954
3	1.265	0	2.745	1.588
4	0.829	0	2.375	1.978
5	0.712	0	2.179	2.257
6	0.562	0	2.055	2.472
7	0.520	0.078	1.967	2.645
8	0.441	0.139	1.901	2.791
9	0.419	0.187	1.850	2.916
10	0.369	0.227	1.809	3.024

Source: Extracted by permission from P. C. Clifford, "Control Without Calculations," *Industrial Quality Control*, Vol. 15, No. 6 (May 1959): 44.

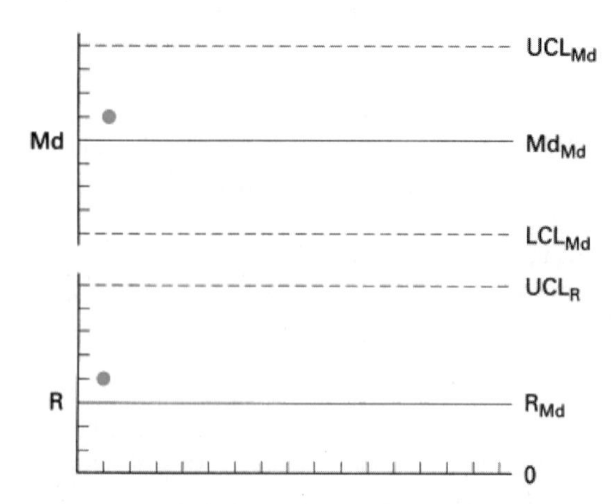

FIGURE 6-31 Control Charts for Median and Range

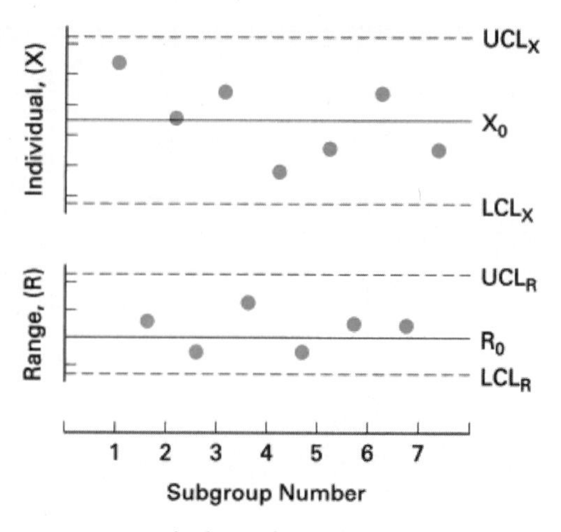

FIGURE 6-32 Control Charts for Individual Values and Moving Range

where D_5 and D_6 are factors for determining the 3σ control limits based on R_{Md} and are found in Table 6-7. An estimate of the population standard deviation can be obtained from $\sigma = R_{Md}/D_3$.

The principal benefits of the median chart are as follows: (1) there is less arithmetic, (2) it is easier to understand, and (3) it can be easily maintained by operators. However, the median chart fails to grant any weight to extreme values in a subgroup.

When these charts are maintained by operating personnel, a subgroup size of 3 is recommended. For example, consider the three values 36, 39, and 35. The Md is 36 and R is 4—all three values are used. Figure 6-31 is an example of a median chart. Subgroup sizes of 5 give a better chart; however, using the manual method, the operator will have to order the data before determining the median. Although these charts are not as sensitive to variation as the $\overline{X}$ and R charts, they can be quite effective, especially after quality improvement has been obtained and the process is in a monitoring phase. An unpublished master's thesis study showed little difference in the effectiveness of the Md and R charts when compared to the $\overline{X}$ and R charts.

Chart for Individual Values

In many situations, only one measurement is taken on a quality characteristic. This may be due to the fact that it is too expensive or too time-consuming, there are too few items to inspect, or it may not be possible. In such cases an X chart will provide some information from limited data, whereas an $\overline{X}$ chart would provide no information or information only after considerable delay to obtain sufficient data. Figure 6-32 illustrates an X chart.

Formulas for the trial central line and control limits are

$$\overline{X} = \frac{\Sigma X}{g} \qquad \overline{R} = \frac{\Sigma R}{g}$$

$$UCL_X = \overline{X} + 2.660\overline{R} \qquad UCL_R = 3.267\overline{R}$$

$$LCL_X = \overline{X} - 2.660\overline{R} \qquad LCL_R = (0)\overline{R}$$

These formulas require the moving-range technique with a subgroup size of 2.[14] To obtain the first range point, the value of X_1 is subtracted from X_2; to obtain the second point, X_2 is subtracted from X_3; and so forth. Each individual value is used for two different points except for the first and last—therefore, the name "moving" range. The range points should be placed between the subgroup number on the R chart because they are obtained from both values, or they can be placed at the second point.

These range points are averaged to obtain $\overline{R}$. Note that g for obtaining $\overline{R}$ will be 1 less than g for obtaining $\overline{X}$.

Formulas for the revised central line and control limits are

$$X_0 = \overline{X}_{new} \qquad \overline{R} = \frac{\Sigma R}{g}$$

$$UCL_X = X_0 + 3\sigma_0 \qquad UCL_R = (3.686)\sigma_0$$

$$LCL_X = X_0 - 3\sigma_0 \qquad LCL_R = (0)\sigma_0$$

where $\sigma_0 = 0.8865R_0$.

The X chart has the advantage of being easier for production personnel to understand and of providing a direct comparison with specifications. It does have the disadvantages of (1) requiring too many subgroups to indicate an out-of-control condition, (2) not summarizing the data as well as $\overline{X}$, and (3) distorting the control limits when the distribution is not normal. To correct for the last disadvantage, tests for normality should be used. Unless there is an insufficient amount of data, the $\overline{X}$ chart is recommended.

Charts with Nonacceptance Limits

Nonacceptance limits have the same relationship to averages as specifications have to individual values. Figure 6-33 shows the relationship of nonacceptance limits, control limits, and specifications for the three cases discussed in the section on specifications. The upper and lower specifications are shown in Figure 6-33 to illustrate the technique and are not included in actual practice.

In Case I the nonacceptance limits are greater than the control limits, which is a desirable situation, because an out-of-control condition will not result in nonconforming product. Case II shows the situation when the nonacceptance limits are equal to the control limits; therefore, any out-of-control

[14] J. M. Juran, *Juran's Quality Control Handbook,* 4th ed. (New York: McGraw-Hill, 1988).

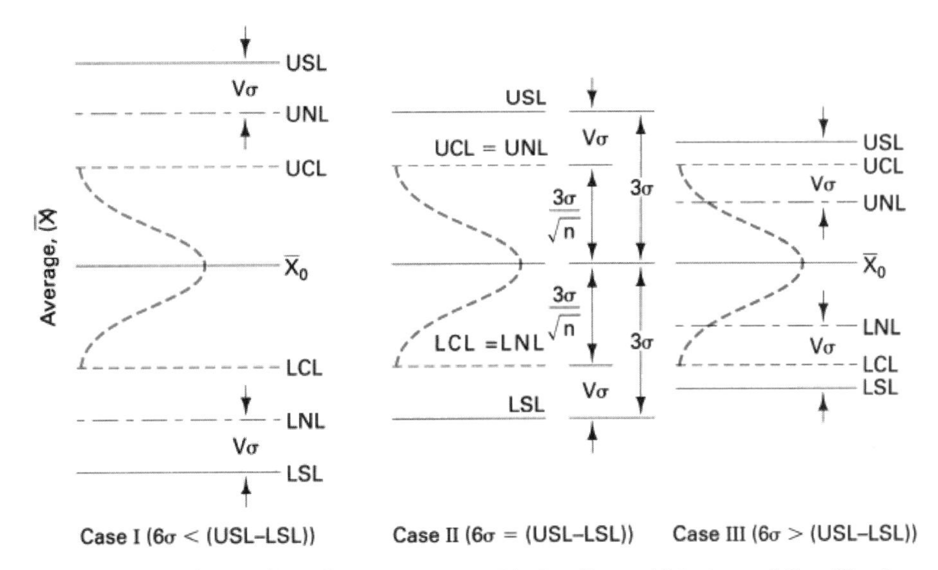

FIGURE 6-33 Relationship of Nonacceptance Limits, Control Limits, and Specifications

situations will result in nonconforming product being manufactured. Case III illustrates the situation when the nonacceptance limits are inside the control limits, and therefore some nonconforming product will be manufactured even when the process is in control.

The figure shows that the nonacceptance limits are a prescribed distance from the specifications. This distance is equal to $V\sigma$, where V varies with the subgroup size and is equal to the value $3 - 3/\sqrt{n}$. The formula for V was derived from Case II, because in that situation the control limits are equal to the nonacceptance limits.

Control limits tell what the process is capable of doing, and reject limits tell when the product is conforming to specifications. This can be a valuable tool for the quality professional and perhaps the first-line supervisor. Posting of nonacceptance limits for operating personnel should be avoided, because they will be confusing and may lead to unnecessary adjustment. Also, the operator is only responsible to maintain the process between the control limits.

Chart for Exponential Weighted Moving-Average

The *exponential weighted moving-average* (EWMA) chart, also called the geometric moving-averge chart, gives the greatest weight to the most recent data and less weight to all previous data. Its primary advantage is the ability to detect small shifts in the process average; however, it does not react as quickly to large shifts as the $\overline{X}$ chart. The advantages of both techniques can be achieved by plotting them on the same chart using different colors.

Another chart that has the ability to detect small shifts in the process average is called the *cusum chart.* It is more difficult to understand and calculate, and it does not react as well as the EWMA chart to large shifts. Details are given in Juran's *Quality Control Handbook.*[15]

[15]Ibid.

The EWMA is defined by the euqation

$$V_t = \lambda \overline{X}_t + (1 - \lambda) V_{t-1}$$

where V_t = the EWMA of the most recent plotted point
V_{t-1} = the EWMA of the previous plotted point
λ = the weight given to the subgroup average or individual value
$\overline{X}_t$ = the subgroup average or individual value

The value of lambda, λ, should be between 0.05 and 0.25, with lower values giving a better ability to detect smaller shifts. Values of 0.08, 0.10, and 0.15 work well. In order to start the sequential calculations, V_{t-1} is given the value $\overline{\overline{X}}$.[16]

Control limits are established by the equations

$$\text{ULC} = \overline{\overline{X}} - A_2\overline{R}\sqrt{\frac{\lambda}{(2 - \lambda)}}$$

$$\text{LCL} = \overline{\overline{X}} - A_2\overline{R}\sqrt{\frac{\lambda}{(2 - \lambda)}}$$

Actually, the control limits for the first few samples use different equations; however, the control limit values increase rapidly to their limiting values as determined by the equations given above.[17]

Example Problem 6-10

Using the information in Table 6-2 with the three out-of-control subgroups removed, determine the control limits and plot the points for an EWMA control chart using a λ value of 0.15. Table 6-8 gives an EXCEL spreadsheet for the calculations, and Figure 6-34 shows the actual control chart.

Calculations for the first few points and the control limits are

$$V_1 = \lambda \overline{X}_1 + (1 - \lambda) V_0 \quad V_0 = \overline{\overline{X}}$$
$$= 0.15(6.36) + (1 - 0.15)(6.394)$$
$$= 6.389$$

[16]Douglas C. Montgomery, *Introduction to Statistical Quality Control, 5e,* Wiley Publishing, Inc., Indianapolis, IN, 2004.
[17]Ibid.

TABLE 6-8 EXCEL Spreadsheet for EWMA Control Chart for Example Problem 6-10

Subgroup Number	X or X-Bar	Range	EWMA
1	6.36	0.08	6.394
2	6.4	0.1	6.389
3	6.36	0.06	6.391
4	6.39	0.1	6.386
5	6.4	0.09	6.387
6	6.43	0.05	6.389
7	6.37	0.08	6.395
8	6.46	0.04	6.391
9	6.42	0.11	6.401
10	6.39	0.03	6.404
11	6.38	0.04	6.402
12	6.4	0.12	6.399
13	6.41	0.07	6.399
14	6.45	0.08	6.401
15	6.34	0.1	6.408
16	6.36	0.12	6.398
17	6.38	0.06	6.392
18	6.4	0.08	6.390
19	6.39	0.07	6.392
20	6.39	0.06	6.392
21	6.38	0.06	6.391
22	6.41	0.06	6.390
			6.393
Sum	140.67	1.68	
X-DBar	6.394		
R-Bar	0.0764		
Lambda	0.15		
UCL	6.410		
LCL	6.378		

$$\begin{aligned} V_2 &= \lambda \bar{X}_2 + (1-\lambda)V_1 \\ &= 0.15(6.40) + (1-0.15)(6.389) \\ &= 6.391 \end{aligned}$$

$$\begin{aligned} V_3 &= \lambda \bar{X}_3 + (1-\lambda)V_2 \\ &= 0.15(6.38) + (1-0.15)(6.391) \\ &= 6.385 \end{aligned}$$

$$\begin{aligned} \text{UCL} &= \bar{\bar{X}} + A_2\bar{R}\sqrt{\frac{\lambda}{2-\lambda}} \\ &= 6.394 + (0.729)(0.0764)\sqrt{\frac{0.15}{(2-0.15)}} \\ &= 6.413 \end{aligned}$$

$$\begin{aligned} \text{LCL} &= \bar{\bar{X}} - A_2\bar{R}\sqrt{\frac{\lambda}{2-\lambda}} \\ &= 6.394 - (0.729)(0.0764)\sqrt{\frac{0.15}{2-0.15}} \\ &= 6.376 \end{aligned}$$

Because the EWMA chart is not sensitive to normality assumptions, it can be used for individual values. It can also be used for attribute charts.

COMPUTER PROGRAM

The EXCEL program files on the website will solve for $\bar{X}$ and R charts, Md and R charts, X and MR charts, EWMA Chart and process capability. Their file names are *X-bar & R Charts, Md & R Charts, X & MR Charts, EWMA Charts,* and *Process Capability.*

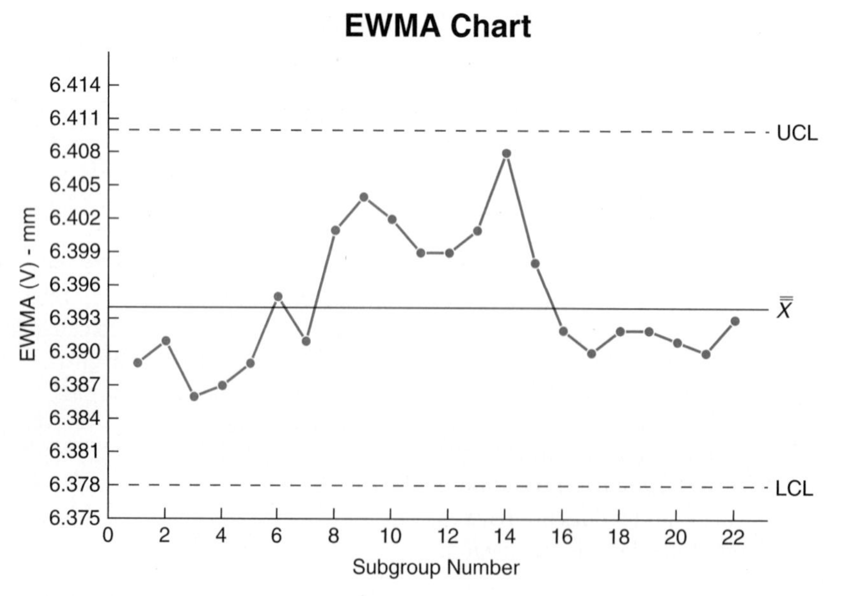

FIGURE 6-34 EWMA Control Chart for Data in Table 6-8

EXERCISES

1. Given is a typical $\overline{X}$ and R chart form with information on acid content in milliliters. Complete the calculations for subgroups 22, 23, 24, and 25. Plot the points to complete the run chart. Construct a control chart by calculating and drawing the trial central line and limits. Analyze the plotted points to determine if the process is stable.

VARIABLES CONTROL CHART DEPT/AREA: CHART ID: *Problem 1*

PART ID:	OPERATION ID:	CHARACTERISTIC: *Acid Content*
CHECK METHOD:	NOMINAL VALUE: *0.70 ml*	TOLERANCE: ± *0.20*

	1	2	3	4	5	6	7	8	9	10	11	12	13	14	15	16	17	18	19	20	21	22	23	24	25
SAMPLE READINGS 1	.85	.75	.80	.65	.75	.60	.80	.70	.75	.60	.80	.75	.70	.65	.85	.80	.70	.70	.65	.65	.55	.75	.80	.65	.65
2	.65	.85	.80	.75	.70	.75	.75	.60	.85	.70	.75	.85	.70	.70	.75	.75	.85	.60	.65	.60	.50	.65	.65	.60	.70
3	.65	.75	.75	.60	.65	.75	.65	.75	.85	.60	.90	.85	.75	.85	.80	.75	.75	.70	.85	.60	.65	.65	.75	.65	.70
4	.70	.85	.70	.70	.80	.70	.75	.75	.80	.80	.50	.65	.70	.75	.80	.80	.70	.70	.65	.65	.80	.80	.65	.60	.60
SUM, ΣX	2.85	3.20	3.05	2.70	2.90	2.80	2.95	2.80	3.25	2.70	2.95	3.10	2.85	2.95	3.20	3.10	3.00	2.70	2.80	2.50	2.50				
AVERAGE, $\overline{X}$	.71	.80	.76	.68	.73	.70	.74	.70	.81	.68	.74	.78	.71	.74	.80	.78	.75	.68	.70	.63	.63				
RANGE, R	.20	.10	.10	.15	.15	.15	.15	.15	.10	.20	.40	.20	.05	.20	.10	.05	.15	.10	.20	.05	.30				

$\overline{\overline{X}}$ = UCL = LCL =

AVERAGES .80 .70 .60

$\overline{R}$ = UCL = LCL =

RANGES .30 .20 .10

2. Control charts for $\overline{X}$ and R are to be established on a certain dimension part, measured in millimeters. Data were collected in subgroup sizes of 6 and are given below. Determine the trial central line and control limits. Assume assignable causes and revise the central line and limits.

Subgroup Number	$\overline{X}$	R	Subgroup Number	$\overline{X}$	R
1	20.35	0.34	14	20.41	0.36
2	20.40	0.36	15	20.45	0.34
3	20.36	0.32	16	20.34	0.36
4	20.65	0.36	17	20.36	0.37
5	20.20	0.36	18	20.42	0.73
6	20.40	0.35	19	20.50	0.38
7	20.43	0.31	20	20.31	0.35
8	20.37	0.34	21	20.39	0.38
9	20.48	0.30	22	20.39	0.33
10	20.42	0.37	23	20.40	0.32
11	20.39	0.29	24	20.41	0.34
12	20.38	0.30	25	20.40	0.30
13	20.40	0.33			

3. The following table gives the average and range in kilograms for tensile tests on an improved plastic cord. The subgroup size is 4. Determine the trial central line and control limits. If any points are out of control, assume assignable causes and calculate revised limits and central line.

Subgroup Number	$\overline{X}$	R	Subgroup Number	$\overline{X}$	R
1	476	32	14	482	22
2	466	24	15	506	23
3	484	32	16	496	23
4	466	26	17	478	25
5	470	24	18	484	24
6	494	24	19	506	23
7	486	28	20	476	25
8	496	23	21	485	29
9	488	24	22	490	25
10	482	26	23	463	22
11	498	25	24	469	27
12	464	24	25	474	22
13	484	24			

4. Rework Exercise 2 assuming subgroup sizes of 3, 4, and 5. How do the control limits compare?

5. Control charts for $\overline{X}$ and R are kept on the weight in kilograms of a color pigment for a batch process. After 25 subgroups with a subgroup size of 4, $\Sigma\overline{X} = 52.08$ kg (114.8 lb), $\Sigma R = 11.82$ kg (26.1 lb). Assuming the process is in a state of control, compute the $\overline{X}$ and R chart central line and control limits for the next production period.

6. Control charts for $\overline{X}$ and s are to be established on the Brinell hardness of hardened tool steel in kilograms per square millimeter. Data for subgroup sizes of 8 are shown below. Determine the trial central line and control limits for the $\overline{X}$ and s charts. Assume that the out-of-control points have assignable causes. Calculate the revised limits and central line.

Subgroup Number	$\overline{X}$	s	Subgroup Number	$\overline{X}$	s
1	540	26	14	551	24
2	534	23	15	522	29
3	545	24	16	579	26
4	561	27	17	549	28
5	576	25	18	508	23
6	523	50	19	569	22
7	571	29	20	574	28
8	547	29	21	563	33
9	584	23	22	561	23
10	552	24	23	548	25
11	541	28	24	556	27
12	545	25	25	553	23
13	546	26			

7. Control charts for $\overline{X}$ and s are maintained on the resistance in ohms of an electrical part. The subgroup size is 6. After 25 subgroups, $\Sigma\overline{X} = 2046.5$ and $\Sigma s = 17.4$. If the process is in statistical control, what are the control limits and central line?

8. Rework Exercise 6 assuming a subgroup size of 3.

9. Copy the s chart of Figure 6-8 on transparent paper. Place this copy on top of the R chart of Figure 6-5 and compare the patterns of variation.

10. In filling bags of nitrogen fertilizer, it is desired to hold the average overfill to as low a value as possible. The lower specification limit is 22.00 kg (48.50 lb), the population mean weight of the bags is 22.73 kg (50.11 lb), and the population standard deviation is 0.80 kg (1.76 lb). What percentage of the bags contains less than 22 kg? If it is permissible for 5% of the bags to be below 22 kg, what would be the average weight? Assume a normal distribution.

11. Plastic strips that are used in a sensitive electronic device are manufactured to a maximum specification of 305.70 mm (approximately 12 in.) and a minimum specification of 304.55 mm. If the strips are less than the minimum specification, they are scrapped; if they are greater than the maximum specification, they are reworked. The part dimensions are normally distributed with a population mean of 305.20 mm and a standard deviation of 0.25 mm. What percentage of the product is scrap? What percentage is rework? How can the process be centered to eliminate all but 0.1% of the scrap? What is the rework percentage now?

12. A company that manufactures oil seals found the population mean to be 49.15 mm (1.935 in.), the population standard deviation to be 0.51 mm (0.020 in.), and the data to be normally distributed. If the internal diameter of the seal is below the lower specification limit of 47.80 mm, the part is reworked. However, if it is above the upper specification limit of 49.80 mm, the seal is scrapped. (a) What percentage of the seals is reworked? What percentage is scrapped? (b) For various reasons, the process average is changed to 48.50 mm. With this new mean or process center, what percentage of the seals is reworked? What percentage is scrapped? If rework is economically feasible, is the change in the process center a wise decision?

13. The historical data of Exercise 37 have a subgroup size of 3. Time is not available to collect data for a process capability study using a subgroup size of 4. Determine the process capability using the first 25 subgroups. Use a D_2 value for $n = 3$.

14. Repeat Exercise 13 using the last 25 subgroups and compare the results.

15. Determine the process capability of the case-hardening process of Exercise 6.

16. Determine the process capability of the tensile tests of the improved plastic cord of Exercise 3.

17. What is the process capability of:

    a. Exercise 2?

    b. Exercise 5?

18. Determine the capability index before ($\sigma_0 = 0.038$) and after ($\sigma_0 = 0.030$) improvement of Example Problem 5-8 using specifications of 6.40 ± 0.15 mm.

19. A new process is started, and the sum of the sample standard deviations for 25 subgroups of size 4 is 750. If the specifications are 700 ± 80, what is the process capability index? What action would you recommend?

20. What is the C_{pk} value after improvement for Exercise 18 when the process center is 6.40? When the process center is 6.30? Explain.

21. What is the C_{pk} value for the information in Exercise 19 when the process average is 700, 740, 780, and 820? Explain.

22. Determine the revised central line and control limits for a subgroup sum chart using the data of:

    a. Exercise 2

    b. Exercise 3

23. Determine the trial central line and control limits for a moving-average and moving-range chart using a time period of 3. Data in liters are as follows: 4.56, 4.65, 4.66, 4.34, 4.65, 4.40, 4.50, 4.55, 4.69, 4.29, 4.58, 4.71, 4.61, 4.66, 4.46, 4.70, 4.65, 4.61, 4.54, 4.55, 4.54, 4.54, 4.47, 4.64, 4.72, 4.47, 4.66, 4.51, 4.43, and 4.34. Are there any out-of-control points?

24. Repeat Exercise 23 using a time period of 4. What is the difference in the central line and control limits? Are there any out-of-control points?

25. The Get-Well Hospital has completed a quality improvement project on the time to admit a patient using $\overline{X}$ and R charts. The hospital now wishes to monitor the activity using median and range charts. Determine the central line and control limits with the latest data in minutes as given below.

Subgroup Number	Observation X_1	X_2	X_3	Subgroup Number	Observation X_1	X_2	X_3
1	6.0	5.8	6.1	13	6.1	6.9	7.4
2	5.2	6.4	6.9	14	6.2	5.2	6.8
3	5.5	5.8	5.2	15	4.9	6.6	6.6
4	5.0	5.7	6.5	16	7.0	6.4	6.1
5	6.7	6.5	5.5	17	5.4	6.5	6.7
6	5.8	5.2	5.0	18	6.6	7.0	6.8
7	5.6	5.1	5.2	19	4.7	6.2	7.1
8	6.0	5.8	6.0	20	6.7	5.4	6.7
9	5.5	4.9	5.7	21	6.8	6.5	5.2
10	4.3	6.4	6.3	22	5.9	6.4	6.0
11	6.2	6.9	5.0	23	6.7	6.3	4.6
12	6.7	7.1	6.2	24	7.4	6.8	6.3

26. Determine the trial central line and control limits for median and range charts for the data of Table 6-2. Assume assignable causes for any out-of-control points and determine the revised central line and control limits. Compare the pattern of variation with the $\overline{X}$ and R charts in Figure 6-4.

27. An X and MR chart is to be maintained on the pH value for the swimming pool water of a leading motel. One reading is taken each day for 30 days. Data are 7.8, 7.9, 7.7, 7.6, 7.4, 7.2, 6.9, 7.5, 7.8, 7.7, 7.5, 7.8, 8.0, 8.1, 8.0, 7.9, 8.2, 7.3, 7.8, 7.4, 7.2, 7.5, 6.8, 7.3, 7.4, 8.1, 7.6, 8.0, 7.4, and 7.0. Plot the data on graph paper, determine the trial central line and limits, and evaluate the variation.

28. Determine upper and lower reject limits for the $\overline{X}$ chart of Exercise 2. The specifications are 20.40 ± 0.25. Compare these limits to the revised control limits.

29. Repeat Exercise 28 for specifications of 20.40 ± 0.30.

30. A new process is starting, and there is the possibility that the process temperature will give problems. Eight readings are taken each day at 8:00 A.M., 10:00 A.M., 12:00 noon, 2:00 P.M., 4:00 P.M., 6:00 P.M., 8:00 P.M., and 10:00 P.M. Prepare a run chart and evaluate the results.

Day	Temperature (0°C)							
Monday	78.9	80.0	79.6	79.9	78.6	80.2	78.9	78.5
Tuesday	80.7	80.5	79.6	80.2	79.2	79.3	79.7	80.3
Wednesday	79.0	80.6	79.9	79.6	80.0	80.0	78.6	79.3
Thursday	79.7	79.9	80.2	79.2	79.5	80.3	79.0	79.4
Friday	79.3	80.2	79.1	79.5	78.8	78.9	80.0	78.8

31. The viscosity of a liquid is checked every half-hour during one 3-shift day. Prepare a histogram with 5 cells and the midpoint value of the first cell equal to 29 and evaluate the distribution. Prepare a run chart and evaluate the distribution again. What does the run chart indicate? Data are 39, 42, 38, 37, 41, 40, 38, 36, 40, 36, 35, 38, 34, 35, 37, 36, 39, 34, 38, 36, 32, 37, 35, 34, 33, 35, 32, 32, 38, 34, 37, 35, 35, 34, 31, 33, 35, 32, 36, 31, 29, 33, 32, 31, 30, 32, 32, and 29.

32. Using the EXCEL program files solve

    a. Exercise 1

    b. Exercise 25

    c. Exercise 27

33. Using EXCEL, write a template for moving-average and moving-range charts for three periods and determine the charts using the data from

    a. Exercise 23

    b. Exercise 30

    c. Exercise 31

34. Using EXCEL, write a template for $\overline{X}$ and s charts and determine the charts for Exercise 1.

35. Using the EXCEL program file, determine an X and MR chart for the data of

    a. Exercise 30

    b. Exercise 31

36. Using the EXCEL program file, determine the process capability of cypress bark bags in kilograms for the data below. Also determine the C_p and C_{pk} for a USL of 130 kg and an LSL of 75 kg.

Subgroup	X_1	X_2	X_3	X_4
1	95	90	93	120
2	76	81	81	83
3	107	80	87	95
4	83	77	87	90
5	105	93	95	103
6	88	76	95	97
7	100	87	100	103
8	97	91	92	94
9	90	91	95	101
10	93	79	91	94
11	106	97	100	90
12	89	91	80	82
13	92	83	95	75
14	87	90	100	98
15	97	95	95	90
16	82	106	99	101
17	100	95	95	90
18	81	94	97	90
19	98	101	87	89
20	78	96	100	72
21	91	91	87	89
22	76	91	106	80
23	95	97	100	93
24	92	99	97	94
25	92	85	90	90

37. Using the EXCEL program file, determine the $\overline{X}$ and R charts for the data on shampoo weights in kilograms given below.

Subgroup Number	X_1	X_2	X_3	Subgroup Number	X_1	X_2	X_3
1	6.01	6.01	5.97	16	6.00	5.98	6.02
2	5.99	6.03	5.99	17	5.97	6.01	5.97
3	6.00	5.96	6.00	18	6.02	5.99	6.02
4	6.01	5.99	5.99	19	5.99	5.98	6.01
5	6.05	6.00	6.00	20	6.01	5.98	5.99
6	6.00	5.94	5.99	21	5.97	5.95	5.99
7	6.04	6.02	6.01	22	6.02	6.00	5.98
8	6.01	5.98	5.99	23	5.98	5.99	6.00
9	6.00	6.00	6.01	24	6.02	6.00	5.98
10	5.98	5.99	6.03	25	5.97	5.99	6.02
11	6.00	5.98	5.96	26	6.00	6.02	5.99
12	5.98	5.99	5.99	27	5.99	5.96	6.01
13	5.97	6.01	6.00	28	5.99	6.02	5.98
14	6.01	6.03	5.99	29	5.99	5.98	5.96
15	6.00	5.98	6.01	30	5.97	6.01	5.98

38. Using the EXCEL program file, determine an EWMA chart for

    a. Exercise 2, using $\lambda = 0.10$ and 0.20.

    b. Exercise 3, using $\lambda = 0.05$ and 0.25.

    Verify your answers with a few hand calculations.

39. Write an EXCEL program for an EWMA chart using the individual data in Exercise 27.

    Hint: Refer to the information on the chart for individual values.

CHAPTER
SEVEN

ADDITIONAL SPC TECHNIQUES FOR VARIABLES

OBJECTIVES

Upon completion of this chapter, the reader is expected to

- explain the difference between discrete, continuous, and batch processes;
- know how to construct and use a group chart;
- be able to construct a multi-vari chart;
- calculate the central line and control limits of a specification chart;
- explain how to use precontrol for setup and run activities;
- calculate the central line and central limits for a $\overline{Z}$ and W and a Z and MW chart;
- be able to perform GR&R.

INTRODUCTION

Chapter 6 covered basic information on variable control charts, which are a fundamental aspect of statistical process control (SPC). For the most part, that discussion concentrated on long production runs of discrete parts. This chapter augments that material by providing information on continuous and batch processes, short runs, and gauge control.

CONTINUOUS AND BATCH PROCESSES

Continuous Processes

One of the best examples of a continuous process is the paper-making process. Paper-making machines are very long, with some exceeding the length of a football field, and wide, over 18 ft, and they may run at speeds of over 3600 ft/min. They operate 24 h/day, 7 days/week, and stop only for scheduled maintenance or emergencies. Briefly, the paper-making process begins with the conversion of wood chips to wood pulp by chemical or mechanical means. The pulp is washed, treated, and refined until it consists of 99% water and 1% pulp. It then flows to the headbox of the paper-making machine, which is shown in Figure 7-1. Pulp flows onto a moving wire screen, and water drains to form a wet mat. The mat passes through pressing rollers and a drying section to remove more water. After calendering to produce a hard, smooth surface, the web is wound into large rolls.

Statistical process control on the web is shown in Figure 7-2. Observed values are taken in the machine direction (md) or cross-machine direction (cd) by either sensors or manually after a roll is complete. Average and range values for machine direction and cross-machine direction are different.

The flow of pulp at the headbox is controlled by numerous valves; therefore, from the viewpoint of SPC, we need a variables control chart for each valve. For example, if the headbox has 48 valves, 48 md control charts are needed to control each valve. This type of activity is referred to as *multiple-stream output.*

In this particular process, a cd control chart would have little value for control of paper caliper. It might have some value for overall moisture control, because a control chart could indicate the need to increase or decrease the temperature of the drying rolls. The customer might be more interested in a cd control chart, because any out-of-control conditions could affect the performance of the paper on the customer's equipment.

It is extremely important for the practitioner to be knowledgeable about the process and to have definite objectives for the control chart. In many continuous processes, it is extremely difficult to obtain samples from a location that can effectively control the process. In such cases, sensors may be helpful to collect data, compare to control limits, and automatically control the process.

Group Chart

A group control chart eliminates the need for a chart for each stream. A single chart controls all the streams; however, it does not eliminate the need for measurements at each stream.

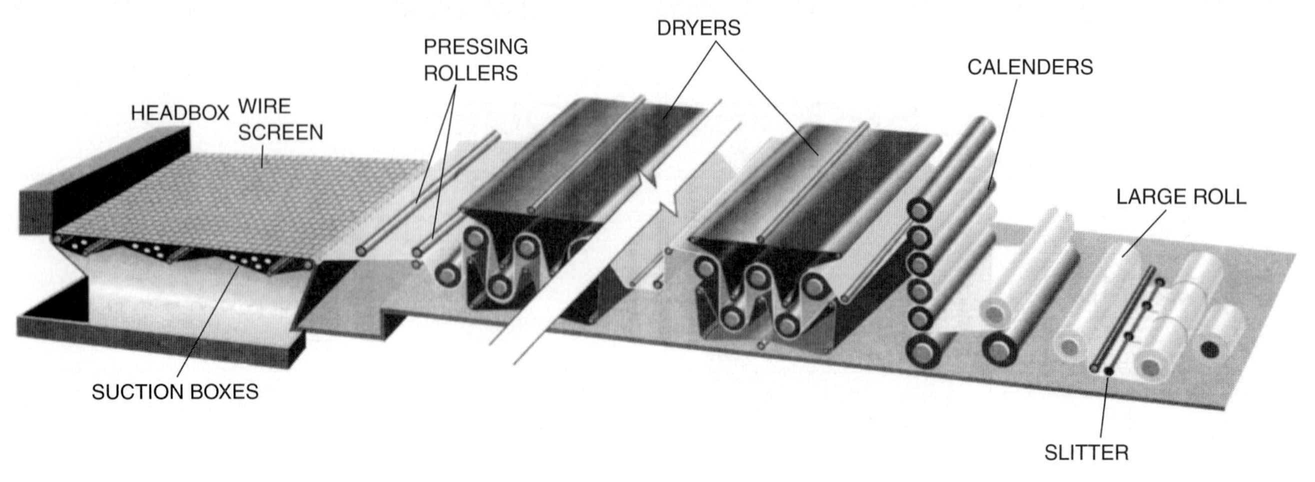

FIGURE 7-1 Paper-Making Machine

(Adapted from *The New Book of Knowledge*, 1969 edition. Copyright 1969 by Grolier Incorporated. Reprinted by permission.)

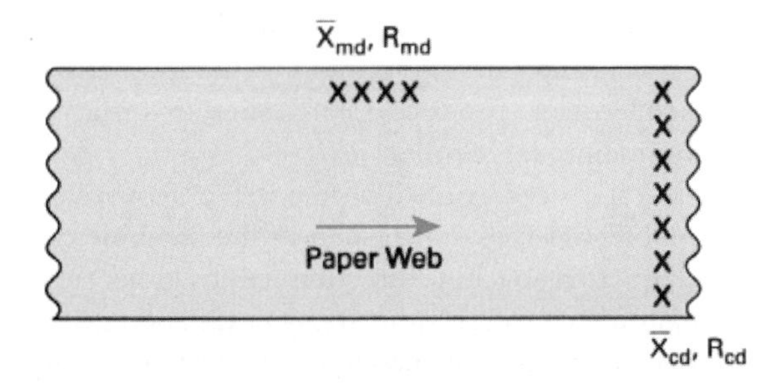

FIGURE 7-2 Paper Web and Observed Valves for md and cd Control Charts

Data are collected in the same manner as outlined in Chapter 6, that is, 25 subgroups for each stream. From this information, the central line and control limits are calculated. The plotted points for the $\overline{X}$ chart are the highest and the lowest averages, $\overline{X}_h$ and $\overline{X}_1$, and for the R chart, the highest range, R_h. Each stream or spindle is given a number, and it is recorded with a plotted point.

Of course, any out-of-control situation will call for corrective action. In addition, we have the out-of-control situation when the same stream gives the highest or lowest value r times in succession. Table 7-1 gives practical r values for the number of streams.

TABLE 7-1 Suggested r Values for the Number of Streams

Number of Streams	r
2	9
3	7
4	6
5–6	5
7–10	4
11–27	3
Over 27	2

Example Problem 7-1

Assume a four-spindle filling machine as shown in Figure 7-3 and a subgroup size of 3. Determine the number of subgroups needed to establish the central lines and control limits. Also determine the number of times in succession one of the spindles can be plotted before an out-of-control situation occurs.

$$25 \text{ per spindle} \times 4 \text{ spindles} = 100 \text{ subgroups of 3 each}$$

From the table, $r = 6$.

This technique is applicable to machines, test equipment, operators, or suppliers as long as the following three criteria are met: Each stream has the same target, same variation, and the variations are as close to normal as required by conventional $\overline{X}$ and R charts.[1]

Batch Processes

Many products are manufactured in batches, such as paint, adhesives, soft drinks, bread, soup, iron, and so forth. Statistical process control of batches has two forms: within-batch variation and between-batch variation.

Within-batch variation can be very minimal for many liquids that are under agitation, heat, pressure, or any combination thereof. For example, the composition of a product such as perfume might be quite uniform throughout the batch. Thus, only one observed value of a particular quality characteristic can be obtained. In this case, an X and R chart for individuals would be an appropriate SPC technique. Each batch in a series of batches would be plotted on the control chart.

Some liquid products, such as soup, will exhibit within-batch variation. Observed values (samples) need to be

[1]For more information, see L. S. Nelson, "Control Chart for Multiple Stream Processes," *Journal of Quality Technology*, Vol. 18, No. 4 (October 1986): 255–256.

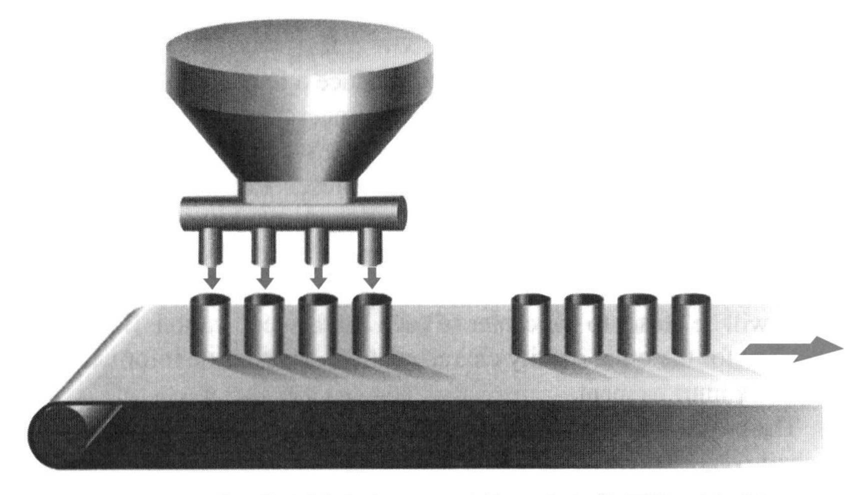

FIGURE 7-3 Example of Multiple Streams: A Four-Spindle Filling Machine

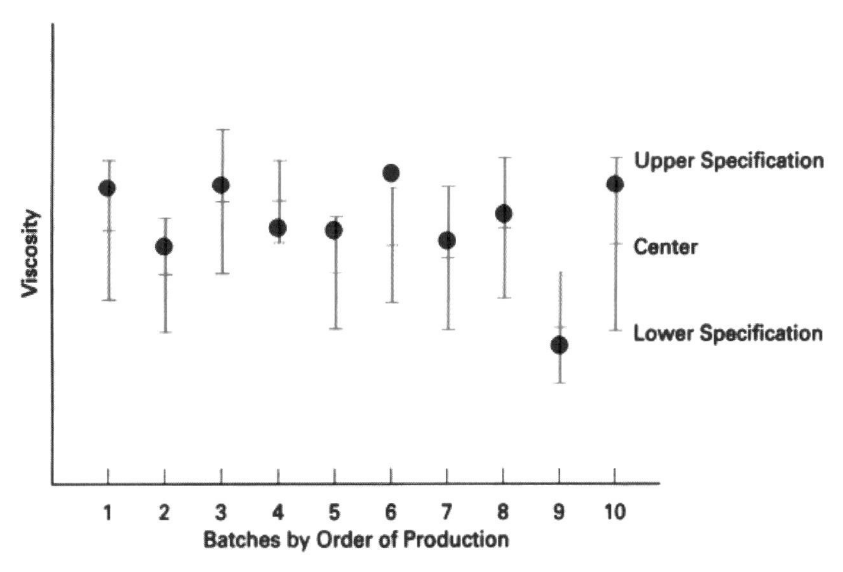

FIGURE 7-4 Batch Chart for Different Batches with Different Specifications

obtained at different locations within the batch, which may be difficult or impossible to accomplish. If samples can be obtained, then $\overline{X}$ and R charts, or similar charts, are appropriate. Sometimes it is necessary to obtain samples from the next operation, which is usually packaging and requires an appropriate location for measuring a volume or weight-fill characteristic. Care must be exercised to ensure that one is measuring within-batch variation, because the volume or weight-fill characteristic is a discrete process.

Batch-to-batch variation does not always occur. Because of the nature of some products, there is only one batch. In other words, a customer orders a product to a particular specification and never repeats the order. When there are repetitive batches of the same product, batch-to-batch variation can be charted in the same manner as for discrete processes.

Many processes do not obey the basic assumptions behind traditional SPC. The nature of the process should be determined before developing control charts. For example, one-sided specifications may indicate nonnormal data. Batch processes may have nested variation sources. Selection of the right model depends on an understanding of the process. The use of analysis of variance and determination of the appropriate distribution would be an initial step.[2]

Many products are manufactured by combinations of continuous, batch, and discrete processes. For example, in the paper-making process that was previously described, the pulping process is by batch in giant pressure cookers called digesters; the actual paper-making process is continuous; and the paper rolls are a discrete process.

Batch Chart

Many processing plants are designed to produce a few basic products to customer specifications. Although the ingredients and process are essentially the same, the specifications will change with each customer's batch. Figure 7-4 shows a run chart for batch viscosity. The solid point represents the

[2]William A. Levinson, "Using SPC in Batch Processes," *Quality Digest* (March 1998): 45–48.

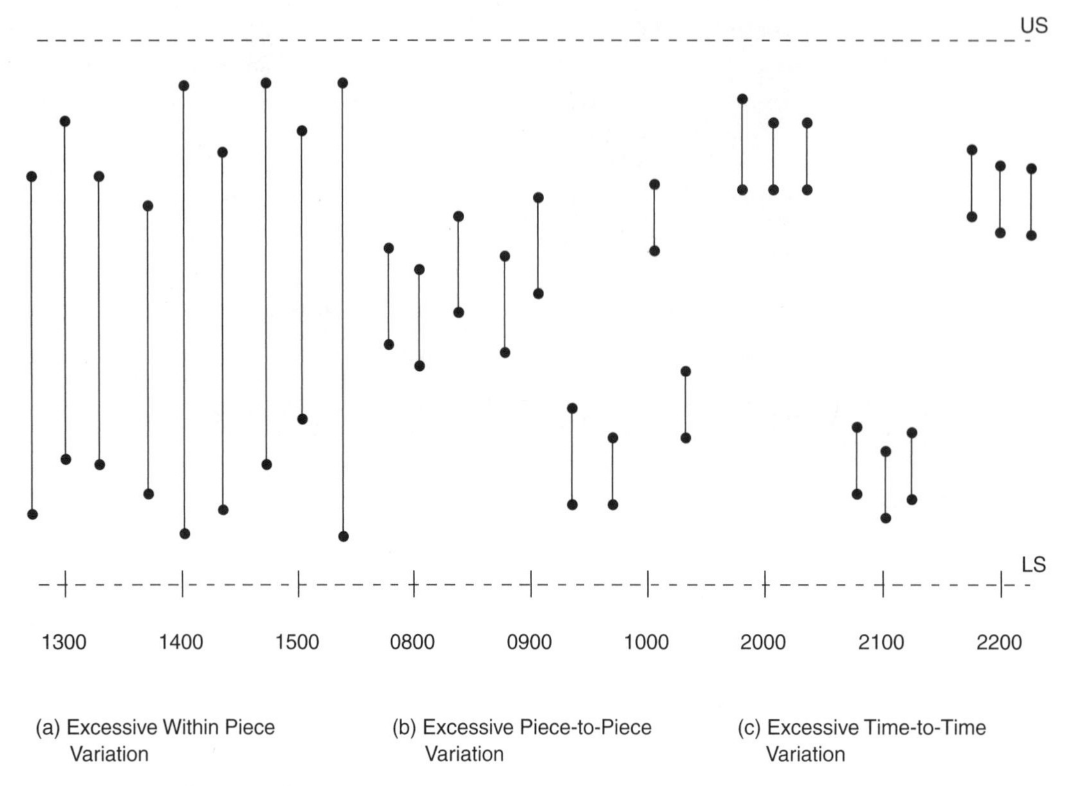

FIGURE 7-5 Multi-Vari Chart

viscosity value, and the vertical line represents the specification range. A cursory analysis of the batches shows that 8 of the 10 plotted points are on the high end of the specification. This information may lead to a minor adjustment so that the viscosity of future batches will be closer to the center of each batch specification. Batch charts such as this one for the other quality characteristics can provide information for effective quality improvement.

The batch chart is not a control chart. It might more appropriately be called a run chart.

MULTI-VARI CHART

The multi-vari chart is a useful tool for detecting different types of variation that are found in products and services. Frequently, the chart will lead to a problem solution much faster than other techniques. Some of the processes that lend themselves to this chart are inside and outside diameters, molds with multiple cavities, and adhesion strength.

The multi-vari chart concept is shown in Figure 7-5. It uses a vertical line to show the range of variation of the observed values within a single piece or service. Types of variation are shown in (a) within a unit, (b) unit to unit, and (c) time to time.

Within-unit variation occurs within a single unit, such as porosity in a casting, surface roughness, or cavities within a mold. Unit to unit variation occurs between consecutive units drawn from a process, batch-to-batch variations, and lot-to-lot variations. Time-to-time variation occurs from hour to hour, shift to shift, day to day, and week to week.

The procedure is to select three to five consecutive units, plot the highest and lowest observed values of each piece, and draw a line between them. After a period of time, usually 1 h or less, the process is repeated until about 80% of the variation of the process is captured.

Another approach to analyzing many variables utilizes Hotelling's T^2 Statistic. This statistic consolidates all the information in a multivariate observation and reduces it to a single value. It is not only a function of how far the observation is from the mean, but of how the variables relate to one another. In addition to control charting with an UCL (the LCL $= 0$), the amount each variable contributes to the T^2 statistic is calculated. The reader is referred to the reference for additional information on this valuable technique.[3]

SHORT-RUN SPC

Introduction

In many processes, the run is completed before the central line and control limits can be calculated. This fact is especially true for a job shop with small lot sizes. Furthermore, as companies practice just-in-time production, short runs are becoming more common.

Possible solutions to this problem are basing the chart on specifications, deviation chart, $\overline{Z}$ and W charts, Z and

[3]Robert L. Mason and John C. Young, "Another Data Mining Tool," *Quality Progress* (February 2003): 76–79.

MW charts, precontrol, and percent tolerance precontrol. This section discusses these charting techniques.

Specification Chart

A specification chart gives some measure of control and a method of quality improvement. The central line and the control limits are established using the specifications.

Assume that the specifications call for 25.00 ± 0.12 mm. Then the central line $\overline{X}_0 = 25.00$. The difference between the upper specification and the lower specification (USL − LSL) is 0.24 mm, which is the spread of the process under the Case II situation ($C_p = 1.00$). Thus,

$$C_p = \frac{\text{USL} - \text{LSL}}{6\sigma}$$

$$\sigma = \frac{\text{USL} - \text{LSL}}{6C_p} = \frac{25.12 - 24.88}{6(1.00)} = 0.04$$

Figure 7-6 shows the relationship between the tolerance (USL − LSL) and the process capability for the Case II situation, which was described in Chapter 6. Thus, for $n = 4$,

$$\text{URL}_{\overline{X}} = \overline{X}_0 + A\sigma = 25.00 + 1.500(0.04) = 25.06$$
$$\text{LRL}_{\overline{X}} = \overline{X}_0 - A\sigma = 25.00 - 1.500(0.04) = 24.94$$
$$R_0 = d_2\sigma = (2.059)(0.04) = 0.08$$
$$\text{URL}_R = D_2\sigma = (4.698)(0.04) = 0.19$$
$$\text{LRL}_R = D_1\sigma = (0)(0.04) = 0$$

These limits represent what we would like the process to do (as a maximum condition) rather than what it is capable of doing. Actually, these limits are reject limits as discussed in the previous chapter; however, the method of calculation is slightly different.

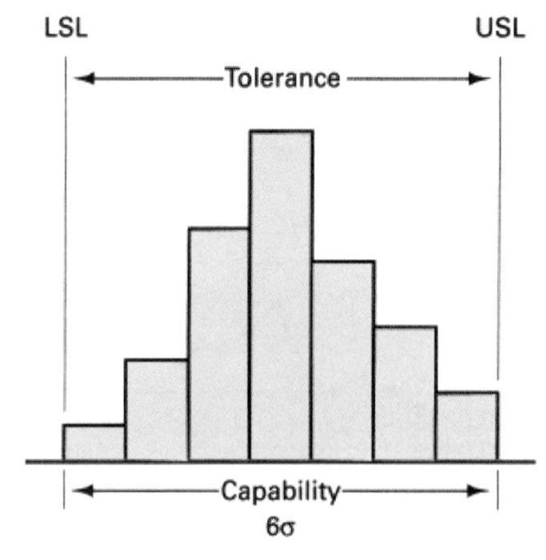

FIGURE 7-6 Relationship of Tolerance and Capability for the Case II Situation

We now have a chart that is ready to use for the first piece produced. Interpretation of the chart is the difficult part. Figure 7-7 shows the plotted point pattern for three situations: in (a) is the Case II situation that was used to determine the reject limits. If the process has a $C_p = 1.00$, the plotted points will form a normal curve within the limits. If the process is quite capable, as illustrated in (b), with a $C_p = 1.33$, then the plotted points will be compact about the central line. The most difficult interpretation occurs in (c), where the process is not capable. For example, if a plotted point falls outside the limits, it could be due to an assignable cause or due to the process not being capable. Because the actual C_p value is unknown until there are sufficient plotted points, personnel need to be well trained in process variation. They must closely observe the pattern to know when to adjust and when not to adjust the machine.

Deviation Chart

Figure 7-8 shows a deviation chart for individuals (*X*'s). It is identical to an *X* chart (see Chapter 6), except the plotted

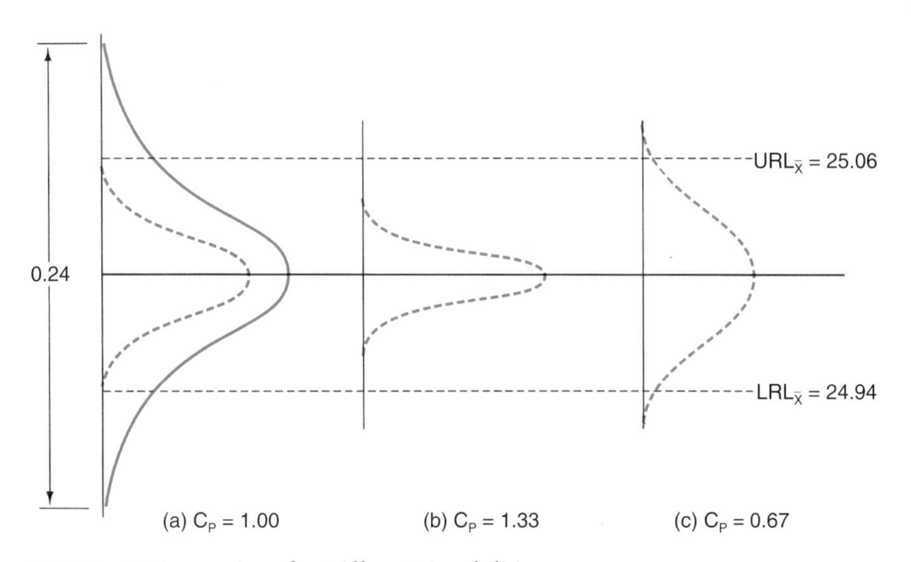

FIGURE 7-7 Comparison for Different Capabilities

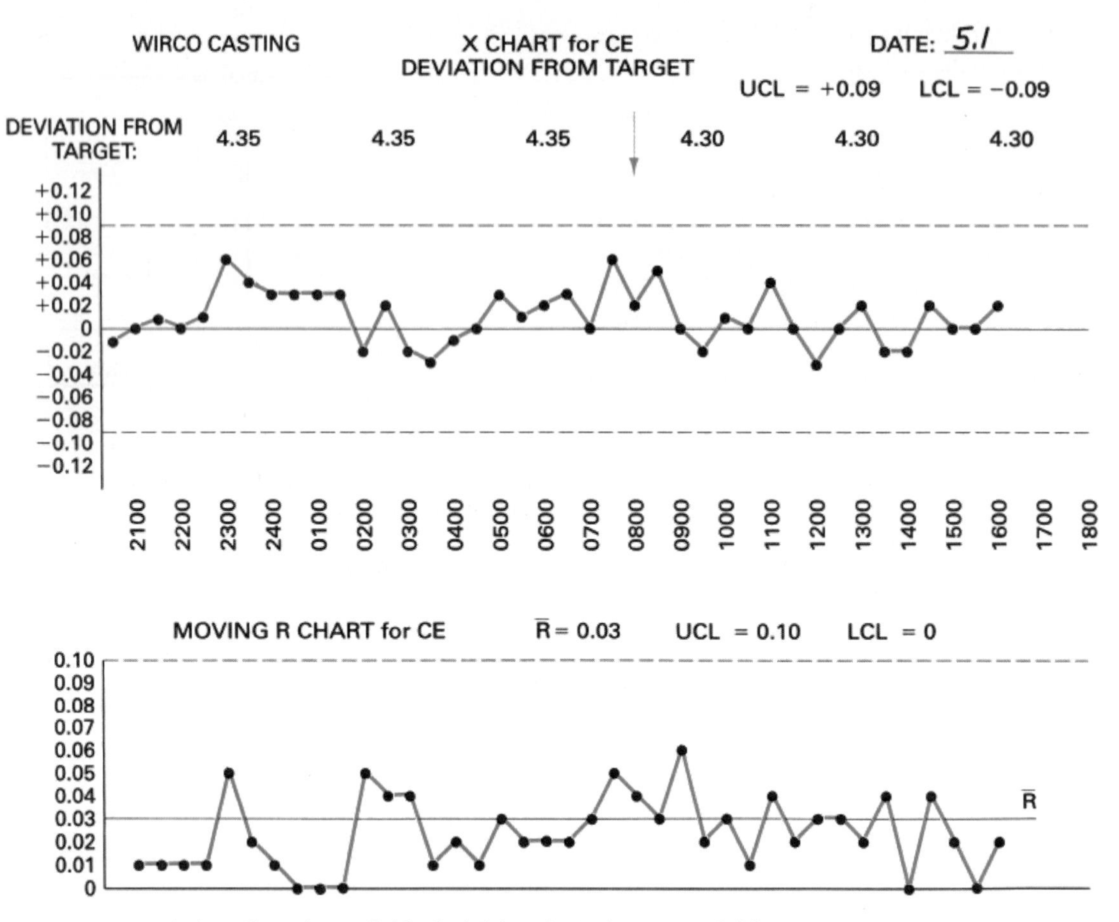

FIGURE 7-8 Deviation Chart for Individuals (X's) and Moving Range (R's)

point is the deviation from the target. For example, at time 0130, a test was taken for the carbon equivalent (CE) of iron melt, with a resulting value of 4.38. The target is 4.35; therefore, the deviation is $4.38 - 4.35 = 0.03$. This value is posted to the chart. There is no change to the R chart; it still uses the moving-range technique.

Even though the target changes, the central line for the X chart is always zero (0). Therefore, the chart can accommodate short runs with different targets. The figure shows that the CE target changes from 4.35 to 4.30. Use of this technique requires the variances (s^2) of the different targets or nominals to be identical. This requirement is verified by an analysis of variance (ANOVA) or using the following rule of thumb:

$$\frac{\overline{R}_{\text{Process}}}{\overline{R}_{\text{Total}}} \leq 1.3$$

where $\overline{R}_{\text{Process}}$ = average range of the process
$\overline{R}_{\text{Total}}$ = average range for all of the processes

Example Problem 7-2

The average range for all of the iron melt processes with different CE targets is 0.03. For the process with a targeted CE value of 4.30, the average range is 0.026. Can this process use the deviation technique? What about the process for a targeted CE value of 4.40 with an average range of 0.038?

$$\frac{\overline{R}_{4.30}}{\overline{R}_{\text{Total}}} = \frac{0.026}{0.03} = 0.87 \quad \text{(ok)}$$

$$\frac{\overline{R}_{4.40}}{\overline{R}_{\text{Total}}} = \frac{0.038}{0.03} = 1.27 \quad \text{(ok)}$$

The deviation technique is also applicable to $\overline{X}$ and R charts. Data are collected as deviations from the target; otherwise the technique is the same as discussed in Chapter 6.

Example Problem 7-3

A lathe turns rough diameters between 5 mm and 50 mm and runs last less than 2 h. Material and depth of cut do not change. Determine the central line and control limits. Data are

Sub Group	Target	X_1	X_2	X_3	X_4	$\overline{X}$	R
1	28.500	0	+.005	−.005	0	0	.010
⋮	⋮	⋮	⋮	⋮	⋮	⋮	⋮
15	45.000	0	−.005	0	−.005	−.0025	.005
⋮	⋮	⋮	⋮	⋮	⋮	⋮	⋮
25	17.000	+.005	0	0	+.005	+.0025	.005
					Σ	+.020	.175

$$\overline{\overline{X}} = \frac{\Sigma \overline{X}}{g} = \frac{0.020}{25} = 0.0008$$

(Note: $\overline{X}_0 = 0$, because the central line must be zero.)

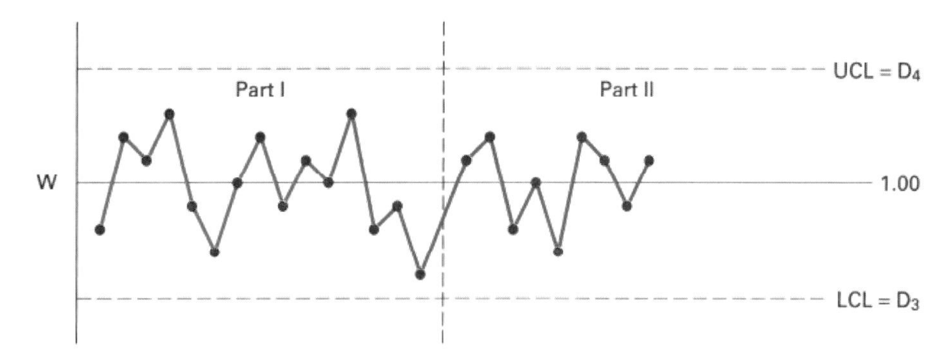

FIGURE 7-9 *W* Chart

$$\overline{R} = \frac{\Sigma R}{g} = \frac{0.175}{25} = 0.007$$

$$UCL_{\overline{X}} = \overline{X}_0 + A_2\overline{R} = 0 + 0.729(0.007) = +0.005$$

$$LCL_{\overline{X}} = \overline{X}_0 - A_2\overline{R} = 0 - 0.729(0.007) = -0.005$$

$$UCL_R = D_4\overline{R} = 2.282(0.007) = 0.016$$

$$LCL_R = D_3\overline{R} = 0(0.007) = 0$$

Deviation charts are also called difference, nominal, or target charts. The disadvantage of this type of chart is the requirement that the variation from process to process be relatively constant. If the variation is too great, as judged by the rule of thumb discussed earlier in this section, then a $\overline{Z}$ or Z chart can be used.

The Boeing Company used a modified deviation chart for process control of a tapered part. On the tapered part the thickness was measured in 13 locations, and although the nominal changed, the tolerance at each location was the same. Thus, each part consisted of a subgroup of 13. From this data, an $\overline{X}$ deviation chart and an s chart were constructed. A normal probability plot of the deviations did not reject the normality assumption.[4]

$\overline{Z}$ and W Charts

Z and W charts are very good for short runs. The central line and control limits are derived from the traditional formulas.

Looking at the R chart first, we have

R chart inequality	$LCL_R < R < UCL_R$
Substituting the formulas	$D_3\overline{R} < R < D_4\overline{R}$
Dividing by $\overline{R}$	$D_3 < \dfrac{R}{\overline{R}} < D_4$

Figure 7-9 shows the UCL $= D_4$ and the LCL $= D_3$. The central line is equal to 1.00 because it occurs when $R = \overline{R}$. This chart is called a W chart, and the plotted point is

$$W = \frac{R}{\text{Target } \overline{R}}$$

The control limits D_3 and D_4 are independent of $\overline{R}$; however, they are functions of n, which must be constant.

Looking at the $\overline{X}$ chart, we have

$\overline{X}$ chart inequality	$LCL_{\overline{X}} < \overline{X} < UCL_{\overline{X}}$
Substituting the formulas	$\overline{\overline{X}} - A_2\overline{R} < \overline{X} < \overline{\overline{X}} + A_2\overline{R}$
Subtract $\overline{\overline{X}}$	$-A_2\overline{R} < \overline{X} - \overline{\overline{X}} < +A_2\overline{R}$
Divide by $\overline{R}$	$-A_2 < \dfrac{\overline{X} - \overline{\overline{X}}}{\overline{R}} < +A_2$

Figure 7-10 shows UCL $= +A_2$ and LCL $= -A_2$. The central line is equal to 0.0 because it occurs when $\overline{X} - \overline{\overline{X}} = 0$, which is the perfect situation. This chart is called a $\overline{Z}$ chart, and the plotted point is

$$\overline{Z} = \frac{(\overline{X} - \text{Target } \overline{\overline{X}})}{\text{Target } \overline{R}}$$

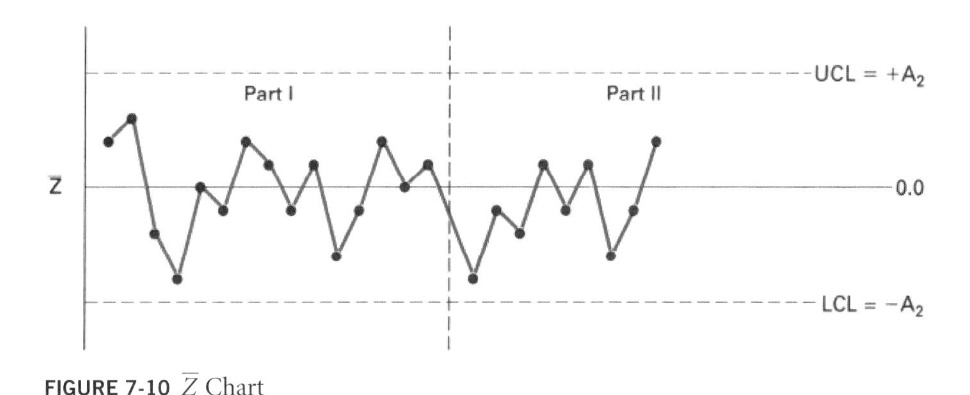

FIGURE 7-10 $\overline{Z}$ Chart

[4]S. K. Vermani, "Modified Nominal/Target Control Charts—A Case Study in Supplier Development," *Quality Management Journal*, Vol. 10, No. 4, ©2003, American Society for Quality.

The control limits $+A_2$ and $-A_2$ are independent of $\overline{R}$; however, they are functions of n, which must be constant.

Target $\overline{\overline{X}}$ and $\overline{R}$ values for a given part are determined by

1. Prior control charts
2. Historical data:
   a. Target $\overline{\overline{X}} = \Sigma\overline{X}/m$
      where m = number of measurements
   b. Target $\overline{R} = s(d_2/c_4)$
      where s = sample standard deviation for m
      d_2 = factor for central line ($\overline{R}$) for n
      c_4 = factor for central line (s) for m
3. Prior experience on similar part numbers
4. Specifications[5]
   a. Target $\overline{\overline{X}}$ = Nominal print specification
   b. Target $\overline{R} = \dfrac{d_2(\text{USL} - \text{LSL})}{6C_p}$

Example Problem 7-4

Determine the central lines and control limits for a $\overline{Z}$ and W chart with a subgroup size of 3. If the target $\overline{\overline{X}}$ is 4.25 and the target $\overline{R}$ is .10, determine the plotted points for three subgroups.

Subgroup	X_1	X_2	X_3	$\overline{X}$	R
1	4.33	4.35	4.32	4.33	.03
2	4.28	4.38	4.22	4.29	.16
3	4.26	4.23	4.20	4.23	.06

From Table B in the Appendix, for $n = 3$, $D_3 = 0$, $D_4 = 2.574$, and $A_2 = 1.023$,

$$\overline{Z}_1 = \frac{\overline{X} - \text{Target}\,\overline{\overline{X}}}{\text{Target}\,\overline{R}} = \frac{4.33 - 4.25}{0.10} = +0.80$$

$$\overline{Z}_2 = \frac{4.29 - 4.25}{0.10} = +0.40$$

$$\overline{Z}_3 = \frac{4.23 - 4.25}{0.10} = -0.20$$

$$W_1 = \frac{R}{\text{Target}\,\overline{R}} = \frac{0.03}{0.10} = 0.3$$

$$W_2 = \frac{0.16}{0.10} = 1.6$$

$$W_3 = \frac{0.06}{0.10} = 0.6$$

[5]*SPC for Short Production Runs*, prepared for U.S. Army Armament Munitions and Chemical Command by Davis R. Bothe, International Quality Institute, Inc., 1988.

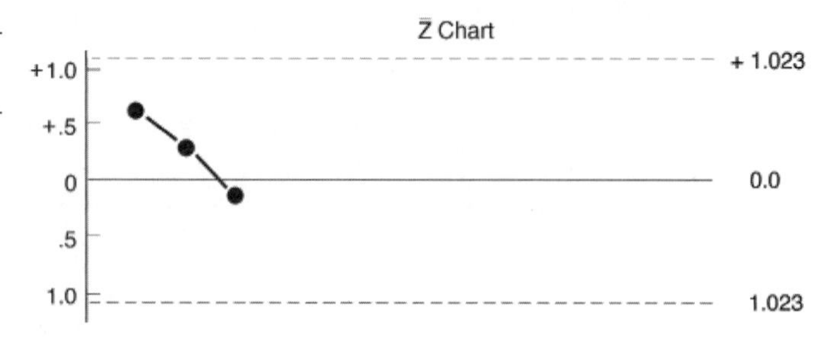

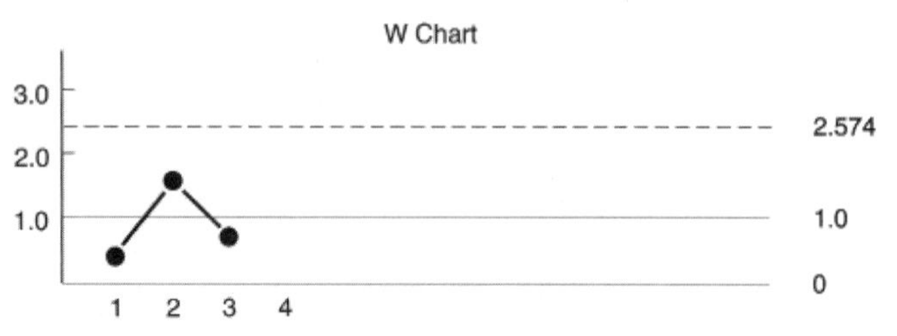

In addition to the advantage of short runs, the $\overline{Z}$ and W charts provide the opportunity to convey enhanced information. On these same charts we can plot

1. Different quality characteristics such as length and width
2. An operator's daily performance
3. The entire part history, thereby providing statistical evidence of the quality for the customer

It must be remembered, however, that the subgroup size must remain constant. The basic disadvantage is the fact that the plotted points are more difficult to calculate.

Z and *MW* Charts

The traditional $\overline{X}$ and R charts have their counterparts in $\overline{Z}$ and W charts. The X (individual) and MR charts, where MR is the moving range of the X values, have their counterparts in Z and MW charts, where MW is the moving range of the Z values. The concept is the same.

Figure 7-11 gives the control limits for the Z and MW charts. These limits and central lines are always the values shown in the figure. The limits are based on a moving range of Z as explained in Chapter 6, under the topic "Chart for Individual Values." The plot points for the Z and MW charts are

$$Z = \frac{X - \text{Target}\,\overline{X}}{\text{Target}\,\overline{R}}$$

$$\text{MW}_{i+1} = Z_i - Z_{i+1}$$

The derivation of the Z and MW chart is left as an exercise. Targets $\overline{X}$ and $\overline{R}$ are found in the same manner as explained in the previous section. Also, the information about including multiple information on the same chart is the same as for $\overline{Z}$ and W charts. The MW chart uses the absolute value.

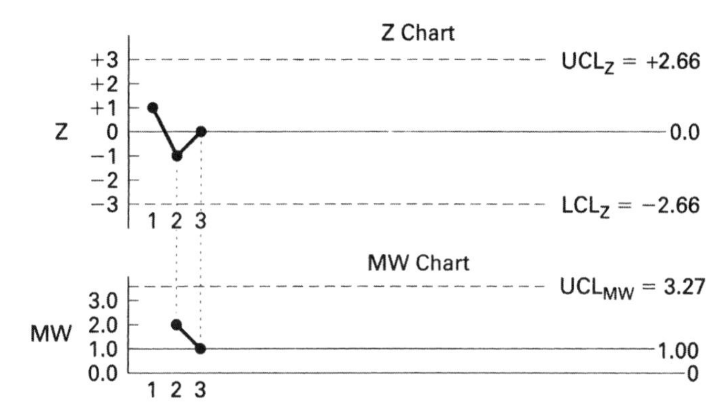

FIGURE 7-11 Central Lines and Control Limits for Z and MW Charts

Example Problem 7-5

Graph the Z and MW charts and plot the points for a target $\overline{X}$ of 39.0 and a target $\overline{R}$ of 0.6. Individual values for four subgroups are 39.6, 40.5, 38.2, and 39.0.

$$Z_1 = \frac{X - \text{Target}\,\overline{X}}{\text{Target}\,\overline{R}} = \frac{39.6 - 39.0}{0.6} = +1.00$$

$$Z_2 = \frac{X - \text{Target}\,\overline{X}}{\text{Target}\,\overline{R}} = \frac{40.5 - 39.0}{0.6} = +2.50$$

$$Z_3 = \frac{X - \text{Target}\,\overline{X}}{\text{Target}\,\overline{R}} = \frac{38.2 - 39.0}{0.6} = -1.33$$

$$Z_4 = \frac{X - \text{Target}\,\overline{X}}{\text{Target}\,\overline{R}} = \frac{39.0 - 39.0}{0.6} = 0$$

$$\text{MW}_2 = |Z_1 - Z_2| = |1.00 - 2.50| = 1.50$$

$$\text{MW}_3 = |Z_2 - Z_3| = |2.50 - (-1.33)| = 3.83$$

$$\text{MW}_4 = |Z_3 - Z_4| = |-1.33 - 0| = 1.33$$

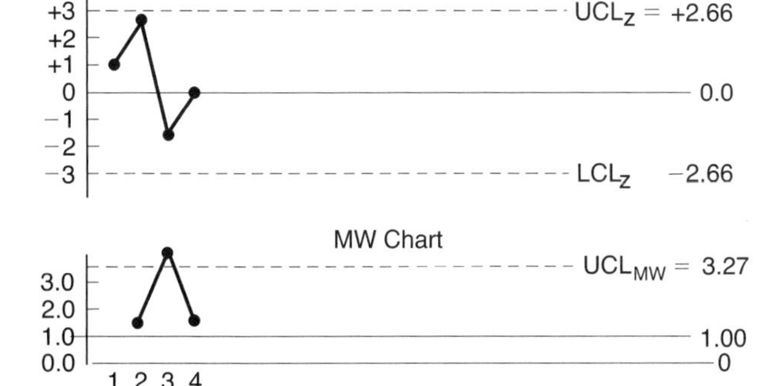

Precontrol

Control charts for variables, notably $\overline{X}$ and R charts, are excellent for problem solving. They do, however, have certain disadvantages when used by operating personnel to monitor a process after a project team has improved the process:

On short runs, the process is often completed before the operators have time to calculate the limits.

Operators may not have the time or ability to make the necessary calculations.

Frequently, operators are confused about specifications and control limits. This fact is especially true when a process is out of control but waste is not being produced.

Precontrol corrects these disadvantages as well as offers some advantages of its own.

The first step in the process is to be sure that the process capability is less than the specifications. Therefore, a capability index, C_p, of 1.00 or more, preferably more, is required. It is management's responsibility to ensure that the process is capable of meeting the specifications. Next, precontrol (PC) lines are established to divide the tolerance into five zones as shown in Figure 7-12(a). These PC lines are located halfway between the nominal value and the outside limits of the tolerance as given by the USL for upper specifications and the LSL for lower specifications. The center zone is one-half the print tolerance and is called the green area. On each side are the yellow zones, and each amounts to one-fourth of the total tolerance. Outside the specifications are the red zones. The colors make the procedure simple to understand and apply.

For a specification of 3.15 ± 0.10 mm, the calculations are

1. Divide tolerance by 4: $0.20/4 = 0.05$
2. Add value to lower specification, 3.05:

$$\text{PC} = 3.05 + 0.05 = 3.10$$

3. Subtract value from upper specification, 3.25:

$$\text{PC} = 3.25 - 0.05 = 3.20$$

Thus, the two PC lines are located at 3.10 and 3.20 mm. These values are shown in (b).

The statistical foundation of precontrol is shown in Figure 7-12(b). First, the process capability is equal to the specifications and is centered as indicated by $C_p = 1.00$ and $C_{pk} = 1.00$. For a normal distribution, 86% of the parts (12 out of 14) will fall between the PC lines, which is the green zone, and 7% of the parts (1 out of 14) will fall between the PC line and the specifications, which are the two yellow zones. As the capability index increases, the

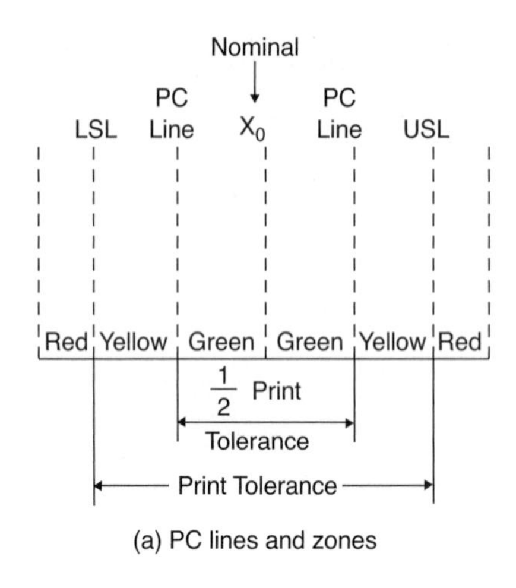

(a) PC lines and zones

(b) Probability when $C_p = 1.00$ and $C_{pk} = 1.00$

FIGURE 7-12 Precontrol Lines

chance of a part falling in the yellow zone decreases. Also, with a large capability index ($C_p = 1.33$ is considered a de facto standard), distributions that depart from normal are easily accommodated.

The precontrol procedure has two stages: startup and run. These stages are shown in Figure 7-13. One part is checked, and the results can fall in one of the three color zones. If the part is outside specifications (red zone), the process is stopped and reset. If the part is between the PC lines and the specifications (yellow zone), a second part is tested; if the second part is in the yellow zone, the process is stopped and reset. If the part falls between the PC lines (green zone), additional parts are tested until five consecutive parts are in the green zone. Operators become quite adept at "nudging" the setting when a reset is required.

Once there are five consecutive parts in the green zone, the run, or frequency testing, stage commences. Frequency testing is the evaluation of pairs of parts. The frequency rule is to sample six pairs between adjustments, and Table 7-2 gives the time between measurements for various adjustment frequencies. As can be seen from the table, there is a linear relationship between the two variables. Thus, if, on the average, an adjustment is made every 6 h, the time between measurement of pairs is 60 min. The time between adjustments is determined by the operator and supervisor based on historical information.

Figure 7-14 shows the decision rules for the measured pairs (designated A, B) for the different color zone possibilities:

1. When a part falls in the red zone, the process is shut down, reset, and the procedure returned to the startup stage.
2. When an A, B pair falls in opposite yellow zones, the process is shut down and help is requested, because this may require a more sophisticated adjustment.
3. When an A, B pair falls in the same yellow zone, the process is adjusted and the procedure returned to the startup stage.
4. When one or both A and B fall in the green zone, the process continues to run.

On the right side of the figure is the probability that a particular A, B pair will occur.

Precontrol is made even easier to use by painting the measuring instrument in green, yellow, and red at the

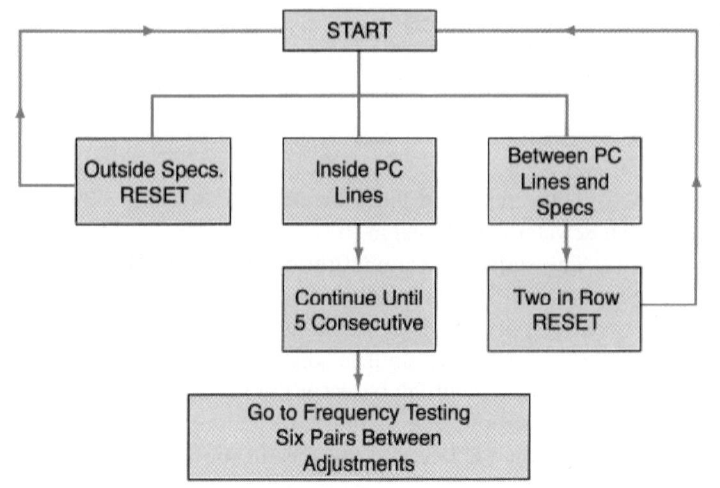

FIGURE 7-13 Precontrol Procedure

TABLE 7-2 Frequency of Measuring

Time Between Adjustments, H	Time Between Measurement, Min
1	10
2	20
3	30
4	40
.	.
.	.
.	.

	Color Zones					
Decision	**Red**	**Yellow**	**Green**	**Yellow**	**Red**	**Probability**
Stop, Go to	A					nil
Start-up					A	nil
Stop, Get		A		B		1/14 * 1/14 = 1/196
help		B		A		1/14 * 1/14 = 1/196
Adjust, Go to		A,B				1/14 * 1/14 = 1/196
Start-up				A,B		1/14 * 1/14 = 1/196
			A,B			12/14 * 12/14 = 144/196
		A	B			1/14 * 12/14 = 12/196
Continue		B	A			1/14 * 12/14 = 12/196
			A	B		12/14 * 1/14 = 12/196
			B	A		12/14 * 1/14 = 12/196
						Total = 196/196

LSL PC X_0 PC USL

Nominal (Target)

FIGURE 7-14 Run Decision and Probability

appropriate places. In this way, the operator knows when to go, apply caution, or stop.

Precontrol can be used for single specifications, as shown by Figure 7-15. In these cases the green area is established at three-fourths of the print tolerance. The figure shows a skewed distribution, which would most likely occur with an out-of-round characteristic when the target is zero.

Precontrol can also be used for attributes. Appropriately colored "go/no-go" gauges that designate the precontrol lines are issued to the operator along with the usual gauges for the upper and lower specifications. Precontrol is also used for visual characteristics by assigning visual standards for the PC lines.

The advantages of precontrol are as follows:

1. It is applicable to short production runs as well as long production runs.
2. No recording, calculating, or plotting of data is involved. A precontrol chart can be used if the consumer desires statistical evidence of process control (see Figure 7-16).
3. It is applicable to startup, so the process is centered on the target.
4. It works directly with the tolerance rather than easily misunderstood control limits.
5. It is applicable to attributes.
6. It is simple to understand, so training is very easy.

Although the precontrol technique has a lot of advantages, we must remember that it is only a monitoring technique. Control charts are used for problem solving, because they have the ability to improve the process by correcting assignable causes and testing improvement ideas. Also, the control chart is more appropriate for process capability and detecting process shifts.

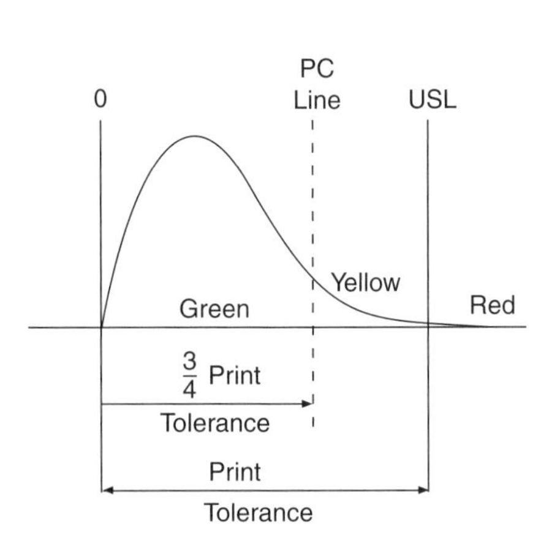

FIGURE 7-15 Precontrol for Single Specifications

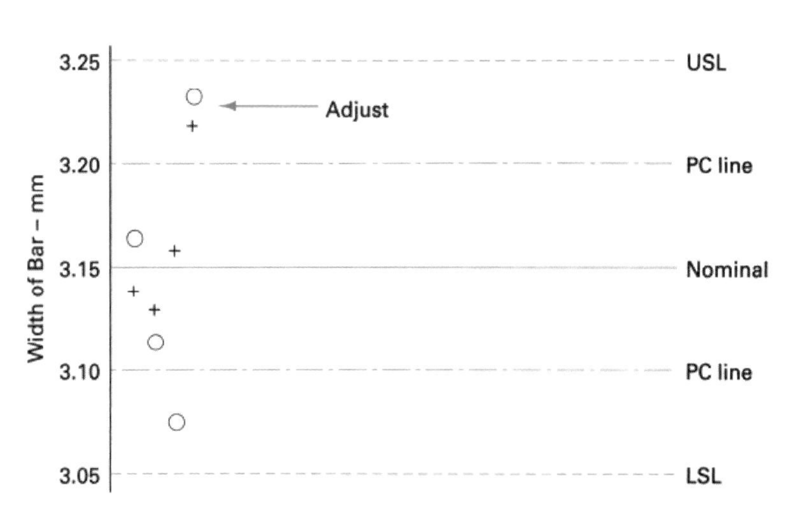

FIGURE 7-16 Precontrol Chart

In summary, precontrol means better management and operator understanding, operator responsibility for quality, reduced rejects, reduced adjustments, reduced operator frustration, and a subsequent increase in morale. These benefits have been realized for many different types of processes.

Percent Tolerance Precontrol Chart[6]

The concept of Z charts, with its ability to accommodate more than one quality characteristic, and precontrol, with its simplicity, can be combined into one technique by use of a percent tolerance precontrol chart (PTPCC). Recall that the plotted point of the Z chart is

$$Z = \frac{X - \text{Target } \overline{X}}{\text{Target } \overline{R}}$$

Using a similar logic, an individual measurement, X, can be transformed to percent deviation from the target or nominal by

$$X^* = \frac{X - \text{Nominal}}{(\text{USL} - \text{LSL})/2}$$

where X^* = deviation from nominal as a percent (decimal) of tolerance

$(\text{USL} - \text{LSL})/2$ = one-half the print tolerance and is the target $\overline{R}$ for the precontrol concept

Some examples will show the use of the formula:

1. The specification of part number 1234 is 2.350 ± 0.005 and an inspected measured value is 2.3485.

$$X^* = \frac{X - \text{nominal}}{(\text{USL} - \text{LSL})/2} = \frac{(2.3485 - 2.350)}{(2.355 - 2.245)/2} = -0.3 \text{ or } -30\%$$

2. The specification of part number 5678 is 0.5000 ± 0.0010 and an inspected measured value is 0.4997.

$$X^* = \frac{X - \text{nominal}}{(\text{USL} - \text{LSL})/2} = \frac{0.4997 - 0.5000}{(0.5010 - 0.4990)/2} = -0.3 \text{ or } -30\%$$

3. The specification of part number 1234 is 2.350 ± 0.005 and an inspected measured value is 2.351.

$$X^* = \frac{X - \text{Nominal}}{(\text{USL} - \text{LSL})/2} = \frac{(2.351 - 2.350)}{(2.355 - 2.345)/2} = 0.2 \text{ or } 20\%$$

Note that both Example 1 and Example 2 have the same percent deviation (−30%) from the nominal, even though the tolerance is much different. The negative value indicates that the observed value is below the nominal. Comparing Example 1 and Example 3, it is seen that while both have the same nominal and tolerance, Example 1 is 30% below the nominal and Example 3 is 20% above the nominal.

A spreadsheet can be set up to calculate the values and generate the plotted points on the PTPCC. Table 7-2 shows the calculations and Figure 7-17 shows the plotted points. Two parts are shown on the PTPCC; however, there could be as many parts or part features as space permits. Each part or part feature can have different nominal values and tolerances. Different parts can be plotted on the same chart. In fact, the entire part history of its different operations can be plotted on one chart.

Percent Tolerance Precontrol Chart Data

Machine Number: *Mill* 21 — Date *24 JAN*

Part No.	Time	USL	LSL	Nominal	Part 1	Part 2	Plot 1 (%)	Plot 2 (%)	Status
1234	0800 h	2.355	2.345	2.350	2.3485	2.3510	−30.0	20.0	ok
	0830 h	2.355	2.345	2.350	2.3480	2.3490	−40.0	−20.0	ok
	0900 h	2.355	2.345	2.350	2.3500	2.3530	0.0	60.0	ok (1Y)
5678	1000 h	0.5010	0.4990	0.5000	0.4997	0.5002	−30.0	20.0	ok
	1030 h	0.5010	0.4990	0.5000	0.5006	0.4997	60.0	−30.0	ok (1Y)
	1100 h	0.5010	0.4990	0.5000	0.5000	0.5003	0.0	30.0	ok
	1130 h	0.5010	0.4990	0.5000	0.4994	0.4992	−60.0	−80.0	2Y

TABLE 7-2 Calculations for Percent Tolerance Precontrol Chart

[6] S. K. Vermani, "SPC Modified with Percent Tolerance Precontrol Charts," *Quality Progress* (October 2000): 43–48.

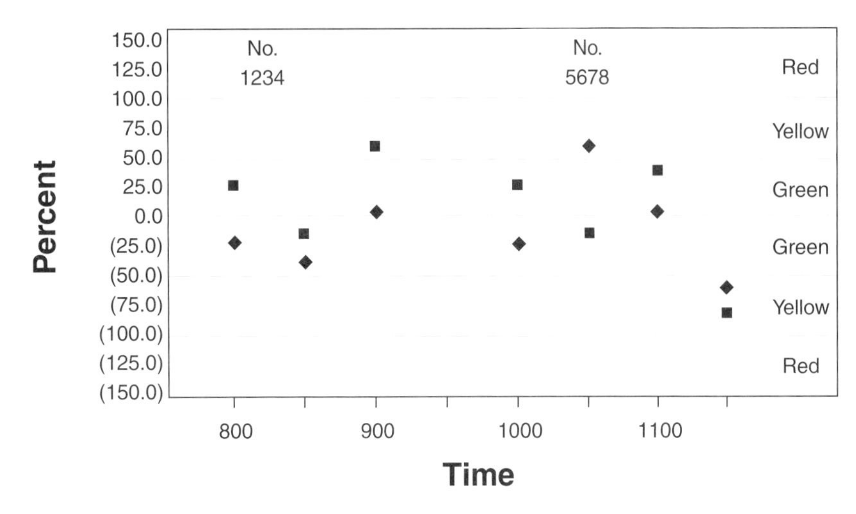

FIGURE 7-17 Percent Tolerance Precontrol Chart

The rules of control, which were defined in the previous section, are applicable. A review of this particular PTPCC shows that the process went out of control at 1130 h with two readings in the yellow zone.

GAUGE CONTROL

SPC requires accurate and precise data (see the section on data collection in Chapter 4); however, all data have measurement errors. Thus, an observed value has two components:

$$\text{Observed Value} = \text{True Value} + \text{Measurement Error}$$

The previous chapter discussed variation that occurs due to the process and the measurement, thus

$$\begin{array}{c}\text{Total}\\ \text{Variation}\end{array} = \begin{array}{c}\text{Product}\\ \text{Variation}\end{array} + \begin{array}{c}\text{Measurement}\\ \text{Variation}\end{array}$$

Measurement variation is further divided into repeatability, which is due to equipment variation, and reproducibility, which is due to appraiser (inspector) variation. It is called GR&R, or gauge repeatability and reproducibility.

Before we perform tests to evaluate GR&R, it is necessary to calibrate the gauge. Calibration must be performed either in-house or by an independent laboratory. It must be accomplished so that it is traceable to reference standards of known accuracy and stability, such as those of the National Institute for Standards and Technology (NIST). For industries or products in which such standards do not exist, the calibration must be traceable to developed criteria.

Analysis of variance (ANOVA), which is discussed in Chapter 13, evaluates the measuring system to determine its viability. The technique determines the amount of variation due to repeatability and due to reproducibility, and combines those values for the measurement variation. A P/T ratio, which compares the precision (P) or measurement variation to the total (T) variation of the process is determined. If the P/T ratio is low, then the effect on the process measurements is small; however, if the ratio is large then measurement variation will have too great an influence on the process measurements. A P/T ratio less than 0.1 is acceptable and a ratio greater than 0.3 is not acceptable. In the later case, the process may produce parts that are declared good when they are actually bad or parts that are declared bad when they are actually good.

The number of parts, appraisers, or trials can vary, but 10 parts, 2 or 3 appraisers, and 2 or 3 trials are considered optimum. Readings are taken by randomizing the part sequence on each trial. For example, on the first trial, each appraiser would measure the part characteristic in the following order: 4, 7, 5, 9, 1, 6, 2, 10, 8, and 3. On the second trial, the order might be 2, 8, 6, 4, 3, 7, 9, 10, 1, and 5. A table of random numbers, as given in Table D in the Appendix, should be used to determine the order on each trial.

Evaluation

If repeatability is large compared to reproducibility, the reasons may be that:

1. The gauge needs maintenance.
2. The gauge should be redesigned to be more rigid.
3. The clamping or location for gaging needs to be improved.
4. There is excessive within-part variation.

If reproducibility is large compared to repeatability, the reasons may be that:

1. The operator needs to be better trained in how to use and read the gauge.

2. Calibrations on the gauge are not legible.
3. A fixture may be needed to help the operator use the gauge consistently.

Guidelines for acceptance of the P/T ratio are—

Under 0.1	Gauge system is satisfactory.
.1 to 0.3	May be acceptable based on the importance of application, cost of gauge, cost of repairs, etc.
Over 0.3	Gauge system is not satisfactory. Identify the causes and take corrective action.

The information given above provides the basic concept. A major modification is given in the literature that adjusts for small sample sizes.[7] Another situation compares testing equipment at two locations.[8]

Environmental conditions such as temperature, humidity, air cleanliness, and electric discharge can influence the GR&R results and should be controlled to minimize variation.

COMPUTER PROGRAM

The EXCEL program files on the website will solve for the $\overline{Z}$ and *W* charts and for the PTPC chart. Their file names are *Z-Bar & W Charts* and *PTPC chart.*

[7] Donald S. Ermer, "Appraiser Variation in Gauge R&R Measurement," *Quality Progress* (May 2006): 75–78.

[8] Neal D. Morchower, "Two-Location Gauge Evaluation," *Quality Progress* (April 1999): 79–86.

EXERCISES

1. Determine the number of subgroups needed to establish the central lines and control limits for $\overline{X}$ and *R* charts with a subgroup size of 2 for an 8-spindle filling machine. How many times in succession can one of the spindles be plotted?
2. Determine the number of subgroups needed to establish the central lines and control limits for *X* and moving *R* charts for a paper towel process that is controlled by 24 valves. How many times in succession can one of the valves be plotted?
3. Given the following data in microinches for the surface roughness of a grinding operation, construct a multi-vari chart and analyze the results.

Time	0700 h			1400 h			2100 h		
Part No.	20	21	22	82	83	84	145	146	147
	38	26	31	32	19	29	10	28	14
Surface	28	08	30	25	29	09	05	11	15
Roughness	30	31	38	16	20	18	26	38	04
Measure	37	20	22	22	21	16	32	30	38
	39	44	35	30	28	24	29	38	10

4. Given below are data for a lathe turning operation. The part is 15 cm long and has a target diameter of 60.000 mm ± 0.012. One measurement is taken at each end and three consecutive parts are measured every 30 min. Construct a multi-vari chart and analyze the process. The data are in deviations from the target of 60.000, thus 60.003 is coded 3 and 59.986 is coded −14.

Time	Part 1	Part 2	Part 3
0800 h	−7/10	−2/13	9/18
0830 h	2/15	5/14	2/14
0900 h	0/14	3/15	−7/15
0930 h	−23/−5	−20/−6	−14/1
1000 h	−20/−8	−22/−7	−11/10
1030 h	−15/9	−18/6	−14/5
1100 h	−9/8	−13/4	−12/8
1130 h	−19/1	−14/9	−13/12

5. Determine the central line and limits for a short production run that will be completed in 3 h. The specifications are 25.0 ± 0.3 Ω. Use $n = 4$.
6. Determine the central line and limits for a short production run that will be completed in 1 h. The specifications are 3.40 ± .05 mm. Use $n = 3$.
7. A five-stage progressive die has four critical dimensions. $\overline{X}$ is 25.30, 14.82, 105.65, and 58.26 mm and $\overline{R}$ is 0.06, 0.05, 0.07, and 0.06 for the dimensions. Can a deviation chart be used? Let $\overline{R}_{Total}$ equal the average of the four range values.
8. Determine the central lines and control limits for $\overline{Z}$ and *W* charts with a subgroup size of 2 and draw the graphs. If the target $\overline{\overline{X}}$ is 1.50 and the target $\overline{R}$ is 0.08, determine the plotted points for three subgroups.

Subgroup	X_1	X_2
1	1.55	1.59
2	1.43	1.53
3	1.30	1.38

9. Determine the central lines and control limits for $\overline{Z}$ and *W* charts with a subgroup size of 3 and draw the graphs. If the target $\overline{\overline{X}}$ is 25.00 and target $\overline{R}$ is 0.05, determine the plotted points for the following three subgroups. Are any points out of control?

Subgroup	X_1	X_2	X_3
1	24.97	25.01	25.00
2	25.08	25.06	25.09
3	25.03	25.04	24.98

10. Using the information in the Chart for Individual Values (Chapter 6), derive the control limits and plotted points for the Z and MW charts.
11. Graph the central lines and control limits for Z and MW charts. The target $\overline{X}$ is 1.15, and the target $\overline{R}$ is 0.03. Plot the points for $X_1 = 1.20$, $X_2 = 1.06$, and $X_3 = 1.14$. Are there any out-of-control points?
12. Graph the central lines and control limits for Z and MW charts. The target $\overline{X}$ is 3.00, and the target $\overline{R}$ is 0.05. Plot the points for $X_1 = 3.06$, $X_2 = 2.91$, and $X_3 = 3.10$. Are any points out of control?
13. What are the PC lines for a process that has a nominal of 32.0°C and a tolerance of ±1.0°C?
14. Determine the PC line for the concentricity of a shaft when the total indicator reading (TIR) tolerance is 0.06 mm and the target is 0. *Hint:* This problem is a one-sided tolerance; however, the green zone is still half the tolerance. Graph the results.
15. What is the probability of an A, B pair being green? Of an A, B pair having one yellow and one green?
16. If a process is adjusted every 3 h, how often should pairs of parts be measured?
17. Determine the PTPC chart plot point for the following:

	USL	LSL	Nominal	Part 1	Part 2
(a)	0.460	0.440	0.450	0.449	0.458
(b)	1.505	1.495	1.500	1.496	1.500
(c)	1.2750	1.2650	1.2700	1.2695	1.2732
(d)	0.7720	0.7520	0.7620	0.7600	0.7590

18. Construct a PTPC chart for the data in Exercise 4. Use the average value for Part 1 and Part 2. For example, the data for 0800 h would be Part 1 $(59.993 + 60.010)/2 = 60.0015$ and Part 2 $(59.998 + 60.013)/2 = 60.0055$. Do not use Part 3.
19. Using the EXCEL program file for $\overline{Z}$ and W charts, solve
    a. Exercise 8
    b. Exercise 9
20. Develop a template for Z and MW charts. (*Hint:* Similar to X and MR charts.)

CHAPTER EIGHT

FUNDAMENTALS OF PROBABILITY

OBJECTIVES

Upon completion of this chapter, the reader is expected to

- define probability using the frequency definition;
- know the seven basic theorems of probability;
- identify the different discrete and continuous probability distributions;
- calculate the probability of nonconforming units occurring using the hypergeometric, binomial, and Poisson distributions;
- know when to use the hypergeometric, binomial, and Poisson distributions.

INTRODUCTION

This chapter covers the fundamentals of probability, including definition, theorems, discrete distributions, continuous distributions, and distribution interrelationship as they pertain to statistical quality control. In particular, the material covers essential information for an understanding of attribute control charts, which are covered in the next chapter.

BASIC CONCEPTS

Definition of Probability

The term *probability* has a number of synonyms, such as likelihood, chance, tendency, and trend. To the layperson, *probability* is a well-known term that refers to the chance that something will happen. "I will probably play golf tomorrow" or "I will probably receive an A in this course" are typical examples. When the commentator on the evening news states, "The probability of rain tomorrow is 25%," the definition has been quantified. It is possible to define probability with extreme mathematical rigor; however, in this text we will define probability from a practical viewpoint as it applies to quality control.

If a nickel is tossed, the probability of a head is $\frac{1}{2}$ and the probability of a tail is $\frac{1}{2}$. A die, which is used in games of chance, is a cube with six sides and spots on each side from one to six. When the die is tossed on the table, the likelihood or probability of one spot is $\frac{1}{6}$, the probability of two spots is $\frac{1}{6}$, . . . , the probability of six spots is $\frac{1}{6}$. Another example of probability is illustrated by the drawing of a card from a deck of cards. The probability of a spade is $\frac{13}{52}$, because there are 13 spades in a deck of cards that contains 52 total cards. For hearts, diamonds, and clubs, the other three suits in the deck, the probability is also $\frac{13}{52}$.

Figure 8-1 shows the probability distributions for the preceding examples. It is noted that the area of each distribution is equal to 1.000 ($\frac{1}{2} + \frac{1}{2} = 1.000$, $\frac{1}{6} + \frac{1}{6} + \frac{1}{6} + \frac{1}{6} + \frac{1}{6} + \frac{1}{6} = 1.000$, and $\frac{13}{52} + \frac{13}{52} + \frac{13}{52} + \frac{13}{52} = 1.000$). Recall that the area under the normal distribution curve, which is a probability distribution, is also equal to 1.000. Therefore, the total probability of any situation will be equal to 1.000. The probability is expressed as a decimal, such as (1) the probability of heads is 0.500, which is expressed in symbols as [$P(h) = 0.500$]; (2) the probability of a 3 on a die is 0.167 [$P(3) = 0.167$]; and (3) the probability of a spade is 0.250 [$P(s) = 0.250$].

The probabilities given in the preceding examples will occur provided sufficient trials are made and provided there is an equal likelihood of the events occurring. In other words, the probability of a head (the event) will be 0.500 provided that the chance of a head or a tail is equal (equally likely). For most coins, the equally likely condition is met; however, the addition of a little extra metal on one side would produce a biased coin and then the equally likely condition could not be met. Similarly, an unscrupulous person might fix a die so that a three appears more often than one out of six times, or he might stack a deck of cards so that all the aces were at the top.

Returning to the example of a six-sided die, there are six possible outcomes (1, 2, 3, 4, 5, and 6). An *event* is a collection of outcomes. Thus, the event of a 2 or 4 occurring on a throw of the die has two outcomes, and the total number of outcomes is 6. The probability is obviously $\frac{2}{6}$, or 0.333.

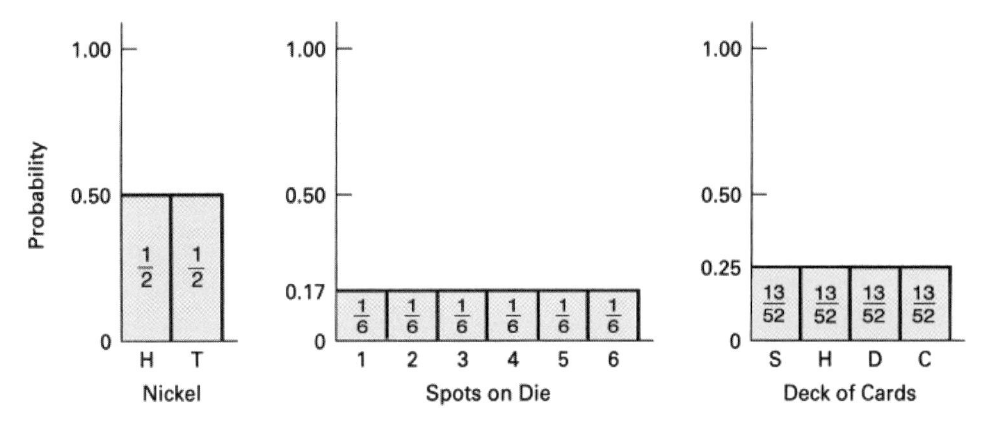

FIGURE 8-1 Probability Distributions

From the preceding discussion, a definition based on a frequency interpretation can be given. If an event A can occur in N_A outcomes out of a total of N possible and equally likely outcomes, then the probability that the event will occur is

$$P(A) = \frac{N_A}{N}$$

where $P(A)$ = probability of an event A occurring to three decimal places
N_A = number of successful outcomes of event A
N = total number of possible outcomes

This definition can be used when the number of outcomes is known or when the number of outcomes is found by experimentation.

Example Problem 8-1

A part is selected at random from a container of 50 parts that are known to have 10 nonconforming units. The part is returned to the container and a record of the number of trials and the number nonconforming is maintained. After 90 trials, 16 nonconforming units were recorded. What is the probability based on known outcomes and on experimental outcomes?

Known outcomes:

$$P(A) = \frac{N_A}{N} = \frac{10}{50} = 0.200$$

Experimental outcomes:

$$P(A) = \frac{N_A}{N} = \frac{16}{90} = 0.178$$

The probability calculated using known outcomes is the true probability, and the one calculated using experimental outcomes is different due to the chance factor. If, say, 900 trials were made, the probability using experimental outcomes would be much closer, because the chance factor would be minimized.

In most cases, the number nonconforming in the container would not be known; therefore, the probability with known outcomes cannot be determined. If we consider the probability using experimental outcomes to represent the sample and known outcomes to represent the population, there is the same relationship between sample and population that was discussed in Chapter 5.

The preceding definition is useful for finite situations where N_A, the number of successful outcomes, and N, total number of outcomes, are known or must be found experimentally. For an infinite situation, where $N = \infty$, the definition will always lead to a probability of zero. Therefore, in the infinite situation the probability of an event occurring is proportional to the population distribution. A discussion of this situation is given in the discussion of continuous and discrete probability distributions.

Theorems of Probability

Theorem 1. Probability is expressed as a number between 1.000 and 0, where a value of 1.000 is a certainty that an event will occur and a value of 0 is a certainty that an event will not occur.

Theorem 2. If $P(A)$ is the probability that event A will occur, then the probability that A will not occur, $P(P(A'))$, is $1.000 - P(A)$.

Example Problem 8-2

If the probability of finding an error on an income tax return is 0.04, what is the probability of finding an error-free or conforming return?

$$\begin{aligned} P(A') &= 1.000 - P(A) \\ &= 1.000 - 0.040 \\ &= 0.960 \end{aligned}$$

Therefore, the probability of finding a conforming income tax return is 0.960.

Before proceeding to the other theorems, it is appropriate to learn where they are applicable. In Figure 8-2 we see that if the probability of only one event is desired, then Theorem 3 or 4 is used, depending on whether the event is

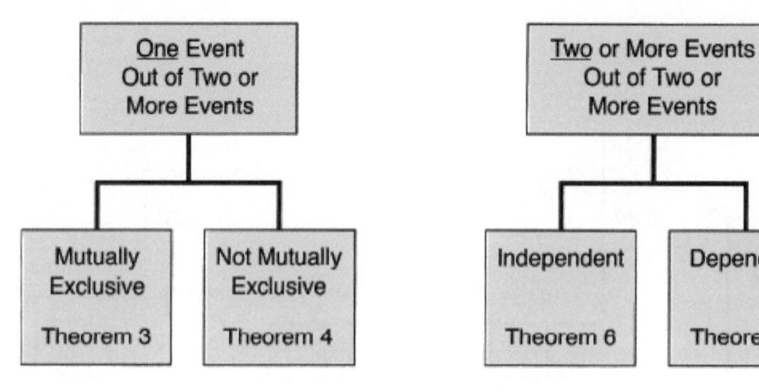

FIGURE 8-2 When to Use Theorems 3, 4, 6, and 7

TABLE 8-1 Inspection Results by Supplier

Supplier	Number Conforming	Number Nonconforming	Total
X	50	3	53
Y	125	6	131
Z	75	2	77
Total	250	11	261

mutually exclusive or not. If the probability of two or more events is desired, then Theorem 6 or 7 is used, depending on whether the events are independent or not. Theorem 5 is not included in the figure, because it pertains to a different concept. Table 8-1 provides the data for the example problems for Theorems 3, 4, 6, and 7.

Theorem 3. If A and B are two mutually exclusive events, then the probability that either event A or event B will occur is the sum of their respective probabilities:

$$P(A \text{ or } B) = P(A) + P(B)$$

Mutually exclusive means that the occurrence of one event makes the other event impossible. Thus, if on one throw of a die a 3 occurred (event A), then event B, say, a 5, could not possibly occur.

Whenever an "or" is verbalized, the mathematical operation is usually addition, or, as we shall see in Theorem 4, it can be subtraction. Theorem 3 was illustrated with two events—it is equally applicable for more than two $[P(A \text{ or } B \text{ or} \ldots \text{ or } F) = P(A) + P(B) + \cdots + P(F)]$.

Example Problem 8-3

If the 261 parts described in Table 8-1 are contained in a box, what is the probability of selecting a random part produced by supplier X or by supplier Z?

$$P(X \text{ or } Z) = P(X) + P(Z) = \frac{53}{261} + \frac{77}{261} = 0.498$$

What is the probability of selecting a nonconforming part from supplier X or a conforming part from supplier Z?

$$P(\text{nc. } X \text{ or co. } Z) = P(\text{nc. } X) + P(\text{co. } Z) = \frac{3}{261} + \frac{75}{261} = 0.299$$

Example Problem 8-4

If the 261 parts described in Table 8-1 are contained in a box, what is the probability that a randomly selected part will be from supplier Z, a nonconforming unit from supplier X, or a conforming part from supplier Y?

$$P(Z \text{ or nc. } X \text{ or co. } Y) = P(Z) + P(\text{nc. } X) + P(\text{co. } Y) = \frac{77}{261} + \frac{3}{261} + \frac{125}{261} = 0.785$$

Theorem 3 is frequently referred to as the *additive law of probability.*

Theorem 4. If event A and event B are not mutually exclusive events, then the probability of either event A or event B or both is given by

$$P(A \text{ or } B \text{ or both}) = P(A) + P(B) - P(\text{both})$$

Events that are not mutually exclusive have some outcomes in common.

Example Problem 8-5

If the 261 parts described in Table 8-1 are contained in a box, what is the probability that a randomly selected part will be from supplier X or a nonconforming unit?

$$P(X \text{ or nc. or both}) = P(X) + P(\text{nc.}) - P(X \text{ and nc.}) = \frac{53}{261} + \frac{11}{261} - \frac{3}{261} = 0.234$$

In Example Problem 8-5, there are three outcomes common to both events. The 3 nonconforming units of

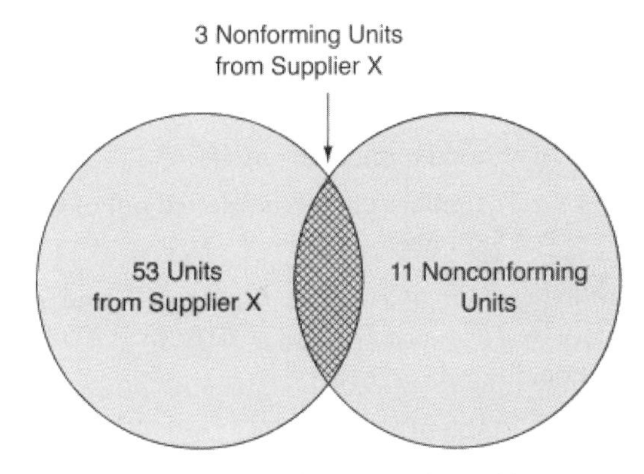

FIGURE 8-3 Venn Diagram for Example Problem 8-5

supplier X are counted twice as outcomes of $P(X)$ and of $P(\text{nc.})$; therefore, one set of three is subtracted out. This theorem is also applicable to more than two events. A Venn diagram is sometimes used to describe the not mutually exclusive concept, as shown in Figure 8-3. The circle on the left contains 53 units from supplier X and the circle on the right contains 11 nonconforming units. The 3 nonconforming units from supplier X are found where the two circles intersect.

Theorem 5. The sum of the probabilities of the events of a situation is equal to 1.000:

$$P(A) + P(B) + \cdots + P(N) = 1.000$$

This theorem was illustrated in Figure 8-1 for the coin-tossing, die-rolling, and card-drawing situations, in which the sum of the events equaled 1.000.

Example Problem 8-6

A health inspector examines 3 products in a subgroup to determine if they are acceptable. From past experience it is known that the probability of finding no nonconforming units in the sample of 3 is 0.990, the probability of 1 nonconforming unit in the sample of 3 is 0.006, and the probability of finding 2 nonconforming units in the sample of 3 is 0.003. What is the probability of finding 3 nonconforming units in the sample of 3?

There are 4, and only 4, events to this situation: 0 nonconforming units, 1 nonconforming unit, 2 nonconforming units, and 3 nonconforming units.

$$P(0) + P(1) + P(2) + P(3) = 1.000$$
$$0.990 + 0.006 + 0.003 + P(3) = 1.000$$
$$P(3) = 0.001$$

Thus, the probability of 3 nonconforming units in the sample of 3 is 0.001.

Theorem 6. If A and B are independent events, then the probability of both A and B occurring is the product of their respective probabilities:

$$P(A \text{ and } B) = P(A) \times P(B)$$

An independent event is one in which its occurrence has no influence on the probability of the other event or events. This theorem is referred to as the *multiplicative law of probability*. Whenever an "and" is verbalized, the mathematical operation is multiplication.

Example Problem 8-7

If the 261 parts described in Table 8-1 are contained in a box, what is the probability that 2 randomly selected parts will be from supplier X and supplier Y? Assume that the first part is returned to the box before the second part is selected (called *with replacement*)

$$P(X \text{ and } Y) = P(X) \times P(Y)$$
$$= \left(\frac{53}{261}\right)\left(\frac{131}{261}\right)$$
$$= 0.102$$

At first thought, the result of Example Problem 8-7 seems too low, but there are other possibilities, such as XX, YY, ZZ, YX, XZ, ZX, YZ, and ZY. This theorem is applicable to more than two events.

Theorem 7. If A and B are *dependent* events, the probability of both A and B occurring is the product of the probability of A and the probability that if A occurred, then B will occur also:

$$P(A \text{ and } B) = P(A) \times P(B|A)$$

The symbol $P(B|A)$ is defined as the probability of event B provided that event A has occurred. A dependent event is one whose occurrence influences the probability of the other event or events. This theorem is sometimes referred to as the *conditional theorem*, because the probability of the second event depends on the result of the first event. It is applicable to more than two events.

Example Problem 8-8

Assume that in Example Problem 8-7 the first part was not returned to the box before the second part was selected. What is the probability?

$$P(X \text{ and } Y) = P(X) \times P(Y|X)$$
$$= \left(\frac{53}{261}\right)\left(\frac{131}{260}\right)$$
$$= 0.102$$

Because the first part was not returned to the box, there was a total of only 260 parts in the box.

What is the probability of choosing both parts from supplier Z?

$$P(Z \text{ and } Z) = P(Z) \times P(Z|Z)$$
$$= \left(\frac{77}{261}\right)\left(\frac{76}{260}\right)$$
$$= 0.086$$

Because the first part was from supplier Z, there are only 76 from supplier Z of the new total of 260 in the box.

To solve many probability problems, it is necessary to use several theorems, as shown by Example Problem 8-9, which uses Theorems 3 and 6.

Example Problem 8-9

If the 261 parts described in Table 8-1 are contained in a box, what is the probability that two randomly selected parts (with replacement) will have one conforming part from supplier *X* and one conforming part from supplier *Y* or supplier *Z*?

$$P[\text{co. } X \text{ and (co. } Y \text{ or co. } Z)] = P(\text{co. } X)[P(\text{co. } Y) + P(\text{co. } Z)]$$

$$= \left(\frac{50}{261}\right)\left(\frac{125}{261} + \frac{75}{261}\right)$$

$$= 0.147$$

Counting of Events

Many probability problems, such as those in which the events are uniform probability distributions, can be solved using counting techniques. There are three counting techniques that are quite often used in the computation of probabilities.

1. **Simple multiplication.** If an event *A* can happen in any of *a* ways or outcomes and, after it has occurred, another event *B* can happen in *b* ways or outcomes, the number of ways that both events can happen is *ab*.

Example Problem 8-10

A witness to a hit-and-run accident remembered the first 3 digits of the license plate out of 5 and noted the fact that the last 2 were numerals. How many owners of automobiles would the police have to investigate?

$$ab = (10)(10) = 100$$

If the last 2 were letters, how many would need to be investigated?

$$ab = (26)(26) = 676$$

2. **Permutations.** A *permutation* is an ordered arrangement of a set of objects. The permutations of the word *cup* are cup, cpu, upc, ucp, puc, and pcu. In this case there are 3 objects in the set and we arranged them in groups of 3 to obtain 6 permutations. This is referred to as a permutation of *n* objects taking *r* at a time, where $n = 3$ and $r = 3$. How many permutations would there be for 4 objects taken 2 at a time? Using the word *fork* to represent the four objects, the permutations are fo, of, fr, rf, fk, kf, or, ro, ok, ko, rk, and kr. As the number of objects, *n*, and the number that are taken at one time, *r*, become larger, it becomes a tedious task to list all the permutations. The formula to find the number of permutations more easily is

$$P_r^n = \frac{n!}{(n-r)!}$$

where P_r^n = number of permutations of *n* objects taken *r* of them at a time (the symbol is sometimes written as $_nP_r$)

n = total number of objects

r = number of objects selected out of the total number

The expression $n!$ is read "*n* factorial" and means $n(n-1)(n-2)\cdots(1)$. Thus, $6! = 6 \cdot 5 \cdot 4 \cdot 3 \cdot 2 \cdot 1 = 720$. By definition, $0! = 1$.

Example Problem 8-11

How many permutations are there of 5 objects taken 3 at a time?

$$P_r^n = \frac{n!}{(n-r)!}$$

$$P_3^5 = \frac{5!}{(5-3)!} = \frac{5 \cdot 4 \cdot 3 \cdot 2 \cdot 1}{2 \cdot 1} = 60$$

Example Problem 8-12

In the license plate example, suppose the witness further remembers that the numerals were not the same.

$$P_r^n = \frac{n!}{(n-r)!}$$

$$P_2^{10} = \frac{10!}{(10-2)!} = \frac{10 \cdot 9 \cdot 8 \cdot 7 \cdots 1}{8 \cdot 7 \cdots 1} = 90$$

Example Problem 8-12 could also have been solved by simple multiplication with $a = 10$ and $b = 9$. In other words, there are 10 ways for the first digit but only 9 ways for the second, because duplicates are not permitted.

The symbol *P* is used for both permutation and probability. No confusion should result from this dual usage, because for permutations the superscript *n* and subscript *r* are used.

3. **Combinations.** If the way the objects are ordered is unimportant, then we have a *combination.* The word *cup* has *six* permutations when the 3 objects are taken 3 at a time. However, there is only *one* combination, because the same three letters are in a different order. The word *fork* has 12 permutations when the 4 letters are taken 2 at a time; but the number of combinations is fo, fr, fk, or, ok, and rk, which gives a total of six. The formula for the number of combinations is

$$C_r^n = \frac{n!}{r!(n-r)!}$$

where C_r^n = number of combinations of *n* objects taken *r* at a time (the symbol is sometimes written $_nC_r$ or $\binom{n}{r}$)

n = total number of objects

r = number of objects selected out of the total number

Example Problem 8-13

An interior designer has five different colored chairs and will use three in a living room arrangement. How many different combinations are possible?

$$C_r^n = \frac{n!}{r!(n-r)!}$$

$$C_3^5 = \frac{5!}{3!(5-3)!} = \frac{5 \cdot 4 \cdot 3 \cdot 2 \cdot 1}{3 \cdot 2 \cdot 1 \cdot 2 \cdot 1}$$

$$= 10$$

There is a symmetry associated with combinations such that $C_3^5 = C_2^5$, $C_1^4 = C_3^4$, $C_2^{10} = C_8^{10}$, and so on. Proof of this symmetry is left as an exercise.

The probability definition, the seven theorems, and the three counting techniques are all used to solve probability problems. Many hand calculators have permutation and combination functional keys that eliminate calculation errors provided that the correct keys are punched.

DISCRETE PROBABILITY DISTRIBUTIONS

When specific values such as the integers 0, 1, 2, 3 are used, then the probability distribution is *discrete*. Typical discrete probability distributions are hypergeometric, binomial, and Poisson.

Hypergeometric Probability Distribution

The *hypergeometric probability distribution* occurs when the population is finite and the random sample is taken without replacement. The formula for the hypergeometric probability distribution is constructed of three combinations (total combinations, nonconforming combinations, and conforming combinations) and is given by

$$P(d) = \frac{C_d^D C_{n-d}^{N-D}}{C_n^N}$$

where $P(d)$ = probability of d nonconforming units in a sample of size n
C_n^N = combinations of all units
C_d^D = combinations of nonconforming units
C_{n-d}^{N-D} = combinations of conforming units
N = number of units in the lot (population)
n = number of units in the sample
D = number of nonconforming units in the lot
d = number of nonconforming units in the sample
$N - D$ = number of conforming units in the lot
$n - d$ = number of conforming units in the sample

The formula is obtained from the application of the probability definition, simple multiplication, and combinations. In other words, the numerator is the ways or outcomes of obtaining nonconforming units times the ways or outcomes of obtaining conforming units, and the denominator is the total possible ways or outcomes. Note that symbols in the combination formula have been changed to make them more appropriate for quality.

An example will make the application of this distribution more meaningful.

Example Problem 8-14

A lot of 9 thermostats located in a container has 3 nonconforming units. What is the probability of drawing 1 nonconforming unit in a random sample of 4?

For instructional purposes, the following is a graphical illustration of the problem.

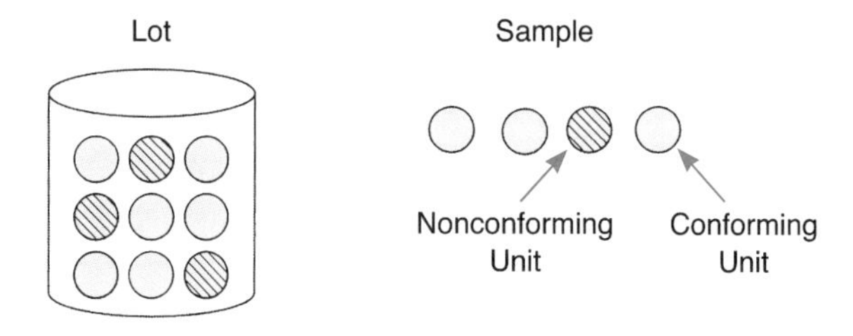

From the picture or from the statement of the problem, $N = 9$, $D = 3$, $n = 4$, and $d = 1$.

$$P(d) = \frac{C_d^D C_{n-d}^{N-D}}{C_n^N}$$

$$P(1) = \frac{C_1^3 C_{4-1}^{9-3}}{C_4^9}$$

$$= \frac{\dfrac{3!}{1!(3-1)!} \cdot \dfrac{6!}{3!(6-3)!}}{\dfrac{9!}{4!(9-4)!}}$$

$$= 0.476$$

Similarly, $P(0) = 0.119$, $P(2) = 0.357$, and $P(3) = 0.048$. Because there are only 3 nonconforming units in the lot, $P(4)$ is impossible. The sum of the probabilities must equal 1000, and this is verified as follows:

$$P(T) = P(0) + P(1) + P(2) + P(3)$$
$$= 0.119 + 0.476 + 0.357 + 0.048$$
$$= 1.000$$

The complete probability distribution is given in Figure 8-4. As the parameters of the hypergeometric distribution change, the shape of the distribution changes, as illustrated by Figure 8-5, where $N = 20$ and $n = 4$. Therefore, each hypergeometric distribution has a unique shape based on N, n, and D. With hand calculators and computers, the efficient calculation of the distribution is not too difficult.

Some solutions require an "or less" probability. In such cases the method is to add up the respective probabilities. Thus,

$$P(2 \text{ or less}) = P(2) + P(1) + P(0)$$

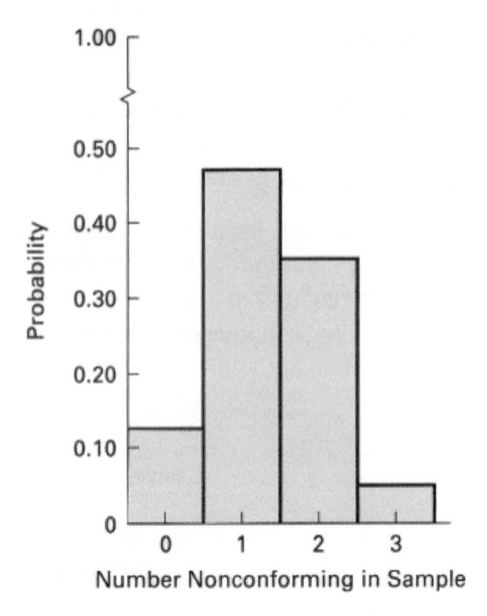

FIGURE 8-4 Hypergeometric Distribution for $N = 9, n = 4$, and $D = 3$

Similarly, some solutions require an "or more" probability and use the formulas

$$P(2 \text{ or more}) = P(\text{T}) - P(1 \text{ or less})$$
$$= P(2) + P(3) + \cdots$$

In the latter series, the number of terms to calculate is determined by the sample size, the number nonconforming in the lot, or when the value is less than 0.001. Thus, if the sample size is, say, 4, you would add $P(2)$, $P(3)$, and $P(4)$; if the number nonconforming in the lot was, say, 6, you would add $P(2)$, $P(3)$, $P(4)$, $P(5)$, and $P(6)$; and if $P(3) = 0.0009$, you would add $P(2)$ and $P(3)$.

The mean and standard deviation of the distribution are

$$\mu = \frac{nD}{N}$$

$$\sigma = \sqrt{\frac{\frac{nD}{N}\left(1 - \frac{D}{N}\right)(N - n)}{N - 1}}$$

Binomial Probability Distribution

The *binomial probability distribution* is applicable to discrete probability problems that have an infinite number of items or that have a steady stream of items coming from a work center. It is applied to problems that have attributes, such as conforming or nonconforming, success or failure, pass or fail, and heads or tails. The binomial distribution is applicable provided the two possible outcomes are constant and the trials are independent. It corresponds to successive terms in the binomial expansion, which is

$$(p + q)^n = p^n + np^{n-1}q + \frac{n(n-1)}{2}p^{n-2}q^2 + \cdots + q^n$$

where p = probability of an event such as a nonconforming unit (proportion nonconforming)
$q = 1 - p$ = probability of a nonevent such as a conforming unit (proportion conforming)
n = number of trials or the sample size

Applying the expansions to the distribution of tails ($p = \frac{1}{2}, q = \frac{1}{2}$) resulting from an infinite number of throws of 11 coins at once, the expansion is

$$(\tfrac{1}{2} + \tfrac{1}{2})^{11} = (\tfrac{1}{2})^{11} + 11(\tfrac{1}{2})^{10}(\tfrac{1}{2}) + 55(\tfrac{1}{2})^{9}(\tfrac{1}{2})^2 + \cdots + (\tfrac{1}{2})^{11}$$
$$= 0.001 + 0.005 + 0.027 + 0.080 + 0.161 + \cdots + 0.001$$

The probability distribution of the number of tails is shown in Figure 8-6. Because $p = q$, the distribution is symmetrical regardless of the value of n; however, when $p \neq q$, the distribution is asymmetrical. In quality work, p is the proportion or fraction nonconforming and is usually less than 0.15.

In most cases in quality work, we are not interested in the entire distribution, only in one or two terms of the binomial expansion. The binomial formula for a single term is

$$P(d) = \frac{n!}{d!(n-d)!}p_0^d q_0^{n-d}$$

where $P(d)$ = probability of d nonconforming units
n = number in the sample
d = number nonconforming in the sample

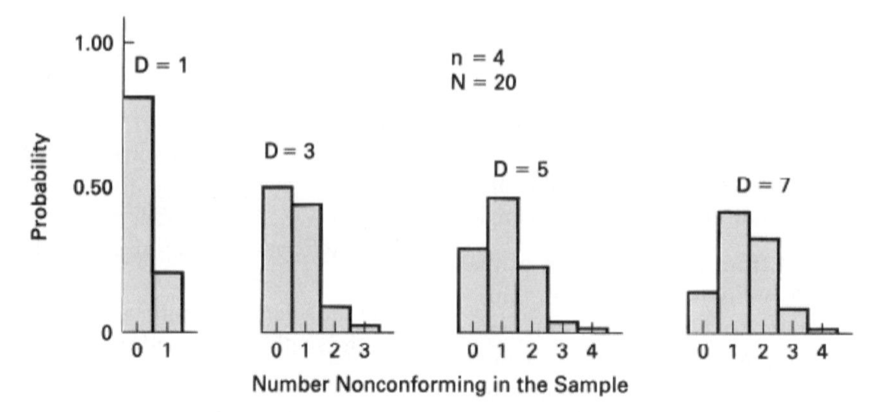

FIGURE 8-5 Comparison of Hypergeometric Distributions with Different Fraction Nonconforming in Lot

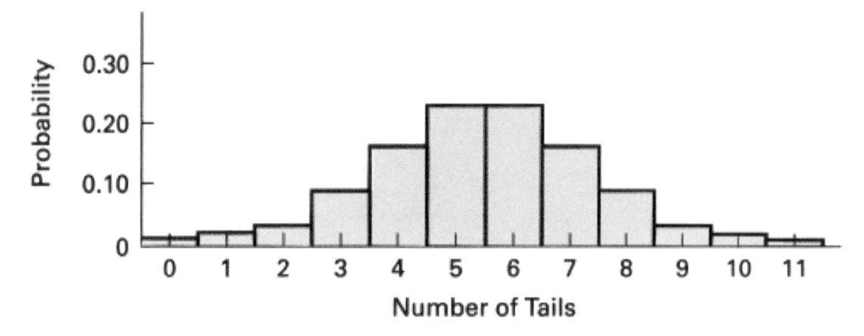

FIGURE 8-6 Distribution of the Number of Tails for an Infinite Number of Tosses of 11 Coins

p_0 = proportion (fraction) nonconforming in the population[1]

q_0 = proportion (fraction) conforming $(1 - p_0)$ in the population

Because the binomial is for the infinite situation, there is no lot size, N, in the formula.

Example Problem 8-15

A random sample of 5 hinges is selected from a steady stream of product from a punch press, and the proportion nonconforming is 0.10. What is the probability of 1 nonconforming unit in the sample? What is the probability of 1 or less? What is the probability of 2 or more?

$$q_0 = 1 - p_0 = 1.00 - 0.10 = 0.90$$

$$P(d) = \frac{n!}{d!(n-d)!} p_0^d q_0^{n-d}$$

$$P(1) = \frac{5!}{1!(5-1)!}(0.10^1)(0.90^{5-1}) = 0.328$$

What is the probability of 1 or less nonconforming units? To solve, we need to use the addition theorem and add $P(1)$ and $P(0)$.

$$P(d) = \frac{n!}{d!(n-d)!} p_0^d q_0^{n-d}$$

$$P(0) = \frac{5!}{0!(5-0)!}(0.10^0)(0.90^{5-0}) = 0.590$$

Thus,

$$P(1 \text{ or less}) = P(0) + P(1) = 0.590 + 0.328 = 0.918$$

What is the probability of 2 or more nonconforming units? Solution can be accomplished using the addition theorem and adding the probabilities of 2, 3, 4, and 5 nonconforming units.

$$P(2 \text{ or more}) = P(2) + P(3) + P(4) + P(5)$$

Or it can be accomplished by using the theorem that the sum of the probabilities is 1.

$$P(2 \text{ or more}) = P(T) - P(1 \text{ or less}) = 1.000 - 0.918 = 0.082$$

Calculations for 2 nonconforming units and 3 nonconforming units for the data in Example Problem 8-15 give $P(2) = 0.073$ and $P(3) = 0.008$. The complete distribution is shown as the graph on the left of Figure 8-7. Calculations for $P(4)$ and $P(5)$ give values less than 0.001, so they are not included in the graph.

Figure 8-7 illustrates the change in the distribution as the sample size increases for the proportion nonconforming of $p = 0.10$, and Figure 8-8 illustrates the change for $p = 0.05$. As the sample size gets larger, the shape of the curve will become symmetrical even though $p \neq q$. Comparing the distribution for $p = 0.10$ and $n = 30$ in

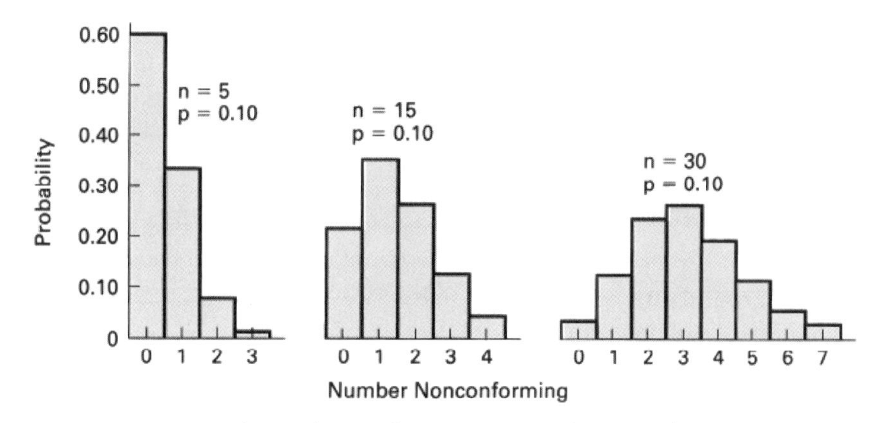

FIGURE 8-7 Binomial Distribution for Various Sample Sizes When $p = 0.10$

[1]Also standard or reference value; see Chapter 6.

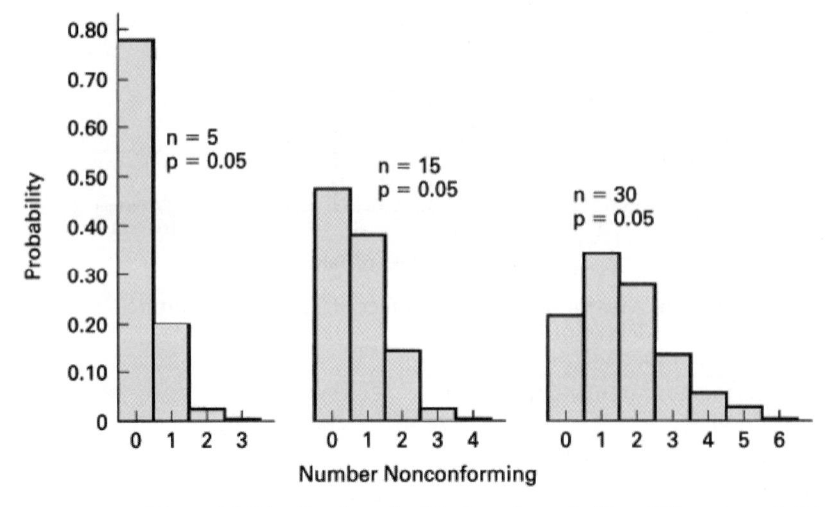

FIGURE 8-8 Binomial Distributions for Various Sample Sizes When $p = 0.05$

Figure 8-7 with the distribution of $p = 0.05$, $n = 30$ in Figure 8-8, it is noted that for the same value of n, the larger the value of the proportion nonconforming p, the greater the symmetry of the distribution.

The shape of the distribution is always a function of the sample size, n, and the proportion nonconforming, p. Change either of these values and a different distribution results.

Tables are available for the binomial distribution. However, because three variables (n, p, and d) are needed, they require a considerable amount of space.

The calculator and computer can make the required calculations quite efficiently; therefore, there is no longer any need for the tables.

The binomial is used for the infinite situation but will approximate the hypergeometric under certain conditions that are discussed later in the chapter. It requires that there be two and only two possible outcomes (a nonconforming or a conforming unit) and that the probability of each outcome does not change. In addition, the use of the binomial requires that the trials are independent; that is, if a nonconforming unit occurs, then the chance of the next one being nonconforming neither increases nor decreases.

The mean and standard deviation of the distribution are

$$\mu = np_0$$
$$\sigma = \sqrt{np_0(1 - p_0)}$$

In addition, the binomial distribution is the basis for one of the control chart groups discussed in Chapter 9.

Poisson Probability Distribution

A third discrete probability distribution is referred to as the *Poisson probability distribution*, named after Simeon Poisson, who described it in 1837. The distribution is applicable to many situations that involve observations per unit of time: for example, the count of cars arriving at a highway toll booth in 1-min intervals, the count of machine breakdowns in 1 day, and the count of shoppers entering a grocery store in 5-min intervals. The distribution is also applicable to situations involving observations per unit of amount: for example, the count of weaving nonconformities in 1000 m^2 of cloth, the count of nonconformities for the number of service calls per week, and the count of rivet nonconformities in a recreational vehicle.

In each of the preceding situations, there are many equal and independent opportunities for the occurrence of an event. Each rivet in a recreational vehicle has an equal opportunity to be a nonconformity; however, there will only be a few nonconformities out of the hundreds of rivets. The Poisson is applicable when n is quite large and p_0 is small.

The formula for the Poisson distribution is

$$P(c) = \frac{(np_0)^c}{c!}e^{-np_0}$$

where c = count, or number, of events of a given classification occurring in a sample, such as count of nonconformities, cars, customers, or machine breakdowns

np_0 = average count, or average number, of events of a given classification occurring in a sample

e = 2.718281

When the Poisson is used as an approximation to the binomial (to be discussed later in the chapter), the symbol c has the same meaning as d has in the binomial and hypergeometric formulas. Because c and np_0 have similar definitions, there is some confusion, which can be corrected by thinking of c as an individual value and np_0 as an average or population value.

Using the formula, a probability distribution can be determined. Suppose that the average count of cars that arrive at a highway toll booth in a 1-min interval is 2; then the calculations are

$$P(c) = \frac{(np_0)^c}{c!}e^{-np_0}$$

$$P(0) = \frac{(2)^0}{0!}e^{-2} = 0.135$$

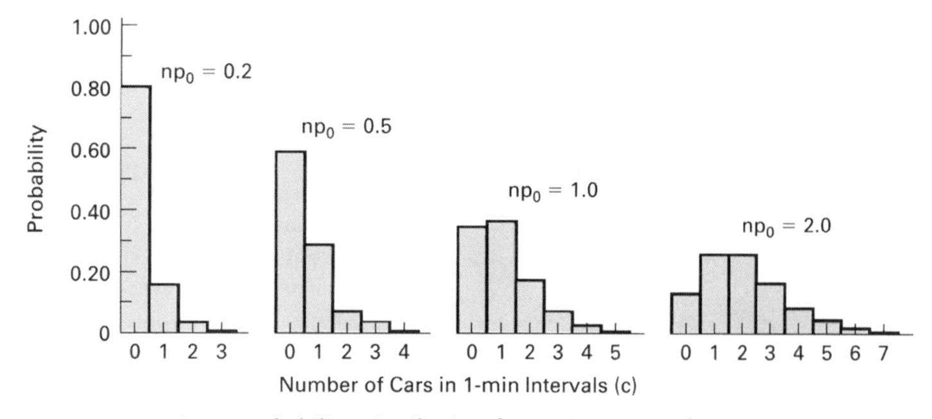

FIGURE 8-9 Poisson Probability Distribution for Various np_0 Values

$$P(1) = \frac{(2)^1}{1!}e^{-2} = 0.271$$

$$P(2) = \frac{(2)^2}{2!}e^{-2} = 0.271$$

$$P(3) = \frac{(2)^3}{3!}e^{-2} = 0.180$$

$$P(4) = \frac{(2)^4}{4!}e^{-2} = 0.090$$

$$P(5) = \frac{(2)^5}{5!}e^{-2} = 0.036$$

$$P(6) = \frac{(2)^6}{6!}e^{-2} = 0.012$$

$$P(7) = \frac{(2)^7}{7!}e^{-2} = 0.003$$

The resulting probability distribution is the one on the right in Figure 8-9. This distribution indicates the probability that a certain count of cars will arrive in any 1-min time interval. Thus, the probability of 0 cars in any 1-min interval is 0.135, the probability of 1 car in any 1-min interval is 0.271, . . . , and the probability of 7 cars in any 1-min interval is 0.003. Figure 8-9 also illustrates the property that as np_0 gets larger, the distribution approaches symmetry.

Probabilities for the Poisson distribution for np_0 values of from 0.1 to 5.0 in intervals of 0.1 and from 6.0 to 15.0 in intervals of 1.0 are given in Table C in the Appendix. Values in parentheses in the table are cumulative probabilities for obtaining "or less" answers. The use of this table simplifies the calculations, as illustrated in Example Problem 8-16.

Example Problem 8-16

The average count of billing errors at a local bank per 8-h shift is 1.0. What is the probability of 2 billing errors? The probability of 1 or less? The probability of 2 or more?

From Table C for an np_0 value of 1.0:

$$P(2) = 0.184$$

$$P(1 \text{ or less}) = 0.736$$

$$\begin{aligned} P(2 \text{ or more}) &= 1.000 - P(1 \text{ or less}) \\ &= 1.000 - 0.736 \\ &= 0.264 \end{aligned}$$

The mean and standard deviation of a distribution are

$$\mu = np_0$$

$$\sigma = \sqrt{np_0}$$

The Poisson probability distribution is the basis for attribute control charts and for acceptance sampling, which are discussed in subsequent chapters. In addition to quality applications, the Poisson distribution is used in other industrial situations, such as accident frequencies, computer simulation, operations research, and work sampling.

From a theoretical viewpoint, a discrete probability distribution should use a bar graph. However, it is a common practice (and the one followed for the figures in this book) to use a rectangle like the histogram.

Other discrete probability distributions are the uniform, geometric, and negative binomial. The uniform distribution was illustrated in Figure 8-1. From an application viewpoint, it is the one used to generate a random number table. The geometric and negative binomial are used in reliability studies for discrete data.

CONTINUOUS PROBABILITY DISTRIBUTIONS

When measurable data such as meters, kilograms, and ohms are used, the probability distribution is continuous. There are many continuous probability distributions, but only the normal is of sufficient importance to warrant a detailed discussion in an introductory text.

Normal Probability Distribution

The *normal curve* is a continuous probability distribution. Solutions to probability problems that involve continuous data can be solved using the normal probability distribution. In Chapter 5, techniques were discussed to determine the percentage of the data that were above a certain value, below a certain value, or between two values. These same techniques are applicable to probability problems, as illustrated in the following example problem.

Example Problem 8-17

If the operating life of an electric mixer, which is normally distributed, has a mean of 2200 h and standard deviation of 120 h, what is the probability that a single electric mixer will fail to operate at 1900 h or less?

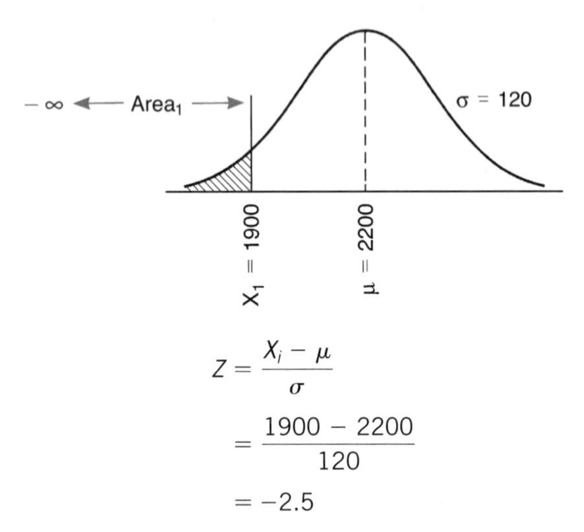

$$Z = \frac{X_i - \mu}{\sigma}$$

$$= \frac{1900 - 2200}{120}$$

$$= -2.5$$

From Table A of the Appendix, for a Z value of -2.5, $\text{area}_1 = 0.0062$. Therefore, the probability of an electric mixer failing is

$$P(\text{failure at 1900 h or less}) = 0.0062$$

The answer in the problem could have been stated as: "The percent of items less than 1900 h is 0.62%." Therefore, the area under the normal curve can be treated as either a probability value or a relative frequency value.

Other Continuous Probability Distributions

Of the many other continuous probability distributions, such as uniform, beta, and gamma, only two are of significant importance to mention their practical applications. The exponential probability distribution is used in reliability studies when there is a constant failure rate, and the Weibull is used when the time to failure is not constant. These two distributions are discussed in Chapter 12.

DISTRIBUTION INTERRELATIONSHIP

With so many distributions, it is sometimes difficult to know when they are applicable. Certainly, because the Poisson can be easily calculated using Table C of the Appendix, it should be used whenever appropriate. Figures 8-5, 8-8, and 8-9 show a similarity among the hypergeometric, binomial, and Poisson distributions.

The hypergeometric is used for finite lots of stated size N. It can be approximated by the binomial when $n/N \leqq 0.10$; or by the Poisson when $n/N \leqq 0.10$, $p_0 \leqq 0.10$, and $np_0 \leqq 5$; or by the normal when $n/N \leqq 0.10$ and the normal approximates the binomial.

The binomial is used for infinite situations or when there is a steady stream of product so that an infinite situation is assumed. It can be approximated by the Poisson when $p_0 \leqq 0.10$ and $np_0 \leqq 5$. The normal curve is an excellent approximation when p_0 is close to 0.5 and $n \geqq 10$. As np_0 deviates from 0.5, the approximation is still good as long as $np_0 \geqq 5$ and n increases to 50 or more for values of p_0 as low as 0.10 and as high as 0.90. Because the binomial calculation time is not too much different than the normal calculation time, there is little advantage to using the normal as an approximation.

The information given above can be considered to provide approximation guidelines rather than absolute laws. Approximations are better the farther the data are from the limiting values. *For the most part, the efficiency of the calculator and computer has made the use of approximations obsolete.*

COMPUTER PROGRAM

Microsoft's EXCEL software has the capability of performing calculations under the Formulas/More Functions/Statistical tabs except combination, which is under Formulas/Math & Trig. These formulas are permutations, combinations, hypergeometric, binomial, and Poisson.

EXERCISES

1. If an event is certain to occur, what is its probability? If an event will not occur, what is its probability?
2. What is the probability that you will live forever? What is the probability that an octopus will fly?
3. If the probability of obtaining a 3 on a six-sided die is 0.167, what is the probability of obtaining any number but a 3?
4. Find an event that has a probability of 1.000.
5. The probability of drawing a pink chip from a bowl of different-colored chips is 0.35, the probability of a blue chip is 0.46, the probability of a green chip is 0.15, and the probability of a purple chip is 0.04. What is the probability of a blue or a purple chip? What is the probability of a pink or a blue chip?
6. At any hour in an intensive care unit, the probability of an emergency is 0.247. What is the probability that there will be tranquility for the staff?
7. If a hotel has 20 king-size beds, 50 queen-size beds, 100 double beds, and 30 twin beds available, what is the probability that you will be given a queen-size or a twin bed when you register?
8. A ball is drawn at random from a container that holds 8 yellow balls numbered 1 to 8, 6 orange balls

numbered 1 to 6, and 10 gray balls numbered 1 to 10. What is the probability of obtaining an orange ball or a ball numbered 5 or an orange ball numbered 5 in a draw of one ball? What is the probability of a gray ball or a ball numbered 8 or a gray ball numbered 8 in a draw of one ball?

9. If the probability of obtaining 1 nonconforming unit in a sample of 2 from a large lot of Neoprene gaskets is 0.18 and the probability of 2 nonconforming units 0.25, what is the probability of 0 nonconforming units?

10. Using the information in Exercise 9, find the probability of obtaining 2 nonconforming units on the first sample of 2 and 1 nonconforming unit on the second sample of 2. What is the probability of 0 nonconforming units on the first sample and 2 nonconforming units on the second? The first gasket selected is returned to the lot before the second one is selected.

11. A basket contains 34 heads of lettuce, 5 of which are spoiled. If a sample of 2 is drawn and not replaced, what is the probability that both will be spoiled?

12. If a sample of 1 can be drawn from an automatic storage and retrieval rack with 3 different storage racks and 6 different trays in each rack, what is the number of different ways of obtaining the sample of 1?

13. A small model-airplane motor has four starting components: key, battery, wire, and glow plug. What is the probability that the system will work if the probability that each component will work is as follows: key (0.998), battery (0.997), wire (0.999), and plug (0.995)?

14. An inspector has to inspect products from three machines in one department, five machines in another, and two machines in a third. The quality manager wants to vary the inspector's route. How many different ways are possible?

15. If, in Example Problem 8-10 of the hit-and-run driver, there were 1 numeral and 1 letter, how many automobiles would need to be investigated?

16. A sample of 3 is selected from 10 people on a Caribbean cruise. How many permutations are possible?

17. From a lot of 90 airline tickets, a sample of 8 is selected. How many permutations are possible?

18. A sample of 4 is selected from a lot of 20 piston rings. How many different sample combinations are possible?

19. From a lot of 100 hotel rooms, a sample of 3 is selected for audit. How many different sample combinations are possible?

20. A sample of 2 is selected from a tray of 20 bolts. How many different sample combinations are possible?

21. In the Illinois lottery, the numbers available are 1 to 54. On Saturday night, 6 numbers are selected. How many different combinations are possible?

22. In the Illinois lottery, each participant selects two sets of 6 numbers. What is the probability of having all 6 numbers?

23. The game of KENO has 80 numbers and you select 15. What is the probability of having all 15 numbers and winning the jackpot?

24. An automatic garage-door opener has 12 switches that can be set on or off. Both the transmitter and receiver are set the same and the owner has the option of setting 1 to 12 switches. What is the probability that another person with the same model transmitter could open the door?

25. Compare the answers of C_3^5 with C_2^5, C_1^4 with C_3^4, and C_2^{10} with C_8^{10}. What conclusion can you draw?

26. Calculate C_0^6, C_0^{10}, and C_0^{25}. What conclusion can you draw?

27. Calculate C_3^3, C_9^9, and C_{35}^{35}. What conclusion can you draw?

28. Calculate C_1^7, C_1^{12}, and C_1^{18}. What conclusion can you draw?

29. A random sample of 4 insurance claims is selected from a lot of 12 that has 3 nonconforming units. Using the hypergeometric distribution, what is the probability that the sample will contain exactly zero nonconforming units? one nonconforming unit? two nonconforming units? three nonconforming units? four nonconforming units?

30. A finite lot of 20 digital watches is 20% nonconforming. Using the hypergeometric distribution, what is the probability that a sample of 3 will contain 2 nonconforming watches?

31. In Exercise 30, what is the probability of obtaining 2 or more nonconforming units? What is the probability of 2 or less nonconforming units?

32. A steady stream of income tax returns has a proportion nonconforming of 0.03. What is the probability of obtaining 2 nonconforming units from a sample of 20? Use the binomial distribution.

33. Find the probability, using the binomial distribution, of obtaining 2 or more nonconforming units when sampling 5 computers from a batch known to be 6% nonconforming.

34. Using the binomial distribution, find the probability of obtaining 2 or less nonconforming restaurants in a sample of 9 when the lot is 15% nonconforming.

35. What is the probability of guessing correctly exactly 4 answers on a true–false examination that has 9 questions? Use the binomial distribution.

36. An injection molder produces golf tees that are 15.0% nonconforming. Using the binomial, find the probability that, in a random sample of 20 golf tees, 1 or less are nonconforming.

37. A random sample of 10 automotive bumpers is taken from a stream of product that is 5% nonconforming.

Using the binomial distribution, determine the probability of 2 nonconforming automotive bumpers.

38. If the average number of nonconforming units is 1.6, what is the probability that a sample will contain 2 or less nonconforming units? Use the Poisson distribution.

39. Using the data from Exercise 38, determine the probability of 2 or more nonconforming units.

40. A sample of 10 washing machines is selected from a process that is 8% nonconforming. What is the probability of 1 nonconforming washing machine in the sample? Use the Poisson distribution.

41. A lot of 15 has 3 nonconforming units. What is the probability that a sample of 3 will have 1 nonconforming unit?

42. A sample of 3 medicine bottles is taken from an injection molding machine that is 10% nonconforming. What is the probability of 1 nonconforming medicine bottle in the sample?

43. A steady stream of light bulbs has a fraction nonconforming of 0.09. If 67 are sampled, what is the probability of 3 nonconforming units?

44. Using the EXCEL software, solve some of the exercises and compare your answers with those obtained using a calculator or by hand.

45. Using EXCEL, construct a graph showing the entire distribution for:

    a. Hypergeometirc, where $n = 4, N = 20$, and $D = 5$

    b. Binomial, where $n = 15$ and $p = 0.05$

    c. Poisson, where $np = 1.0$

CHAPTER
NINE

CONTROL CHARTS FOR ATTRIBUTES[1]

OBJECTIVES

Upon completion of this chapter, the reader is expected to

- know the limitations of variable control charts and the different types of attribute charts;
- know the objectives of the *p*-chart group and the applicable distribution;
- construct a

  Fraction defective chart—fixed subgroup size.

  Fraction defective chart—variable subgroup size.

  Percent defective chart.

  Number defective chart.
- know how to minimize the effect of variable subgroup size;
- know the applications of the *c*-chart group, the applicable distribution, and two conditions;
- construct a *c* chart and *u* chart and know the difference between them;
- know the three classes of defect severity.

INTRODUCTION

Attributes

An attribute was defined in Chapter 5, and the definition is repeated to refresh the reader's memory. The term *attribute*, as used in quality, refers to those quality characteristics that conform to specifications or do not conform to specifications.

Attributes are used

1. Where measurements are not possible—for example, visually inspected items such as color, missing parts, scratches, and damage.
2. Where measurements can be made but are not made because of time, cost, or need. In other words, although the diameter of a hole can be measured with an inside micrometer, it may be more convenient to use a "go–no go" gauge and determine if it conforms or does not conform to specifications.

Where an attribute does not conform to specifications, various descriptive terms are used. A *nonconformity* is a departure of a quality characteristic from its intended level or state that occurs with a severity sufficient to cause an associated product or service not to meet a specification requirement. The definition of a *defect* is similar, except it is concerned with satisfying intended normal, or reasonably foreseeable, usage requirements. *Defect* is appropriate for use when evaluation is in terms of usage, and *nonconformity* is appropriate for conformance to specifications.

The term *nonconforming unit* is used to describe a unit of product or service containing at least one nonconformity. *Defective* is analogous to defect and is appropriate for use when a unit of product or service is evaluated in terms of usage rather than conformance to specifications.

In this book we use the terms *nonconformity* and *nonconforming unit*. This practice avoids the confusion and misunderstanding that occurs with *defect* and *defective* in product-liability lawsuits.

Limitations of Variable Control Charts

Variable control charts are excellent means for controlling quality and subsequently improving it; however, they do have limitations. One obvious limitation is that these charts cannot be used for quality characteristics that are attributes. The converse is not true, because a variable can be changed to an attribute by stating that it conforms or does not conform to specifications. In other words, nonconformities such as missing parts, incorrect color, and so on, are not measureable, and a variable control chart is not applicable.

Another limitation concerns the fact that there are many variables in a manufacturing entity. Even a small manufacturing plant could have as many as 1000 variable quality characteristics. Because an $\overline{X}$ and R chart is needed for each characteristic, 1000 charts would be required. Clearly, this would be too expensive and impractical. Note that in Chapter 7 some advanced techniques were discussed

[1]The information in this chapter is based on ANSI/ASQ B1–B3.

to minimize this limitation. A control chart for attributes can also minimize this limitation by providing overall quality information at a fraction of the cost.

Types of Attribute Charts

There are two different groups of control charts for attributes. One group of charts is for nonconforming units. It is based on the binomial distribution. A proportion, p, chart shows the proportion nonconforming in a sample or subgroup. The proportion is expressed as a fraction or a percent. Similarly, we could have charts for proportion conforming, and they too could be expressed as a fraction or a percent. Another chart in the group is for the number nonconforming, an np chart,[2] and it too could be expressed as number conforming.

The other group of charts is for nonconformities. It is based on the Poisson distribution. A c chart shows the count of nonconformities in an inspected unit such as an automobile, a bolt of cloth, or a roll of paper. Another closely related chart is the u chart, which is for the count of nonconformities per unit.

Much of the information on control charts for attributes is similar to that given in Chapter 6. The reader is referred to the sections titled "State of Control" and "Analysis of Out-of-Control Condition."

CONTROL CHARTS FOR NONCONFORMING UNITS

Introduction

The p chart is used for data that consist of the proportion of the number of occurrences of an event to the total number of occurrences. It is used in quality to report the fraction or percent nonconforming in a product, quality characteristic, or group of quality characteristics. As such, the *fraction* nonconforming is the proportion of the number nonconforming in a sample or subgroup to the total number in the sample or subgroup. In symbolic terms, the formula is

$$p = \frac{np}{n}$$

where p = proportion or fraction nonconforming in the sample or subgroup
n = number in the sample or subgroup
np = number nonconforming in the sample or subgroup

Example Problem 9-1

During the first shift, 450 inspections are made of book-of-the-month shipments and 5 nonconforming units are found. Production during the shift was 15,000 units. What is the fraction nonconforming?

$$p = \frac{np}{n} = \frac{5}{450} = 0.011$$

[2] The ANSI/ASQ B1–B3 standard uses the symbol pn; however, current practice uses np.

The fraction nonconforming, p, is usually small, say, 0.10 or less. Except in unusual circumstances, values greater than 0.10 indicate that the organization is in serious difficulty, and that measures more drastic than a control chart are required. Because the fraction nonconforming is very small, the subgroup sizes must be quite large to produce a meaningful chart.

The p chart is an extremely versatile control chart. It can be used to control one quality characteristic, as is done with the $\overline{X}$ and R charts; to control a group of quality characteristics of the same type or of the same part; or to control the entire product. The p chart can be established to measure the quality produced by a work center, by a department, by a shift, or by an entire plant. It is frequently used to report the performance of an operator, group of operators, or management as a means of evaluating their quality performance.

The subgroup size of the p chart can be either variable or constant. A constant subgroup size is preferred; however, there may be many situations, such as changes in mix and 100% automated inspection, where the subgroup size changes.

Objectives

The objectives of nonconforming charts are to

1. **Determine the average quality level.** Knowledge of the quality average is essential as a benchmark. This information provides the process capability in terms of attributes.
2. **Bring to the attention of management any changes in the average.** Once the average quality (proportion nonconforming) is known, changes, either increasing or decreasing, become significant.
3. **Improve the product quality.** In this regard, a p chart can motivate operating and management personnel to initiate ideas for quality improvement. The chart will tell whether the idea is an appropriate or inappropriate one. A continual and relentless effort must be made to improve the quality.
4. **Evaluate the quality performance of operating and management personnel.** Supervisors of activities and especially the chief executive officer (CEO) should be evaluated by a chart for nonconforming units. Other functional areas, such as engineering, sales, finance, and so on, may find a chart for nonconformities more applicable for evaluation purposes.
5. **Suggest places to use $\overline{X}$ and R charts.** Even though the cost of computing and charting $\overline{X}$ and R charts is more than the chart for nonconforming units, the $\overline{X}$ and R charts are much more sensitive to variations and are more helpful in diagnosing causes. In other words, the chart for nonconforming units suggests the source of difficulty, and the $\overline{X}$ and R charts find the cause.
6. **Determine acceptance criteria of a product before shipment to the customer.** Knowledge of the proportion nonconforming provides management with information on whether to release an order.

These objectives indicate the scope and value of a nonconforming chart.

p-Chart Construction for Constant Subgroup Size

The general procedures that apply to variable control charts also apply to the p chart.

1. **Select the quality characteristic(s).** The first step in the procedure is to determine the use of the control chart. A p chart can be established to control the proportion nonconforming of (a) a single quality characteristic, (b) a group of quality characteristics, (c) a part, (d) an entire product, or (e) a number of products. This establishes a hierarchy of utilization so that any inspections applicable for a single quality characteristic also provide data for other p charts, which represent larger groups of characteristics, parts, or products.

A p chart can also be established for performance control of an (a) operator, (b) work center, (c) department, (d) shift, (e) plant, or (f) corporation. Using the chart in this manner, comparisons may be made between like units. It is also possible to evaluate the quality performance of a unit. A hierarchy of utilization exists so that data collected for one chart can also be used on a more all-inclusive chart.

The use for the chart or charts will be based on securing the greatest benefit for a minimum of cost. One chart should measure the CEO's quality performance.

2. **Determine the subgroup size and method.** The size of the subgroup is a function of the proportion nonconforming. If a part has a proportion nonconforming, p, of 0.001 and a subgroup size, n, of 1000, then the average number nonconforming, np, would be one per subgroup. This would *not* make a good chart, because a large number of values posted to the chart would be 0. If a part has a proportion nonconforming of 0.15 and a subgroup size of 50, the average number of nonconforming would be 7.5, which would make a good chart.

Therefore, the selection of the subgroup size requires some preliminary observations to obtain a rough idea of the proportion nonconforming and some judgment as to the average number of nonconforming units that will make an adequate graphic chart. A minimum size of 50 is suggested as a starting point. Inspection can either be by audit or online. Audits are usually done in a laboratory under optimal conditions. Online provides immediate feedback for corrective action.

A precise method of determining the sample size is given by the formula

$$n = p(1 - p)\left(\frac{Z_{\alpha/2}}{E}\right)^2$$

where n = sample size.

p = estimate of the population proportion nonconforming. If no estimate is available, assume "worst case," where $p = 0.50$. To be safe, one should estimate on the high side.

$Z_{\alpha/2}$ = normal distribution coefficient (Z value) for area between the two tails. This area represents the decimal equivalent of the confidence limit.

$Z_{\alpha/2}$	Confidence Limit
1.036	70%
1.282	80%
1.645	90%
1.96	95%
2.575	99%
3.00	99.73%

E = maximum allowable error in the estimate of p, which is also referred to as the desired precision.

An example problem will illustrate the use of the technique.

Example Problem 9-2

An insurance manager wishes to determine the proportion of automobile insurance claims that are incorrectly filled out (nonconforming). Based on some preliminary data, she estimates the percent nonconforming as 20% ($p = 0.20$). She desires a precision of 10% and a confidence level of 95%. Determine the sample size.

$$E = 10\% \text{ of } p = 0.10(0.20) = 0.02$$

$$Z_{\alpha/2} = 1.96 \text{ from table}$$

$$n = p(1 - p)\left(\frac{Z_{\alpha/2}}{E}\right)^2$$

$$= 0.20(1 - 0.20)\left(\frac{1.96}{0.02}\right)^2$$

$$= 1537$$

If, after taking a sample of 1537, the value of $p = 0.17$, then the true value will be between 0.15 and 0.19 ($p \pm E$) 95% of the time.

It is suggested that the calculation of n be repeated periodically during the study to obtain a better estimate of p. The normal distribution is a good approximation of the binomial distribution as long as $np \geq 5$ and n increases to 50 or more.

3. **Collect the data.** The quality technician will need to collect sufficient data for at least 25 subgroups, or the data may be obtained from historical records. Perhaps the best source is from a check sheet designed by a project team. Table 9-1 gives the inspection results for the blower motor in an electric hair dryer for the motor department. For each subgroup, the proportion nonconforming is calculated by the formula $p = np/n$. The quality technician reported that subgroup 19 had an abnormally large number of nonconforming units, owing to faulty contacts.

The data can be plotted as a run chart, as shown in Figure 9-1. A run chart shows the variation in the data; however, we need statistical limits to determine if the process is stable.

This type of chart is very effective during the start-up phase of a new item or process, when the process is very erratic. Also, many organizations prefer to use this type of chart to measure quality performance, rather than a control chart.

Because the run chart does not have limits, it is not a control chart. This fact does not reduce its effectiveness in many situations.

TABLE 9-1 Inspection Results of Hair Dryer Blower Motor, Motor Department, May

Subgroup Number	Number Inspected n	Number Nonconforming np	Proportion Nonconforming p
1	300	12	0.040
2	300	3	0.010
3	300	9	0.030
4	300	4	0.013
5	300	0	0.0
6	300	6	0.020
7	300	6	0.020
8	300	1	0.003
9	300	8	0.027
10	300	11	0.037
11	300	2	0.007
12	300	10	0.033
13	300	9	0.030
14	300	3	0.010
15	300	0	0.0
16	300	5	0.017
17	300	7	0.023
18	300	8	0.027
19	300	16	0.053
20	300	2	0.007
21	300	5	0.017
22	300	6	0.020
23	300	0	0.0
24	300	3	0.010
25	300	2	0.007
Total	7500	138	

4. **Calculate the trial central line and control limits.** The formula for the trial control limits is given by

$$\text{UCL} = \bar{p} + 3\sqrt{\frac{\bar{p}(1-\bar{p})}{n}}$$

$$\text{LCL} = \bar{p} - 3\sqrt{\frac{\bar{p}(1-\bar{p})}{n}}$$

where $\bar{p}$ = average proportion nonconforming for many subgroups

n = number inspected in a subgroup

The average proportion nonconforming, $\bar{p}$, is the central line and is obtained by the formula $\bar{p} = \Sigma np/\Sigma n$. Calculations for the 3 σ trial control limits using the data on the electric hair dryer are as follows:

$$\bar{p} = \frac{\Sigma np}{\Sigma n} = \frac{138}{7500} = 0.018$$

$$\text{UCL} = \bar{p} + 3\sqrt{\frac{\bar{p}(1-\bar{p})}{n}}$$

$$= 0.018 + 3\sqrt{\frac{0.018(1-0.018)}{300}}$$

$$= 0.041$$

$$\text{LCL} = \bar{p} - 3\sqrt{\frac{\bar{p}(1-\bar{p})}{n}}$$

$$= 0.018 - 3\sqrt{\frac{0.018(1-0.018)}{300}}$$

$$= -0.005 \text{ or } 0.0$$

Calculations for the lower control limit resulted in a *negative* value, which is a theoretical result. In practice, a negative proportion nonconforming would be impossible. Therefore, the lower control limit value of –0.005 is changed to 0.

When the lower control limit is positive, it may in some cases be changed to 0. If the p chart is to be viewed by operating personnel, it would be difficult to explain why a proportion nonconforming that is below the lower control limit is out of control. In other words, performance of exceptionally good quality would be classified as out of control. To avoid the need to explain this situation to operating personnel, the lower control limit is changed from a positive value to 0. When the p chart is to be used by quality personnel and by management, a positive lower control limit is left unchanged. In this manner, exceptionally good performance (below the lower control limit) will be treated as an out-of-control situation and investigated for an assignable cause. It is hoped that the assignable cause will indicate how the situation can be repeated.

The central line, $\bar{p}$, and the control limits are shown in Figure 9-2; the proportion nonconforming, p, from Table 9-1

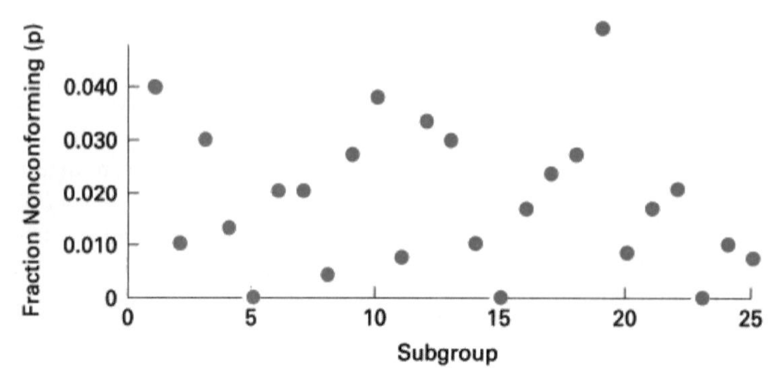

FIGURE 9-1 Run Chart for the Data of Table 9-1

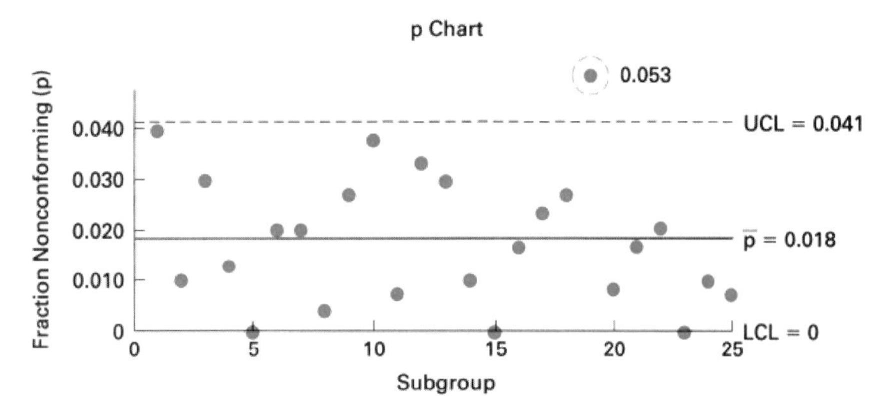

FIGURE 9-2 A p Chart to Illustrate the Trial Central Line and Control Limits Using the Data from Table 9-1

is also posted to that chart. This chart is used to determine if the process is stable and is not posted. It is important to recognize that the central line and control limits were determined from the data.

5. **Establish the revised central line and control limits.** In order to determine the revised 3σ control limits, the standard or reference value for the proportion nonconforming, p_0, needs to be determined. If an analysis of the chart of Step 4 shows good control (a stable process), then $\bar{p}$ can be considered to be representative of that process. Therefore, the best estimate of p_0 at this time is $\bar{p}$, and $p_0 = \bar{p}$.

Most industrial processes, however, are not in control when they are first analyzed, and this fact is illustrated in Figure 9-2 by subgroup 19, which is above the upper control limit and therefore out of control. Because subgroup 19 has an assignable cause, it can be discarded from the data and a new $\bar{p}$ computed with all of the subgroups except 19. The calculations can be simplified by using the formula

$$\bar{p}_{\text{new}} = \frac{\Sigma np - np_d}{\Sigma n - n_d}$$

where np_d = number nonconforming in the discarded subgroups

n_d = number inspected in the discarded subgroups

In discarding data, it must be remembered that only those subgroups with assignable causes are discarded. Those subgroups without assignable causes are left in the data. Also, out-of-control points below the lower control limit do not need to be discarded, because they represent exceptionally good quality. If the out-of-control point on the low side is due to an inspection error, it should, however, be discarded.

With an adopted standard or reference value for the proportion nonconforming, p_0, the revised control limits are given by

$$p_0 = \bar{p}_{\text{new}}$$

$$\text{UCL} = p_0 + 3\sqrt{\frac{p_0(1 - p_0)}{n}}$$

$$\text{LCL} = p_0 - 3\sqrt{\frac{p_0(1 - p_0)}{n}}$$

where p_0, the central line, represents the reference or standard value for the fraction nonconforming. These formulas are for the control limits for 3 standard deviations from the central line p_0.

Thus, for the preliminary data in Table 9-1, a new $\bar{p}$ is obtained by discarding subgroup 19.

$$\bar{p}_{\text{new}} = \frac{\Sigma np - np_d}{\Sigma n - n_d} = \frac{138 - 16}{7500 - 300} = 0.017$$

Because $\bar{p}_{\text{new}}$ is the best estimate of the standard or reference value, $p_0 = 0.017$. The revised control limits for the p chart are obtained as follows:

$$\text{UCL} = p_0 + 3\sqrt{\frac{p_0(1 - p_0)}{n}} = 0.017 + 3\sqrt{\frac{0.017(1 - 0.017)}{300}} = 0.039$$

$$\text{LCL} = p_0 - 3\sqrt{\frac{p_0(1 - p_0)}{n}} = 0.017 - 3\sqrt{\frac{0.017(1 - 0.017)}{300}} = -0.005 \text{ or } 0.0$$

The revised control limits and the central line, p_0, are shown in Figure 9-3. This chart, without the plotted points, is posted in an appropriate place, and the proportion nonconforming, p, for each subgroup is plotted as it occurs.

6. **Achieve the objective.** The first five steps are planning. The last step involves action and leads to the achievement of the objective. The revised control limits were based on data collected in May. Some representative values of inspection results for the month of June are shown in Figure 9-3. Analysis of the June results shows that the quality improved. This improvement is expected, because the posting of a quality control chart usually results in improved quality. Using the

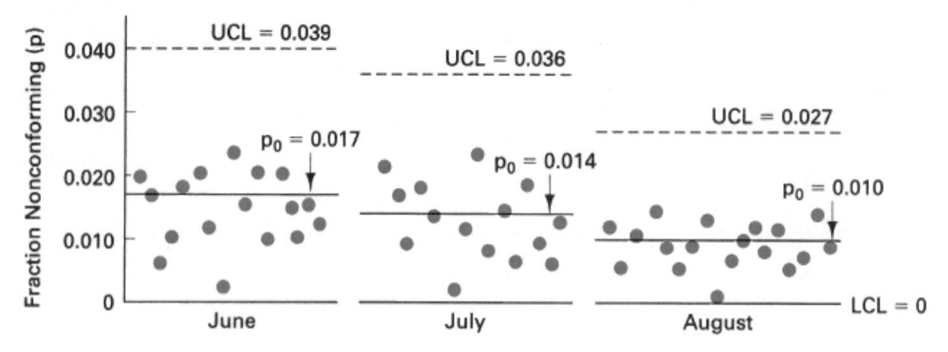

FIGURE 9-3 Continuing Use of the p Chart for Representative Values of the Proportion Nonconforming, p

June data, a better estimate of the proportion nonconforming is obtained. The new value ($p_0 = 0.014$) is used to obtain the UCL of 0.036.

During the latter part of June and the entire month of July, various quality improvement ideas generated by a project team are tested. These ideas are new shellac, change in wire size, stronger spring, $\overline{X}$ and R charts on the armature, and so on. In testing ideas, there are three criteria: A minimum of 25 subgroups is required. They can be compressed in time as long as no sampling bias occurs, and only one idea can be tested at one time. The control chart will tell whether the idea improves the quality, reduces the quality, or has no effect on the quality. The control chart should be located in a conspicuous place so operating personnel can view it.

Data from July are used to determine the central line and control limits for August. The pattern of variation for August indicates that no further improvement resulted. However, a 41% improvement occurred from June (0.017) to August (0.010). At this point, we have obtained considerable improvement by testing the ideas of the project team. Although this improvement is very good, we must continue our relentless pursuit of quality improvement—1 out of every 100 is still nonconforming. Perhaps a detailed failure analysis or technical assistance from product engineering will lead to additional ideas that can be evaluated. A new project team may help.

Quality improvement is never terminated. Efforts may be redirected to other areas based on need and/or resources available.

Some Comments on p Charts

Like the $\overline{X}$ and R chart, the p chart is most effective if it is posted where operating and quality personnel can view it. Also, like the $\overline{X}$ and R chart, the control limits are usually 3 standard deviations from the central value. Therefore, approximately 99% of the plotted points, p, will fall between the upper and lower control limits.

A state of control for a p chart is treated in a manner similar to that described in Chapter 6. The reader may wish to briefly review that section. A control chart for subgroup values of p will aid in disclosing the occasional presence of assignable causes of variation in the process. The elimination of these assignable causes will lower p_0 and, therefore, have a positive effect on spoilage, efficiency, and cost per unit. A p chart will also indicate long-range trends in quality, which will help to evaluate changes in personnel, methods, equipment, tooling, materials, and inspection techniques.

If the population proportion nonconforming, ϕ, is known, it is not necessary to calculate the trial control limits. This is a considerable timesaver, because $p_0 = \phi$, which allows the p chart to be introduced immediately. Also, p_0 may be assigned a desired value—in which case the trial control limits are not necessary.

Because the p chart is based on the binomial distribution, there must be a constant chance of selecting a nonconforming unit. In some operations, if one nonconforming unit occurs, all units that follow will be nonconforming until the condition is corrected. This type of condition also occurs in batch processes when the entire batch is nonconforming or when an error is made in dimensions, color, and so on. In such cases a constant chance of obtaining a nonconforming unit does not occur, and therefore, the p chart is not suitable.

Presentation Techniques

The information in the preceding example is presented as a fraction nonconforming. It could also be presented as percent nonconforming, fraction conforming, or percent conforming. All four techniques convey the same information, as shown by Figure 9-4. The two lower figures show opposite information from the respective upper figures.

Table 9-2 shows the equations for calculating the central line and control limits for the four techniques as a function of p_0.

Many organizations are taking the positive approach and using either of the two conforming presentation techniques. The use of the chart and the results will be the same no matter which chart is used.

p-Chart Construction for Variable Subgroup Size

Whenever possible, p charts should be developed and used with a constant subgroup size. This situation is not possible when the p chart is used for 100% inspection of

TABLE 9-2 Calculating Central Line and Limits for the Various Presentation Techniques

	Fraction Nonconforming	Percent Nonconforming	Fraction Conforming	Percent Conforming
Central line	p_0	$100p_0$	$q_0 = 1 - p_0$	$100q_0 = 100(1 - p_0)$
Upper control limit	UCL_p	$100(UCL_p)$	$UCLq = 1 - LCL_p$	$100(UCLq)$
Lower control limit	LCL_p	$100(LCL_p)$	$LCLq = 1 - UCL_p$	$100(LCLq)$

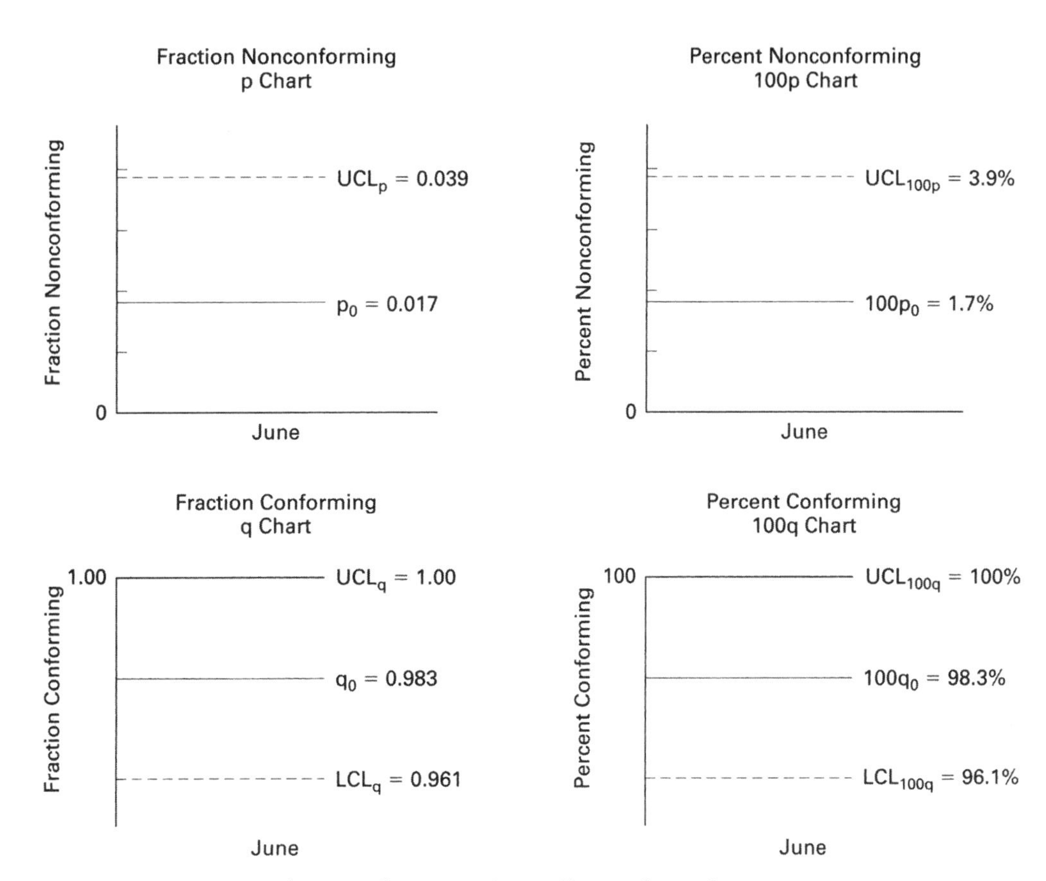

FIGURE 9-4 Various Techniques for Presenting p-Chart Information

output that varies from day to day. Also, data for p-chart use from sampling inspection might vary for a variety of reasons. Because the control limits are a function of the subgroup size, n, the control limits will vary with the subgroup size. Therefore, they need to be calculated for each subgroup.

Although a variable subgroup size is undesirable, it does exist and must be handled. The procedures of data collection, trial central line and control limits, and revised central line and control limits are the same as those for a p chart with constant subgroup size. An example without Steps 1 and 2 will be used to illustrate the procedure.

Step 3. Collect the data. A computer modem manufacturer has collected data from the final test of the product for the end of March and all of April. Subgroup size was 1 day's inspection results. The inspection results for 25 subgroups are shown in the first three columns of Table 9-3: subgroup designation, number inspected, and number nonconforming. A fourth column for the fraction nonconforming is calculated using the formula $p = np/n$. The last two columns are for the upper and lower control limit calculations, which are discussed in the next section.

The variation in the number inspected per day can be due to a number of reasons. Machines may have breakdowns or not be scheduled. Product models may have different production requirements, which will cause day-to-day variations.

For the data in Table 9-3, there was a low on April 9 of 1238 inspections because the second shift did not work, and a high on April 22 of 2678 inspections because of overtime in one work center.

Step 4. Determine the trial central line and control limits. Control limits are calculated using the same procedures and formulas as for a constant subgroup. However, because the subgroup size changes each day, limits must be calculated for each day. First, the average fraction

TABLE 9-3 Preliminary Data of Computer Modem Final Test and Control Limits for Each Subgroup

Subgroup		Number Inspected n	Number Nonconforming np	Fraction Nonconforming p	Limit UCL	Limit LCL
March	29	2,385	55	0.023	0.029	0.011
	30	1,451	18	0.012	0.031	0.009
	31	1,935	50	0.026	0.030	0.010
April	1	2,450	42	0.017	0.028	0.012
	2	1,997	39	0.020	0.029	0.011
	5	2,168	52	0.024	0.029	0.011
	6	1,941	47	0.024	0.030	0.010
	7	1,962	34	0.017	0.030	0.010
	8	2,244	29	0.013	0.029	0.011
	9	1,238	53	0.043	0.032	0.008
	12	2,289	45	0.020	0.029	0.011
	13	1,464	26	0.018	0.031	0.009
	14	2,061	47	0.023	0.029	0.011
	15	1,667	34	0.020	0.030	0.010
	16	2,350	31	0.013	0.029	0.011
	19	2,354	38	0.016	0.029	0.011
	20	1,509	28	0.018	0.031	0.009
	21	2,190	30	0.014	0.029	0.011
	22	2,678	113	0.042	0.028	0.012
	23	2,252	58	0.026	0.029	0.011
	26	1,641	34	0.021	0.030	0.010
	27	1,782	19	0.011	0.030	0.010
	28	1,993	30	0.015	0.030	0.010
	29	2,382	17	0.007	0.029	0.011
	30	2,132	46	0.022	0.029	0.011
		50,515	1,015			

nonconforming, which is the central line, must be determined, and it is

$$\bar{p} = \frac{\Sigma np}{\Sigma n} = \frac{1{,}015}{50{,}515} = 0.020$$

Using $\bar{p}$, the control limits for each day can be obtained. For March 29, the limits are

$$\begin{aligned} UCL_{29} &= \bar{p} + 3\sqrt{\frac{\bar{p}(1-\bar{p})}{n_{29}}} \\ &= 0.020 + 3\sqrt{\frac{0.020(1-0.020)}{2385}} \\ &= 0.029 \end{aligned}$$

$$\begin{aligned} LCL_{29} &= \bar{p} - 3\sqrt{\frac{\bar{p}(1-\bar{p})}{n_{29}}} \\ &= 0.020 - 3\sqrt{\frac{0.020(1-0.020)}{2385}} \\ &= 0.011 \end{aligned}$$

For March 30, the control limits are

$$\begin{aligned} UCL_{30} &= \bar{p} + 3\sqrt{\frac{\bar{p}(1-\bar{p})}{n_{30}}} \\ &= 0.020 + 3\sqrt{\frac{0.020(1-0.020)}{1451}} \\ &= 0.031 \end{aligned}$$

$$\begin{aligned} LCL_{30} &= \bar{p} - 3\sqrt{\frac{\bar{p}(1-\bar{p})}{n_{30}}} \\ &= 0.020 - 3\sqrt{\frac{0.020(1-0.020)}{1451}} \\ &= 0.009 \end{aligned}$$

The control limit calculations above are repeated for the remaining 23 subgroups. Because n is the only variable that is changing, it is possible to simplify the calculations as follows:

$$CL\text{'s} = \bar{p} \pm \frac{3\sqrt{\bar{p}(1-\bar{p})}}{\sqrt{n}}$$

$$= 0.020 \pm \frac{3\sqrt{0.020(1 - 0.020)}}{\sqrt{n}}$$

$$= 0.020 \pm \frac{0.42}{\sqrt{n}}$$

Using this technique, the calculations are much faster. The control limits for all 25 subgroups are shown in columns 5 and 6 of Table 9-3. A graphical illustration of the trial control limits, central line, and subgroup values is shown in Figure 9-5.

Note that as the subgroup size gets larger, the control limits are closer together; as the subgroup size gets smaller, the control limits become wider apart. This fact is apparent from the formula and by comparing the subgroup size, n, with its UCL and LCL.

Step 5. Establish revised central line and control limits. A review of Figure 9-5 shows that an out-of-control situation was present on April 9, April 22, and April 29. There was a problem with the wave solder on April 9 and April 22. Also, it was found that on April 29 the testing instrument was out of calibration. Because all these out-of-control points have assignable causes, they are discarded. A new $\bar{p}$ is obtained as follows:

$$\bar{p}_{new} = \frac{\Sigma np - np_d}{\Sigma n - n_d}$$

$$= \frac{1{,}015 - 53 - 113 - 17}{50{,}515 - 1{,}238 - 2{,}678 - 2{,}382}$$

$$= 0.019$$

Because this value represents the best estimate of the standard or reference value of the fraction nonconforming, $p_0 = 0.019$.

The fraction nonconforming, p_0, is used to calculate upper and lower control limits for the next period, which is the month of May. However, the limits cannot be calculated until the end of each day, when the subgroup size, n, is known. This means that the control limits are never known ahead of time. Table 9-4 shows the inspection results for the first three working days in May. Control limits and the fraction nonconforming for May 3 are as follows:

TABLE 9-4 Inspection Results for May 3, 4, and 5

Subgroup	Number Inspected	Number Nonconforming
May 3	1,535	31
4	2,262	28
5	1,872	45

$$p_{May\,3} = \frac{np}{n} = \frac{31}{1{,}535} = 0.020$$

$$UCL_{May\,3} = p_0 + 3\sqrt{\frac{p_0(1 - p_0)}{n_{May\,3}}}$$

$$= 0.019 + 3\sqrt{\frac{0.019(1 - 0.019)}{1535}}$$

$$= 0.029$$

$$LCL_{May\,3} = p_0 - 3\sqrt{\frac{p_0(1 - p_0)}{n_{May\,3}}}$$

$$= 0.019 - 3\sqrt{\frac{0.019(1 - 0.019)}{1535}}$$

$$= 0.009$$

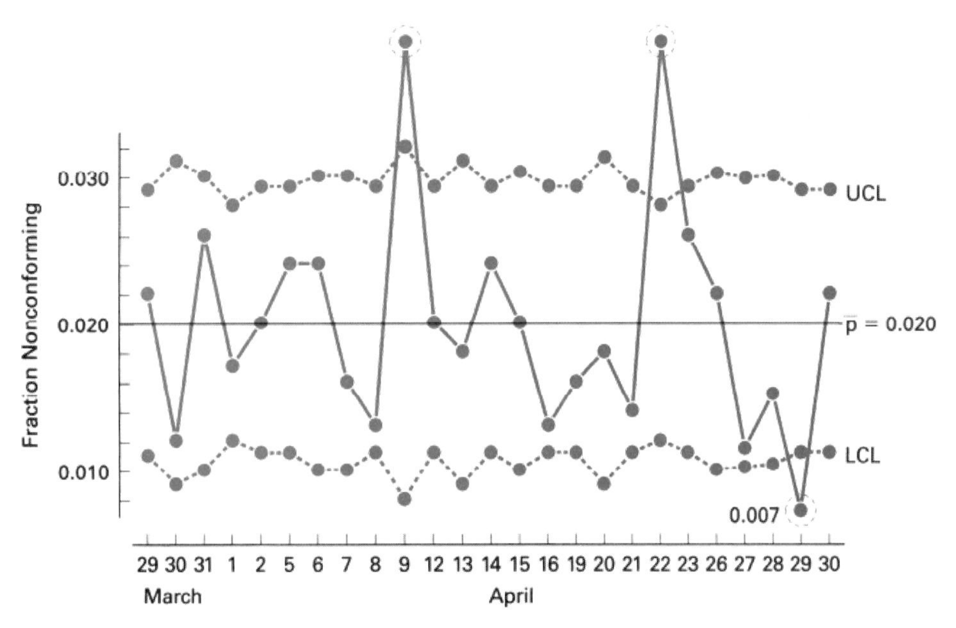

FIGURE 9-5 Preliminary Data, Central Line, and Trial Control Limits

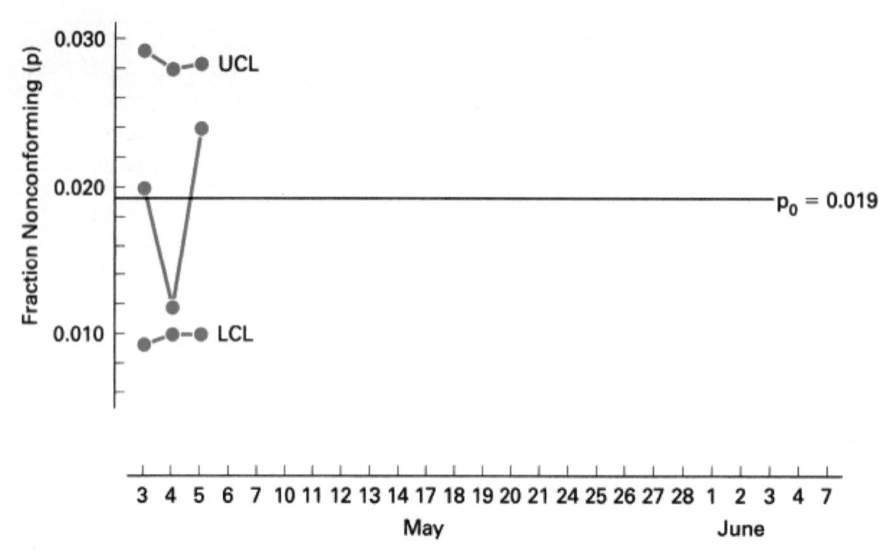

FIGURE 9-6 Control Limits and Fraction Nonconforming for First Three Working Days in May

The upper and lower control limits and the fraction nonconforming for May 3 are posted to the p chart as illustrated in Figure 9-6. In a similar manner, calculations are made for May 4 and 5 and the results posted to the chart.

The chart is continued until the end of May, using $p_0 = 0.019$. Because an improvement usually occurs after introduction of a chart, a better estimate of p_0 will probably be obtained at the end of May using that month's data. In the future, the value of p_0 should be evaluated periodically.

If p_0 is known, the process of data collection and trial control limits is not necessary. This saves considerable time and effort.

Because some confusion occurs among $p_0, \bar{p}$, and p, their definitions will be repeated:

1. p is the proportion (fraction) nonconforming in a single subgroup. It is posted to the chart but is *not* used to calculate the control limits.
2. $\bar{p}$ is the average proportion (fraction) nonconforming of many subgroups. It is the sum of the number nonconforming divided by the sum of the number inspected and is used to calculate the trial control limits.
3. p_0 is the standard or reference value of the proportion (fraction) nonconforming based on the best estimate of $\bar{p}$. It is used to calculate the revised control limits. It can be specified as a desired value.
4. ϕ is the population proportion (fraction) nonconforming. When this value is known, it can be used to calculate the limits, because $p_0 = \phi$.

Minimizing the Effect of Variable Subgroup Size

When the control limits vary from subgroup to subgroup, it presents an unattractive chart that is difficult to explain to operating personnel. It is also difficult to explain that control limits are calculated at the end of each day or time period rather than ahead of time. There are two techniques that minimize the effect of the variable subgroup size.

1. **Control limits for an average subgroup size.** By using an average subgroup size, one limit can be calculated and placed on the control chart. The average group size can be based on the anticipated production for the month or the previous month's inspections. As an example, the average number inspected for the preliminary data in Table 9-3 would be

$$n_{av} = \frac{\Sigma n}{g} = \frac{50{,}515}{25} = 2020.6, \quad \text{say, } 2000$$

Using a value of 2000 for the subgroup size, n, and $p_0 = 0.019$, the upper and lower control limits become

$$\begin{aligned} \text{UCL} &= p_0 + 3\sqrt{\frac{p_0(1-p_0)}{n_{av}}} \\ &= 0.019 + 3\sqrt{\frac{0.019(1-0.019)}{2000}} \\ &= 0.028 \\ \text{LCL} &= p_0 - 3\sqrt{\frac{p_0(1-p_0)}{n_{av}}} \\ &= 0.019 - 3\sqrt{\frac{0.019(1-0.019)}{2000}} \\ &= 0.010 \end{aligned}$$

These control limits are shown in the p chart of Figure 9-7 along with the fraction nonconforming, p, for each day in May.

When an average subgroup size is used, there are four situations that occur between the control limits and the individual fraction nonconforming values.

Case I. This case occurs when a point (subgroup fraction nonconforming) falls inside the limits and its subgroup size is smaller than the average subgroup size.

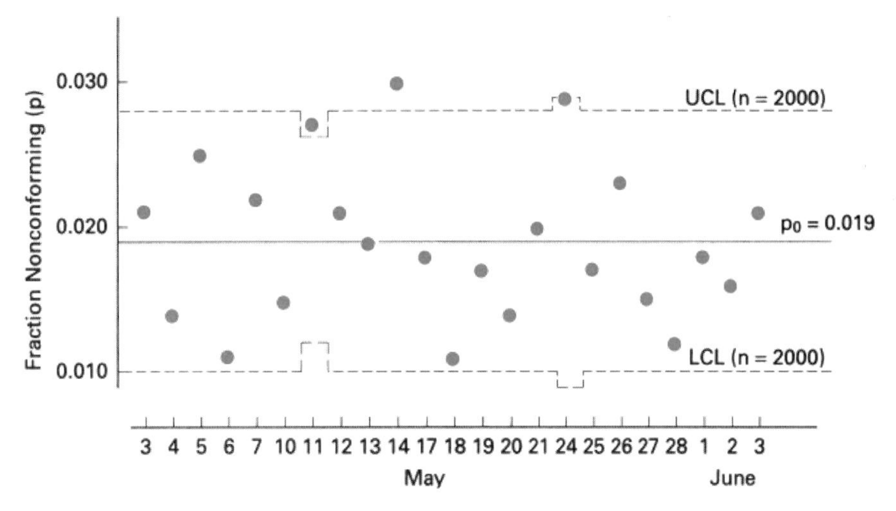

FIGURE 9-7 Chart for May Data Illustrating Use of an Average Subgroup Size

The data for May 6, $p = 0.011$ and $n = 1828$, represent this case. Because the May 6 subgroup size (1828) is less than the average of 2000, the control limits for May 6 will be farther apart than the control limits for the average subgroup size. Therefore, in this case individual control limits are not needed. If p is in control when $n = 2000$, it must also be in control when $n = 1828$.

Case II. This case occurs when a point (subgroup fraction nonconforming) falls inside the average limits and its subgroup size is larger than the average subgroup size. The data for May 11, $p = 0.027$ and $n = 2900$, illustrate this case. Because the May 11 subgroup size is greater than the average subgroup size, the control limits for May 11 will be closer together than the control limits for the average subgroup size. Therefore, when there is a substantial difference in the subgroup size, individual control limits are calculated. For May 11 the values for the upper and lower control limits are 0.026 and 0.012, respectively. These individual control limits are shown in Figure 9-7. It is seen that the point is beyond the individual control limit and so we have an out-of-control situation.

Case III. This case occurs when a point (subgroup fraction nonconforming) falls outside the limits and its subgroup size is larger than the average subgroup size: The data for May 14, $p = 0.030$ and $n = 2365$, illustrate this case. Because the May 14 subgroup size (2365) is greater than the average of 2000, the control limits for May 14 will be narrower than the control limits for the average subgroup size. Therefore, in this case individual control limits are not needed. If p is out of control when $n = 2000$, it must also be out of control when $n = 2365$.

Case IV. This case occurs when a point (subgroup fraction nonconforming) falls outside the limits and its subgroup size is less than the average subgroup size. The data for May 24, $p = 0.029$ and $n = 1590$, illustrate this case. Because the May 24 subgroup size (1590) is less than the average of 2000, the control limits for May 24 will be wider apart than the control limits for the average subgroup size. Therefore, when there is a substantial difference in the subgroup size, individual control limits are calculated. For May 24 the values for the upper and lower control limits are 0.029 and 0.009, respectively. These individual control limits are shown in Figure 9-7. It is seen that the point is on the individual control limit and is assumed to be in control.

It is not always necessary to calculate the individual control limits in cases II and IV. Only when the value of p is close to the control limits is it necessary to determine the individual limits. For this example problem, p values within, say, ± 0.002 of the original limits should be checked. Because approximately 5% of the p values will be close to the control limits, few p values will need to be evaluated.

In addition, it is not necessary to calculate individual control limits as long as the subgroup size does not deviate substantially from the average, say, 15%. For this example, subgroup sizes from 1700 to 2300 would be satisfactory and not need to have individual limit calculations.

Actually, when the average subgroup size is used, individual control limits are determined infrequently—about once every 3 months.

2. **Control limits for different subgroup sizes.** Another technique, which has been found to be effective, is to establish control limits for different subgroup sizes. Figure 9-8 illustrates such a chart. Using the different control limits and the four cases described previously, the need to calculate individual control limits would be rare. For example, the subgroup for July 16 with 1150 inspections is in control, and the subgroup for July 22 with 3500 inspections is out of control.

An analysis of Figure 9-8 shows that the relationship of the control limits to the subgroup size, n, is exponential

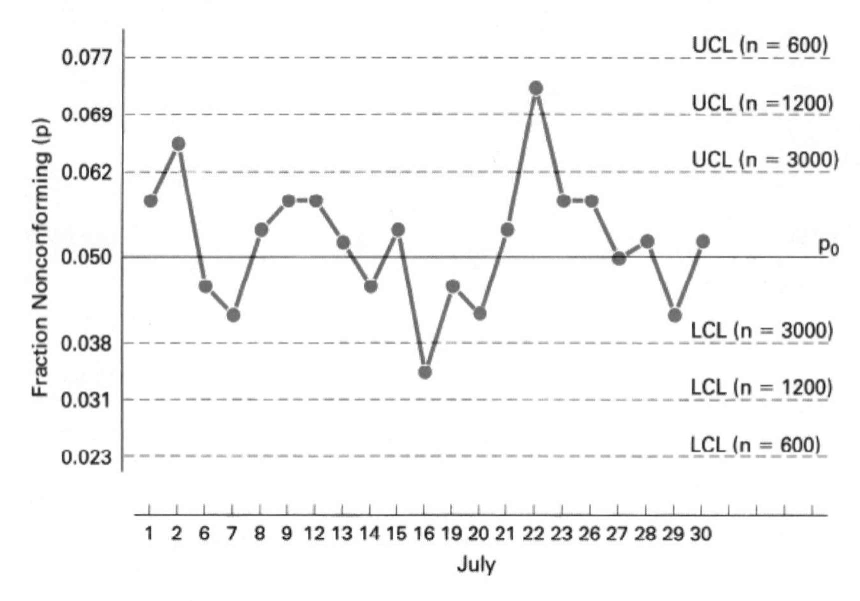

FIGURE 9-8 *p* Chart Illustrating Central Line and Control Limits for Different Subgroup Sizes

rather than linear. In other words, the control limit lines are not equally spaced for equal subdivisions of the subgroup size, n. This type of chart can be effective when there are extreme variations in subgroup size.

Number Nonconforming Chart

The number nonconforming chart (np chart) is almost the same as the p chart; however, you would not use both for the same objective.

The np chart is easier for operating personnel to understand than the p chart. Also, inspection results are posted directly to the chart without any calculations.

If the subgroup size is allowed to vary, the central line and the control limits will vary, which presents a chart that is almost meaningless. Therefore, one limitation of an np chart is the requirement that the subgroup size be constant. The sample size should be shown on the chart so viewers have a reference point.

Because the number nonconforming chart is mathematically equivalent to the proportion nonconforming chart, the central line and control limits are changed by a factor of n. Formulas are

$$\text{Central line} = np_0$$
$$\text{Control limits} = np_0 \pm 3\sqrt{np_0(1 - p_0)}$$

If the fraction nonconforming p_0 is unknown, then it must be determined by collecting data, calculating trial control limits, and obtaining the best estimate of p_0. The trial control limits formulas are obtained by substituting $\bar{p}$ for p_0 in the formulas. An example problem illustrates the technique.

Example Problem 9-3

A government agency samples 200 documents per day from a daily lot of 6000. From past records, the standard or reference value for the fraction nonconforming p_0, is 0.075.

Central line and control limit calculations are

$$np_0 = 200(0.075) = 15.0$$
$$\begin{aligned} \text{UCL} &= np_0 + 3\sqrt{np_0(1 - p_0)} \\ &= 15 + 3\sqrt{15(1 - 0.075)} \\ &= 26.2 \end{aligned}$$
$$\begin{aligned} \text{LCL} &= np_0 - 3\sqrt{np_0(1 - p_0)} \\ &= 15 - 3\sqrt{15(1 - 0.075)} \\ &= 3.8 \end{aligned}$$

Because the number nonconforming is a whole number, the limit values should be whole numbers; however, they can be left as fractions. This practice prevents a plotted point from falling on a control limit. Of course the central line is a fraction. The control chart is shown in Figure 9-9 for 4 weeks in October.

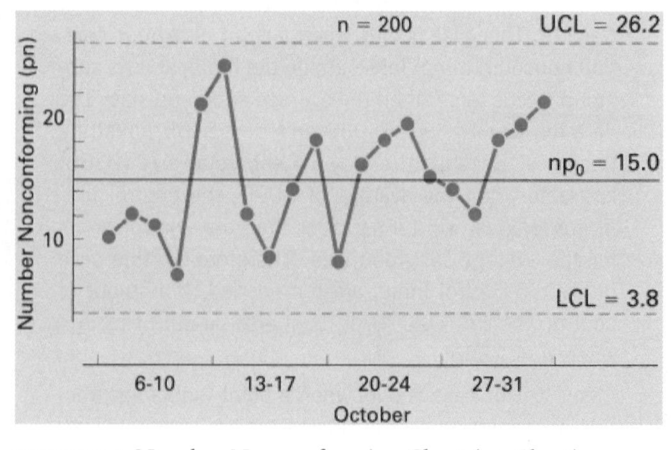

FIGURE 9-9 Number Nonconforming Chart (np Chart)

Process Capability

The process capability of a variable was described in Chapter 6. For an attribute, this process is much simpler. In fact, the process capability is the central line of the control chart.

Figure 9-10 shows a percent nonconforming chart for first-run automobile water leaks with a central line of 5.0%. The 5.0% value is the process capability, and the plotted points vary from the capability within the control limits. This variation occurs in a random manner but follows the binomial distribution.

Although the control limits show the limits of the variation of the capability, it should be understood that the limits are a function of the subgroup size. This fact is shown in Figure 9-10 for subgroup sizes of 500, 200, and 50. As the subgroup size increases, the control limits become closer to the central line.

Management is responsible for the capability. If the 5% value is not satisfactory, then management must initiate the procedures and provide the resources to take the necessary corrective action. As long as operating personnel (operators, first-line supervisors, and maintenance workers) are maintaining the plotted points within the control limits, they are doing what the process is capable of doing. When the plotted point is outside the control limit, operating personnel are usually responsible. A plotted point below the lower control limit is due to exceptionally good quality. It should be investigated to determine the assignable cause so that, if it is not due to an inspection error, it can be repeated.

CONTROL CHARTS FOR COUNT OF NONCONFORMITIES

Introduction

The other group of attribute charts is the nonconformity charts. Whereas a p chart controls the proportion nonconforming of the product or service, the nonconformities chart controls the count of nonconformities within the product or service. Remember, an item is classified as a nonconforming unit whether it has one or many nonconformities. There are two types of charts: count of nonconformities (c) charts and count of nonconformities per unit (u) charts.

Because these charts are based on the Poisson distribution, two conditions must be met. First, the average count of nonconformities must be much less than the total possible count of nonconformities. In other words, the opportunity for nonconformities is large, whereas the chance of a nonconformity at any one location is very small. This situation is typified by the rivets on a commercial airplane, where there are a large number of rivets but a small chance of any one rivet being a nonconformity. The second condition specifies that the occurrences are independent. In other words, the occurrence of one nonconformity does not increase or decrease the chance of the next occurrence being a nonconformity. For example, if a typist types an incorrect letter, there is an equal likelihood of the next letter being incorrect. Any beginning typist knows that this is not always the case because if the hands are not on the home keys, the chance of the second letter being incorrect is almost a certainty.

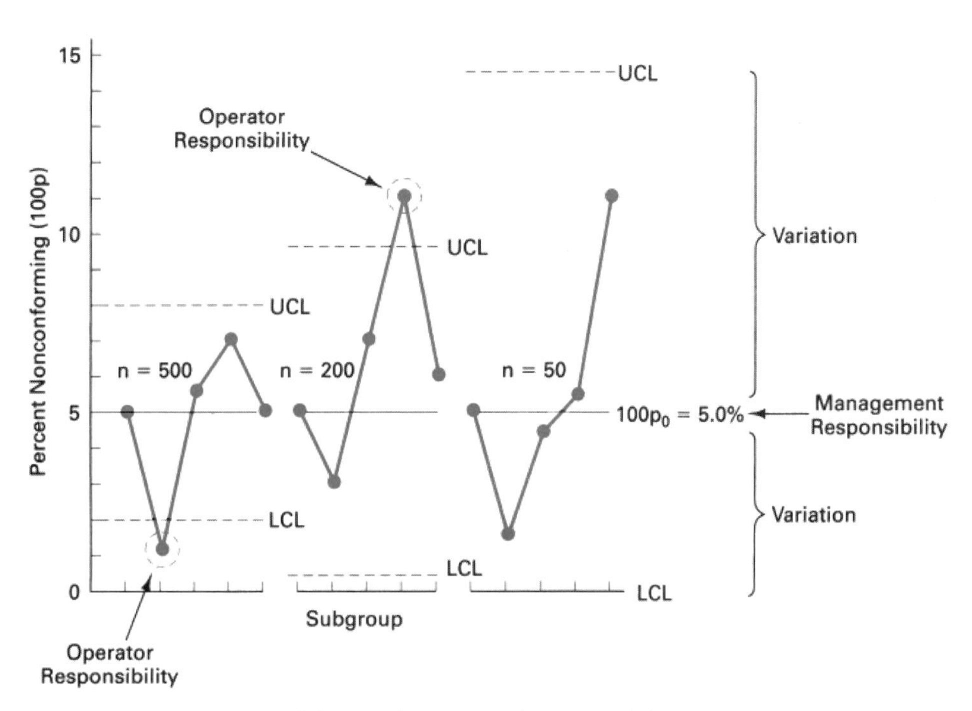

FIGURE 9-10 Process Capability Explanation and Responsibility

Other places where a chart of nonconformities meets the two conditions are imperfections in a large roll of paper, typographical errors on a printed page, rust spots on steel sheets, seeds or air pockets in glassware, adhesion defects per 1000 square feet of corrugated board, mold marks on fiberglass canoes, billing errors, and errors on forms.

Like nonconforming unit charts, the control limits for charts for nonconformities are usually based on 3σ from the central line. Therefore, approximately 99% of the subgroup values will fall within the limits. It is suggested that the reader review the section "State of Control" in Chapter 6, because much of that information is applicable to the nonconformity charts.

Objectives

The charts for count of nonconformities are not as inclusive as the $\overline{X}$ and R charts or the p charts, but they still have a number of applications, some of which have been mentioned.

The objectives of charts for count of nonconformities are to

1. Determine the average quality level as a benchmark or starting point. This information gives the initial process capability.
2. Bring to the attention of management any changes in the average. Once the average quality is known, any change becomes significant.
3. Improve the product or service quality. In this regard a chart for count of nonconformities can motivate operating and management personnel to initiate ideas for quality improvement. The chart will tell whether the idea is an appropriate or an inappropriate one. A continual and relentless effort must be made to improve the quality.
4. Evaluate the quality performance of operating and management personnel. As long as the chart is in control, operating personnel are performing satisfactorily. Because the charts for count of nonconformities are usually applicable to errors, they are very effective in evaluating the quality of the functional areas of finance, sales, customer service, and so on.
5. Suggest places to use the $\overline{X}$ and R charts. Some applications of the charts for count of nonconformities lend themselves to more detailed analysis by $\overline{X}$ and R charts.
6. Provide information concerning the acceptability of the product prior to shipment.

These objectives are almost identical to those for nonconforming charts. Therefore, the reader is cautioned to be sure that the appropriate group of charts is being used.

Because of the limitations of the charts for count of nonconformities, many organizations do not have occasion for their use.

c-Chart Construction

The procedures for the construction of a c chart are the same as those for the p chart. If the count of nonconformities, c_0, is unknown, it must be found by collecting data, calculating trial control limits, and obtaining the best estimate.

1. **Select the quality characteristic(s).** The first step in the procedure is to determine the use of the control chart. Like the p chart, it can be established to control (a) a single quality characteristic, (b) a group of quality characteristics, (c) a part, (d) an entire product, or (e) a number of products. It can also be established for performance control of (a) an operator, (b) a work center, (c) a department, (d) a shift, (e) a plant, or (f) a corporation. The use for the chart or charts will be based on securing the greatest benefit for a minimum of cost.

2. **Determine the subgroup size and method.** The size of a c chart is one inspected unit. An inspected unit could be one airplane, one case of soda cans, one gross of pencils, one bundle of Medicare applications, one stack of labels, and so forth. The method of obtaining the sample can either be by audit or online.

3. **Collect the data.** Data were collected on the count of nonconformities of a blemish nature for fiberglass canoes. These data were collected during the first and second weeks of May by inspecting random samples. Data are shown in Table 9-5 for 25 canoes, which is the minimum number of subgroups needed for trial control limit calculations. Note that canoes MY132 and MY278 both had production difficulties.

4. **Calculate the trial central line and control limits.** The formulas for the trial control limits are

$$\text{UCL} = \bar{c} + 3\sqrt{\bar{c}}$$
$$\text{LCL} = \bar{c} - 3\sqrt{\bar{c}}$$

where $\bar{c}$ is the average count of nonconformities for a number of subgroups. The value of $\bar{c}$ is obtained from the formula $\bar{c} = \Sigma c/g$, where g is the number of subgroups and c is the count of nonconformities. For the data in Table 9-5, the calculations are

$$\bar{c} = \frac{\Sigma c}{g} = \frac{141}{25} = 5.64$$

$$\begin{aligned} \text{UCL} &= \bar{c} + 3\sqrt{\bar{c}} \\ &= 5.64 + 3\sqrt{5.64} \\ &= 12.76 \end{aligned} \qquad \begin{aligned} \text{LCL} &= \bar{c} - 3\sqrt{\bar{c}} \\ &= 5.64 - 3\sqrt{5.64} \\ &= -1.48, \text{ or } 0 \end{aligned}$$

Because a lower control limit of -1.48 is impossible, it is changed to 0. The upper control limit of 12.76 is left as a fraction so that the plotted point, which is a whole number, cannot lie on the control limit. Figure 9-11 illustrates the central line, $\bar{c}$, the control limits, and the count of nonconformities, c, for each canoe of the preliminary data.

5. **Establish the revised central line and control limits.** In order to determine the revised 3σ control limits, the standard or reference value for the count of defects, c_0, is needed. If an analysis of the preliminary data shows good

TABLE 9-5 Count of Blemish Nonconformities (c) by Canoe Serial Number

Serial Number	Count of Nonconformities	Comment	Serial Number	Count of Nonconformities	Comment
MY102	7		MY198	3	
MY113	6		MY208	2	
MY121	6		MY222	7	
MY125	3		MY235	5	
MY132	20	Mold sticking	MY241	7	
MY143	8		MY258	2	
MY150	6		MY259	8	
MY152	1		MY264	0	
MY164	0		MY267	4	
MY166	5		MY278	14	Fell off skid
MY172	14		MY281	4	
MY184	3		MY288	5	
MY185	1		Total	$\Sigma c = 141$	

control, then $\bar{c}$ can be considered to be representative of that process, $c_0 = \bar{c}$.

Usually, however, an analysis of the preliminary data does not show good control, as illustrated in Figure 9-11. A better estimate of $\bar{c}$ (one that can be adopted for c_0) can be obtained by discarding out-of-control values with assignable causes. Low values that do not have an assignable cause represent exceptionally good quality. The calculations can be simplified by using the formula

$$\bar{c}_{\text{new}} = \frac{\Sigma c - c_d}{g - g_d}$$

where c_d = count of nonconformities in the discarded subgroups

g_d = number of discarded subgroups

Once an adopted standard or reference value is obtained, the revised 3σ control limits are found using the formulas

$$\text{UCL} = c_0 + 3\sqrt{c_0}$$
$$\text{LCL} = c_0 - 3\sqrt{c_0}$$

where c_0 = the reference or standard value for the count of nonconformities

The count of nonconformities, c_0, is the central line of the chart and is the best estimate using the available data. It equals $\bar{c}_{\text{new}}$.

Using the information from Figure 9-11 and Table 9-5, revised limits can be obtained. An analysis of Figure 9-11 shows that canoe numbers 132, 172, and 278 are out of control. Because canoes 132 and 278 have an assignable cause (see Table 9-5), they are discarded; however, canoe 172 may

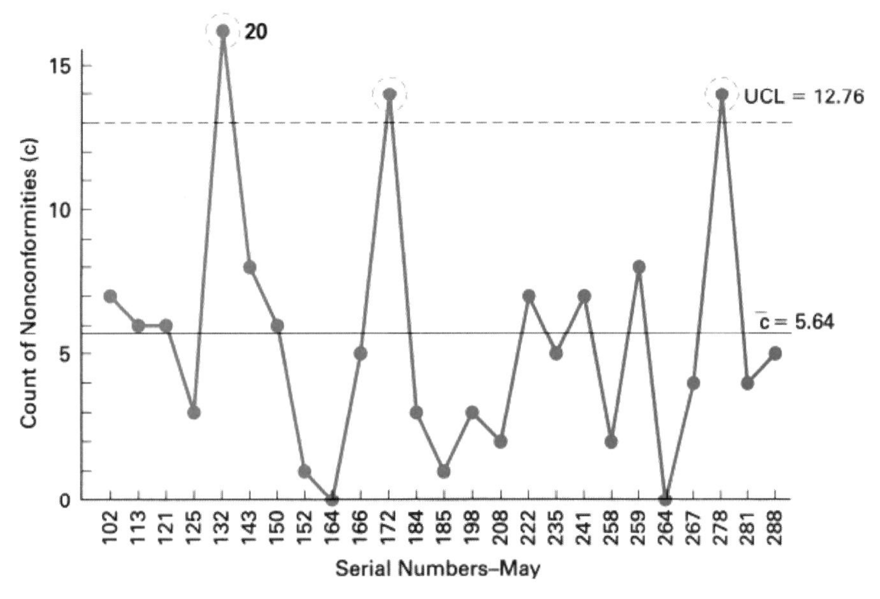

FIGURE 9-11 Control Chart for Count of Nonconformities (c Chart), Using Preliminary Data

Chart for Canoe Blemish Nonconformities
Model–17S

Type of Nonconformity																														
Scratches	1		2		2		3			1						2					1	2	1		1			1		
Paint Imperfections						1	2	1						1			3				1				1					3
Indentations	1		2						2					1					1		1	2							1	
Scuff Marks	1	1	5		3	4	3	5	2	2				4	1	2		1	2		1	3	1	5	2	1	2	2		3
Total	3	1	9	0	5	5	8	6	4	3			0	6	1	4	3	1	3	0	4	7	2	5	4	1	2	3	1	6
Serial Number	305	310	321	354	373	409	441	469	485	487			129	150	178	185	209	230	260	283	303	321	347	359	407	471	485	493	564	589

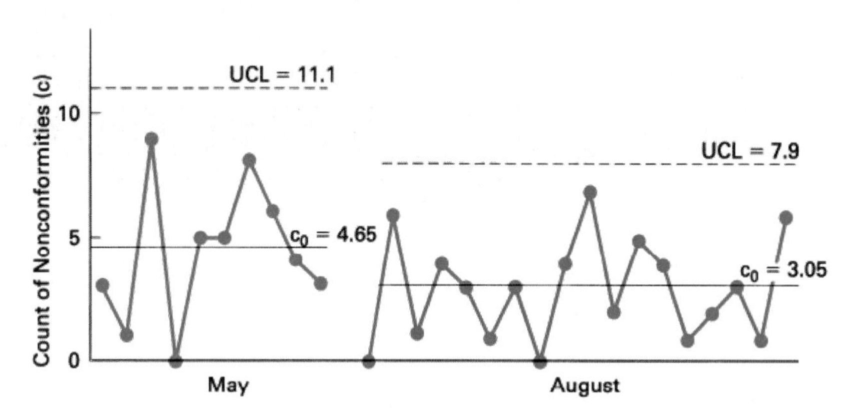

FIGURE 9-12 *c* Chart for Canoe Blemish Nonconformities

be due to a chance cause, it is not discarded. Therefore, $\bar{c}_{\text{new}}$ is obtained as follows:

$$\bar{c}_{\text{new}} = \frac{\Sigma c - c_d}{g - g_d}$$
$$= \frac{141 - 20 - 14}{25 - 2}$$
$$= 4.65$$

Because $\bar{c}_{\text{new}}$ is the best estimate of the central line, $c_0 = 4.65$. The revised control limits for the c chart are

$$\text{UCL} = c_0 + 3\sqrt{c_0} \qquad \text{LCL} = c_0 - 3\sqrt{c_0}$$
$$= 4.65 + 3\sqrt{4.65} \qquad = 4.65 - 3\sqrt{4.65}$$
$$= 11.1 \qquad = -1.82, \text{ or } 0$$

These control limits are used to start the chart beginning with canoes produced during the third week of May and are shown in Figure 9-12.

If c_0 had been known, the data collection and trial control limit phase would have been unnecessary.

6. **Achieve the objective.** The reason for the control chart is to achieve one or more of the previously stated objectives. Once the objective is reached, the chart is discontinued or inspection activity is reduced and resources are allocated to another quality problem. Some of the objectives, however, such as the first one, can be ongoing.

As with the other types of control charts, an improvement in the quality is expected after the introduction of a chart. At the end of the initial period, a better estimate of the number of nonconformities can be obtained. Figure 9-12 illustrates the change in c_0 and in the control limits for August as the chart is continued in use. Quality improvement resulted from the evaluation of ideas generated by the project team, such as attaching small pieces of carpet to the skids, faster-drying ink, worker training programs, and so on. The control chart shows whether the idea improves the quality, reduces the quality, or does not change the quality. A minimum of 25 subgroups is needed to evaluate each idea. The subgroups can be taken as often as practical, as long as they are representative of the process. Only one idea should be evaluated at a time.

Figure 9-12 also illustrates a technique for reporting the number of nonconformities of individual quality characteristics, and the graph reports the total. This is an excellent technique for presenting the total picture and one that is accomplished with little additional time or cost. It is interesting to note that the serial numbers of the canoes that were selected for inspection were obtained from a random-number table.

The control chart should be placed in a conspicuous place where it can be viewed by operating personnel.

Chart for Count of Nonconformities per Unit (*u* Chart)[3]

The c chart is applicable where the subgroup size is an inspected unit of one, such as a canoe, an airplane, 1000 square feet of cloth, a ream of paper, 100 income tax forms,

[3] The chart is not included in ANSI/ASQ B1–B3.

and a keg of nails. The inspected unit can be any size that meets the objective; however, it must be constant. Recall that the subgroup size, n, is not in the calculations because its value is 1. When situations arise where the subgroup size varies, then the u chart (count of nonconformities/unit) is the appropriate chart. The u chart can also be used when the subgroup size is constant.

The u chart is mathematically equivalent to the c chart. It is developed in the same manner as the c chart, with the collection of 25 subgroups, calculation of trial central line and control limits, acquisition of an estimate of the standard or reference count of nonconformities per unit, and calculation of the revised limits. Formulas used for the procedure are

$$u = \frac{c}{n} \qquad \bar{u} = \frac{\sum c}{\sum n}$$

$$\text{UCL} = \bar{u} + 3\sqrt{\frac{\bar{u}}{n}} \qquad \text{LCL} = \bar{u} - 3\sqrt{\frac{\bar{u}}{n}}$$

where c = count of nonconformities in a subgroup
n = number inspected in a subgroup
u = count of nonconformities/unit in a subgroup
$\bar{u}$ = average count of nonconformities/unit for many subgroups

Revised control limits are obtained by substituting u_0 in the trial control limit formula. The u chart will be illustrated by an example.

Each day a clerk inspects the waybills of a small overnight air freight company for errors. Because the number of waybills varies from day to day, a u chart is the appropriate technique. If the number of waybills was constant, either the c or u chart would be appropriate. Data are collected as shown in Table 9-6. The date, number inspected, and count of nonconformities are obtained and posted to the table. The count of nonconformities per unit, u, is calculated and posted. Also, because the subgroup size varies, the control limits are calculated for each subgroup.

TABLE 9-6 Count of Nonconformities per Unit for Waybills

Date	Number Inspected n	Count of Nonconformities c	Nonconformities per unit u	UCL	LCL
Jan. 30	110	120	1.09	1.51	0.89
31	82	94	1.15	1.56	0.84
Feb. 1	96	89	.93	1.53	0.87
2	115	162	1.41	1.50	0.90
3	108	150	1.39	1.51	0.89
4	56	82	1.46	1.64	0.76
6	120	143	1.19	1.50	0.90
7	98	134	1.37	1.53	0.87
8	102	97	.95	1.53	0.87
9	115	145	1.26	1.50	0.90
10	88	128	1.45	1.55	0.85
11	71	83	1.16	1.59	0.81
13	95	120	1.26	1.54	0.86
14	103	116	1.13	1.52	0.88
15	113	127	1.12	1.51	0.89
16	85	92	1.08	1.56	0.84
17	101	140	1.39	1.53	0.87
18	42	60	1.19	1.70	0.70
20	97	121	1.25	1.53	0.87
21	92	108	1.17	1.54	0.86
22	100	131	1.31	1.53	0.87
23	115	119	1.03	1.50	0.90
24	99	93	.94	1.53	0.87
25	57	88	1.54	1.64	0.76
27	89	107	1.20	1.55	0.85
28	101	105	1.04	1.53	0.87
Mar. 1	122	143	1.17	1.49	0.91
2	105	132	1.26	1.52	0.88
3	98	100	1.02	1.53	0.87
4	48	60	1.25	1.67	0.73
	Total 2823	3389			

Data for 5 weeks at 6 days per week are collected for a total of 30 subgroups. Although only 25 subgroups are required, this approach eliminates any bias that could occur from the low activity that occurs on Saturday. The calculation for the trial central line is

$$\bar{u} = \frac{\Sigma c}{\Sigma n} = \frac{3389}{2823} = 1.20$$

Calculations for the trial control limits and the plotted point, u, must be made for each subgroup. For January 30 they are

$$UCL_{\text{Jan 30}} = \bar{u} + 3\sqrt{\frac{\bar{u}}{n}} \qquad LCL_{\text{Jan 30}} = \bar{u} - 3\sqrt{\frac{\bar{u}}{n}}$$

$$= 1.20 + 3\sqrt{\frac{1.20}{110}} \qquad = 1.20 - 3\sqrt{\frac{1.20}{110}}$$

$$= 1.51 \qquad = 0.89$$

$$u_{\text{Jan 30}} = \frac{c}{n} = \frac{120}{110} = 1.09$$

These calculations must be repeated for 29 subgroups and the values posted to the table.

A comparison of the plotted points with the upper and lower control limits in Figure 9-13 shows that there are no out-of-control values. Therefore, $\bar{u}$ can be considered the best estimate of u_0 and $u_0 = 1.20$. A visual inspection of the plotted points indicates a stable process. This situation is somewhat unusual at the beginning of control-charting activities.

To determine control limits for the next 5-week period, we can use an average subgroup size in the same manner as the variable subgroup size of the p chart. A review of the chart shows that the control limits for Saturday are much farther apart than for the rest of the week. This condition is due to the smaller subgroup size. Therefore, it appears appropriate to establish separate control limits for Saturday. Calculations are as follows:

$$n_{\text{Sat. avg.}} = \frac{\Sigma n}{g} = \frac{(56 + 71 + 42 + 57 + 48)}{5} = 55$$

$$UCL = u_0 + 3\sqrt{\frac{u_0}{n}} \qquad LCL = u_0 - 3\sqrt{\frac{u_0}{n}}$$

$$= 1.20 + 3\sqrt{\frac{1.20}{55}} \qquad = 1.20 - 3\sqrt{\frac{1.20}{55}}$$

$$= 1.64 \qquad = 0.76$$

$$n_{\text{daily avg.}} = \frac{\Sigma n}{g} = \frac{2823 - 274}{25} = 102, \quad \text{say, } 100$$

$$UCL = u_0 + 3\sqrt{\frac{u_0}{n}} \qquad LCL = u_0 - 3\sqrt{\frac{u_0}{n}}$$

$$= 1.20 + 3\sqrt{\frac{1.20}{100}} \qquad = 1.20 - 3\sqrt{\frac{1.20}{100}}$$

$$= 1.53 \qquad = 0.87$$

The control chart for the next period is shown in Figure 9-14. When the subgroup is a day's inspections, the true control limits will need to be calculated about once every 3 months.

The control chart can now be used to achieve the objective. If a project team is involved, it can test ideas for quality improvement.

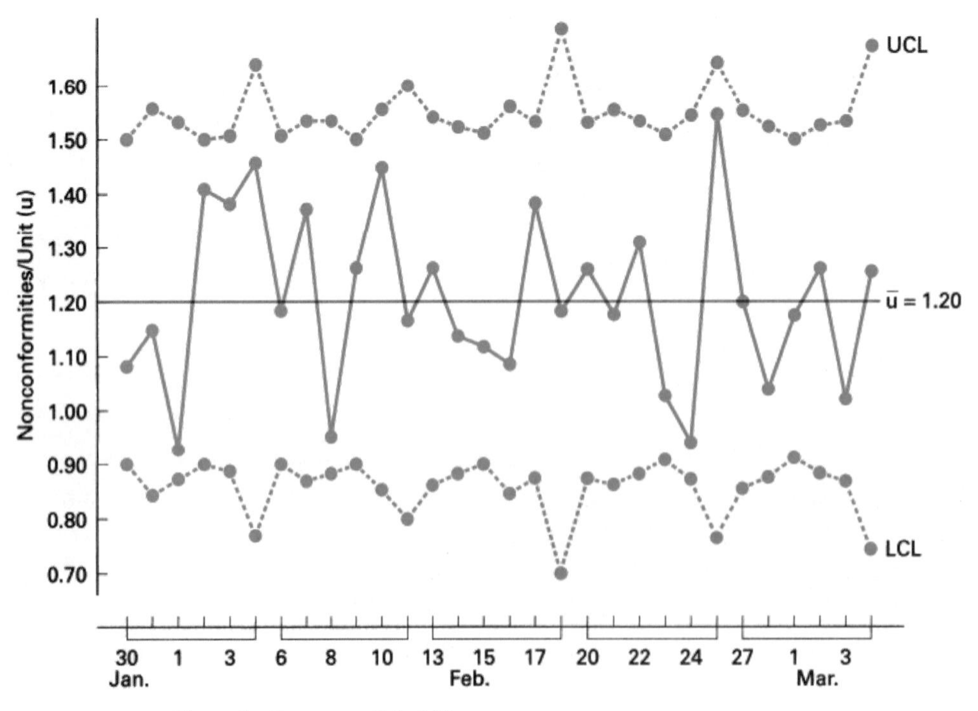

FIGURE 9-13 u Chart for Errors on Waybills

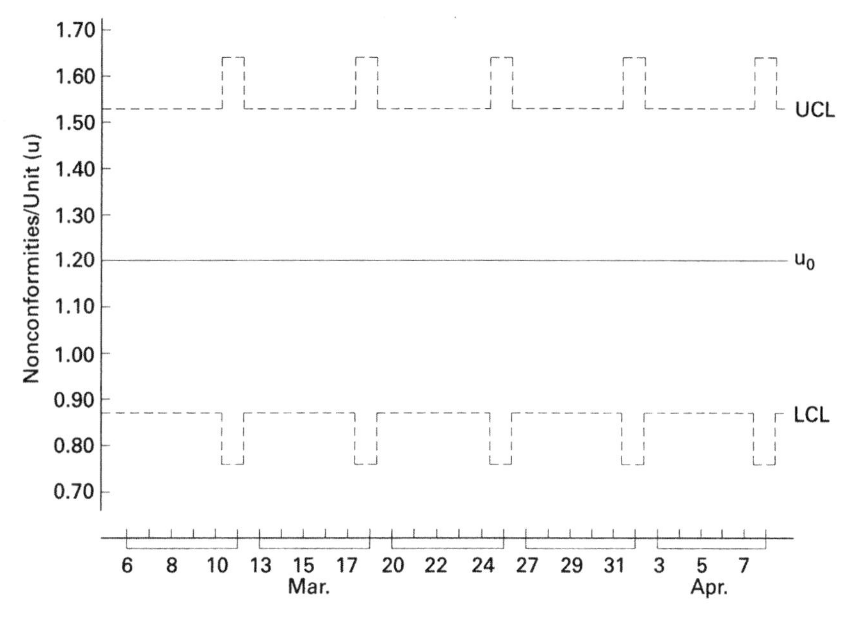

FIGURE 9-14 *u* Chart for Next Period

The *u* chart is identical to the *c* chart in all aspects except two. One difference is the scale, which is continuous for a *u* chart but discrete for the *c* chart. This difference provides more flexibility for the *u* chart, because the subgroup size can vary. The other difference is the subgroup size, which is 1 for the *c* chart.

The *u* chart is limited in that we do not know the location of the nonconformities. For example, in Table 9-6, February 4 has 82 nonconformities out of 56 inspected for a value of 1.46. All 82 nonconformities could have been counted on one unit.

Final Comments

Process capability for nonconformities is treated in a manner similar to nonconforming units. The reader is referred to Figure 9-10.

Figure 9-15 shows when to use the various attribute charts. First you need to decide whether to chart nonconformities or nonconforming units. Next you need to determine whether the subgroup size will be constant or will vary. These two decisions give the appropriate chart.

		Attribute Chart	
		Nonconforming Units	Nonconformities
Sample Size	Constant	np	c (n = 1)
	Constant or Varies	p	u

FIGURE 9-15 When to Use the Various Attribute Charts

A QUALITY RATING SYSTEM

Introduction

In the attribute charts of the preceding section, all nonconformities and nonconforming units had the same weight, regardless of their seriousness. For example, in the evaluation of desk chairs, one chair might have 5 nonconformities, all related to the surface finish, while another chair might have 1 nonconformity, a broken leg. The usable chair with 5 trivial nonconformities has 5 times the influence on the attribute chart as the unusable chair with 1 serious nonconformity. This situation presents an incorrect evaluation of the product quality. A quality rating system will correct this deficiency.

There are many situations in which it is desirable to compare the performance of operators, shifts, plants, or vendors. In order to compare quality performance, a quality rating system is needed to classify, weigh, and evaluate nonconformities.

Nonconformity Classification

Nonconformities and, for that matter, nonconforming units, are classified according to their severity. One system groups nonconformities into three classes:

1. **Critical nonconformities.** A critical nonconformity is a nonconformity that judgment and experience indicate is likely to result in hazardous or unsafe conditions for individuals using, maintaining, or depending on the product or service, or a nonconformity that judgment and experience indicate is likely to prevent performance of the function of the product or service.
2. **Major nonconformities.** A major nonconformity is a nonconformity, other than critical, that is likely to result in

failure or to reduce materially the usability of the product or service for its intended purpose.

3. **Minor nonconformities.** A minor nonconformity is a nonconformity that is not likely to reduce materially the usability of the product or service for its intended purpose. Minor nonconformities are usually associated with appearance.

To summarize, a critical nonconformity *will* affect usability; a major nonconformity *might* affect usability; and a minor nonconformity *will not* affect usability of the unit.

Other classification systems use four classes or two classes, depending on the complexity of the product. A catastrophic class is sometimes used.

Once the classifications are determined, the weights to assign to each class can be established. Although any weights can be assigned to the classifications, 9 points for a critical, 3 points for a major, and 1 point for a minor are usually considered to be satisfactory, because a major is three times as important as a minor and a critical is three times as important as a major.

Control Chart[4]

Control charts are established and plotted for count of demerits per unit. A demerit per unit is given by the formula

$$D = w_c u_c + w_{ma} u_{ma} + w_{mi} u_{mi}$$

where D = demerits per unit

w_c, w_{ma}, w_{mi} = weights for the three classes—critical, major, and minor

u_c, u_{ma}, u_{mi} = count of nonconformities per unit in each of the three classes—critical, major, and minor

When w_c, w_{ma}, and w_{mi} are 9, 3, and 1, respectively, the formula is

$$D = 9u_c + 3u_{ma} + 1u_{mi}$$

The D values calculated from the formula are posted to the chart for each subgroup.

The central line and the 3σ control limits are obtained from the formulas

$$D_0 = 9u_{0c} + 3u_{0ma} + 1u_{0mi}$$

$$\sigma_{0u} = \sqrt{\frac{9^2 u_{0c} + 3^2 u_{0ma} + 1^2 u_{0mi}}{n}}$$

$$\text{UCL} = D_0 + 3\sigma_{0u} \qquad \text{LCL} = D_0 - 3\sigma_{0u}$$

where u_{0c}, u_{0ma}, and u_{0mi} represent the standard nonconformities per unit for the critical, major, and minor classifications, respectively. The nonconformities per unit for the critical, major, and minor classifications are obtained by separating the nonconformities into the three classifications and treating each as a separate u chart.

[4] The demerit chart is not included in ANSI/ASQ/B1–B3.

Example Problem 9-4

Assuming that a 9:3:1 three-class weighting system is used, determine the central line and control limits when $u_{0c} = 0.08$, $u_{0ma} = 0.5$, $u_{0mi} = 3.0$, and $n = 40$. Also calculate the demerits per unit for May 25 when critical nonconformities are 2, major nonconformities are 26, and minor nonconformities are 160 for the 40 units inspected on that day. Is the May 25 subgroup in control or out of control?

$$D_0 = 9u_{0c} + 3u_{0ma} + 1u_{0mi}$$
$$= 9(0.08) + 3(0.5) + 1(3.0)$$
$$= 5.2$$

$$\sigma_{0u} = \sqrt{\frac{9^2 u_{0c} + 3^2 u_{0ma} + 1^2 u_{0mi}}{n}}$$
$$= \sqrt{\frac{81(0.08) + 9(0.5) + 1(3.0)}{40}}$$
$$= 0.59$$

$$\text{UCL} = D_0 + 3\sigma_{0u} \qquad \text{LCL} = D_0 - 3\sigma_{0u}$$
$$= 5.2 + 3(0.59) \qquad = 5.2 - 3(0.59)$$
$$= 7.0 \qquad = 3.4$$

The central line and control limits are illustrated in Figure 9-16. Calculations for the May 25 subgroup are

$$D_{\text{May 26}} = 9u_c + 3u_{ma} + 1u_{mi}$$
$$= 9\left(\frac{2}{40}\right) + 3\left(\frac{26}{40}\right) + 1\left(\frac{160}{40}\right)$$
$$= 6.4 \text{ (in control)}$$

Quality rating systems based on demerits per unit are useful for performance control and can be an important feature of a total quality system.

COMPUTER PROGRAM

The EXCEL program files on the website will solve for the four charts in this chapter. Their file names are *p-chart, np-chart, c-chart,* and *u-chart.*

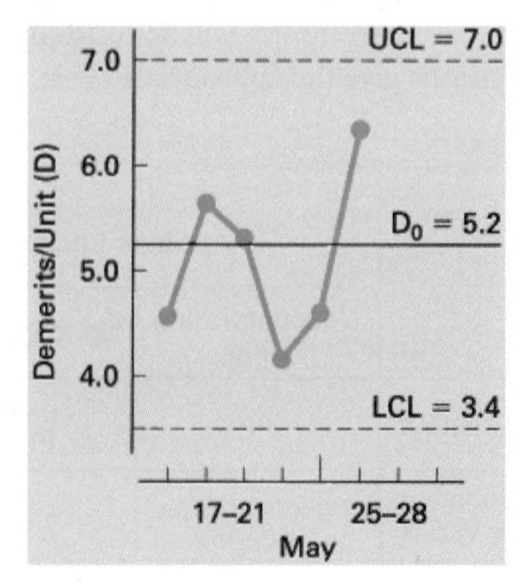

FIGURE 9-16 Demerit-Per-Unit Chart (D Chart)

EXERCISES

1. On page 144 is a typical attribute chart form with information concerning 2-L soda bottles.
   a. Calculate the proportion nonconforming for subgroups 21, 22, 23, 24, and 25. Construct a run chart.
   b. Calculate the trial central line and control limits. Draw these values on the chart.
   c. If it is assumed that any out-of-control points have assignable causes, what central line and control limits should be used for the next period?
2. Determine the trial central line and control limits for a *p* chart using the following data, which are for the payment of dental insurance claims. Plot the values on graph paper and determine if the process is stable. If there are any out-of-control points, assume an assignable cause and determine the revised central line and control limits.

Subgroup Number	Number Inspected	Number Nonconforming	Subgroup Number	Number Inspected	Number Nonconforming
1	300	3	14	300	6
2	300	6	15	300	7
3	300	4	16	300	4
4	300	6	17	300	5
5	300	20	18	300	7
6	300	2	19	300	5
7	300	6	20	300	0
8	300	7	21	300	2
9	300	3	22	300	3
10	300	0	23	300	6
11	300	6	24	300	1
12	300	9	25	300	8
13	300	5			

3. The supervisor is not sure about the best way to display the quality performance determined in Exercise 2. Calculate the central line and limits for the other methods of presentation.
4. After achieving the objective in the example problem concerning the hair dryer motor, it is decided to reduce the sample size to 80. What are the central line and control limits?
5. Fifty motor generators are inspected per day from a stable process. The best estimate of the fraction nonconforming is 0.076. Determine the central line and control limits. On a particular day, 5 nonconforming generators were discovered. Is this in control or out of control?
6. Inspection results of video-of-the-month shipments to customers for 25 consecutive days are given in the following table. What central line and control limits should be established and posted if it is assumed that any out-of-control points have assignable causes? The number of inspections each day is constant and equals 1750.

Date	Number Nonconforming	Date	Number Nonconforming
July 6	47	July 23	37
7	42	26	39
8	48	27	51
9	58	28	44
12	32	29	61
13	38	30	48
14	53	Aug. 2	56
15	68	3	48
16	45	4	40
19	37	5	47
20	57	6	25
21	38	9	35
22	53		

ATTRIBUTES CONTROL CHART p ☒ np ☐ u ☐ c ☐ CHART ID:

PART ID: *2 LITER-BOTTLE* **OPERATION ID:** *NEW PACKAGING LINE* **DEPT/AREA:** *PACKAGING*

CHECK METHOD: *VISUAL* **CHARACTERISTIC:** *CASE PACKING DEFECTS*

DAY:	1	2	3	4	5	6	7	8	9	10	11	12	13	14	15	16	17	18	19	20	21	22	23	24	25					
SAMPLE (n)	400	400	400	400	400	400	400	400	400	400	400	400	400	400	400	400	400	400	400	400	400	400	400	400	400					
NUMBER (np, c)	43	21	14	20	15	16	8	12	18	4	6	12	5	4	3	8	7	31	8	6	4	7	9	6	10					
PROPORTION (p, u)	.108	.053	.035	.050	.038	.040	.020	.030	.045	.010	.015	.030	.013	.010	.008	.020	.018	.078	.020	.015										

AVG = UCL = LCL =

7. The performance of the first shift is reflected in the inspection results of electric carving knives. Determine the trial central line and control limits for each subgroup. Assume that any out-of-control points have assignable causes and determine the standard value for the fraction nonconforming for the next production period.

Date	Number Inspected	Number Nonconforming	Date	Number Inspected	Number Nonconforming
Sept. 6	500	5	Sept. 23	525	10
7	550	6	24	650	3
8	700	8	27	675	8
9	625	9	28	450	23
10	700	7	29	500	2
13	550	8	30	375	3
14	450	16	Oct. 1	550	8
15	600	6	4	600	7
16	475	9	5	700	4
17	650	6	6	660	9
20	650	7	7	450	8
21	550	8	8	500	6
22	525	7	11	525	1

8. Daily inspection results for the model 305 electric range assembly line are given in the following table. Determine trial control limits for each subgroup. Assume that any out-of-control points have assignable causes, and determine the standard value for the fraction nonconforming for December.

Date and Shift	Number Inspected	Number Nonconforming	Date and Shift	Number Inspected	Number Nonconforming
Nov. 8 I	171	31	Nov. 17 I	165	16
II	167	6	II	170	35
9 I	170	8	18 I	175	12
II	135	13	II	167	6
10 I	137	26	19 I	141	50
II	170	30	II	159	26
11 I	45	3	22 I	181	16
II	155	11	II	195	38
12 I	195	30	23 I	165	33
II	180	36	II	140	21
15 I	181	38	24 I	162	18
II	115	33	II	191	22
16 I	165	26	25 I	139	16
II	189	15	II	181	27

9. Control limits are to be established based on the average number inspected from the information in Exercise 8. What are these control limits and the central line? Describe the cases where individual control limits will need to be calculated.

10. Control charts are to be established on the manufacture of backpack frames. The revised fraction nonconforming is 0.08. Determine control limit lines for inspection rates of 1000 per day, 1500 per day, and 2000 per day. Draw the control chart. Why are the control limits not spaced equally?

11. Determine the revised central line and control limits for a percent nonconforming chart for the information in:

    a. Exercise 2

    b. Exercise 6

12. From the information of Exercise 2, determine the revised central line and control limits for an *np* chart.

13. From the information of Exercise 6, determine the revised central line and control limits for an *np* chart. Which chart is more meaningful to operating personnel?

14. An *np* chart is to be established on a painting process that is in statistical control. If 35 pieces are to be inspected every 4 h and the fraction nonconforming is 0.06, determine the central line and control limits.

15. Determine the revised central line and control limits for *fraction conforming, percent conforming,* and *number conforming* charts for the information in:

    a. Exercise 2

    b. Exercise 6

16. Find the process capability for

    a. Exercise 6

    b. Exercise 7

    c. Exercise 10

17. A credit card manager wishes to determine the proportion of customer calls that result in a dissatisfied customer. Based on some preliminary data, she estimates the percentage to be 10% ($p = 0.10$). A precision of 15% and confidence level of 90% are desired. What is the sample size?

18. The sample size of a *p* chart for a fertilizer packing line needs to be determined. Preliminary data indicate that 8% of the bags are outside the weight specifications. What is the sample size for a precision of 10% and a 70% confidence level? For a precision of 10% and a confidence level of 99%? For a precision of 10% and a confidence level of 95%? What conclusions can you make concerning the precision and confidence level?

19. The count of surface nonconformities in 1000 m^2 of 20-kg kraft paper is given in the following table. Determine the trial central line and control limits and the revised central line and control limits, assuming that out-of-control points have assignable causes.

Lot Number	Count of Nonconformities	Lot Number	Count of Nonconformities
20	10	36	2
21	8	37	12
22	6	38	0
23	6	39	6
24	2	40	14
25	10	41	10
26	8	42	8
27	10	43	6
28	0	44	2
29	2	45	14
30	8	46	16
31	2	47	10
32	20	48	2
33	10	49	6
34	6	50	3
35	30		

20. A bank has compiled the data in the following table showing the count of nonconformities for 100,000 accounting transactions per day during December and January. What control limits and central line are recommended for the control chart for February? Assume any out-of-control points have assignable causes.

Count of Nonconformities	Count of Nonconformities
8	17
19	14
14	9

Count of Nonconformities	Count of Nonconformities
18	7
11	15
16	22
8	19
15	38
21	12
8	13
23	5
10	2
9	16

21. A quality technician has collected data on the count of rivet nonconformities in 4-m travel trailers. After 30 trailers, the total count of nonconformities is 316. Trial control limits have been determined, and a comparison with the data shows no out-of-control points. What is the recommendation for the central line and the revised control limits for a count of nonconformities chart?

22. One hundred product labels are inspected every day for surface nonconformities. Results for the past 25 days are 22, 29, 25, 17, 20, 16, 34, 11, 31, 29, 15, 10, 33, 23, 27, 15, 17, 17, 19, 22, 23, 27, 29, 33, and 21. Plot the points on graph paper (run chart) and determine if the process is stable. Determine the trial central line and control limits.

23. Determine the trial control limits and revised control limits for a *u* chart using the data in the table for the surface finish of rolls of white paper. Assume any out-of-control points have assignable causes.

Lot Number	Sample Size	Total Nonconformities	Lot Number	Sample Size	Total Nonconformities
1	10	45	15	10	48
2	10	51	16	11	35
3	10	36	17	10	39
4	9	48	18	10	29
5	10	42	19	10	37
6	10	5	20	10	33
7	10	33	21	10	15
8	8	27	22	10	33
9	8	31	23	11	27
10	8	22	24	10	23
11	12	25	25	10	25
12	12	35	26	10	41
13	12	32	27	9	37
14	10	43	28	10	28

24. A warehouse distribution activity has been in statistical control, and control limits are needed for the next period. If the subgroup size is 100, the total count of nonconformities is 835, and the number of subgroups is 22, what are the new control limits and central line?

25. Construct a control chart for the data in the table for empty bottle inspections of a soft drink manufacturer. Assume assignable causes for any points that are out of control.

Number of Bottles	Chips, Scratches, Other	Foreign Material on Sides	Foreign Material on Bottom	Total Nonconformities
40	9	9	27	45
40	10	1	29	40
40	8	0	25	33
40	8	2	33	43
40	10	6	46	62
52	12	16	51	79
52	15	2	43	60
52	13	2	35	50
52	12	2	59	73
52	11	1	42	54
52	15	15	25	55
52	12	5	57	74
52	14	2	27	43
52	12	7	42	61
40	11	2	30	43
40	9	4	19	32

Number of Bottles	Chips, Scratches, Other	Foreign Material on Sides	Foreign Material on Bottom	Total Nonconformities
40	5	6	34	45
40	8	11	14	33
40	3	9	38	50
40	9	9	10	28
52	13	8	37	58
52	11	5	30	46
52	14	10	47	71
52	12	3	41	56
52	12	2	28	42

26. Assuming that a 10:5:1 demerit weighting system is used, determine the central line and control limits when $u_c = 0.11$, $u_{ma} = 0.70$, $u_{mi} = 4.00$, and $n = 50$. If the subgroup inspection results for a particular day are 1 critical, 35 major, and 110 minor nonconformities, determine if the results are in control or out of control.

27. Solve the following problems using the EXCEL program files on the website.

    a. Exercise 2

    b. Exercise 13

    c. Exercise 17

    d. Exercise 21

28. Prepare an EXCEL template for the four charts to ensure that the LCL is always 0.

29. Write an EXCEL template for a D chart.

30. Determine the formulas for a number conforming chart, nq.

CHAPTER
TEN

ACCEPTANCE SAMPLING[1]

OBJECTIVES

Upon completion of this chapter, the reader is expected to

- know the advantages and disadvantages of sampling; the types of sampling plans and selection factors; criteria for formation of lots; criteria for sample selection; and decisions concerning rejected lots;
- determine the OC curve for a single sampling plan;
- determine the equations needed to graph the OC curve for a double sampling plan;
- know the properties of OC curves;
- know the consumer–producer relationships of risk, AQL, and LQ;
- determine the AOQ curve and the AOQL for a single, sampling plan;
- design single sampling plans for stipulated producer's risk and for stipulated consumer's risk.

INTRODUCTION

In recent years, acceptance sampling has declined in importance as statistical process control has assumed a more prominent role in the quality function. However, acceptance sampling still has a place in the entire body of knowledge that constitutes quality science. In addition to statistical acceptance sampling, discussed in this chapter, there are several other acceptance sampling practices, such as fixed percentage, occasional random checks, and 100% inspection.

FUNDAMENTAL CONCEPTS

Description

Lot-by-lot acceptance sampling by attributes is the most common type of sampling. With this type of sampling, a predetermined number of units (sample) from each lot is inspected by attributes. If the number of nonconforming units is less than the prescribed minimum, the lot is accepted; if not, the lot is not accepted. Acceptance sampling can be used either for the number of nonconforming units or for nonconformities per unit. To simplify the presentation in this chapter, the number of nonconforming units is used; however, it is understood that the information is also applicable to nonconformities per unit. Sampling plans are established by severity (critical, major, minor) or on a demerit-per-unit basis.

A single sampling plan is defined by the lot size, N, the sample size, n, and the acceptance number, c. Thus, the plan

$$N = 9000$$
$$n = 300$$
$$c = 2$$

means that a lot of 9000 units has 300 units inspected. If two or fewer nonconforming units are found in the 300-unit sample, the lot is accepted. If three or more nonconforming units are found in the 300-unit sample, the lot is not accepted.

Acceptance sampling can be performed in a number of different situations where there is a consumer–producer relationship. The consumer and producer can be from two different organizations, two plants within the same organization, or two departments within the same organization's facility. In any case, there is always the problem of deciding whether to accept or not accept the product.

Acceptance sampling is most likely to be used in one of five situations:

1. When the test is destructive (such as a test on an electrical fuse or a tensile test), sampling is necessary; otherwise, all of the units will be destroyed by testing.
2. When the cost of 100% inspection is high in relation to the cost of passing a nonconforming unit.
3. When there are many similar units to be inspected, sampling will frequently produce as good, if not better, results than 100% inspection. This is true because, with manual inspection, fatigue and boredom cause a higher percentage of nonconforming material to be passed than would occur on the average using a sampling plan.

[1]This chapter is based on ANSI/ASQ S2.

4. When information concerning producer's quality, such as $\overline{X}$ and R, p or c charts, and C_{pk}, is not available.
5. When automated inspection is not available.

Advantages and Disadvantages of Sampling

When sampling is compared with 100% inspection, it has the following advantages:

1. It places responsibility for quality in the appropriate place rather than on inspection, thereby encouraging rapid improvement.
2. It is more economical, owing to fewer inspections (fewer inspectors) and less handling damage during inspection.
3. It upgrades the inspection job from monotonous piece-by-piece decisions to lot-by-lot decisions.
4. It applies to destructive testing.
5. Entire lots are not accepted, rather than the return of a few nonconforming units, thereby giving stronger motivation for improvement.

Inherent disadvantages of acceptance sampling are that

1. There are certain risks of not accepting conforming lots and accepting nonconforming lots.
2. More time and effort is devoted to planning and documentation.
3. Less information is provided about the product, although there is usually enough.
4. There is no assurance that the entire lot conforms to specifications.

Types of Sampling Plans

There are four types of sampling plans: single, double, multiple, and sequential. In the single sampling plan, one sample is taken from the lot and a decision to accept or not accept the lot is made based on the inspection results of that sample. This type of sampling plan was described earlier in the chapter.

Double sampling plans are somewhat more complicated. On the initial sample, a decision, based on the inspection results, is made whether (1) to accept the lot, (2) not to accept the lot, or (3) to take another sample. If the quality is very good, the lot is accepted on the first sample and a second sample is not taken; if the quality is very poor, the lot is not accepted on the first sample and a second sample is not taken. Only when the quality level is neither very good nor very bad is a second sample taken.

If a second sample is required, the results of that inspection and the first inspection are used to make a decision. A double sampling plan is defined by

N = lot size

n_1 = sample size of first sample

c_1 = acceptance number for the first sample (sometimes the symbol *Ac* is used)

r_1 = nonacceptance number for the first sample (sometimes the symbol *Re* is used)

n_2 = sample size of second sample

c_2 = acceptance number for both samples

r_2 = nonacceptance number for both samples

If values are not given for r_1 and r_2, they are equal to $c_2 + 1$.

An illustrative example will help to clarify the double sampling plan: $N = 9000$, $n_1 = 60$, $c_1 = 1$, $r_1 = 5$, $n_2 = 150$, $c_2 = 6$, and $r_2 = 7$. An initial sample (n_1) of 60 is selected from the lot (N) of 9000 and inspected. One of the following judgments is made:

1. If there is 1 or fewer nonconforming unit (c_1), the lot is accepted.
2. If there are 5 or more nonconforming units (r_1), the lot is not accepted.
3. If there are 2, 3, or 4 nonconforming units, no decision is made and a second sample is taken.

A second sample of 150 (n_2) from the lot (N) is inspected, and one of the following judgments is made:

1. If there are 6 or fewer nonconforming units (c_2) in both samples, the lot is accepted. This number (6 or fewer) is obtained by 2 in the first sample and 4 or fewer in the second sample, by 3 in the first sample and 3 or fewer in the second sample, or by 4 in the first sample and 2 or fewer in the second sample.
2. If there are 7 or more nonconforming units (r_2) in both samples, the lot is not accepted. This number (7 or more) is obtained by 2 in the first sample and 5 or more in the second sample, by 3 in the first sample and 4 or more in the second sample, or by 4 in the first sample and 3 or more in the second sample.

A multiple sampling plan is a continuation of double sampling in that three, four, five, or as many samples as desired can be established. Sample sizes are much smaller. The technique is the same as that described for double sampling; therefore, a detailed description is not given. Multiple sampling plans of ANSI/ASQ Z1.4 use seven samples. An example of a multiple sampling plan with four samples is illustrated later in this chapter.

In sequential sampling, items are sampled and inspected one after another. A cumulative record is maintained, and a decision is made to accept or not accept the lot as soon as there is sufficient cumulative evidence.

All four types of sampling plans can give the same results; therefore, the chance of a lot being accepted under a single sampling plan is the same under the appropriate double, multiple, or sequential sampling plan. Thus, the type of plan for a particular unit is based on factors other than effectiveness. These factors are simplicity, administrative costs, quality information, number of units inspected, and psychological impact.

Perhaps the most important factor is simplicity. In this regard, single sampling is the best and sequential sampling the poorest.

TABLE 10-1 Random Number

74972	38712	36401	45525	40640	16281	13554	79945
75906	91807	56827	30825	40113	08243	08459	28364
29002	46453	25653	06543	27340	10493	60147	15702
80033	69828	88215	27191	23756	54935	13385	22782
25348	04332	18873	96927	64953	99337	68689	03263

Administrative costs for training, inspection, record keeping, and so on, are least for single sampling and greatest for sequential sampling.

Single sampling provides more information concerning the quality level in each lot than double sampling and much more than multiple or sequential sampling.

In general, the number of units inspected is greatest under single sampling and least under sequential. An ASN curve, shown later in the chapter, illustrates this concept.

A fifth factor concerns the psychological impact of the four types of sampling plans. Under single sampling there is no second chance; however, in double sampling, if the first sample is borderline, a second chance is possible by taking another sample. Many producers like the second-chance psychology provided by the double sample. In multiple and sequential sampling, there are a number of "second chances"; therefore, the psychological impact is less than with double sampling.

Careful consideration of the five factors is necessary to select a type of sampling plan that will be best for the particular situation.

Formation of Lots

Lot formation can influence the effectiveness of the sampling plan. Guidelines are as follows:

1. Lots should be homogeneous, which means that all product in the lot is produced by the same machine, same operator, same input material, and so on. When units from different sources are mixed, the sampling plan does not function properly. Also, it is difficult to take corrective action to eliminate the source of nonconforming units.

2. Lots should be as large as possible. Because sample sizes do not increase as rapidly as lot sizes, a lower inspection cost results with larger lot sizes. For example, a lot of 2000 would have a sample size of 125 (6.25%), but an equally effective sampling plan for a lot of 4000 would have a sample size of 200 (5.00%). When an organization starts a just-in-time procurement philosophy, the lot sizes are usually reduced to a 2- or 3-day supply. Thus, the relative amount inspected and the inspection costs will increase. The benefits to just-in-time are far greater than the increase in inspection costs; therefore, smaller lot sizes are to be expected.

The reader is cautioned not to confuse the packaging requirements for shipment and materials handling with the concept of a homogeneous lot. In other words, a lot may consist of a number of packages and may also consist of a number of shipments. If two different machines and/or two different operators are included in a shipment, they are separate lots and should be so identified. The reader should also be aware that partial shipments of a homogeneous lot can be treated as if they are homogeneous lots.

Sample Selection

The sample units selected for inspection should be representative of the entire lot. All sampling plans are based on the premise that each unit in the lot has an equal likelihood of being selected. This is referred to as *random sampling*.

The basic technique of random sampling is to assign a number to each unit in the lot. Then a series of random numbers is generated that tells which of the numbered units are to be sampled and inspected. Random numbers can be generated from a computer, an electronic hand calculator, a 20-sided random-number die, numbered chips in a bowl, and so on. They may be used to select the sample or to develop a table of random numbers.

A random-number table is shown in Table D of the Appendix. A portion of Table D is reproduced here as Table 10-1. To use the table, it is entered at any location and numbers are selected sequentially from one direction, such as up, down, left, or right. Any number that is not appropriate is discarded. For locating convenience, this table is established with 5 digits per column. It could have been established with 2, 3, 6, or any number per column. In fact, the digits could have run across the page with no spaces, but that format would make the table difficult to read. Any number of digits can be used for a random number.

An example will help to illustrate the technique. Assume that a lot of 90 units has been assigned numbers from 1 to 90 and it is desired to select a sample of 9. A two-digit number is selected at random, as indicated by the number 53. Numbers are selected downward on the right of the third column, and the first three numbers are 53, 15, and 73. Starting at the top of the next column, the numbers 45, 30, 06, 27, and 96 are obtained. The number 96 is too high and is discarded. The next numbers are 52 and 82. Units with the numbers 53, 15, 73, 45, 30, 06, 27, 52, and 82 comprise the sample.

Many units have serial numbers that can be used as the assigned number. This practice avoids the difficult process of assigning numbers to each unit. In many situations, units are systematically packed in a container and the assigned number can be designated by the location. A three-digit number would represent the width, height, and depth in a container

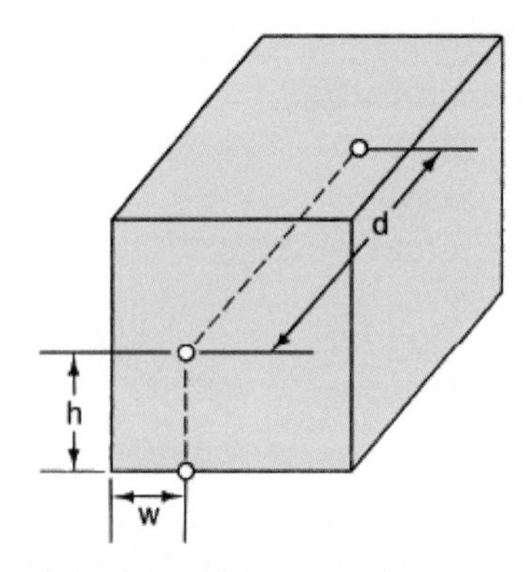

FIGURE 10-1 Location and Random Numbers

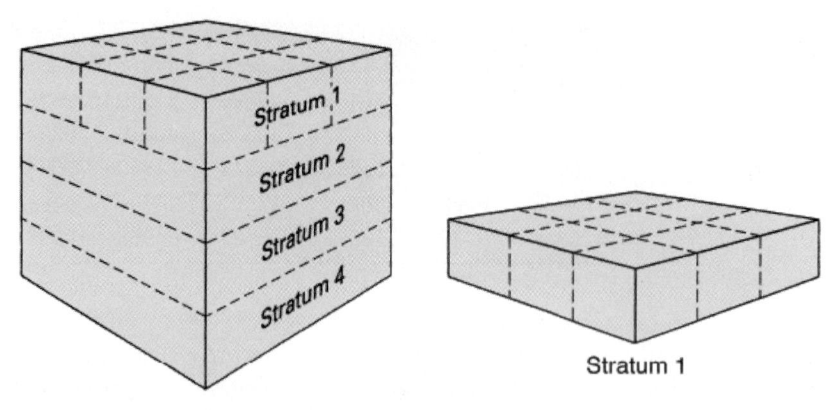

FIGURE 10-2 Dividing a Lot for Stratified Sampling

as shown in Figure 10-1. Thus, the random number 328 could specify the unit located at the third row, second level, and eighth unit from the front. For fluid or other well-mixed products, the sample can be taken from any location, because the product is presumed to be homogeneous.

It is not always practical to assign a number to each unit, utilize a serial number, or utilize a locational number. Stratification of the lot or package with samples drawn from each stratum can be an effective substitute for random sampling. The technique is to divide the lot or package into strata or layers as shown in Figure 10-2. Each stratum is further subdivided into cubes, as illustrated by stratum 1. Within each cube, samples are drawn from the entire volume. The dividing of the lot or package into strata and cubes within each stratum is an imaginary process done by the inspector. By this technique, pieces are selected from all locations in the lot or package.

Unless an adequate sampling method is used, a variety of biases can occur. An example of a biased sample occurs when the operator makes sure that units on the top of a lot are the best quality, and the inspector selects the sample from the same location. Adequate supervision of operators and inspectors is necessary to ensure that no bias occurs.

Nonaccepted Lots

Once a lot has not been accepted, there are a number of courses of action that can be taken.

1. The nonaccepted lot can be passed to the production facilities and the nonconforming units sorted by production personnel. This action is not a satisfactory alternative, because it defeats the purpose of sampling inspection and slows production. However, if the units are badly needed, there may be no other choice.

2. The nonaccepted lot can be rectified at the consumer's plant by personnel from either the producer's or the consumer's plant. Although shipping costs are saved, there is a psychological disadvantage, because all the consumer's personnel are aware that producer *X* had product that was not accepted. This fact may be used as a crutch to explain poor performance when using producer *X*'s material at a future time. In addition, space at the consumer's plant must be provided for personnel to perform the sorting operation.

3. The nonaccepted lot can be returned to the producer for rectification. This is the only appropriate course of action, because it results in long-run improvement in the quality. Because shipping costs are paid in both directions, cost becomes a motivating factor to improve the quality. Also, when the lot is sorted in the producer's plant, all the employees are aware that consumer *Y* expects to receive quality units. This, too, is a motivating factor for quality improvement the next time an order is produced for consumer *Y*. This course of action may require the production line to be shut down, which would be a loud and clear signal to the supplier and operating personnel that quality is important.

It is assumed that nonaccepted lots will receive 100% inspection and the nonconforming units discarded. A resubmitted lot is not normally reinspected, but if it is, the inspection should be confined to the original nonconformity. Because the nonconforming units are discarded, a resubmitted lot will have fewer units than the original.

STATISTICAL ASPECTS

OC Curve for Single Sampling Plans

An excellent evaluation technique is an *operating characteristic* (OC) *curve*. In judging a particular sampling plan, it is desirable to know the probability that a lot submitted with a certain percent nonconforming, $100p_0$, will be accepted. The OC curve will provide this information, and a typical OC curve is shown in Figure 10-3. When the percent nonconforming is low, the probability of the lot being accepted is large and decreases as the percent nonconforming increases.

The construction of an OC curve can be illustrated by a concrete example. A single sampling plan has a lot size $N = 3000$, a sample size $n = 89$, and an acceptance number $c = 2$. It is assumed that the lots are from a steady stream of product that can be considered infinite, and therefore the binomial probability distribution can be used for the calculations. Fortunately, the Poisson is an excellent approximation to the binomial for almost all sampling plans; therefore, the Poisson is used for determining the probability of the acceptance of a lot.

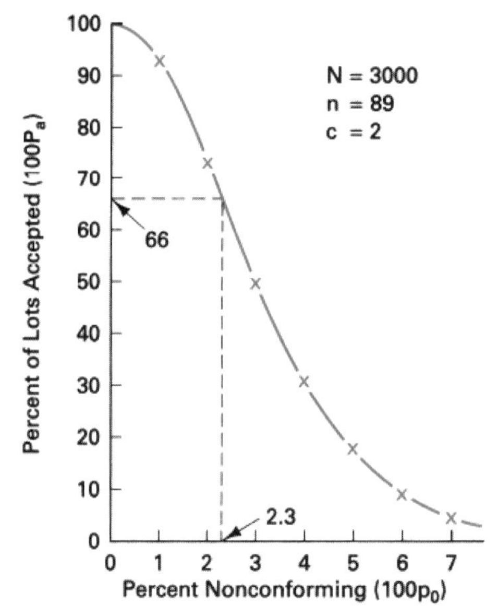

FIGURE 10-3 OC Curve for the Single Sampling Plan $N = 3000$, $n = 89$, and $c = 2$

In graphing the curve with the variables $100P_a$ (percent of lots accepted) and $100p_0$ (percent nonconforming), one value, $100p_0$, will be assumed and the other calculated. For illustrative purposes, we will assume a $100p_0$ value of, say, 2%, which gives an np_0 value of

$$P_0 = 0.02$$
$$np_0 = (89)(0.02) = 1.8$$

Acceptance of the lot is based on the acceptance number $c = 2$ and is possible when there are 0 nonconforming units in the sample, 1 nonconforming unit in the sample, or 2 nonconforming units in the sample. Thus,

$$\begin{aligned} P_a &= P_0 + P_1 + P_2 \\ &= P_2 \text{ or less} \\ &= 0.731 \text{ or } 100P_a = 73.1\% \end{aligned}$$

The P_a value is obtained from Table C in the Appendix for $c = 2$ and $np_0 = 1.8$.

A table can be used to assist with the calculations, as shown in Table 10-2. The curve is terminated when the P_a value is close to 0.05. Because $P_a = 0.055$ for $100p_0 = 7\%$, it is not necessary to make any calculations for values greater than 7%. Approximately 7 points are needed to describe the curve with a greater concentration of points where the curve changes direction.

Information from the table is plotted to obtain the OC curve shown in Figure 10-3. The steps are (1) assume p_0 value, (2) calculate np_0 value, (3) attain P_a values from the Poisson table using the applicable c and np_0 values, (4) plot point $(100p_0, 100P_a)$, and (5) repeat steps 1, 2, 3, and 4 until a smooth curve is obtained.

To make the curve more readable, the label Percent of Lots (expected to be) Accepted is used rather than Probability of Acceptance.

Once the curve is constructed, it shows the chance of a lot being accepted for a particular incoming quality. Thus, if the incoming process quality is 2.3% nonconforming, the percent of the lots that are expected to be accepted is 66%. Similarly, if 55 lots from a process that is 2.3% nonconforming are inspected using this sampling plan, 36 $[(55)(0.66) = 36]$ will be accepted and 19 $[55 - 36 = 19]$ will be unacceptable.

This OC curve is unique to the sampling plan defined by $N = 3000$, $n = 89$, and $c = 2$. If this sampling plan does not give the desired effectiveness, then the sampling plan should be changed and a new OC curve constructed and evaluated.

OC Curve for Double Sampling Plans

The construction of an OC curve for double sampling plans is somewhat more involved, because two curves must be determined. One curve is for the probability of acceptance on the first sample; the second curve is the probability of acceptance on the combined samples.

A typical OC curve is shown in Figure 10-4 for the double sampling plan $N = 2400$, $n_1 = 150$, $c_1 = 1$, $r_1 = 4$, $n_2 = 200$, $c_2 = 5$, and $r_2 = 6$. The first step in the construction of the OC curve is to determine the equations. If

TABLE 10-2 Probabilities of Acceptance for the Single Sampling Plan: $n = 89$, $c = 2$

ASSUMED PROCESS QUALITY P_0	$100P_0$	SAMPLE SIZE, n	np_0	PROBABILITY OF ACCEPTANCE P_a	PERCENT OF LOTS ACCEPTED $100P_a$
0.01	1.0	89	0.9	0.938	93.8
0.02	2.0	89	1.8	0.731	73.1
0.03	3.0	89	2.7	0.494	49.4
0.04	4.0	89	3.6	0.302	30.2
0.05	5.0	89	4.5	0.174	17.4
0.06	6.0	89	5.3	0.106*	10.6
0.07	7.0	89	6.2	0.055*	5.5

*By interpolation.

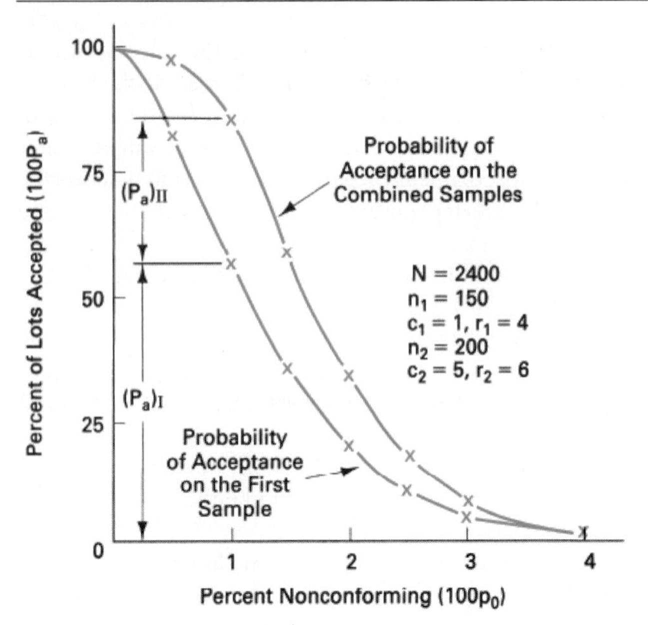

FIGURE 10-4 OC Curve for Double Sampling Plan

there is one or fewer nonconforming unit on the first sample, the lot is accepted. Symbolically, the equation is

$$(P_a)_{\mathrm{I}} = (P_{1 \text{ or less}})_{\mathrm{I}}$$

To obtain the equation for the second sample, the number of different ways in which the lot can be accepted is determined. A second sample is taken only if there are 2 or 3 nonconforming units on the first sample. If there is 1 or less, the lot is accepted; if there are 4 or more, the lot is not accepted. Therefore, the lot can be accepted by obtaining

1. Two nonconforming units on the first sample *and* 3 or less nonconforming units on the second sample, *or*
2. Three nonconforming units on the first sample *and* 2 or less nonconforming units on the second sample.

The *and's* and *or's* are emphasized to illustrate the use of the additive and multiplicative theorems, which were discussed in Chapter 8. Where an *and* occurs, multiply, and where an *or* occurs, add, and the equation becomes

$$(P_a)_{\mathrm{II}} = (P_2)_{\mathrm{I}}(P_{3 \text{ or less}})_{\mathrm{II}} + (P_3)_{\mathrm{I}}(P_{2 \text{ or less}})_{\mathrm{II}}$$

Roman numerals are used as subscripts for the sample number. The equations derived above are applicable only to this double sampling plan; another plan will require a different set of equations. Figure 10-5 graphically illustrates the technique. Note that the number of nonconforming units in each term in the second equation is equal to or less than the acceptance number, c_2. By combining the equations, the probability of acceptance for the combined samples is obtained:

$$(P_a)_{\text{combined}} = (P_a)_{\mathrm{I}} + (P_a)_{\mathrm{II}}$$

Once the equations are obtained, the OC curves are found by assuming various P_0 values and calculating the respective first and second sample P_a values. For example, using Table C of the Appendix and assuming a p_0 value of 0.01 $(100p_0 = 1.0)$,

$$(np_0)_{\mathrm{I}} = (150)(0.01) = 1.5 \quad (np_0)_{\mathrm{II}} = (200)(0.01) = 2.0$$

$$(P_a)_{\mathrm{I}} = (P_{1 \text{ or less}})_1 = 0.558$$

$$(P_a)_{\mathrm{II}} = (P_2)_{\mathrm{I}}(P_{3 \text{ or less}})_{\mathrm{II}} + (P_3)_{\mathrm{I}}(P_{2 \text{ or less}})_{\mathrm{II}}$$

$$(P_a)_{\mathrm{II}} = (0.251)(0.857) + (0.126)(0.677)$$

$$(P_a)_{\mathrm{II}} = 0.300$$

$$(P_a)_{\text{combined}} = (P_a)_{\mathrm{I}} + (P_a)_{\mathrm{II}}$$

$$(P_a)_{\text{combined}} = 0.558 + 0.300$$

$$(P_a)_{\text{combined}} = 0.858$$

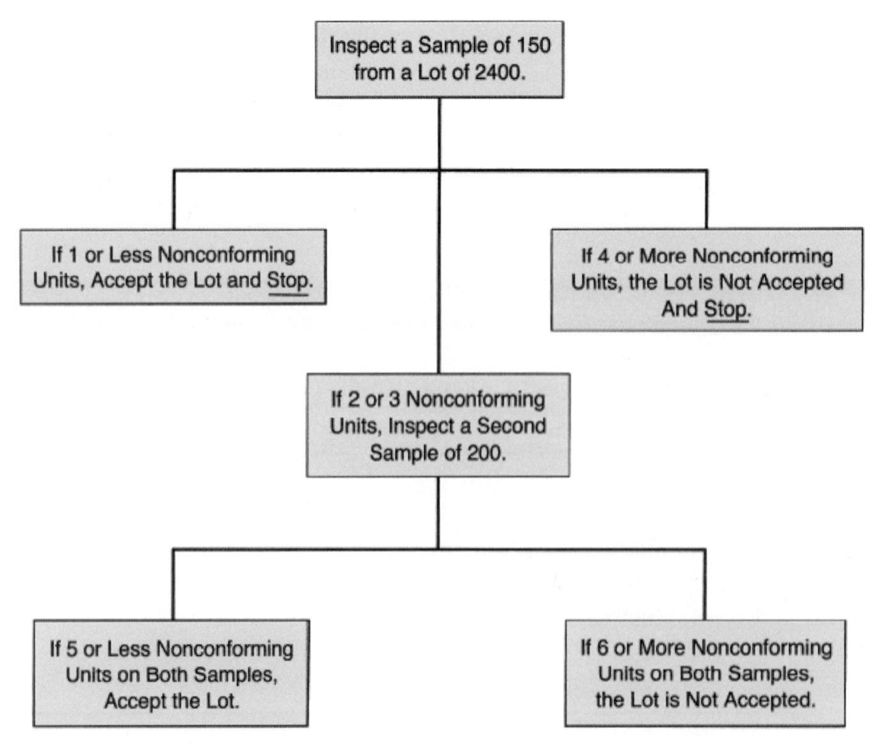

FIGURE 10-5 Graphical Description of the Double Sampling Plan: $N = 2400$, $n_1 = 150, c_1 = 1, r_1 = 4, n_2 = 200, c_2 = 5$, and $r_2 = 6$

These results are illustrated in Figure 10-4. When the two sample sizes are different, the np_0 values are different, which can cause a calculating error. Another source of error is neglecting to use the "or less" probabilities. Calculations are usually to three decimal places. The remaining calculations for other points on the curve are

For $P_0 = 0.005$ $(100P_0 = 0.5)$,

$$(np_0)_{\text{I}} = (150)(0.005) = 0.75$$
$$(np_0)_{\text{II}} = (200)(0.005) = 1.00$$
$$(P_a)_{\text{I}} = 0.826$$
$$(P_a)_{\text{II}} = (0.133)(0.981) + (0.034)(0.920) = 0.162$$
$$(P_a)_{\text{combined}} = 0.988$$

For $p_0 = 0.015$ $(100p_0 = 1.5)$,

$$(np_0)_{\text{I}} = (150)(0.015) = 2.25$$
$$(np_0)_{\text{II}} = (200)(0.015) = 3.00$$
$$(P_a)_{\text{I}} = 0.343$$
$$(P_a)_{\text{II}} = (0.266)(0.647) + (0.200)(0.423) = 0.257$$
$$(P_a)_{\text{combined}} = 0.600$$

For $p_0 = 0.020$ $(100p_0 = 2.0)$,

$$(np_0)_{\text{I}} = (150)(0.020) = 3.00$$
$$(np_0)_{\text{II}} = (200)(0.020) = 4.00$$
$$(P_a)_{\text{I}} = 0.199$$
$$(P_a)_{\text{II}} = (0.224)(0.433) + (0.224)(0.238) = 0.150$$
$$(P_a)_{\text{combined}} = 0.349$$

For $p_0 = 0.025$ $(100p_0 = 2.5)$

$$(np_0)_{\text{I}} = (150)(0.025) = 3.75$$
$$(np_0)_{\text{II}} = (200)(0.025) = 5.00$$
$$(P_a)_{\text{I}} = 0.112$$
$$(P_a)_{\text{II}} = (0.165)(0.265) + (0.207)(0.125) = 0.070$$
$$(P_a)_{\text{combined}} = 0.182$$

For $p_0 = 0.030$ $(100p_0 = 3.0)$

$$(np_0)_{\text{I}} = (150)(0.030) = 4.5$$
$$(np_0)_{\text{II}} = (200)(0.030) = 6.0$$
$$(P_a)_{\text{I}} = 0.061$$
$$(P_a)_{\text{II}} = (0.113)(0.151) + (0.169)(0.062) = 0.028$$
$$(P_a)_{\text{combined}} = 0.089$$

For $p_0 = 0.040$ $(100p_0 = 4.0)$,

$$(np_0)_{\text{I}} = (150)(0.040) = 6.0$$
$$(np_0)_{\text{II}} = (200)(0.040) = 8.0$$
$$(P_a)_{\text{I}} = 0.017$$
$$(P_a)_{\text{II}} = (0.045)(0.043) + (0.089)(0.014) = 0.003$$
$$(P_a)_{\text{combined}} = 0.020$$

Similar to the construction of the OC curve for single sampling, points are plotted as they are calculated, with the last few calculations used for locations where the curve changes direction. Whenever possible, both sample sizes should be the same value to simplify the calculations and the inspector's job. Also, if r_1 and r_2 are not given, they are equal to $c_2 + 1$.

The steps are (1) assume p_0 value, (2) calculate $(np_0)_{\text{I}}$ and $(np_0)_{\text{II}}$ values, (3) determine P_a value using the three equations and Table C, (4) plot points, and (5) repeat steps 1, 2, 3, and 4 until a smooth curve is obtained.

OC Curve for Multiple Sampling Plans

The construction of an OC curve for multiple sampling plans is more involved than for double or single sampling plans; however, the technique is the same. A multiple sampling plan with four levels is illustrated in Figure 10-6 and is specified as

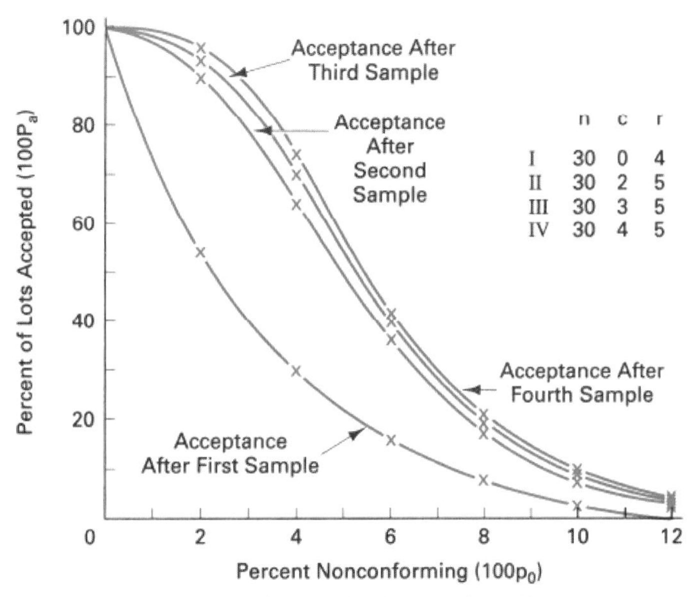

FIGURE 10-6 OC Curve for a Multiple Sampling Plan

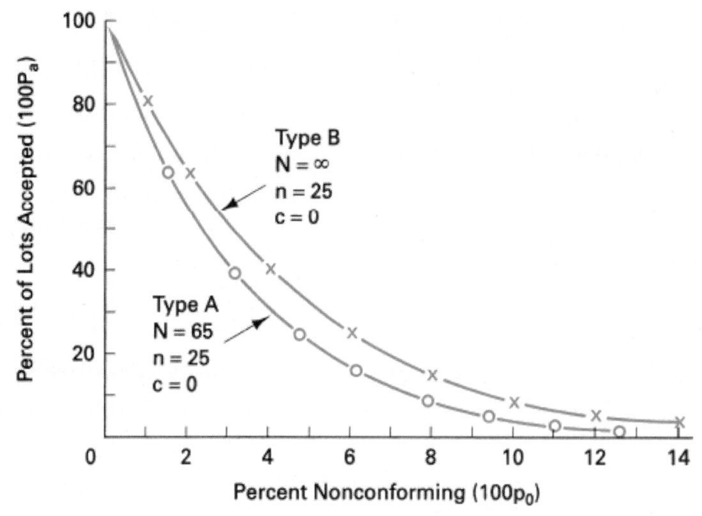

FIGURE 10-7 Types A and B OC Curves

$$N = 3000$$
$$n_1 = 30 \quad c_1 = 0 \quad r_1 = 4$$
$$n_2 = 30 \quad c_2 = 2 \quad r_2 = 5$$
$$n_3 = 30 \quad c_3 = 3 \quad r_3 = 5$$
$$n_4 = 30 \quad c_4 = 4 \quad r_4 = 5$$

Equations for this multiple sampling plan are

$$(P_a)_{\text{I}} = (P_0)_{\text{I}}$$

$$(P_a)_{\text{II}} = (P_1)_{\text{I}}(P_{1 \text{ or less}})_{\text{II}} + (P_2)_{\text{I}}(P_0)_{\text{II}}$$

$$(P_a)_{\text{III}} = (P_1)_{\text{I}}(P_2)_{\text{II}}(P_0)_{\text{III}} + (P_2)_{\text{I}}(P_1)_{\text{II}}(P_0)_{\text{III}} + (P_3)_{\text{I}}(P_0)_{\text{II}}(P_0)_{\text{III}}$$

$$(P_a)_{\text{IV}} = (P_1)(P_2)_{\text{II}}(P_1)_{\text{III}}(P_0)_{\text{IV}} + (P_1)_{\text{I}}(P_3)_{\text{II}}(P_0)_{\text{III}}(P_0)_{\text{IV}} + (P_2)_{\text{I}}(P_1)_{\text{II}}(P_1)_{\text{III}}(P_0)_{\text{IV}} + (P_2)_{\text{I}}(P_2)_{\text{II}}(P_0)_{\text{III}}(P_0)_{\text{IV}} + (P_3)_{\text{I}}(P_0)_{\text{II}}(P_1)_{\text{III}}(P_0)_{\text{IV}} + (P_3)_{\text{I}}(P_1)_{\text{II}}(P_0)_{\text{III}}(P_0)_{\text{IV}}$$

Using the equations above and varying the fraction nonconforming, p_0, the OC curve of Figure 10-6 is constructed. This is a tedious task and one that is ideally suited for the computer.

Comment

An operating characteristic curve evaluates the effectiveness of a particular sampling plan. If that sampling plan is not satisfactory, as shown by the OC curve, another one should be selected and its OC curve constructed.

Because the process quality or lot quality is usually not known, the OC curve, as well as other curves in this chapter, is a "what if" curve. In other words, if the quality is a particular percent nonconforming, the percent of lots accepted can be obtained from the curve.

Difference Between Type A and Type B OC Curves

The OC curves that were constructed in the previous sections are type B curves. It was assumed that the lots came from a continuous stream of product, and therefore the calculations are based on an infinite lot size. The binomial is the exact distribution for calculating the acceptance probabilities; however, the Poisson was used because it is a good approximation. Type B curves are continuous.

Type A curves give the probability of accepting an isolated finite lot. With a finite situation, the hypergeometric is used to calculate the acceptance probabilities. As the lot size of a type A curve increases, it approaches the type B curve and will become almost identical when the lot size is at least 10 times the sample size ($n/N \le 0.10$). A type A curve is shown in Figure 10-7, with the small open circles representing the discrete data and a discontinuous curve; however, the curve is drawn as a continuous one. Thus, a 4% value is impossible, because it represents 2.6 nonconforming units in the lot of 65 [(0.04)(65) = 2.6], but 4.6% nonconforming units are possible, as it represents 3 nonconforming units in the lot of 65 [(0.046)(65) = 3.0]. Therefore, the "curve" exists only where the small open circles are located.

In comparing the type A and type B curves of Figure 10-7, the type A curve is always lower than the type B curve. When the lot size is small in relation to the sample size, the difference between the curves is significant enough to construct the type A curve.

Unless otherwise stated, all discussion of OC curves will be in terms of type B curves.

OC Curve Properties

Acceptance sampling plans with similar properties can give different OC curves. Four of these properties and the OC curve information are given in the information that follows.

1. **Sample size as a fixed percentage of lot size.** Prior to the use of statistical concepts for acceptance sampling, inspectors were frequently instructed to sample a fixed percentage of the lot. If this value is, say, 10% of the lot size, plans for lot sizes of 900, 300, and 90 are

$$N = 900 \quad n = 90 \quad c = 0$$
$$N = 300 \quad n = 30 \quad c = 0$$
$$N = 90 \quad n = 9 \quad c = 0$$

Figure 10-8 shows the OC curves for the three plans, and it is evident that they offer different levels of protection. For

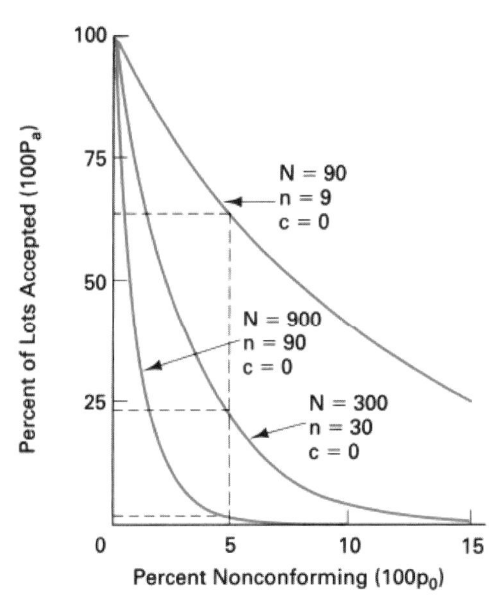

FIGURE 10-8 OC Curves for Sample Sizes that are 10% of the Lot Size

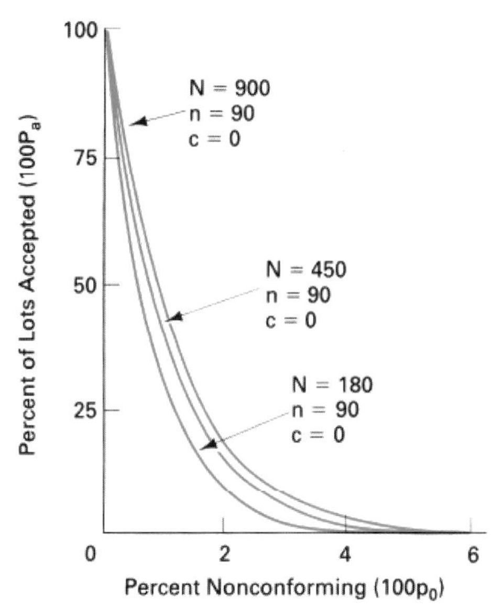

FIGURE 10-9 OC Curves for Fixed Sample Size (Type A)

example, for a process that is 5% nonconforming, $100P_a = 2\%$ for lot sizes of 900, $100P_a = 22\%$ for lot sizes of 300, and $100P_a = 63\%$ for lot sizes of 90.

2. **Fixed sample size.** When a fixed or constant sample size is used, the OC curves are very similar. Figure 10-9 illustrates this property for the type A situation where $n = 10\%$ of N. Naturally, for type B curves or when $n < 10\%$ of N, the curves are identical. The sample size has more to do with the shape of the OC curve and the resulting quality protection than does the lot size.

3. **As sample size increases, the curve becomes steeper.** Figure 10-10 illustrates the change in the shape of the OC curve. As the sample size increases, the slope of the curve becomes steeper and approaches a straight vertical line. Sampling plans with large sample sizes are better able to discriminate between acceptable and unacceptable quality. Therefore, the consumer has fewer lots of unacceptable quality accepted and the producer fewer lots of acceptable quality that are not accepted.

4. **As the acceptance number decreases, the curve becomes steeper.** The change in the shape of the OC curve as the acceptance number changes is shown in Figure 10-11. As the acceptance number decreases, the curve becomes steeper. This fact has frequently been used to justify the use of sampling plans with acceptance numbers of 0. However, the OC curve for $N = 2000$, $n = 300$, and $c = 2$, which is shown by the dashed line, is steeper than the plan with $c = 0$.

A disadvantage of sampling plans with $c = 0$ is that their curves drop sharply down rather than have a horizontal plateau before descending. Because this is the area of the producer's risk

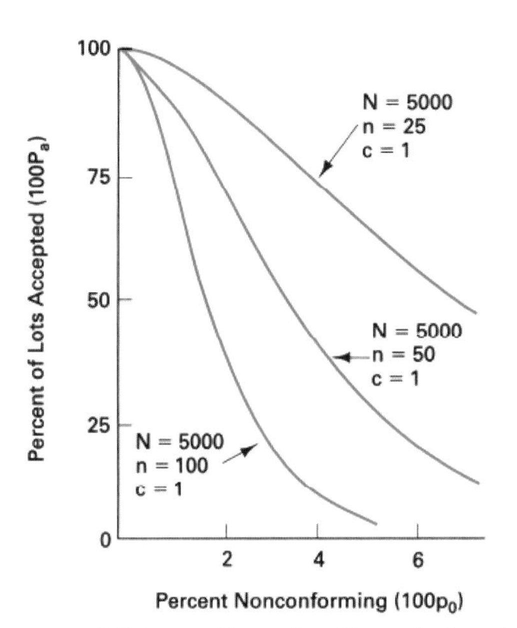

FIGURE 10-10 OC Curves Illustrating Change in Sample Size

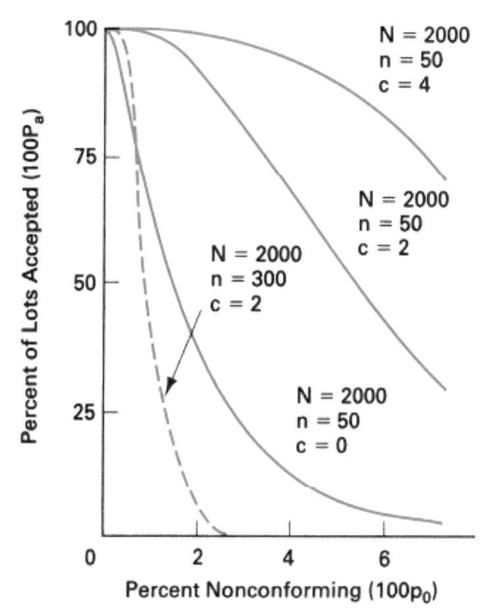

FIGURE 10-11 OC Curves Illustrating Change in Acceptance Number

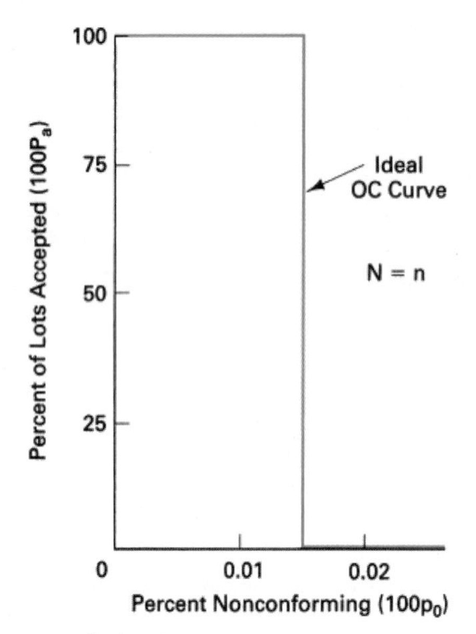

FIGURE 10-12 Ideal OC Curve

(discussed in the next section), sampling plans with $c = 0$ are more demanding of the producer. Sampling plans with acceptance numbers greater than 0 can actually be superior to those with 0; however, these require a larger sample size, which is more costly. In addition, many producers have a psychological aversion to plans that do not accept lots when only one nonconforming unit is found in the sample. The primary advantage of sampling plans with $c = 0$ is the perception that nonconforming units will not be tolerated and should be used for critical nonconformities. For major and minor nonconformities, acceptance numbers greater than 0 should be considered.

Consumer–Producer Relationship

When acceptance sampling is used, there is a conflicting interest between the consumer and the producer. The producer wants all conforming lots accepted, and the consumer wants all nonconforming lots not accepted. Only an ideal sampling plan that has an OC curve that is a vertical line can satisfy both the producer and consumer. An "ideal" OC curve, as shown in Figure 10-12, can be achieved only with 100% inspection, and the pitfalls of this type of inspection were mentioned earlier in the chapter. Therefore, sampling carries risks of not accepting lots that are acceptable and of accepting lots that are unacceptable. Because of the seriousness of these risks, various terms and concepts have been standardized.

The *producer's risk,* which is represented by the symbol α, is the probability of nonacceptance of a conforming lot. This risk is frequently given as 0.05, but it can range from 0.001 to 0.10 or more. Because α is expressed in terms of the probability of nonacceptance, it cannot be located on an OC curve unless specified in terms of the probability of acceptance. This conversion is accomplished by subtracting from 1. Thus, $P_a = 1 - \alpha$, and for $\alpha = 0.05, P_a = 1 - 0.05 = 0.95$. Figure 10-13 shows the producer's risk, α, or 0.05 on an imaginary axis labeled "Percent of Lots Not Accepted."

Associated with the producer's risk is a numerical definition of an acceptable lot, which is called acceptance quality limit (AQL). *The AQL is the quality level that is the worst tolerable process average when a continuing series of lots is submitted for acceptance sampling. It is a reference point on the OC curve and is not meant to convey to the producer that any percent nonconforming is acceptable. It is a statistical term and is not meant to be used by the general public.* The only way the producer can be guaranteed that a lot will be accepted is to have 0% nonconforming or to have the number nonconforming in the lot less than or equal to the acceptance number. In other words, the producer's quality goal is to meet or exceed the specifications so that no nonconforming units are present in the lot.

For the sampling plan $N = 4000$, $n = 300$, and $c = 4$, the AQL $= 0.7\%$ for $100 = 5\%$, as shown in Figure 10-13. In other words, units that are 0.7% nonconforming will have

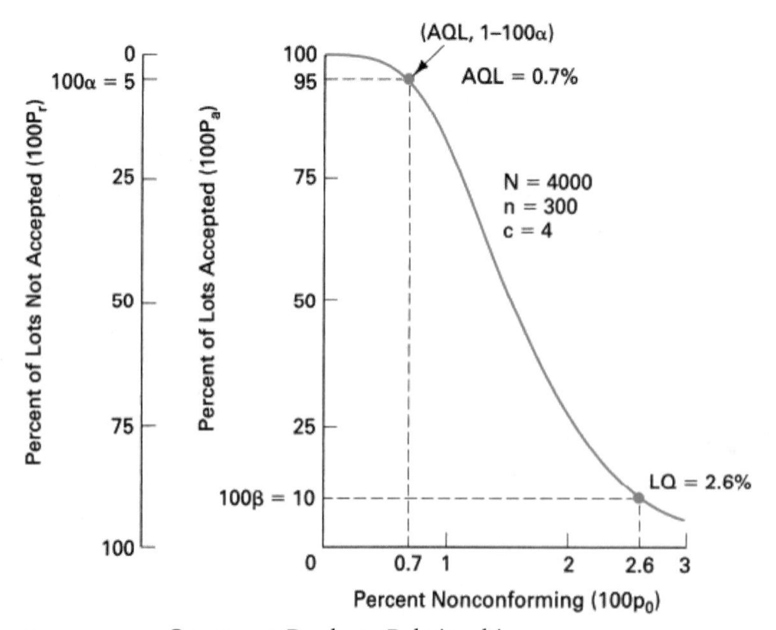

FIGURE 10-13 Consumer–Producer Relationship

a nonacceptance probability of 0.05, or 5%. Or, stated another way, 1 out of 20 lots that are 0.7% nonconforming will not be accepted by the sampling plan.

The *consumer's risk,* represented by the symbol β, is the probability of acceptance of a nonconforming lot. This risk is frequently given as 0.10. Because β is expressed in terms of probability of acceptance, no conversion is necessary.

Associated with the consumer's risk is a numerical definition of a nonconforming lot, called *limiting quality* (LQ). *The LQ is the percent nonconforming in a lot or batch for which, for acceptance sampling purposes, the consumer wishes the probability of acceptance to be low.* For the sampling plan in Figure 10-13, the LQ = 2.6% for 100β = 10%. In other words, lots that are 2.6% nonconforming will have a 10% chance of being accepted. Or, stated another way, 1 out of 10 lots that are 2.6% nonconforming will be accepted by this sampling plan.

Average Outgoing Quality

The *average outgoing quality* (AOQ) is another technique for the evaluation of a sampling plan. Figure 10-14 shows an AOQ curve for the sampling plan $N = 3000$, $n = 89$, and $c = 2$. This is the same plan as the one for the OC curve shown in Figure 10-3.

The information for the construction of an average outgoing quality curve is obtained by adding one column (an AOQ column) to the table used to construct an OC curve. Table 10-3 shows the information for the OC curve and the additional column for the AOQ curve. The average outgoing quality in percent nonconforming is determined by the formula AOQ = $(100p_0)(P_a)$. This formula does not account for the discarded nonconforming units; however, it is close enough for practical purposes and is simpler to use.

Note that to present a more readable graph, the AOQ scale is much larger than the incoming process quality scale. The curve is constructed by plotting the percent nonconforming $(100p_0)$ with its corresponding AOQ value.

TABLE 10-3 Average Outgoing Quality (AOQ) for the Sampling Plan $N = 3000$, $n = 89$, and $c = 2$

PROCESS QUALITY $100p_0$	SAMPLE SIZE n	np_0	PROBABILITY OF ACCEPTANCE P_a	AOQ $(100P_0)(P_a)$
1.0	89	0.9	0.938	0.938
2.0	89	1.8	0.731	1.462
3.0	89	2.7	0.494	1.482
4.0	89	3.6	0.302	1.208
5.0	89	4.5	0.174	0.870
6.0	89	5.3	0.106	0.636
7.0	89	6.2	0.055	0.385
2.5*	89	2.2	0.623	1.558

*Additional point where curve changes direction.

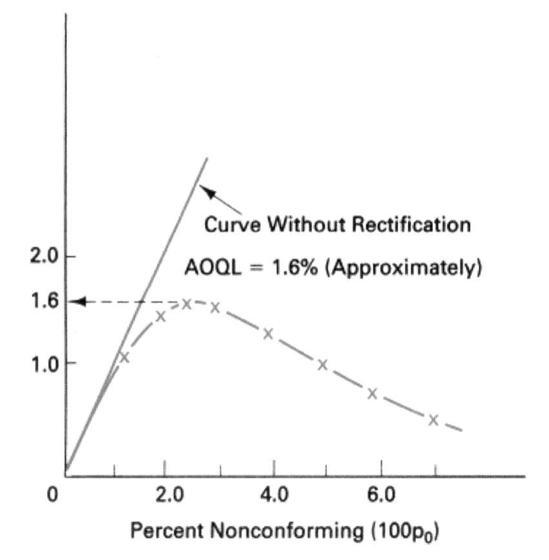

FIGURE 10-14 Average Outgoing Quality Curve for the Sampling Plan $N = 3000$, $n = 89$, and $c = 2$

The AOQ is the quality that leaves the inspection operation. It is assumed that any nonaccepted lots have been rectified or sorted and returned with 100% good product. When rectification does not occur, the AOQ is the same as the incoming quality, and this condition is represented by the straight line in Figure 10-14.

Analysis of the curve shows that when the incoming quality is 2.0% nonconforming, the average outgoing quality is 1.46% nonconforming, and when the incoming quality is 6.0% nonconforming, the average outgoing quality is 0.64% nonconforming. Therefore, because nonaccepted lots are rectified, the average outgoing quality is always better than the incoming quality. In fact, there is a limit that is given the name average outgoing quality limit (AOQL). Thus, for this sampling plan, as the percent nonconforming of the incoming quality changes, the average outgoing quality never exceeds the limit of approximately 1.6% nonconforming.

A better understanding of the concept of acceptance sampling can be obtained from an example. Suppose that, over a period of time, 15 lots of 3000 each are shipped by the producer to the consumer. The lots are 2% nonconforming and a sampling plan of $n = 89$ and $c = 2$ is used to determine acceptance. Figure 10-15 shows this information by a solid line. The OC curve for this sampling plan (Figure 10-3) shows that the percent of lots accepted for a 2% nonconforming lot is 73.1%. Thus, 11 lots ($15 \times 0.731 = 10.97$) are accepted by the consumer, as shown by the wavy line. Four lots are not accepted by the sampling plan and are returned to the producer for rectification, as shown by the dashed line. These four lots receive 100% inspection and are returned to the consumer with 0% nonconforming, as shown by a dashed line.

A summary of what the consumer actually receives is shown at the bottom of the figure. Two percent, or 240, of the four rectified lots are discarded by the producer, which gives 11,760 rather than 12,000. The calculations show that the consumer actually receives 1.47% nonconforming, whereas the producer's quality is 2% nonconforming.

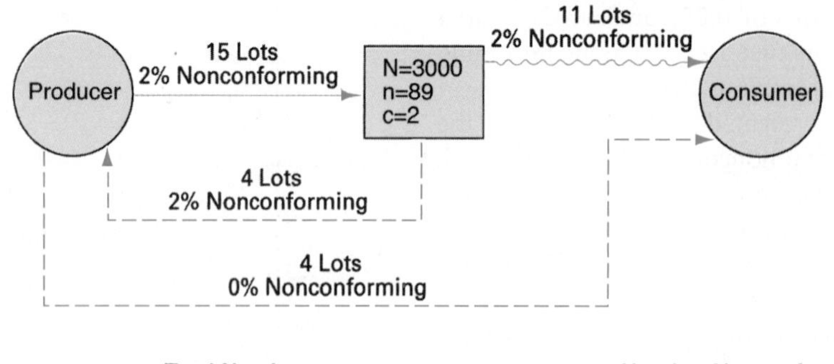

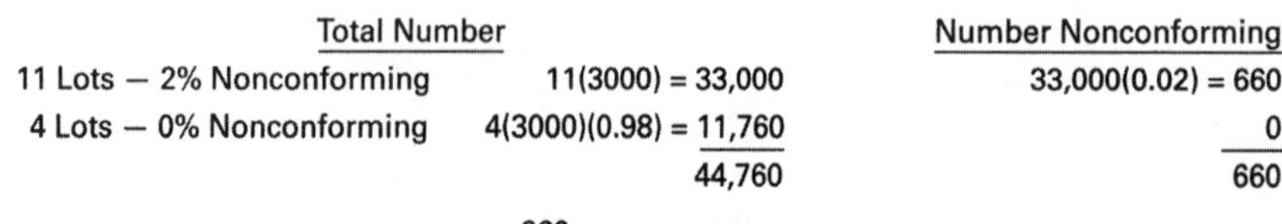

	Total Number	Number Nonconforming
11 Lots – 2% Nonconforming	11(3000) = 33,000	33,000(0.02) = 660
4 Lots – 0% Nonconforming	4(3000)(0.98) = 11,760	0
	44,760	660

$$\text{Percent Nonconforming (AOQ)} = \frac{660}{44{,}760} \times 100 = 1.47\%$$

FIGURE 10-15 How Acceptance Sampling Works

It should be emphasized that the acceptance sampling system works only when nonaccepted lots are returned to the producer and rectified. The AQL for this particular sampling plan at $\alpha = 0.05$ is 0.9%; therefore, the producer at 2% nonconforming is not achieving the desired quality level.

The AOQ curve, in conjunction with the OC curve, provides two powerful tools for describing and analyzing acceptance sampling plans.

Average Sample Number

The *average sample number* (ASN) is a comparison of the average amount inspected per lot by the consumer for single, double, multiple, and sequential sampling. Figure 10-16 shows the comparison for the four different but equally effective sampling plan types. In single sampling the ASN is constant and equal to the sample size, n. For double sampling the process is somewhat more complicated because a second sample may or may not be taken.

The formula for double sampling is

$$\text{ASN} = n_1 + n_2(1 - P_1)$$

where P_I is the probability of a decision on the first sample. An example problem will illustrate the concept.

Example Problem 10-1

Given the single sampling plan $n = 80$ and $c = 2$ and the equally effective double sampling plan $n_1 = 50$, $c_1 = 0$, $r_1 = 3$, $n_2 = 50$, $c_2 = 3$, and $r_2 = 4$, compare the ASN of the two by constructing their curves.

For single sampling, the ASN is the straight line at $n = 80$. For double sampling, the solution is

$$P_1 = P_0 + P_{3\text{ or more}}$$

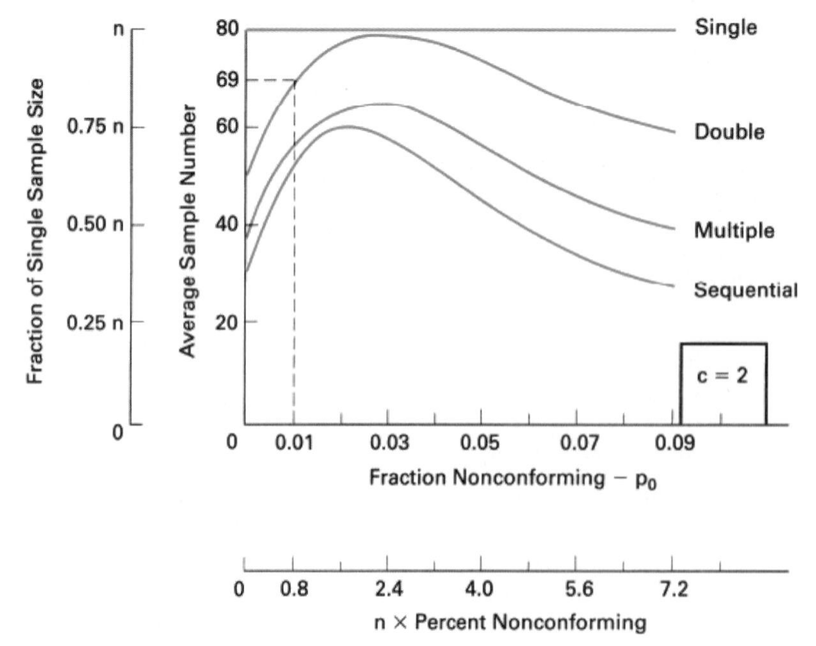

FIGURE 10-16 ASN Curves for Single, Double, Multiple, and Sequential Sampling

Assume that $p_0 = 0.01$; then $np_0 = 50(0.01) = 0.5$. From Appendix Table C:

$$P_0 = 0.607$$

$$P_{3 \text{ or more}} = 1 - P_{2 \text{ or less}} = 1 - 0.986 = 0.014$$

$$\begin{aligned} \text{ASN} &= n_1 + n_2(1 - [P_0 + P_{3 \text{ or more}}]) \\ &= 50 + 50(1 - [0.607 + 0.014]) \\ &= 69 \end{aligned}$$

Repeating for different values of p_0, the double sampling plan is plotted as shown in Figure 10-16.

The formula assumes that inspection continues even after the nonacceptance number is reached. It is frequently the practice to discontinue inspection after the nonacceptance number is reached on either the first or second sample. This practice is called *curtailed inspection,* and the formula is much more complicated. Thus, the ASN curve for double sampling is somewhat lower than what actually occurs.

An analysis of the ASN curve for double sampling in Figure 10-16 shows that at a fraction nonconforming of 0.03, the single and double sampling plans have about the same amount of inspection. For fraction nonconforming less than 0.03, double sampling has less inspection because a decision to accept on the first sample is more likely. Similarly, for fraction nonconforming greater than 0.03, double sampling has less inspection because a decision not to accept on the first sample is more likely and a second sample is not required. It should be noted that in most ASN curves, the double sample curve does not get close to the single sample one.

Calculation of the ASN curve for multiple sampling is much more difficult than for double sampling. The formula is

$$\text{ASN} = n_1P_\text{I} + (n_1 + n_2)P_\text{II} + \cdots + (n_1 + n_2 + \cdots + n_k)P_k$$

where n_k is the sample size of the last level and P_k is the probability of a decision at the last level.

Determining the probabilities of a decision at each level is quite involved—more so than for the OC curve, because the conditional probabilities must also be determined.

Figure 10-16 shows the ASN curve for an equivalent multiple sampling plan with seven levels. As expected, the average amount inspected is much less than single or double sampling.

The reader may be curious about the two extra scales in Figure 10-16. Because we are comparing equivalent sampling plans, the double and multiple plans can be related to the single sampling plans where $c = 2$ and n is the equivalent single sample size by the additional scales. To use the horizontal scale, multiply the single sample size n by the fraction nonconforming. The ASN value is found from the vertical scale by multiplying the scale fraction with the single sample size.

ANSI/ASQ Z1.4 shows a number of ASN curve comparisons indexed by the acceptance number, c. These curves can be used to find the amount inspected per lot for different percent nonconforming without having to make the calculations. When inspection costs are great because of inspection time, equipment costs, or equipment availability, the ASN curves are a valuable tool for justifying double or multiple sampling.

Average Total Inspection

The *average total inspection* (ATI) is another technique for evaluating a sampling plan. ATI is the amount inspected by both the consumer and the producer. Like the ASN curve, it is a curve that provides information on the amount inspected and not on the effectiveness of the plan. For single sampling, the formula is

$$\text{ATI} = n + (1 - P_a)(N - n)$$

It assumes that rectified lots will receive 100% inspection. If lots are submitted with 0% nonconforming, the amount inspected is equal to n, and if lots are submitted that are 100% nonconforming, the amount inspected is equal to N. Because neither of these possibilities is likely to occur, the amount inspected is a function of the probability of nonacceptance $(1 - P_a)$. An example problem will illustrate the calculation.

Example Problem 10-2

Determine the ATI curve for the single sampling plan $N = 3000$, $n = 89$, and $c = 2$.

Assume that $p_0 = 0.02$. From the OC curve (Figure 10-3), $P_a = 0.731$.

$$\begin{aligned} \text{ATI} &= n + (1 - P_a)(N - n) \\ &= 89 + (1 - 0.731)(3000 - 89) \\ &= 872 \end{aligned}$$

Repeat for other p_0 values until a smooth curve is obtained, as shown in Figure 10-17.

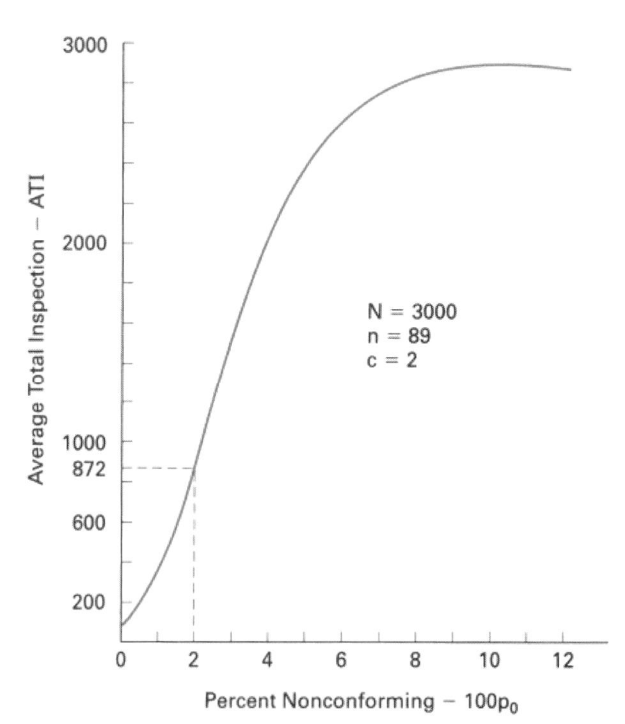

FIGURE 10-17 ATI Curve for $N = 3000$, $n = 89$, and $c = 2$

Examination of the curve shows that when the process quality is close to 0% nonconforming, the average total amount inspected is close to the sample size n. When process quality is very poor, at, say, 9% nonconforming, most of the lots are not accepted, and the ATI curve becomes asymptotic to 3000. As the percent nonconforming increases, the amount inspected by the producer dominates the curve.

Double sampling and multiple sampling formulas for the ATI curves are more complicated. These ATI curves will be slightly below the one for single sampling. The amount below is a function of the ASN curve, which is the amount inspected by the consumer, and this amount is usually very small in relation to the ATI, which is dominated by the amount inspected by the producer. From a practical viewpoint, the ATI curves for double and multiple sampling are not necessary, because the equivalent single sampling curve will convey a good estimate.

SAMPLING PLAN DESIGN

Sampling Plans for Stipulated Producer's Risk

When the producer's risk α and its corresponding AQL are specified, a sampling plan or, more precisely, a family of sampling plans can be determined. For a producer's risk, α, of, say, 0.05 and an AQL of 1.2%, the OC curves for a family of sampling plans as shown in Figure 10-18 are obtained. Each of the plans passes through the point defined by $100P_a = 95\%$ ($100\alpha = 5\%$) and $p_{0.95} = 0.012$. Therefore, each of the plans will ensure that product 1.2% nonconforming will not be accepted 5% of the time or, conversely, accepted 95% of the time.

The sampling plans are obtained by assuming a value for c and finding its corresponding np_0 value from Table C. When np_0 and p_0 are known, the sample size n is obtained. In order to find the np_0 values using Table C, interpolation is required. To eliminate the interpolation operation, np_0 values for various α and β values are reproduced in Table 10-4. In this table, c is cumulative, which means that a c value of 2 represents 2 or less.

Calculations to obtain the three sampling plans of Figure 10-18 are as follows:

$$P_a = 0.95 \qquad p_{0.95} = 0.012$$

For $c = 1$, $np_{0.95} = 0.355$ (from Table 10-4) and

$$n = \frac{np_{0.95}}{p_{0.95}} = \frac{0.355}{0.012} = 29.6 \text{ or } 30$$

For $c = 2$, $np_{0.95} = 0.818$ (from Table 10-4) and

$$n = \frac{np_{0.95}}{p_{0.95}} = \frac{0.818}{0.012} = 68.2 \text{ or } 68$$

For $c = 6$, $np_{0.95} = 3.286$ (from Table 10-4) and

$$n = \frac{np_{0.95}}{p_{0.95}} = \frac{3.286}{0.012} = 273.9 \text{ or } 274$$

The sampling plans for $c = 1$, $c = 2$, and $c = 6$ were arbitrarily selected to illustrate the technique. Construction of the OC curves is accomplished by the methods given at the beginning of the chapter.

While all the plans provide the same protection for the producer, the consumer's risk, at, say, $\beta = 0.10$, is quite different. From Figure 10-18 for the plan $c = 1$, $n = 30$, product that is 13% nonconforming will be accepted 10% ($\beta = 0.10$) of the time; for the plan $c = 2$, $n = 68$, the product that is 7.8% nonconforming will be accepted 10% ($\beta = 0.10$) of the time; and, for the plan $c = 6$, $n = 274$, product that is 3.8% nonconforming will be accepted 10% ($\beta = 0.10$) of the time. From the consumer's viewpoint, the latter plan provides better protection; however, the sample size is greater, which increases the inspection cost. The selection of the appropriate plan to use is a matter of judgment,

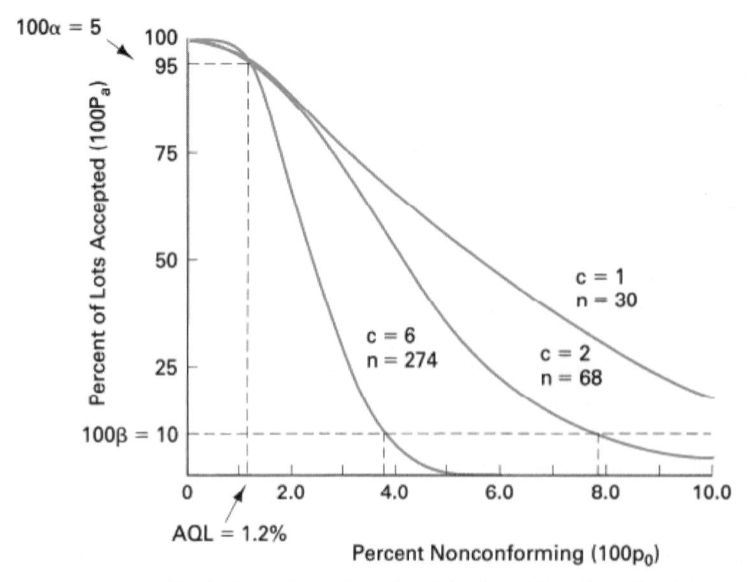

FIGURE 10-18 Single Sampling Plans for Stipulated Producer's Risk and AQL

TABLE 10-4 *np* Values for Corresponding *c* Values and Typical Producer's and Consumer's Risks

c	$P_a = 0.99$ ($\alpha = 0.01$)	$P_a = 0.95$ ($\alpha = 0.05$)	$P_a = 0.90$ ($\alpha = 0.10$)	$P_a = 0.10$ ($\beta = 0.10$)	$P_a = 0.05$ ($\beta = 0.05$)	$P_a = 0.01$ ($\beta = 0.01$)	RATIO OF $p_{0.10}/p_{0.95}$
0	0.010	0.051	0.105	2.303	2.996	4.605	44.890
1	0.149	0.355	0.532	3.890	4.744	6.638	10.946
2	0.436	0.818	1.102	5.322	6.296	8.406	6.509
3	0.823	1.366	1.745	6.681	7.754	10.045	4.890
4	1.279	1.970	2.433	7.994	9.154	11.605	4.057
5	1.785	2.613	3.152	9.275	10.513	13.108	3.549
6	2.330	3.286	3.895	10.532	11.842	14.571	3.206
7	2.906	3.981	4.656	11.771	13.148	16.000	2.957
8	3.507	4.695	5.432	12.995	14.434	17.403	2.768
9	4.130	5.426	6.221	14.206	15.705	18.783	2.618
10	4.771	6.169	7.021	15.407	16.962	20.145	2.497
11	5.428	6.924	7.829	16.598	18.208	21.490	2.397
12	6.099	7.690	8.646	17.782	19.442	22.821	2.312
13	6.782	8.464	9.470	18.958	20.668	24.139	2.240
14	7.477	9.246	10.300	20.128	21.886	25.446	2.177
15	8.181	10.035	11.135	21.292	23.098	26.743	2.122

Source: Extracted by permission from J. M. Cameron, "Tables for Constructing and for Computing the Operating Characteristics of Single-Sampling Plans," *Industry Quality Control,* Vol. 9, No. 1 (July 1952): 39.

which usually involves the lot size. This selection would also include plans for $c = 0, 3, 4, 5, 7$, and so forth.

Sampling Plans for Stipulated Consumer's Risk

When the consumer's risk β and its corresponding LQ are specified, a family of sampling plans can be determined. For a consumer's risk, β, of, say, 0.10 and a LQ of 6.0%, the OC curves for a family of sampling plans as shown in Figure 10-19 are obtained. Each of the plans passes through the point defined by $P_a = 0.10$ ($\beta = 0.10$) and $p_{0.10} = 0.060$. Therefore, each of the plans will ensure that product 6.0% nonconforming will be accepted 10% of the time.

The sampling plans are determined in the same manner as used for a stipulated producer's risk. Calculations are as follows:

$$P_a = 0.10 \quad p_{0.10} = 0.060$$

For $c = 1$, $np_{0.10} = 3.890$ (from Table 10-4) and

$$n = \frac{np_{0.10}}{p_{0.10}} = \frac{3.890}{0.060} = 64.8, \text{ or } 65$$

For $c = 3$, $np_{0.10} = 6.681$ (from Table 10-4) and

$$n = \frac{np_{0.10}}{p_{0.10}} = \frac{6.681}{0.060} = 111.4, \text{ or } 111$$

For $c = 7$, $np_{0.10} = 11.771$ (from Table 10-4) and

$$n = \frac{np_{0.10}}{p_{0.10}} = \frac{11.771}{0.060} = 196.2, \text{ or } 196$$

The sampling plans for $c = 1$, $c = 3$, and $c = 7$ were arbitrarily selected to illustrate the technique. Construction of the OC curves is accomplished by the method given at the beginning of the chapter.

Although all the plans provide the same protection for the consumer, the producer's risk, at, say, $\alpha = 0.05$, is quite different. From Figure 10-19 for the plan $c = 1$, $n = 65$, product that is 0.5% nonconforming will not be accepted 5% ($100\alpha = 5\%$) of the time; for the plan $c = 3$, $n = 111$,

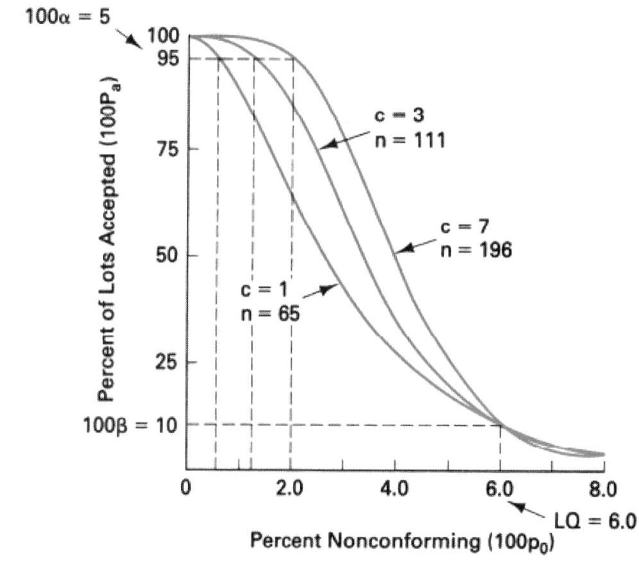

FIGURE 10-19 Single Sampling Plans for Stipulated Consumer's Risk and LQ

product that is 1.2% nonconforming will not be accepted 5% ($100\alpha = 5\%$) of the time; and for the plan $c = 7, n = 196$, product that is 2.0% nonconforming will not be accepted 5% ($\alpha = 0.05$) of the time. From the producer's viewpoint, the latter plan provides better protection; however, the sample size is greater, which increases the inspection costs. The selection of the appropriate plan is a matter of judgment, which usually involves the lot size. This selection would also include plans for $c = 0, 2, 4, 5, 6, 8$, and so forth.

Sampling Plans for Stipulated Producer's and Consumer's Risk

Sampling plans are also stipulated for both the consumer's risk and the producer's risk. It is difficult to obtain an OC curve that will satisfy both conditions. More than likely there will be four sampling plans that are close to meeting the consumer's and producer's stipulations. Figure 10-20 shows four plans that are close to meeting the stipulations of $\alpha = 0.05$, AQL $= 0.9$ and $\beta = 0.10$, LQ $= 7.8$. The OC curves of two plans meet the consumer's stipulation that product that is 7.8% nonconforming (LQ) will be accepted 10% ($\beta = 0.10$) of the time and comes close to the producer's stipulation. These two plans are shown by the dashed lines in Figure 10-20 and are $c = 1, n = 50$ and $c = 2, n = 68$. The two other plans exactly meet the producer's stipulation that product that is 0.9% nonconforming (AQL) will not be accepted 5% ($\alpha = 0.05$) of the time. These two plans are shown by the solid lines and are $c = 1, n = 39$ and $c = 2, n = 91$.

In order to determine the plans, the first step is to find the ratio of $p_{0.10}/p_{0.95}$, which is

$$\frac{p_{0.10}}{p_{0.95}} = \frac{0.078}{0.009} = 8.667$$

From the ratio column of Table 10-4, the ratio of 8.667 falls between the row for $c = 1$ and the row for $c = 2$. Thus, plans that exactly meet the consumer's stipulation of LQ $= 7.8\%$ for $\beta = 0.10$ are

For $c = 1$,

$$p_{0.10} = 0.078$$
$$np_{0.10} = 3.890 \text{ (from Table 10-4)}$$
$$n = \frac{np_{0.10}}{p_{0.10}} = \frac{3.890}{0.078} = 49.9, \text{ or } 50$$

For $c = 2$,

$$p_{0.10} = 0.078$$
$$np_{0.10} = 5.322 \text{ (from Table 10-4)}$$
$$n = \frac{np_{0.10}}{p_{0.10}} = \frac{5.322}{0.078} = 68.2, \text{ or } 68$$

Plans that exactly meet the producer's stipulation of AQL $= 0.9\%$ for $\alpha = 0.05$ are

For $c = 1$,

$$p_{0.95} = 0.009$$
$$np_{0.95} = 0.355 \text{ (from Table 10-4)}$$
$$n = \frac{np_{0.95}}{p_{0.95}} = \frac{0.355}{0.009} = 39.4, \text{ or } 39$$

For $c = 2$,

$$p_{0.95} = 0.009$$
$$np_{0.95} = 0.818 \text{ (from Table 10-4)}$$
$$n = \frac{np_{0.95}}{p_{0.95}} = \frac{0.818}{0.009} = 90.8, \text{ or } 91$$

Construction of the OC curve follows the method given at the beginning of the chapter.

Which of the four plans to select is based on one of four additional criteria? The first additional criterion is the stipulation that the plan with the lowest sample size be selected. The plan with the lowest sample size is one of the two with the lowest acceptance number. Thus, for the example problem, only the two plans for $c = 1$ are calculated, and $c = 1, n = 39$ is the sampling plan selected. A second addi-

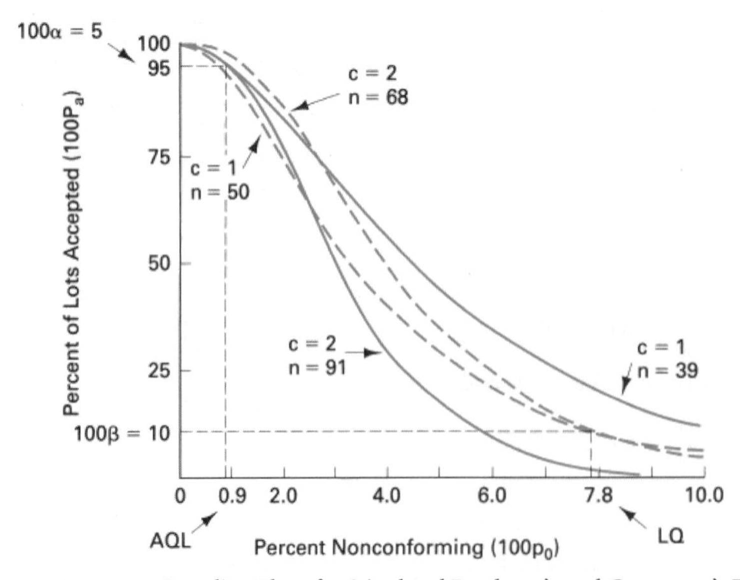

FIGURE 10-20 Sampling Plans for Stipulated Producer's and Consumer's Risk

tional criterion is the stipulation that the plan with the greatest sample size be selected. The plan with the greatest sample size is one of two with the largest acceptance number. Thus, for the example problem, only the two plans for $c = 2$ are calculated, and $c = 2, n = 91$ is the sampling plan selected.

A third additional criterion is the stipulation that the plan exactly meets the consumer's stipulation and comes as close as possible to the producer's stipulation. The two plans that exactly meet the consumer's stipulation are $c = 1, n = 50$ and $c = 2, n = 68$. Calculations to determine which plan is closest to the producer's stipulation of $\text{AQL} = 0.9, \alpha = 0.05$ are

For $c = 1, n = 50$,

$$p_{0.95} = \frac{np_{0.95}}{n} = \frac{0.355}{50} = 0.007$$

For $c = 2, n = 68$,

$$p_{0.95} = \frac{np_{0.95}}{n} = \frac{0.818}{68} = 0.012$$

Because $p_{0.95} = 0.007$ is closest to the stipulated value of 0.009, the plan of $c = 1, n = 50$ is selected.

The fourth additional criterion for the selection of one of the four sampling plans is the stipulation that the plan exactly meet the producer's stipulation and comes as close as possible to the consumer's stipulation. The two plans that are applicable are $c = 1, n = 39$ and $c = 2, n = 91$. Calculations to determine which is the closest to the consumer's stipulation of $\text{LQ} = 7.8, \beta = 0.10$ are

For $c = 1, n = 39$,

$$p_{0.10} = \frac{np_{0.10}}{n} = \frac{0.3890}{39} = 0.100$$

For $c = 2, n = 91$,

$$p_{0.10} = \frac{np_{0.10}}{n} = \frac{5.322}{91} = 0.058$$

Because $p_{0.10} = 0.058$ is closest to the stipulated value of 0.078, the plan of $c = 2, n = 91$ is selected.

Some Comments

The previous discussions have concerned single sampling plans. Double and multiple sampling plan design, although more difficult, will follow similar techniques.

In the previous discussion, a producer's risk of 0.05 and a consumer's risk of 0.10 were used to illustrate the technique. The producer's risk is usually set at 0.05 but can be as small as 0.01 or as high as 0.15. The consumer's risk is usually set at 0.10 but can be as low as 0.01 or as high as 0.20.

Sampling plans can also be specified by the AOQL. If an AOQL of 1.5% for an incoming quality of, say, 2.0% is stipulated, the probability of acceptance is

$$\text{AOQL} = (100p_0)(P_a)$$
$$1.5 = 2.0P_a$$
$$P_a = 0.75 \text{ or } 100P_a = 75\%$$

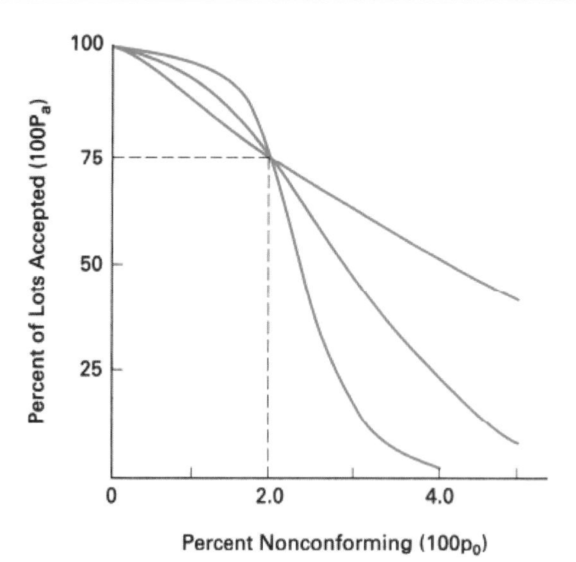

FIGURE 10-21 AOQL Sampling Plans

Figure 10-21 shows a family of OC curves for various sampling plans that satisfy the AOQL criteria.

To design a sampling plan, some initial stipulations are necessary by the producer, consumer, or both. These stipulations are decisions based on historical data, experimentation, or engineering judgment. In some cases the stipulations are negotiated as part of the purchasing contract.

SAMPLING PLAN SYSTEMS

The task of designing a sampling plan system is a tedious one. Fortunately, sampling plan systems are available.

ANSI/ASQ Z1.4

One such system that is almost universally used for the acceptance of product is ANSI/ASQ Z1.4. This system is an AQL, or producer's risk system. The standard is applicable, but not limited, to attribute inspection of the following: (1) end items, (2) components and raw materials, (3) operations, (4) materials in process, (5) supplies in storage, (6) maintenance operations, (7) data or records, and (8) administrative procedures. Sampling plans of this standard are intended to be used for a continuing series of lots, but plans may be designed for isolated lots by consulting the operating characteristic (OC) curve to determine the plan with the desired protection.

The standard provides for three types of sampling: single, double, and multiple. For each type of sampling plan, provision is made for normal, tightened, or reduced inspection. Tightened inspection is used when the producer's recent quality history has deteriorated. Acceptance requirements under tightened inspection are more stringent than under normal inspection. Reduced inspection is used when the producer's recent quality history has been exceptionally good.

ANSI/ASQ Standard Q3

This standard is to be used for inspection of isolated lots by attributes. It complements ANSI/ASQ Z1.4 that is appropriate for a continuous stream of lots. This standard indexes tables by LQ values and is applicable to type A or type B lots or batches. The LQ values are determined by the same techniques used to determine AQL values.

Dodge-Romig Tables

In the 1920s, H. F. Dodge and H. G. Romig developed a set of inspection tables for the lot-by-lot acceptance of product by sampling for attributes. These tables are based on LQ and AOQL. For each of these concepts there are tables for single and double sampling. No provision is made for multiple sampling. The principal advantage of the Dodge–Romig tables is a minimum amount of inspection for a given inspection procedure. This advantage makes the tables desirable for in-house inspection.

Chain Sampling[2]

A special type of lot-by-lot acceptance sampling plan for attributes was developed by H. F. Dodge called Chain Sampling. It is applicable to quality characteristics that involve destructive or costly tests where the acceptance number is zero and sample size is small. These plans make use of the cumulative results of previous samples.

Sequential Sampling[3]

Sequential sampling is similar to multiple sampling except that sequential sampling can theoretically continue indefinitely. In practice, the plan is truncated after the number inspected is equal to three times the number inspected by a corresponding single sampling plan. Sequential sampling, which is used for costly or destructive tests, usually has a subgroup size of 1, thereby making it an item-by-item plan.

Skip-Lot Sampling

Skip-lot sampling was devised by H. F. Dodge in 1955.[4] It is a single sampling plan based on the AOQL for minimizing inspection costs when there is a continuing supply of lots from the same source. It is particularly applicable to chemical and physical characteristics that require laboratory analyses. As companies emphasize statistical process control (SPC) and just-in-time (JIT) procurement, this type of sampling has become more applicable.

[2]For more information, see H. F. Dodge, "Chain Sampling Inspection Plan," *Industrial Quality Control*, Vol. 11, No. 4 (January 1955): 10–13.

[3]For more information, see Abraham Wald, *Sequential Analysis* (New York: John Wiley & Sons, 1947)

[4]H. F. Dodge, "Skip-Lot Sampling Plans," *Industrial Quality Control*, Vol. 11, No. 5 (February 1955): 3–5.

ANSI/ASQ S1

The purpose of this standard is to provide procedures to reduce the inspection effort when the supplier's quality is superior. It is a skip-lot scheme used in conjunction with the attribute lot-by-lot plans given in ANSI/ASQZ1.4; it is not to be confused with Dodge's skip-lot scheme described in the previous section. This sampling plan is an alternate to the reduced inspection of ANSI/ASQZ1.4, which permits smaller sizes than normal inspection.

MIL-STD-1235B

The standard is composed of five different continuous sampling plans. Inspection is by attributes for nonconformities or nonconforming units using three classes of severity: critical, major, and minor. The continuous sampling plans are designed based on the AOQL. In order to be comparable with ANSI/ASQ Z1.4 and other standards, the plans are also indexed by the AQL. The AQL is merely an index to the plans and has no other meaning.

Shainin Lot Plot Plan

The Shainin lot plot plan is a variable sampling plan used in some industries. It was developed by Dorian Shainin while he was chief inspector at Hamilton Standard Division of United Aircraft Corporation.[5] The plan uses a plotted frequency distribution (histogram) to evaluate a sample for decisions concerning acceptance or nonacceptance of a lot. The most significant feature of the plan is the fact that it is applicable to both normal and nonnormal frequency distributions. Another feature is its simplicity. It is a practical plan for in-house inspection as well as receiving inspection.

ANSI/ASQ Z1.9

This standard is a lot-by-lot acceptance sampling plan by variables and closely matches ANSI'ASQ Z1.4 for attributes. The standard is indexed by numerical values of the AQL that range from 0.10 to 10.0%. Provision is made for single and double sampling with normal, tightened, and reduced inspections. Sample sizes are a function of the lot size and the inspection level. The standard assumes a normally distributed random variable.

COMPUTER PROGRAM

The EXCEL program file on the website will solve for OC and AOQ curves. Its file name is *OC Curve*.

[5]Dorian Shainin, "The Hamilton Standard Lot Plot Method of Acceptance Sampling by Variables," *Industrial Quality Control*, Vol. 7, No. 1 (July 1950): 15–34.

EXERCISES

1. A real estate firm evaluates incoming selling agreement forms using the single sampling plan $N = 1500$, $n = 110$, and $c = 3$. Construct the OC curve using about 7 points.
2. A doctor's clinic evaluates incoming disposable cotton-tipped applicators using the single sampling plan $N = 8000$, $n = 62$, and $c = 1$. Construct the OC curve using about 7 points.
3. Determine the equation for the OC curve for the sampling plan $N = 10{,}000$, $n_1 = 200$, $c_1 = 2$, $r_1 = 6$, $n_2 = 350$, $c_2 = 6$, and $r_2 = 7$. Construct the curve using about 5 points.
4. Determine the equation for the OC curve for the following sampling plans:
   a. $N = 500$, $n_1 = 50$, $c_1 = 0$, $r_1 = 3$, $n_2 = 70$, $c_2 = 2$, and $r_2 = 3$
   b. $N = 6000$, $n_1 = 80$, $c_1 = 2$, $r_1 = 4$, $n_2 = 160$, $c_2 = 5$, and $r_2 = 6$
   c. $N = 22{,}000$, $n_1 = 260$, $c_1 = 5$, $r_1 = 9$, $n_2 = 310$, $c_2 = 8$, and $r_2 = 9$
   d. $N = 10{,}000$, $n_1 = 300$, $c_1 = 4$, $r_1 = 9$, $n_2 = 300$, and $c_2 = 8$
   e. $N = 800$, $n_1 = 100$, $c_1 = 0$, $r_1 = 5$, $n_2 = 100$, and $c_2 = 4$
5. For the sampling plan of Exercise 1, determine the AOQ curve and the AOQL.
6. For the sampling plan of Exercise 2, determine the AOQ curve and the AOQL.
7. A major U.S. automotive manufacturer is using a sampling plan of $n = 200$ and $c = 0$ for all lot sizes. Construct the OC and AOQ curves. Graphically determine the AQL value for $\alpha = 0.05$ and the AOQL value.
8. A leading computer firm uses a sampling plan of $n = 50$ and $c = 0$ regardless of lot sizes. Construct the OC and AOQ curves. Graphically determine the AQL value for $\alpha = 0.05$ and the AOQL value.
9. Construct the ASN curves for the single sampling plan $n = 200$, $c = 5$ and the equally effective double sampling plan $n_1 = 125$, $c_1 = 2$, $r_1 = 5$, $n_2 = 125$, $c_2 = 6$, and $r_2 = 7$.
10. Construct the ASN curves for the single sampling plan $n = 80$, $c = 3$ and the equally effective double sampling plan $n_1 = 50$, $c_1 = 1$, $r_1 = 4$, $n_2 = 50$, $c_2 = 4$, and $r_2 = 5$.
11. Construct the ATI curve for $N = 500$, $n = 80$, and $c = 0$.
12. Construct the ATI curve for $N = 10{,}000$, $n = 315$, and $c = 5$.
13. Determine the AOQ curve and the AOQL for the single sampling plan $N = 16{,}000$, $n = 280$, and $c = 4$.
14. Using $c = 1$, $c = 5$, and $c = 8$, determine three sampling plans that ensure product 0.8% nonconforming will be rejected 5.0% of the time.
15. For $c = 3$, $c = 6$, and $c = 12$, determine the sampling plans for AQL $= 1.5$ and $\alpha = 0.01$.
16. A bedsheet supplier and a large motel system have decided to evaluate units in lots of 1000 using an AQL of 1.0% with a probability of nonacceptance of 0.10. Determine sampling plans for $c = 0, 1, 2$, and 4. How would you select the most appropriate plan?
17. For a consumer's risk of 0.10 and an LQ of 6.5%, determine the sampling plans for $c = 2, 6$, and 14.
18. If product that is 8.3% nonconforming is accepted 5% of the time, determine three sampling plans that meet this criterion. Use $c = 0, 3$, and 7.
19. A manufacturer of loudspeakers has decided that product 2% nonconforming will be accepted with a probability of 0.01. Determine single sampling plans for $c = 1, 3$, and 5.
20. Construct the OC and AOQ curves for the $c = 3$ plan of Exercise 19.
21. A single sampling plan is desired with a consumer's risk of 0.10 of accepting 3.0% nonconforming units and a producer's risk of 0.05 of not accepting 0.7% nonconforming units. Select the plan with the lowest sample size.
22. The producer's risk is defined by $\alpha = 0.05$ for 1.5% nonconforming units, and the consumer's risk is defined by $\beta = 0.10$ for 4.6% nonconforming units. Select a sampling plan that exactly meets the producer's stipulation and comes as close as possible to the consumer's stipulation.
23. For the information of Exercise 21, select the plan that exactly meets the consumer's stipulation and comes as close as possible to the producer's stipulation.
24. For the information of Exercise 22, select the plan with the smallest sample size.
25. Given $p_{0.10} = 0.053$ and $p_{0.95} = 0.014$, determine the single sampling plan that exactly meets the consumer's stipulation and comes as close as possible to the producer's stipulation.
26. For the information of Exercise 25, select the plan that meets the producer's stipulation and comes as close as possible to the consumer's stipulation.
27. If a single sampling plan is desired with an AOQL of 1.8% at an incoming quality of 2.6%, what is the common point on the OC curves for a family of sampling plans that meet the AOQL and $100p_0$ stipulation?

28. Using the EXCEL program file, solve
    a. Exercises 1 and 5
    b. Exercises 2 and 6
29. Using the EXCEL program file, copy the template to a new sheet and change the increment for the data points from 0.0025 to 0.002. Resolve Exercises 28(a) and 28(b) and compare results.
30. Using EXCEL, write a program for
    a. OC curve for double sampling.
    b. AOQ curve for double sampling.
    c. ASN curve for single and double sampling.
    d. ATI curve for single and double sampling.

CHAPTER
ELEVEN

RELIABILITY

OBJECTIVES

Upon completion of this chapter, the reader is expected to

- know the definition of reliability and the factors associated with it;
- know the various techniques to obtain reliability;
- understand the probability distributions, failure curves, and reliability curves as factors of time;
- calculate the failure rate under different conditions;
- construct the life history curve and describe its three phases;
- calculate the normal, exponential, and Weibull failure rates;
- calculate the OC curve;
- determine life and reliability test plans;
- understand the different types of test design;
- understand the concepts of availability and maintainability.

INTRODUCTION

This chapter covers basic information on reliability. Advanced topics are covered by means of information on recent technical articles and are linked by footnotes.

FUNDAMENTAL ASPECTS

Definition

Simply stated, reliability is quality over the long run. It is the ability of the product or service to perform its intended function over a period of time. A product that "works" for a long period of time is a reliable one. Because all units of a product will fail at different times, reliability is a probability.

A more precise definition is—*Reliability is the probability that a product will perform its intended function satisfactorily for a prescribed life under certain stated environmental conditions.* From the definition, there are four factors associated with reliability: (1) numerical value, (2) intended function, (3) life, and (4) environmental conditions.

The numerical value is the probability that the product will function satisfactorily during a particular time. Thus, a value of 0.93 would represent the probability that 93 of 100 products will function after a prescribed period of time and 7 products will not function after the prescribed period of time. Particular probability distributions can be used to describe the failure[1] rate of units of product.

The second factor concerns the intended function of the product. Products are designed for particular applications and are expected to be able to perform those applications. For example, an electric hoist is expected to lift a certain design load; it is not expected to lift a load that exceeds the design specification. A screwdriver is designed to turn screws, not open paint cans.

The third factor in the definition of reliability is the intended life of the product—in other words, how long the product is expected to last. Thus, the life of automobile tires is specified by different values, such as 36 months or 70,000 km, depending on the construction of the tire. Product life is specified as a function of usage, time, or both.

The last factor in the definition involves environmental conditions. A product that is designed to function indoors, such as an upholstered chair, cannot be expected to function reliably outdoors in the sun, wind, and precipitation. Environmental conditions also include the storage and transportation aspects of the product. These aspects may be more severe than actual use.

Achieving Reliability

Emphasis Increased emphasis is being given to product reliability. One of the reasons for this emphasis is the Consumer Protection Act. Another reason is the fact that products

[1] The word *failure* is used in this chapter in its limited technical sense and refers to the testing activity rather than usage.

are more complicated. At one time the washing machine was a simple device that agitated clothes in a hot, soapy solution. Today, a washing machine has different agitating speeds, different rinse speeds, different cycle times, different water temperatures, different water levels, and provisions to dispense a number of washing ingredients at precise times in the cycle. An additional reason for the increased emphasis on reliability is automation; people are, in many cases, not able to manually operate the product when an automated component does not function.

System Reliability As products become more complex (have more components), the chance that they will not function increases. The method of arranging the components affects the reliability of the entire system. Components can be arranged in series, parallel, or a combination. Figure 11-1 illustrates the various arrangements.

When components are arranged in series, the reliability of the system is the product of the individual components (multiplicative theorem). Thus, for the series arrangement of Figure 11-1(a), the multiplicative theorem is applicable and the series reliability, R_S, is calculated as follows:

$$\begin{aligned} R_S &= (R_A)(R_B)(R_C) \\ &= (0.955)(0.750)(0.999) \\ &= 0.716 \end{aligned}$$

Note that R_A, R_B, and R_C are the probabilities (P_A, P_B, and P_C) that components A, B, and C will work. As components

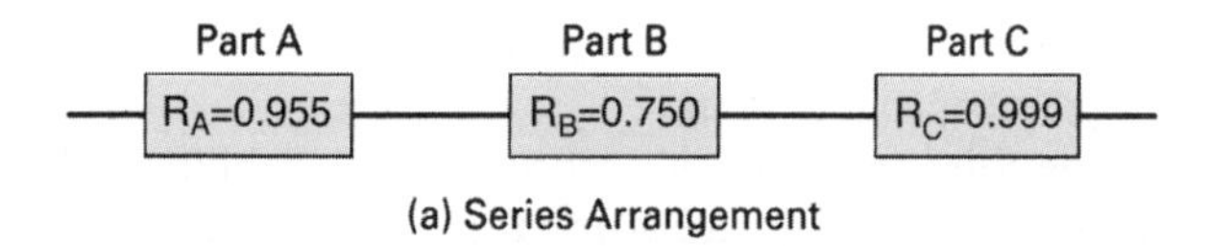

(a) Series Arrangement

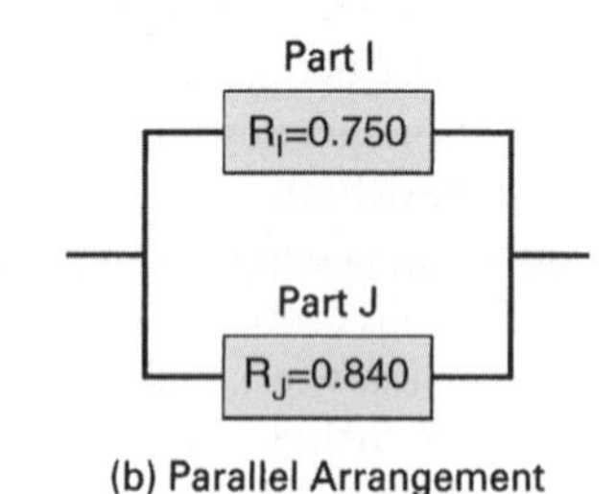

(b) Parallel Arrangement

Part I
RI=0.750
Part A
RA=0.955
Part C
RC=0.999
Part J
RJ=0.840

(c) Combination Arrangement

FIGURE 11-1 Methods of Arranging Components

are added to the series, the system reliability decreases. Also, the system reliability is always less than its lowest value. The cliché that a chain is only as strong as its weakest link is a mathematical fact.

Example Problem 11-1

A system has 5 components, A, B, C, D, and E, with reliability values of 0.985, 0.890, 0.985, 0.999, and 0.999, respectively. If the components are in series, what is the system reliability?

$$\begin{aligned} R_S &= (R_A)(R_B)(R_C)(R_D)(R_E) \\ &= (0.985)(0.890)(0.985)(0.999)(0.999) \\ &= 0.862 \end{aligned}$$

When components are arranged in series and a component does not function, then the entire system does not function. This is not the case when the components are arranged in parallel. When a component does not function, the product continues to function using another component until all parallel components do not function. Thus, for the parallel arrangement in Figure 11-1(b), the parallel system, R_S, is calculated as follows:

$$\begin{aligned} R_S &= 1 - (1 - R_I)(1 - R_J) \\ &= 1 - (1 - 0.750)(1 - 0.840) \\ &= 0.960 \end{aligned}$$

Note that $(1 - R_I)$ and $(1 - R_J)$ are the probabilities that components I and J will not function. As the number of components in parallel increases, the reliability increases. The reliability for a parallel arrangement of components is greater than the reliability of the individual components.

Example Problem 11-2

Determine the system reliability of 3 components—A, B, and C—with individual reliabilities of 0.989, 0.996, and 0.994 when they are arranged in parallel.

$$\begin{aligned} R_S &= 1 - (1 - R_A)(1 - R_B)(1 - R_C) \\ &= 1 - (1 - 0.989)(1 - 0.996)(1 - 0.994) \\ &= 0.999999736 \end{aligned}$$

Note that 9 significant figures are used in the answer to emphasize the parallel component principle.

Most complex products are a combination of series and parallel arrangements of components. This is illustrated in Figure 11-1(c), where part B is replaced by the parallel components, part I and part J. The reliability of the system, R_S, is calculated as follows:

$$\begin{aligned} R_S &= (R_A)(R_{I,J})(R_C) \\ &= (0.95)(0.96)(0.99) \\ &= 0.90 \end{aligned}$$

Example Problem 11-3

Find the reliability of the system below, where components 1, 2, 3, 4, 5, and 6 have reliabilities of 0.900, 0.956, 0.982, 0.999, 0.953, and 0.953.

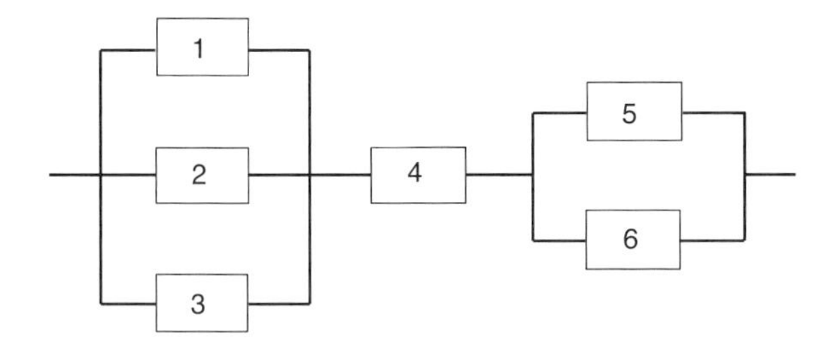

$$
\begin{aligned}
R_S &= (R_{1,2,3})(R_4)(R_{5,6}) \\
&= [1 - (1 - R_1)(1 - R_2)(1 - R_3)][R_4] \\
&\quad [1 - (1 - R_5)(1 - R_6)] \\
&= [1 - (1 - 0.900)(1 - 0.956)(1 - 0.982)] \\
&\quad [0.999][1 - (1 - 0.953)(1 - 0.953)] \\
&= .997
\end{aligned}
$$

Although most products comprise series and parallel systems, there are complex systems, such as a wheatstone bridge or standby redundancy, that are more difficult to analyze.

Design The most important aspect of reliability is the design. It should be as simple as possible. As pointed out earlier, the fewer the number of components, the greater the reliability. If a system has 50 components in series and each component has a reliability of 0.990, the system reliability is

$$R_S = R^n = 0.990^{50} = 0.605$$

If the system has 20 components in series, the system reliability is

$$R_S = R^n = 0.990^{20} = 0.818$$

Although this example may not be realistic, it does support the fact that the fewer the components, the greater the reliability.

Another way of achieving reliability is to have a backup or redundant component. When the primary component does not function, another component is activated. This concept was illustrated by the parallel arrangement of components. It is frequently cheaper to have inexpensive redundant components to achieve a particular reliability than to have a single expensive component.

Reliability can also be achieved by overdesign. The use of large factors of safety can increase the reliability of a product. For example, a 1-in. rope may be substituted for a $\frac{1}{2}$-in. rope, even though the $\frac{1}{2}$-in. rope would have been sufficient.

When an unreliable product can lead to a fatality or substantial financial loss, a fail-safe type of device should be used. Thus, disabling extremity injuries from power-press operations are minimized by the use of a clutch. The clutch must be engaged for the ram and die to descend. If there is a malfunction of the clutch-activation system, the press will not operate.

The maintenance of the system is an important factor in reliability. Products that are easy to maintain will likely receive better maintenance. In some situations it may be more practical to eliminate the need for maintenance. For example, oil-impregnated bearings do not need lubrication for the life of the product.

Environmental conditions such as dust, temperature, moisture, and vibration can be the cause of an unreliable product. The designer must protect the product from these conditions. Heat shields, rubber vibration mounts, and filters are used to increase reliability under adverse environmental conditions.

There is a definite relationship between investment in reliability (cost) and reliability. After a certain point, there is only a slight improvement in reliability for a large increase in product cost. For example, assume that a \$50 component has a reliability of 0.750. If the cost is increased to \$100, the reliability becomes 0.900; if the cost is increased to \$150, the reliability becomes 0.940; and if the cost is increased to \$200, the reliability becomes 0.960. As can be seen by this hypothetical example, there is a diminishing reliability return for the investment dollar.

Production The production process is the second most important aspect of reliability. Basic quality techniques that have been described in previous chapters will minimize the risk of product unreliability. Emphasis should be placed on those components that are least reliable.

Production personnel can take action to ensure that the equipment used is right for the job and investigate new equipment as it becomes available. In addition, they can experiment with process conditions to determine which conditions produce the most reliable product.

Transportation The third aspect of reliability is the transportation of the product to the customer. No matter how well conceived the design or how carefully the product is produced, the actual performance of the product by the customer is the final evaluation. The reliability of the product at the point of use can be greatly affected by the type of handling the product receives in transit. Good packaging techniques and shipment evaluation are essential.

Maintenance Although designers try to eliminate the need for customer maintenance, there are many situations where it is not practical or possible. In such cases, the customer should be given ample warning—for example, a warning light or buzzer when a component needs a lubricant. Maintenance should be simple and easy to perform.

ADDITIONAL STATISTICAL ASPECTS

Distributions Applicable to Reliability

Types of continuous probability distributions used in reliability studies are exponential, normal, and Weibull.[2] Their

[2] A fourth type, the gamma distribution, is not given because of its limited application. Also, the discrete probability distributions, geometric and negative binomial, are not given for the same reason.

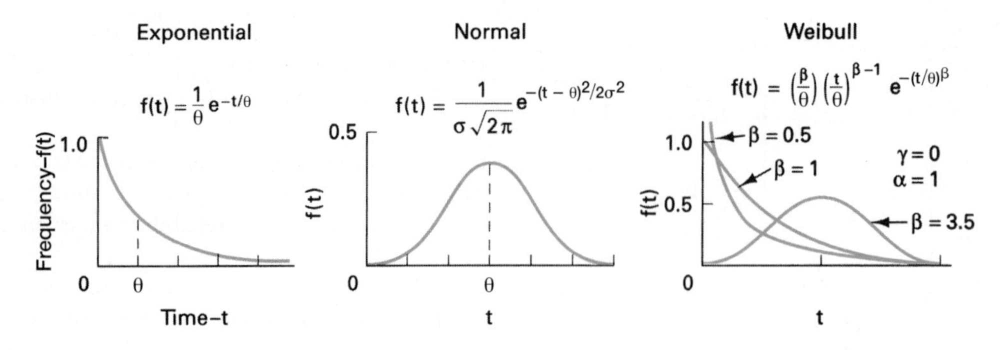

(a) Frequency Distribution as a Function of Time

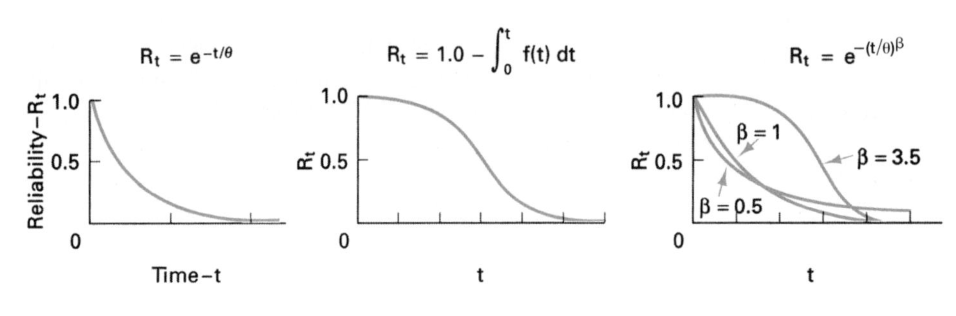

(b) Reliability as a Function of Time

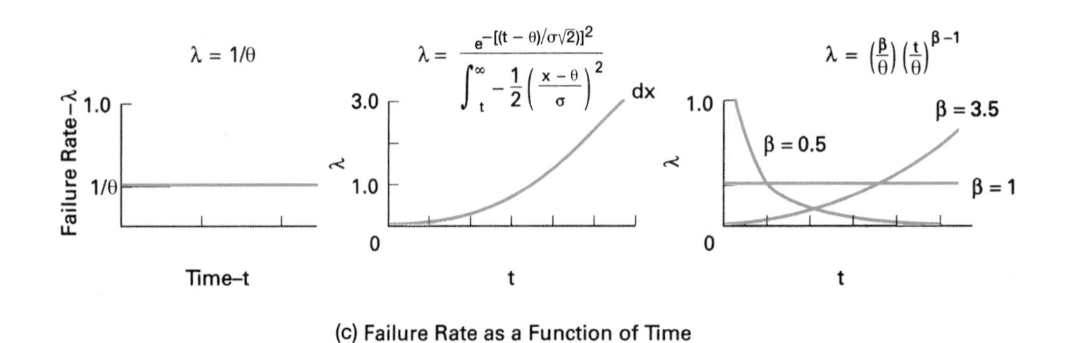

(c) Failure Rate as a Function of Time

FIGURE 11-2 Probability Distributions, Failure-Rate Curves, and Reliability Curves as a Function of Time

frequency distributions as a function of time are given in Figure 11-2(a) .

Reliability Curves

Reliability curves for the exponential, normal, and Weibull distributions as a function of time are given in Figure 11-2(b). The formulas for these distributions are also given in the figure. For the exponential and Weibull curves, the formulas are $R_t = e^{-t/\theta}$, and $R_t = e^{-\alpha t^{\beta}}$, respectively. The formula for the normal distribution is

$$R_t = 1.0 - \int_0^t f(t)dt,$$

which requires integration. However, Table A in the Appendix can be used to find the area under the curve, which is the $\int_0^t f(t)dt$.

Failure-Rate Curve

Failure rate is important in describing the life history of a product. Failure-rate curves and formulas for the exponential, normal, and Weibull distributions as a function of time are shown in Figure 11-2(c).

Failure rate can be estimated from test data by use of the formula

$$\lambda_{\text{est}} = \frac{\text{number of test failures}}{\text{sum of test times or cycles}} = \frac{r}{\Sigma t + (n - r)T}$$

where λ = failure rate, which is the probability that a unit will fail in a stated unit of time or cycles
r = number of test failures
t = test time for a failed item
n = number of items tested
T = termination time

The formula is applicable for the time terminated without a replacement situation. It is modified for the time-terminated with-replacement and failure-terminated situations. The following examples illustrate the difference.

Example Problem 11-4 (Time Terminated without Replacement)

Determine the failure rate for an item that has the test of 9 items terminated at the end of 22 h. Four of the items failed after 4, 12, 15, and 21 h, respectively. Five items were still operating at the end of 22 h.

$$\lambda_{est} = \frac{r}{\Sigma t + (n - r)T}$$
$$= \frac{4}{(4 + 12 + 15 + 21) + (9 - 4)22}$$
$$= 0.025$$

Example Problem 11-5 (Time Terminated with Replacement)

Determine the failure rate for 50 items that are tested for 15 h. When a failure occurs, the item is replaced with another unit. At the end of 15 h, 6 of the items had failed.

$$\lambda_{est} = \frac{r}{\Sigma t}$$
$$= \frac{6}{50(15)}$$
$$= 0.008$$

Note that the formula was simplified because the total test time is equal to Σt.

Example Problem 11-6 (Failure Terminated)

Determine the failure rate for 6 items that are tested to failure. Test cycles are 1025, 1550, 2232, 3786, 5608, and 7918.

$$\lambda = \frac{r}{\Sigma t}$$
$$= \frac{6}{1025 + 1550 + 2232 + 3786 + 5608 + 7918}$$
$$= 0.00027$$

Note that the formula was simplified because the total test time is equal to Σt.

For the exponential distribution and for the Weibull distribution when β, the shape parameter, equals 1, there is a constant failure rate. When the failure rate is constant, the relationship between mean life and failure rate is as follows:[3]

$$\theta = \frac{1}{\lambda} \quad \text{(for constant failure rate)}$$

where θ = mean life or mean time between failures (MTBF)

Example Problem 11-7

Determine the mean life for the three previous example problems. Assume that there is a constant failure rate.

$$\theta = \frac{1}{\lambda} = \frac{1}{0.025} = 40 \text{ h}$$
$$\theta = \frac{1}{\lambda} = \frac{1}{0.008} = 125 \text{ h}$$
$$\theta = \frac{1}{\lambda} = \frac{1}{0.00027} = 3704 \text{ cycles}$$

Life-History Curve

Figure 11-3 shows a typical life-history curve of a complex product for an infinite number of items. The curve, sometimes referred to as the "bathtub" curve, is a comparison of failure rate with time. It has three distinct phases: the debugging phase, the chance failure phase, and the wear-out phase. The probability distributions shown in Figure 11-2(c) are used to describe these phases.

The *debugging phase*, which is also called the burn-in or infant-mortality phase, is characterized by marginal and short-life parts that cause a rapid decrease in the failure rate. Although the shape of the curve will vary somewhat with the type of product, the Weibull distribution with shaping parameters less than 1, $\beta < 1$, is used to describe the occurrence of failures. For some products, the debugging phase may be part of the testing activity prior to shipment. For other products, this phase is usually covered by the warranty period. In either case, it is a significant quality cost.

The *chance failure phase* is shown in the figure as a horizontal line, thereby making the failure rate constant. Failures occur in a random manner due to the constant failure rate. The assumption of a constant failure rate is valid for most products; however, some products may have a failure rate that increases with time. In fact, a few products show a slight decrease, which means that the product is actually improving over time. The exponential distribution and the Weibull distribution with shape parameter equal to 1 are used to describe this phase of the life history. When the curve increases or decreases, a Weibull shape parameter greater or less than 1 can be used. Reliability studies and sampling plans are, for the most part, concerned with the chance failure phase. The lower the failure rate, the better the product.

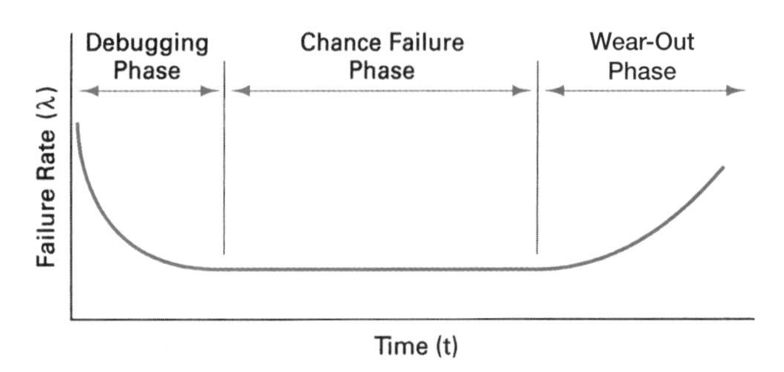

FIGURE 11-3 Typical Life History of a Complex Product for an Infinite Number of Items

[3] Failure rate is also equal to $f(t)/R_t$.

The third phase is the *wear-out phase,* which is depicted by a sharp rise in the failure rate. Usually the normal distribution is the one that best describes the wear-out phase. However, the Weibull distribution with shape parameter greater than 1, $\beta > 1$, can be used depending on the type of wear-out distribution.

The curve shown in Figure 11-3 is the type of failure pattern exhibited by most products; however, some products will deviate from this curve. It is important to know the type of failure pattern so that known probability distributions can be used for analysis and prediction of product reliability. By changing the shape parameter, β, all three phases can be modeled using the Weibull distribution. Test results from samples are used to determine the appropriate probability distribution. An example will illustrate the construction of the life-history curve.

Example Problem 11-8

Determine the life-history curve for the test data in cycles for 1000 items. Assume that failure occurred at $\frac{1}{2}$ the cycle range and survivors went to the end of the cycle range. Data are

Number of Cycles	Number of Failures	Number of Survivors	Calculations $\lambda = r/\sigma t$
0–10	347	653	347/((5)(347) + (10)(653)) = 0.0420
11–20	70	583	70/((15)(70) + (20)(583)) = 0.0055
21–30	59	524	59/((25)(59) + (30)(524)) = 0.0034
31–40	53	471	53/((35)(53) + (40)(471)) = 0.0026
41–50	51	420	51/((45)(51) + (50)(420)) = 0.0022
51–60	60	360	60/((55)(60) + (60)(360)) = 0.0024
61–70	79	281	79/((65)(79) + (70)(281)) = 0.0032
71–80	92	189	92/((75)(92) + (80)(189)) = 0.0043
81–90	189	0	189/((85)(189) + (90)(0)) = 0.0118

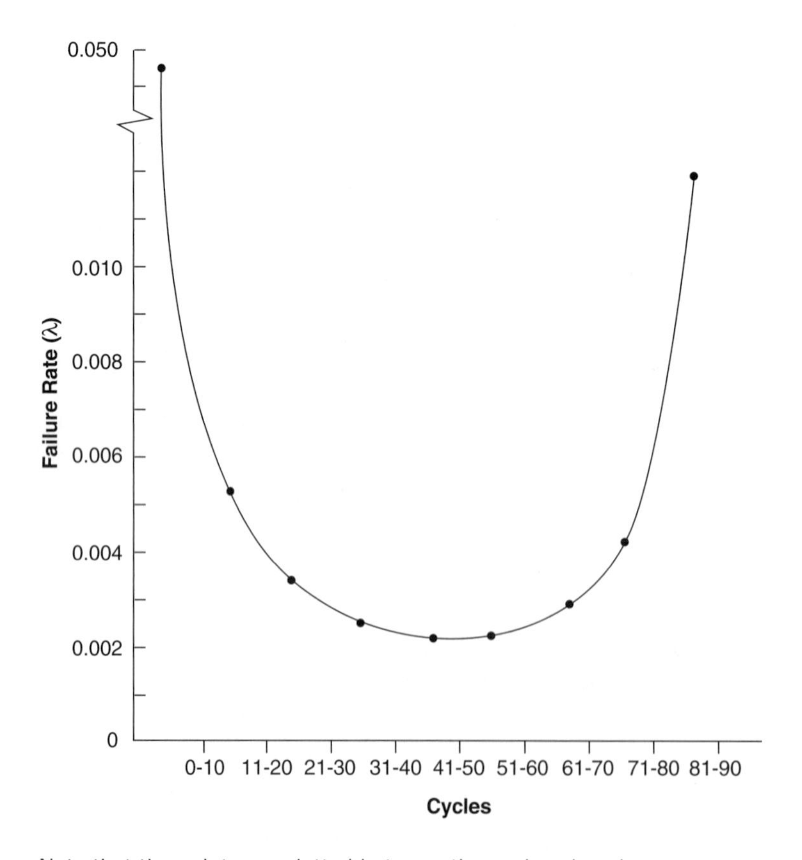

Note that the points are plotted between the cycle values because the failure rate is for the entire cell.

Normal Failure Analysis

Although the normal curve is applicable to the wear-out phase, the Weibull is usually used. The normal curve is introduced first because the reader will be familiar with its use. From Figure 11-2(b), the formula for reliability is

$$R_t = 1.0 - \int_0^t f(t)dt$$

However, the integral, $\int_0^t f(t)dt$ [see Figure 11-2(a)], is the area under the curve to the left of time, t, and is obtained from Appendix Table A. Thus, our equation becomes

$$R_t = 1.0 - P(t)$$

where R_t = reliability at time t
$P(t)$ = probability of failure or area of the normal curve to the left of time t

The process is the same as that learned in Chapter 5. An example problem will illustrate the technique.

Example Problem 11-9

A 25-W light bulb has a mean life of 750 h with a standard deviation of 50 h. What is the reliability at 850 h?

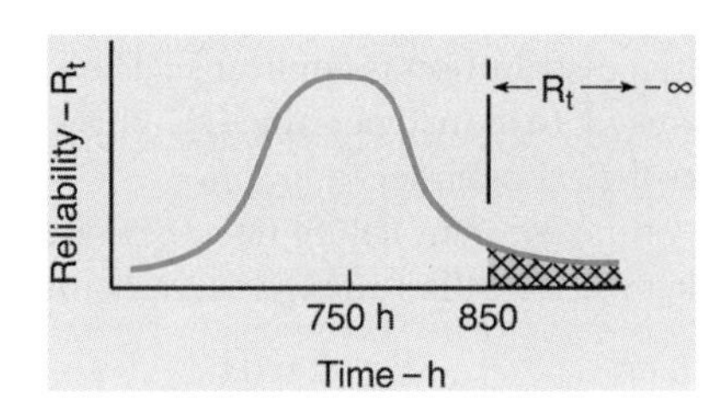

$$Z = \frac{X - \theta}{\sigma} = \frac{850 - 750}{50} = 2.0$$

From Table A, $P(t) = 0.9773$

$$R_{t=850} = 1.0 - P(t)$$
$$= 1.0 - 0.9773$$
$$= 0.0127 \text{ or } 1.27\%$$

On the average a light bulb will last 8.50 h 1.27% of the time. Stated another way, 127 light bulbs out of 10,000 will last 850 h or more.

[a]Note that θ is substituted for μ in the equation for z.

Exponential Failure Analysis

As previously stated, the exponential distribution and the Weibull distribution with shape parameter 1 are used to describe the constant failure rate. Knowing the failure rate and its reciprocal, the mean life, we can calculate the reliability using the formula

$$R_t = e^{-t/\theta}$$

where t = time or cycles
θ = mean life

Example Problem 11-10

Determine the reliability at $t = 30$ where the mean life for a constant failure rate was 40 h.

$$R_t = e^{-t/\theta}$$
$$= e^{-30/40}$$
$$= 0.472$$

What is the reliability at 10 h?

$$R_t = e^{-t/\theta}$$
$$= e^{-10/40}$$
$$= 0.453$$

What is the reliability at 50 h?

$$R_t = e^{-t/\theta}$$
$$= e^{-50/40}$$
$$= 0.287$$

The example problem shows that the reliability of the item is less as the time increases. This fact is graphically illustrated by Figure 11-2(b).

Weibull Failure Analysis

The Weibull distribution can be used for the debugging phase ($\beta < 1$), the chance failure phase ($\beta = 1$), and the wear-out phase ($\beta > 1$). By setting $\beta = 1$, the Weibull equals the exponential; by setting $\beta = 3.4$, the Weibull approximates the normal.

From Figure 11-2(b) the formula for the reliability is

$$R_t = e^{-(t/\theta)^\beta}$$

where β = the Weibull slope

The estimation of the parameters θ and β can be accomplished either graphically, analytically, or with an electronic spreadsheet such as EXCEL. Graphical analysis uses special Weibull probability paper. The data is plotted on the paper and a best-fit line is eyeballed. From this information, θ and β are determined. The computer has made this technique obsolete. An electronic spreadsheet can accomplish the same goal as probability paper, and it is more accurate.[4]

Example Problem 11-11

The failure pattern of a new type of battery fits the Weibull distribution with slope 4.2 and mean life 103 h. Determine its reliability at 120 h.

$$R_t = e^{-(t/\theta)^\beta}$$
$$= e^{-(120/103)^{42}}$$
$$= 0.150$$

[4]For more information, see D. L. Grosh, *A Primer of Reliability Theory* (New York: John Wiley & Sons, 1989), pp. 67–69; and Mitchell O. Locks, "How to Estimate the Parameters of a Weibull Distribution," *Quality Progress* (August 2002): 59–64.

OC Curve Construction

The operating characteristic (OC) curve is constructed in a manner similar to that discussed in Chapter 10. However, the fraction nonconforming, p_0, is replaced by the mean life θ. The shape of the OC curve as shown in Figure 11-4 is different than those of Chapter10. If lots are submitted with a mean life of 5000 h, the probability of acceptance is 0.697 using the sampling plan described by the OC curve of Figure 11-4.

An example problem for a constant failure rate will be used to illustrate the construction. A lot-by-lot acceptance sampling plan with replacement is as follows.

Select a sample of 16 units from a lot and test each item for 600 h. If 2 or less items fail, accept the lot; if 3 or more items fail, do not accept the lot. In symbols, the plan is $n = 16, T = 600\ h, c = 2$, and $r = 3$. When an item fails, it is replaced by another one from the lot. The first step in the construction of the curve is to assume values for the mean life θ. These values are converted to the failure rate, λ, as shown in the second column of Table 11-1. The expected average number of failures for this sampling plan is obtained by multiplying nT [$nT = (16)(600)$] by the failure rate as shown in the third column of the table.

The value $nT\lambda$ performs the same function as the value of np_0, which was previously used for the construction of an OC curve. Values for the probability of acceptance of the lot are found in Table C of the Appendix for $c = 2$. Typical calculations are as follows (assume that $\theta = 2000$):

$$\lambda = \frac{1}{\theta} = \frac{1}{2000} = 0.0005$$

$$nT\lambda = (16)(600)(0.0005) = 4.80$$

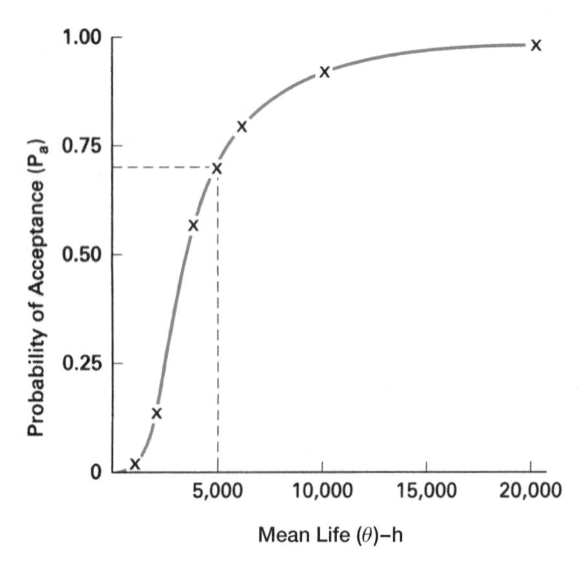

FIGURE 11-4 OC Curve for the Sampling Plan $n = 16$, $T = 600$ h, $c = 2$, and $r = 3$

TABLE 11-1 Calculations for the OC Curve for the Sampling Plan $n = 16$, $T = 600$ h, $c = 2$, $r = 3$

Mean Life θ	Failure Rate $\lambda = 1/\theta$	Expected Average Number of Failures $nT\lambda$	P_a $c = 2$
20,000	0.00005	0.48	0.983[a]
10,000	0.0001	0.96	0.927[a]
5,000	0.0002	1.92	0.698[a]
2,000	0.0005	4.80	0.142
1,000	0.0010	9.60	0.004
4,000	0.00025	2.40	0.570
6,000	0.00017	1.60	0.783

[a] By interpolation.

From Table C of the Appendix, for $nT\lambda = 4.80$ and $c = 2$,

$$P_a = 0.142$$

Additional calculations for other assumed values of θ are shown in Table 11-1.

Because this OC curve assumes a constant failure rate, the exponential distribution is applicable. The Poisson[5] distribution is used to construct the OC curve because it approximates the exponential.

Because of the constant failure rate, there are other sampling plans that will have the same OC curve. Some of these are

$n = 4$	$T = 2400$ h	$c = 2$
$n = 8$	$T = 1200$ h	$c = 2$
$n = 24$	$T = 450$ h	$c = 2$

Any combination of n and T values that give 9600 with $c = 2$ will have the same OC curve.

OC curves for reliability sampling plans are also plotted as a function of θ/θ_0, which is the actual mean life/acceptable mean life. When the OC curve is constructed in this manner, all OC curves for life tests with or without replacement have one point in common. This point is the producer's risk α and $\theta/\theta_0 = 1.0$.

LIFE AND RELIABILITY TESTING PLANS

Types of Tests

Because reliability testing requires the use of the product and sometimes its destruction, the type of test and the amount of testing is usually an economic decision. Testing is normally done on the end product; however, components and parts can be tested if they are presenting problems.

[5] $P(c) = \frac{(np_0)^c}{c!}e^{-np_0}$ (Poisson formula). Substituting $\lambda T = np_0$ and $c = 0$, then $R_t = P(0) = \frac{\lambda T^0}{0!}e^{-\lambda T} = e^{-\lambda T}$

Because testing is usually done in the laboratory, every effort should be made to simulate the real environment under controlled conditions.

Life tests are of the following three types:

Failure-Terminated. These life-test sample plans are terminated when a preassigned number of failures occurs to the sample. Acceptance criteria for the lot are based on the accumulated item test times when the test is terminated.

Time-Terminated. This type of life-test sampling plan is terminated when the sample obtains a predetermined test time. Acceptance criteria for the lot are based on the number of failures in the sample during the test time.

Sequential. A third type of life-testing plan is a sequential life-test sampling plan whereby neither the number of failures nor the time required to reach a decision is fixed in advance. Instead, decisions depend on the accumulated results of the life test. The sequential life-test plans have the advantage that the expected test time and the expected number of failures required to reach a decision as to lot acceptability are less than the failure-terminated or the time-terminated types.

Testing may be conducted with replacement of a failed unit or without replacement. *With replacement* occurs when a failure is replaced with another unit. Test time continues to be accumulated with the new sample unit. This situation is possible when there is a constant failure rate and the replaced unit has an equal chance of failure. The *without-replacement* situation occurs when the failure is not replaced.

Tests are based on one or more of the following characteristics:

1. **Mean Life** the average life of the product.
2. **Failure Rate** the percentage of failures per unit time or number of cycles.
3. **Hazard Rate** the instantaneous failure rate at a specified time. This varies with age except in the special case of a constant failure rate, where the failure rate and hazard rate are the same. The Weibull distribution is applicable and the hazard rate increases with age if the shape parameter β is greater than 1 and decreases with age if the shape parameter is less than 1.
4. **Reliable Life** the life beyond which some specified portion of the items in the lot will survive. The Weibull distribution and the normal distribution as they pertain to the wear-out phase are applicable.

There are a number of life-testing and reliability plans. A description is given in Juran's *Quality Control Handbook*. The most comprehensive plan is Handbook H108, which is provided in the next section. It uses time-terminated tests with mean-life criteria, which is the most common plan.

Handbook H108

Quality Control and Reliability Handbook H108[6] gives sampling procedures and tables for life and reliability testing. Sampling plans in the handbook are based on the exponential distribution. The handbook provides for the three different types of tests: failure-terminated, time-terminated, and sequential. For each of these types of tests, provision is made for the two situations: with replacement of failed units during the test or without replacement. Essentially, the plans are based on the mean-life criterion, although failure rate is used in one part of the handbook.

Because the handbook is over 70 pages long, only one of the plans will be illustrated. This plan is a time-terminated, with-replacement, mean-life plan, which is a common plan. There are three methods of obtaining this plan. Example problems will be used to illustrate the methods.

1. **Stipulated Producer's Risk, Consumer's Risk, and Sample Size.** Determine the time-terminated, with-replacement, mean-life sampling plan, where the producer's risk, α, of rejecting lots with mean life $\theta_0 = 900$ h is 0.05 and the customer's risk, β, of accepting lots with mean life $\theta_1 = 300$ h is 0.10. The ratio θ_1/θ_0 is

$$\frac{\theta_1}{\theta_0} = \frac{300}{900} = 0.333$$

From Table 11-2, for $\alpha = 0.05, \beta = 0.10$, and $\theta_1/\theta_0 = 0.333$, the code letter B-8 is obtained. Because the calculated ratio will rarely equal the one in the table, the next larger one is used.

For each code letter, A, B, C, D, and E, there is a table to determine the nonacceptance number and the value of the ratio T/θ_0, where T is the test time. Table 11-3 gives the value for code letter B. Thus, for code B-8, the nonacceptance number r is 8. The value of T/θ_0 is a function of the sample size.

The sample size is selected from one of the multiples of the nonacceptance number: $2r, 3r, 4r, 5r, 6r, 7r, 8r, 9r, 10r$, and $20r$. For the life-test plans, the sample size depends on the relative cost of placing large numbers of units of products on test and on the expected length of time the life tests must continue in order to determine acceptability of the lots. Increasing the sample size will, on one hand, cut the average time required to determine acceptability but, on the other hand, will increase the cost due to placing more units of product on test. For this example problem, the multiple $3r$ is selected, which gives a sample size $n = 3(8) = 24$. The corresponding value of $T/\theta_0 = 0.166$, which gives a test time T of

$$\begin{aligned} T &= 0.166(\theta_0) \\ &= 0.166(900) \\ &= 149.4 \text{ or } 149 \text{ h} \end{aligned}$$

A sample of 24 items is selected from a lot and all are tested simultaneously. If the eighth failure occurs before

[6]U.S. Department of Defense, *Quality Control and Reliability Handbook H108* (Washington, DC: U.S. Government Printing Office, 1960).

TABLE 11-2 Life-Test Sampling Plan Code Designation[a] (Table 2A-1 of H108)

$\alpha = 0.01$, $\beta = 0.01$		$\alpha = 0.05$, $\beta = 0.10$		$\alpha = 0.10$, $\beta = 0.10$		$\alpha = 0.25$, $\beta = 0.10$		$\alpha = 0.50$, $\beta = 0.10$	
Code	θ_1/θ_0	Code	θ_1/θ_0	Code	θ_1/θ_0	Code	θ_1/θ_0	Code	θ_1/θ_0
A-1	0.004	B-1	0.022	C-1	0.046	D-1	0.125	E-1	0.301
A-2	0.038	B-2	0.091	C-2	0.137	D-2	0.247	E-2	0.432
A-3	0.082	B-3	0.154	C-3	0.207	D-3	0.325	E-3	0.502
A-4	0.123	B-4	0.205	C-4	0.261	D-4	0.379	E-4	0.550
A-5	0.160	B-5	0.246	C-5	0.304	D-5	0.421	E-5	0.584
A-6	0.193	B-6	0.282	C-6	0.340	D-6	0.455	E-6	0.611
A-7	0.221	B-7	0.312	C-7	0.370	D-7	0.483	E-7	0.633
A-8	0.247	B-8	0.338	C-8	0.396	D-8	0.506	E-8	0.652
A-9	0.270	B-9	0.361	C-9	0.418	D-9	0.526	E-9	0.667
A-10	0.291	B-10	0.382	C-10	0.438	D-10	0.544	E-10	0.681
A-11	0.371	B-11	0.459	C-11	0.512	D-11	0.608	E-11	0.729
A-12	0.428	B-12	0.512	C-12	0.561	D-12	0.650	E-12	0.759
A-13	0.470	B-13	0.550	C-13	0.597	D-13	0.680	E-13	0.781
A-14	0.504	B-14	0.581	C-14	0.624	D-14	0.703	E-14	0.798
A-15	0.554	B-15	0.625	C-15	0.666	D-15	0.737	E-15	0.821
A-16	0.591	B-16	0.658	C-16	0.695	D-16	0.761	E-16	0.838
A-17	0.653	B-17	0.711	C-17	0.743	D-17	0.800	E-17	0.865
A-18	0.692	B-18	0.745	C-18	0.774	D-18	0.824	E-18	0.882

[a]Producer's risk, α, is the probability of not accepting lots with mean life θ_2; consumer's risk, β, is the probability of accepting lots with mean life θ_1.

TABLE 11-3 Values of T/θ_0 for $\alpha = 0.05$—Time-Terminated, with Replacement Code Letter B [Table 2C-2(b) of H108]

		Sample Size									
Code	r	$2r$	$3r$	$4r$	$5r$	$6r$	$7r$	$8r$	$9r$	$10r$	$20r$
B-1	1	0.026	0.017	0.013	0.010	0.009	0.007	0.006	0.006	0.005	0.003
B-2	2	0.089	0.059	0.044	0.036	0.030	0.025	0.022	0.020	0.018	0.009
B-3	3	0.136	0.091	0.068	0.055	0.045	0.039	0.034	0.030	0.027	0.014
B-4	4	0.171	0.114	0.085	0.068	0.057	0.049	0.043	0.038	0.034	0.017
B-5	5	0.197	0.131	0.099	0.079	0.066	0.056	0.049	0.044	0.039	0.020
B-6	6	0.218	0.145	0.109	0.087	0.073	0.062	0.054	0.048	0.044	0.022
B-7	7	0.235	0.156	0.117	0.094	0.078	0.067	0.059	0.052	0.047	0.023
B-8	8	0.249	0.166	0.124	0.100	0.083	0.071	0.062	0.055	0.050	0.025
B-9	9	0.261	0.174	0.130	0.104	0.087	0.075	0.065	0.058	0.052	0.026
B-10	10	0.271	0.181	0.136	0.109	0.090	0.078	0.068	0.060	0.054	0.027
B-11	15	0.308	0.205	0.154	0.123	0.103	0.088	0.077	0.068	0.062	0.031
B-12	20	0.331	0.221	0.166	0.133	0.110	0.095	0.083	0.074	0.066	0.033
B-13	25	0.348	0.232	0.174	0.139	0.116	0.099	0.087	0.077	0.070	0.035
B-14	30	0.360	0.240	0.180	0.144	0.120	0.103	0.090	0.080	0.072	0.036
B-15	40	0.377	0.252	0.189	0.151	0.126	0.108	0.094	0.084	0.075	0.038
B-16	50	0.390	0.260	0.195	0.156	0.130	0.111	0.097	0.087	0.078	0.039
B-17	75	0.409	0.273	0.204	0.164	0.136	0.117	0.102	0.091	0.082	0.041
B-18	100	0.421	0.280	0.210	0.168	0.140	0.120	0.105	0.093	0.084	0.042

the termination time of 149 h, the lot is not accepted; if the eighth failure still has not occurred after 149 test hours, the lot is accepted.

2. **Stipulated Producer's Risk, Rejection Number, and Sample Size.** Determine the time-terminated, with-replacement, mean-life sampling plan where the producer's risk of not accepting lots with mean life $\theta_0 = 1200$ h is 0.05, the nonacceptance number is 5, and the sample size is 10, or $2r$. The same set of tables is used for this method as for the previous one. Table 11-3 is the table for the code letter B designation as well as for $\alpha = 0.05$. Thus, using Table 11-3, the value for $T/\theta_0 = 0.197$ and the value for T is

$$\begin{aligned} T &= 0.197(\theta_0) \\ &= 0.197(1200) \\ &= 236.4 \text{ or } 236 \text{ h} \end{aligned}$$

A sample of 10 items is selected from a lot and all are tested simultaneously. If the fifth failure occurs before the termination time of 236 h, the lot is not accepted; if the fifth failure still has not occurred after 236 h, the lot is accepted.

3. **Stipulated Producer's Risk, Consumer's Risk, and Test Time.** Determine the time-terminated, with-replacement, mean-life sampling plan that is not to exceed 500 h and that will accept a lot with mean life of 10,000 h (θ_0) at least 90% of the time ($\beta = 0.10$) but will not accept a lot with mean life of 2000 h (θ_1) about 95% of the time ($\alpha = 0.05$). The first step is to calculate the two ratios, θ_1/θ_0 and T/θ_0:

$$\frac{\theta_1}{\theta_0} = \frac{2{,}000}{10{,}000} = \frac{1}{5}$$

$$\frac{T}{\theta_0} = \frac{500}{10{,}000} = \frac{1}{20}$$

Using the values of θ_1/θ_0, T/θ_0, α, and β, the values of r and n are obtained from Table 11-4 and are $n = 27$ and $r = 4$.

The sampling plan is to select a sample of 27 items from a lot. If the fourth failure occurs before the termination time of 500 h, the lot is not accepted; if the fourth failure still has not occurred after 500 h, the lot is accepted.

TABLE 11-4 Sampling Plans for Specified α, β, θ_1/θ_0, and T/θ_0 (Table 2C-4 of H108)

		T/θ_0					T/θ_0			
		1/3	1/5	1/10	1/20		1/3	1/5	1/10	1/20
θ_1/θ_0	r	n	n	n	n	r	n	n	n	n
		$\alpha = 0.01$		$\beta = 0.01$			$\alpha = 0.05$		$\beta = 0.01$	
2/3	136	331	551	1103	2207	95	238	397	795	1591
1/2	46	95	158	317	634	33	72	120	241	483
1/3	19	31	51	103	206	13	25	38	76	153
1/5	9	10	17	35	70	7	9	16	32	65
1/10	5	4	6	12	25	4	4	6	13	27
		$\alpha = 0.01$		$\beta = 0.05$			$\alpha = 0.05$		$\beta = 0.05$	
2/3	101	237	395	790	1581	67	162	270	541	1082
1/2	35	68	113	227	454	23	47	78	157	314
1/3	15	22	37	74	149	10	16	27	54	108
1/5	8	8	14	29	58	5	6	10	19	39
1/10	4	3	4	8	16	3	3	4	8	16
		$\alpha = 0.01$		$\beta = 0.10$			$\alpha = 0.05$		$\beta = 0.10$	
2/3	83	189	316	632	1265	55	130	216	433	867
1/2	30	56	93	187	374	19	37	62	124	248
1/3	13	18	30	60	121	8	11	19	39	79
1/5	7	7	11	23	46	4	4	7	13	27
1/10	4	2	4	8	16	3	3	4	8	16
		$\alpha = 0.01$		$\beta = 0.25$			$\alpha = 0.05$		$\beta = 0.25$	
2/3	60	130	217	434	869	35	77	129	258	517
1/2	22	37	62	125	251	13	23	38	76	153
1/3	10	12	20	41	82	6	7	13	26	52
1/5	5	4	7	13	25	3	3	4	8	16
1/10	3	2	2	4	8	2	1	2	3	7

(continued)

TABLE 11-4 (continued)

		T/θ_0					T/θ_0			
		1/3	1/5	1/10	1/20		1/3	1/5	1/10	1/20
θ_1/θ_0	r	n	n	n	n	r	n	n	n	n
		$\alpha = 0.10$		$\beta = 0.01$			$\alpha = 0.25$		$\beta = 0.01$	
2/3	77	197	329	659	1319	52	140	234	469	939
1/2	26	59	98	197	394	17	42	70	140	281
1/3	11	21	35	70	140	7	15	25	50	101
1/5	5	7	12	24	48	3	5	8	17	34
1/10	3	3	5	11	22	2	2	4	9	19
		$\alpha = 0.10$		$\beta = 0.05$			$\alpha = 0.25$		$\beta = 0.05$	
2/3	52	128	214	429	859	32	84	140	280	560
1/2	18	38	64	128	256	11	25	43	86	172
1/3	8	13	23	46	93	5	10	16	33	67
1/5	4	5	8	17	34	2	3	5	10	19
1/10	2	2	3	5	10	2	2	4	9	19
		$\alpha = 0.10$		$\beta = 0.10$			$\alpha = 0.25$		$\beta = 0.10$	
2/3	41	99	165	330	660	23	58	98	196	392
1/2	15	30	51	102	205	8	17	29	59	119
1/3	6	9	15	31	63	4	7	12	25	50
1/5	3	4	6	11	22	2	3	4	9	19
1/10	2	2	2	5	10	1	1	2	3	5
		$\alpha = 0.10$		$\beta = 0.25$			$\alpha = 0.25$		$\beta = 0.25$	
2/3	25	56	94	188	376	12	28	47	95	190
1/2	9	16	27	54	108	5	10	16	33	67
1/3	4	5	8	17	34	2	2	4	9	19
1/5	3	3	5	11	22	1	1	2	3	6
1/10	2	1	2	5	10	1	1	1	2	5

When using this technique, the tables provide for values of $\alpha = 0.01, 0.05, 0.10$, and 0.25; $\beta = 0.01, 0.05, 0.10$, and 0.25; $\theta_1/\theta_0 = \frac{2}{3}, \frac{1}{2}, \frac{1}{3}, \frac{1}{5}, \frac{1}{10}$; and $T/\theta_0 = \frac{1}{3}, \frac{1}{5}, \frac{1}{10}$, and $\frac{1}{20}$.

The method to use for obtaining the desired life-test sampling plan is determined by the available information. For example, large sample sizes provide greater confidence; however, they may be impractical from a cost or time viewpoint.

TEST DESIGN

Systems require individual components to be highly reliable. Warranty data provides one of the best sources for repairable products.[7,8] However, not all products are repairable, and this approach is reactive. It should be used for fine-tuning a reliable product.

Proactive effort is needed to build high reliability into products at the design stage. However, despite our best efforts, field failures occur, especially in newly released products. Usually the time for development is very short, which imposes severe constraints on life and reliability testing.

Demonstrating high statistical confidence in a reliability measure with tests of reasonable size and length is difficult. To minimize this difficulty, the time period must be compressed. Accelerated life testing (ALT) should provide statistical assurance that reliability goals can be met or an early warning that they cannot. There are three types of ALT: use-rate acceleration, product aging acceleration, and product street acceleration.[9]

Use-rate acceleration refers to products that are not in continuous use; therefore, test units can be run more frequently than is normal. For example, a newly designed coffee maker,

[7] Necip Doganaksoy, Gerald J. Hahn, and William G. Meeker, "Improving Reliability Through Warranty Data Analysis," *Quality Progress* (November 2006): 63–67.

[8] Necip Doganaksoy, Gerald J. Hahn, and William G. Meeker, "How to Analyze Reliability Data for Repairable Products," *Quality Progress* (June 2006): 93–95.

[9] Gerald J. Hahn, William G. Meeker, and Necip Doganaksoy, "Speedier Reliability Analysis," *Quality Progress* (June 2003): 58–64.

which usually operates one cycle per day, could be run 50 cycles per day, which would result in one year's use time occurring in 7 days. A sampling plan from the previous section would be used with the desired statistical confidence to test the required number of prototype coffee makers. The product must return to a steady state before the next cycle begins.[10]

Product aging acceleration refers to exposing test units to severe temperature, moisture, air quality, or other environmental conditions. These test conditions accelerate the physical or chemical degradation process that causes certain failure modes. For example, as a telecommunications laser ages, more current is needed to maintain the light output. The product needs to operate at least 200,000 h over 20 years at a temperature of 20°C. From experience, it was conservatively estimated that an accelerated temperature of 80°C would provide an accelerated factor of 40, thereby requiring 5000 test hours.[11]

Product stress acceleration refers to the application of increased stress such as vibration, voltage, pressure, or other types of stress. For example, degradation of a new insulation for a generator armature causes a reduction in voltage strength. The normal use condition of 120 V/mm was accelerated using five voltages of 170 to 220 V/mm.[12]

Many failure mechanisms can be traced to an underlying degradation process that leads eventually to product failure. Degradation measurement data often provide more information than failure-time data for improving reliability. Effective use of degradation data is based on identifying a degradation measure that is a true precursor of the failure mode. The advantages of using degradation data are that it provides more information and requires less time than traditional failure-time testing.

Another type of acceleration testing is highly accelerated life tests (HALTs). Both ALTs and HALTs involve testing under accelerated conditions that are carefully chosen to obtain relevant failure modes. However, the goal of HALTs is to identify and eliminate reliability problems in order to make design changes by causing the test units to fail quickly. ALTs, on the other hand, are designed to estimate reliability under use conditions.[13]

AVAILABILITY AND MAINTAINABILITY

For long-lasting products and services such as refrigerators, electric power lines, and front-line service, the time-related factors of availability, reliability, and maintainability are interrelated. For example, when a water line breaks (reliability), it is no longer available to provide water to customers and must be repaired or maintained.

Availability is a time-related factor that measures the ability of a product, process, or service to perform its designated function. The product, process, or service is available when it is in the operational state, which includes active and standby use. A calculator is operational (uptime) when it is being used to perform calculations and when it is being carried in a book bag. Availability can be quantified by the ratio

$$A = \frac{\text{Uptime}}{\text{Uptime} + \text{Downtime}} = \frac{\text{MTBF}}{\text{MTBF} + \text{MTDT}}$$

where MTBF = mean time between failures
MTDT = mean total downtime

For a repairable item, mean downtime (MTDT) is the mean time to repair (MTTR); for items that are not repairable, MTDT is the time to obtain a replacement. For the calculator example, the downtime may just require the time to change the batteries, the time to send the calculator back to the manufacturer for repair (not cost-effective), or the time to purchase a new calculator. For a manufacturing process such as a steel rolling mill, mean time to repair is critical and may even require an overnight shipment of repair parts. Downtime by a person waiting on a telephone for a service representative to become available to respond to an inquiry may require the organization to increase the number of service representatives or the number of telephone lines. Downtime takes on many different aspects, depending on the product or service.

Maintainability is the ease with which preventative and corrective maintenance on a product or service can be achieved. One of the best times to improve maintainability is in the design phase of a product or service. Improvements in design have resulted in 100,000 miles between automotive tune-ups, self-lubricating bearings, expert systems for service activities, and so forth. Production processes rely on total productivity maintenance to improve maintainability. Maintainability uses a number of different figures of merit, such as mean time to repair, mean time to service, repair hours per 1000 number of operating hours, preventative maintenance cost, and downtime probability.

According to David Mulder, keeping maintainability low may be a more cost-effective method of keeping availability high than concentrating on reliability. For example, a Rolls Royce automobile is an extremely reliable automobile; however, when it does break down, the wait time to find a dealer, obtain parts, and make repairs may be many days.[14]

COMPUTER PROGRAM

Microsoft's EXCEL has the capability of performing calculations under the Formulas/More Functions/Statistical tabs. These formulas are exponential distribution and Weibull distribution.

The EXCEL program file on the website will solve for β and θ of the Weibull distribution. Its file name is *Weibull*.

[10]Necip Doganaksoy, Gerald J. Hahn, and William G. Meeker, "Reliability Assessment by Use-Rate Acceleration," *Quality Progress* (June 2007): 74–76.
[11]William G. Meeker, Necip Doganaksoy, and Gerald J. Hahn, "Using Degradation Data for Product Reliability Analysis," *Quality Progress* (June 2001): 60–65.
[12]Gerald J. Hahn, William G. Meeker, and Necip Doganaksoy, "Speedier Reliability Analysis," *Quality Progress* (June 2003): 58–64.
[13]Ibid.

[14]Statement and example from David C. Mulder, "Comparing the Quality Measurements," unpublished manuscript.

EXERCISES

1. A system has 4 components, A, B, C, and D, with reliability values of 0.98, 0.89, 0.94, and 0.95, respectively. If the components are in series, what is the system reliability?
2. A flashlight has 4 components: 2 batteries with reliability of 0.998, a light bulb with reliability of 0.999, and a switch with reliability of 0.997. Determine the reliability of this series system.
3. Christmas tree light bulbs used to be manufactured in series—if one bulb went out, they all did. What would be the reliability of this system if each bulb had a reliability of 0.999 and there were 20 bulbs in the system?
4. What is the reliability of the system below?

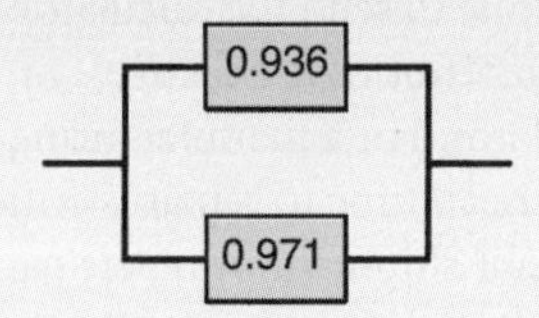

5. If component B of Exercise 1 is changed to 3 parallel components and each has the same reliability, what is the system reliability now?
6. What is the reliability of the system below, where the reliabilities of components A, B, C, and D are 0.975, 0.985, 0.988, and 0.993, respectively?

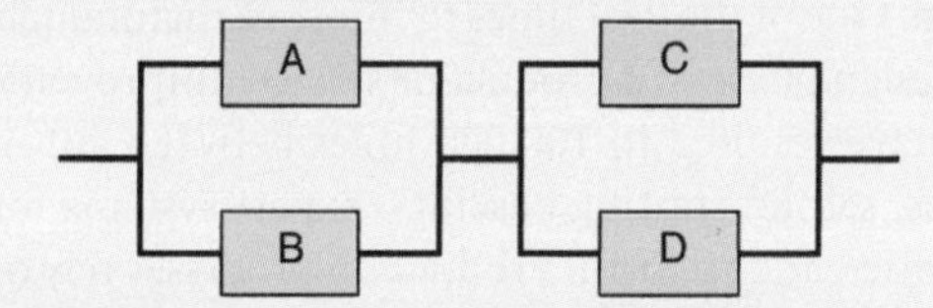

7. Using the same reliabilities as in Exercise 6, what is the reliability of the system below?

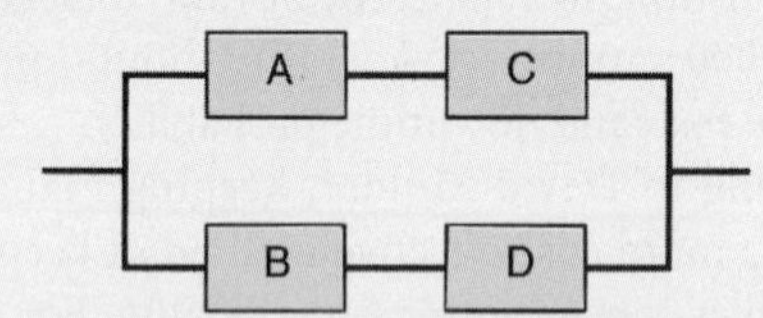

8. A system is composed of 5 components in series, and each has a reliability of 0.96. If the system can be changed to 3 components in series, what is the change in the reliability?
9. Determine the failure rate for 5 items that are tested to failure. Test data in hours are 184, 96, 105, 181, and 203.
10. Twenty-five parts are tested for 15 h. At the end of the test, 3 parts had failed at 2, 5, and 6 h. What is the failure rate?
11. Fifty parts are tested for 500 cycles each. When a part fails, it is replaced by another one. At the end of the test, 5 parts had failed. What is the failure rate?
12. Assume a constant failure rate and determine the mean life for Exercises 9, 10, and 11.
13. Determine the failure rate for a 150-h test of 9 items, where 3 items failed without replacement at 5, 76, and 135 h. What is the mean life for a constant failure rate?
14. If the mean life for a constant failure rate is 52 h, what is the failure rate?
15. Construct the life-history curve for the following test data:

Test Hours	Failures	Survivors
0–69	150	350
70–139	75	275
140–209	30	245
210–279	27	218
280–349	23	194
350–419	32	163
420–489	53	110
490–559	62	48
560–629	32	16
630–699	16	0
	500	500

16. Using normal distribution, determine the reliability at 6000 cycles of a switch with a mean life of 5500 cycles and a standard deviation of 165 cycles.
17. Determine the reliability at $t = 80$ h for the Example Problem 11-10, where $\theta = 125$ and there is a constant failure rate. What is the reliability at $t = 125$ h? At $t = 160$ h?
18. Determine the reliability at $t = 3500$ cycles for Example Problem 11-10, where the mean life of a constant failure rate is 3704 cycles. What is the reliability at $t = 3650$ cycles? At 3900 cycles?
19. Using the Weibull distribution for Exercise 16 rather than the normal distribution, determine the reliability when $\beta = 3.5$.
20. The failure pattern of an automotive engine water pump fits the Weibull distribution with $\beta = 0.7$. If the mean life during the debugging phase is 150 h, what is the reliability at 50 h?
21. Construct the OC curve for a sampling plan specified as $n = 24$, $T = 149$, $c = 7$, and $r = 8$.
22. Construct the OC curve for a sampling plan specified as $n = 10$, $T = 236$, $c = 4$, and $r = 5$.
23. Determine the time-terminated, with-replacement, mean-life sampling plan where the producer's risk of rejecting lots with mean life of 800 h is 0.05 and the consumer's risk of accepting lots with mean life $\theta_1 = 220$ is 0.10. The sample size is 30.
24. Determine the time-terminated, with-replacement sampling plan that has the following specifications: $T = 160$, $\theta_1 = 400$, $\beta = 0.10$, $\theta_0 = 800$, and $\alpha = 0.05$.

25. Determine the time-terminated, with-replacement sampling plan where the producer's risk of rejecting lots with mean life $\theta_0 = 900$ h is 0.05, the rejection number is 3, and the sample size is 9.
26. Find a replacement life-test sampling plan of 300 h that will accept a lot with mean life of 3000 h 95% of the time but will reject a lot with mean life of 1000 h 90% of the time.
27. If the probability of accepting a lot with a mean life of 1100 cycles is 0.95 and the probability of not accepting a lot with mean life of 625 cycles is 0.90, what is the sampling plan for a sample size of 60?
28. Find a life-test, time-terminated sampling plan with replacement that will accept a lot with a mean life of 900 h with probability of 0.95 ($\alpha = 0.05$). The test is to be stopped after the occurrence of the second failure, and 12 units of product are to be placed on test.
29. Using EXCEL, design a template for the construction of an OC curve and test it by solving Exercise 21.
30. Using the EXCEL program file for the Weibull distribution, determine β and θ for the ordered data set 20, 32, 40, 46, 54, 62, 73, 85, 89, 99, 102, 118, 140, 151.

CHAPTER
TWELVE

MANAGEMENT AND PLANNING TOOLS[1]

OBJECTIVES

Upon completion of this chapter, the reader is expected to

- describe the why why, forced field, and nominal group techniques;
- know how to develop and utilize the following tools:
  1. Affinity diagram
  2. Interrelationship diagram
  3. Tree diagram
  4. Matrix diagram
  5. Prioritization matrices
  6. Process decision program chart
  7. Activity network diagram

INTRODUCTION

Although the statistical process control (SPC) tools are excellent problem-solving tools, there are many situations where they are not appropriate. This chapter discusses some additional tools that can be very effective for teams and, in some cases, for individuals. They do not use hard data but rely on subjective information.

The first three tools are quite simple. The next seven are more complicated and are generally referred to as the "Seven Management and Planning Tools." Application of these tools has been proven useful in process improvement, cost reduction, policy deployment, and new-product development. The first three applications only sustain a level of competitive advantage, whereas product or service innovation is the actual source of survival in a global market. Using quality improvement tools during the innovation process will result in a higher quality product or service at a lower cost with a shorter lead time. These tools are useful in conceptualization and idea generation that is provided by a structured approach to problem solving.[2]

[1] Reproduced, with permission, from Besterfield et al., *Total Quality Management*, 3rd ed. (Upper Saddle River, NJ: Prentice Hall, 2003).

[2] Justin Levesque and H. Fred Walker, "The Innovation Process and Quality Tools," *Quality Progress* (July 2007): 18–22.

WHY, WHY

Although the "why, why" tool is very simple, it is effective. It can be a key to finding the root cause of a problem by focusing on the process rather than on people. The procedure is to describe the problem in specific terms and then ask why. You may have to ask why three or more times to obtain the root cause. An example will help illustrate the concept.

Why did we miss the delivery date?
It wasn't scheduled in time.
Why?
There were a lot of engineering changes.
Why?
Customer requested them.
The team suggested changing the delivery date whenever engineering changes occurred.

This tool is very beneficial in developing critical thinking. It is frequently a quick method of solving a problem.

FORCED FIELD ANALYSIS

Forced field analysis is used to identify the forces and factors that may influence the problem or goal. It helps an organization to better understand promoting or driving and restraining or inhibiting forces so that the positives can be reinforced and the negatives reduced or eliminated. The procedure is to define the objective, determine criteria for evaluating the effectiveness of the improvement action, brainstorm the forces that promote and inhibit achieving the goal, prioritize the forces from greatest to least, and take action to strengthen the promoting forces and weaken the inhibiting forces. An example will illustrate the tool.

Objective: Stop Smoking

Promoting Forces	→	←	Inhibiting Forces
Poor health	→	←	Habit
Smelly clothing	→	←	Addiction
Poor example	→	←	Taste
Cost	→	←	Stress
Impact on others	→	←	Advertisement

The benefits are the determination of the positives and negatives of a situation, encouraging people to agree and prioritize the competing forces, and identify the root causes.

NOMINAL GROUP TECHNIQUE

The nominal group technique provides for issue/idea input from everyone on the team and for effective decisions. An example will illustrate the technique. Let's assume that the team wants to decide which problem to work on. Everyone writes on a piece of paper the problem they think is most important. The papers are collected, and all problems are listed on a flip chart. Then each member of the team uses another piece of paper to rank the problems from least important to most important. The rankings are given a numerical value, starting at 1 for least important and continuing to the most important. Points for each problem are totaled, and the item with the highest number of points is considered to be the most important.

AFFINITY DIAGRAM

The affinity diagram allows the team to creatively generate a large number of issues/ideas and then logically group them for problem understanding and possible breakthrough solution. The procedure is to state the issue in a full sentence, brainstorm using short sentences on self-adhesive notes, post them for the team to see, sort ideas into logical groups, and create concise descriptive headings for each group. Figure 12-1 illustrates the technique.

Large groups should be divided into smaller groups with appropriate headings. Notes that stand alone could become

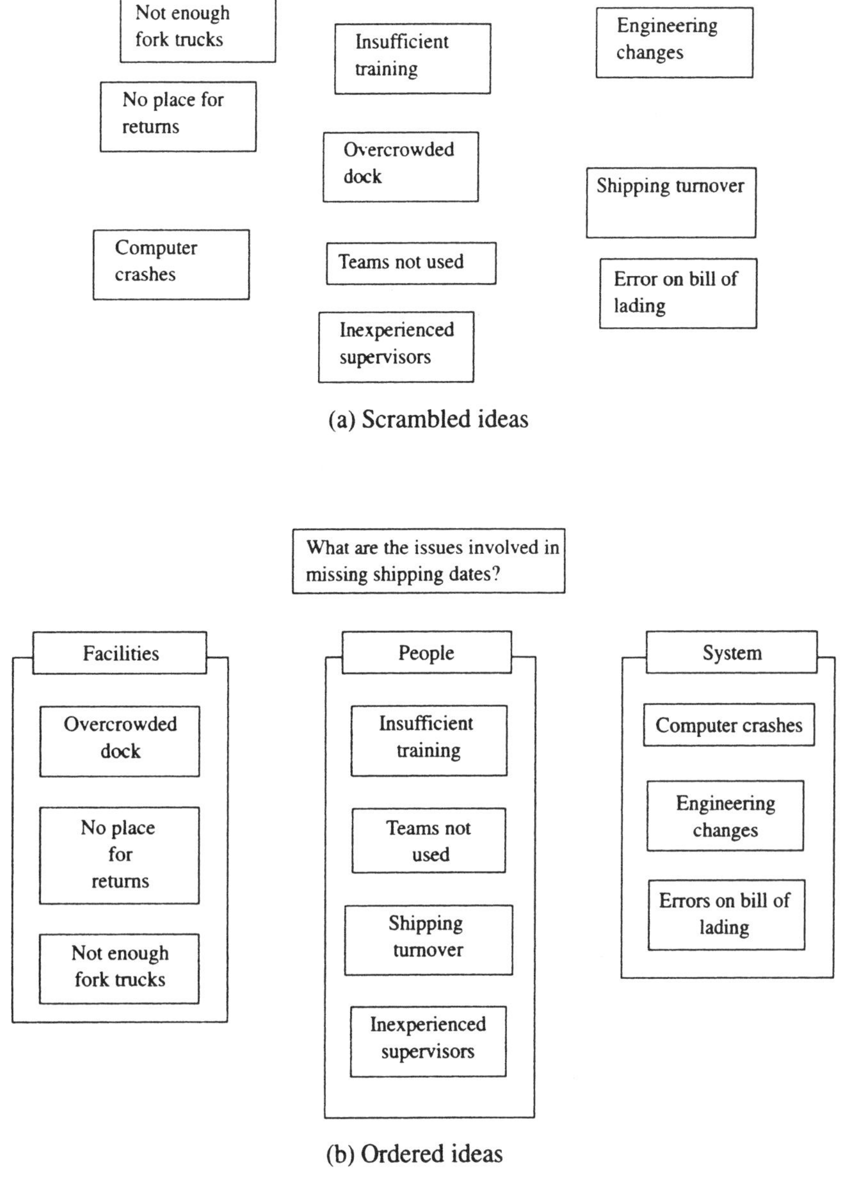

FIGURE 12-1 Affinity Diagram

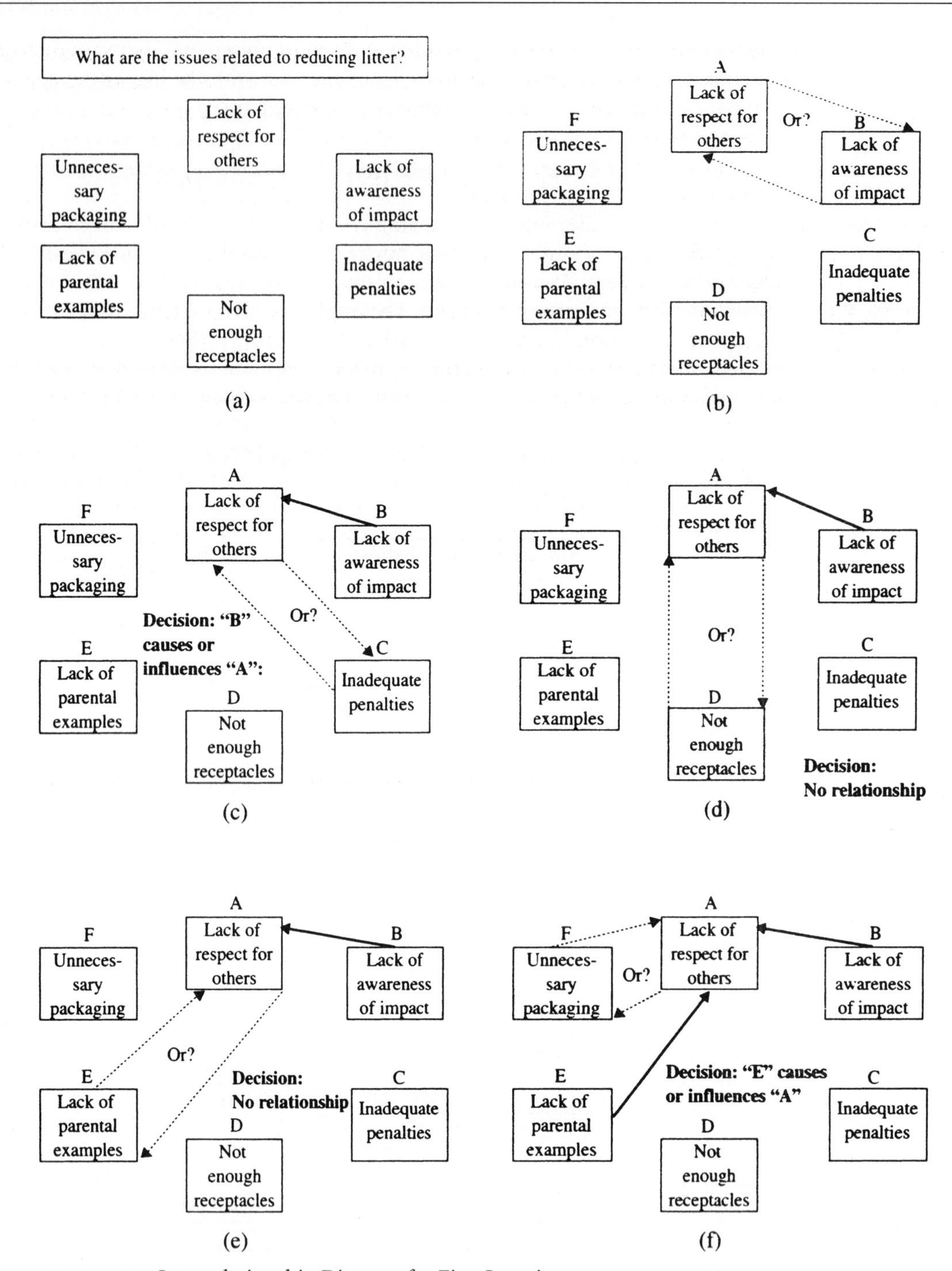

FIGURE 12-2 Interrelationship Diagram for First Iteration

headers or be placed in a miscellaneous category. Affinity diagrams encourage team creativity, break down barriers, facilitate breakthroughs, and stimulate ownership of the process.

INTERRELATIONSHIP DIAGRAM[3]

The interrelationship diagram (ID) clarifies the interrelationship of many factors of a complex situation. It allows the team to classify the cause-and-effect relationships among all the factors so that the key drivers and outcomes can be used to solve the problem. The procedure is somewhat more complicated than the previous tools; thus, it will be itemized.

1. The team should agree on the issue or problem statement.
2. All of the ideas or issues from other techniques or from brainstorming should be laid out, preferably in a circle as shown in Figure 12-2(a).
3. Start with the first issue, "Lack of Respect for Others" (A), and evaluate the cause-and-effect relationship with "Lack of Awareness of Impact" (B). In this situation, Issue B is stronger than Issue A; therefore, the arrow is drawn from Issue B to Issue A as shown in Figure 12-2(c). Each issue in the circle is compared to

[3]This section adapted, with permission, from GOAL/QPC, Salem, NH, 1989, www.MemoryJogger.org.

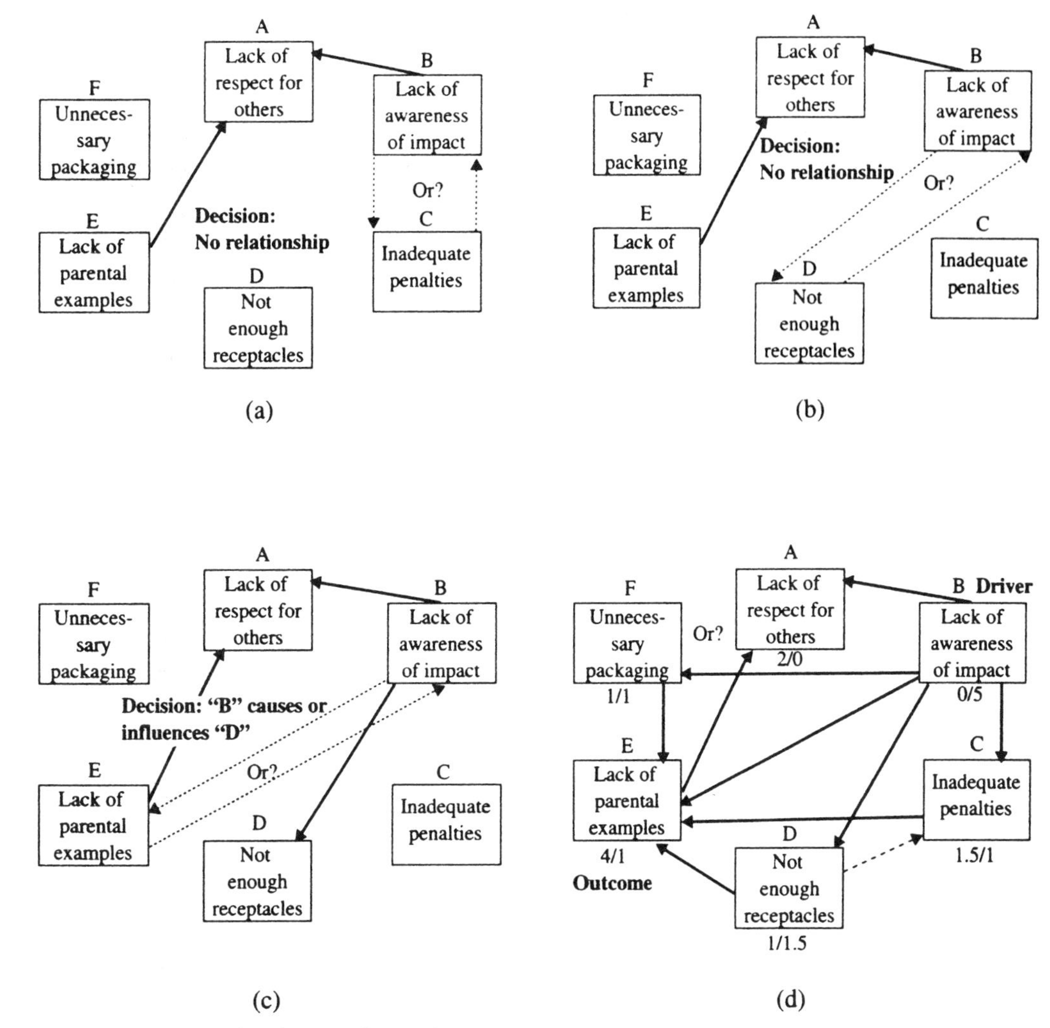

FIGURE 12-3 Completed Interrelationship Diagram

Issue A as shown in Figure 12-2(c), (d), (e), and (f). Only Issues B and E have a relationship with Issue A. The first iteration is complete.

4. The second iteration is to compare Issue B with Issues C Figure 12-3(a), D Figure 12-3(b), E Figure 12-3(c), and F. The third iteration is to compare Issue C with Issues D, E, and F. The fourth iteration is to compare Issue D with Issues E and F. The fifth iteration is to compare Issue E with Issue F.
5. The entire diagram should be reviewed and revised where necessary. It is a good idea to obtain information from other people on upstream and downstream processes.
6. The diagram is completed by tallying the incoming and outgoing arrows and placing this information below the box. Figure 12-3(d) shows a completed diagram. Issue B is the "driver" because it has zero incoming arrows and five outgoing arrows. It is usually the root cause. The issue with the highest incoming arrows is Issue E. It is a meaningful measure of success.

A relationship diagram allows a team to identify root causes from subjective data, systematically explores cause-and-effect relationships, encourages members to think multi-directionally, and develops team harmony and effectiveness.

TREE DIAGRAM

The tree diagram is used to reduce any broad objective to increasing levels of detail in order to achieve the objective. The procedure is to first choose an action-oriented objective statement from the interrelationship diagram, affinity diagram, brainstorming, team mission statement, and so forth. Second, using brainstorming, choose the major headings as shown in Figure 12-4(a) under "Means."

The third step is to generate the next level by analyzing the major headings. Ask, "What needs to be addressed to achieve the objective?" Repeat this question at each level. Three levels below the objective are usually sufficient to complete the diagram and make appropriate assignments. The diagram should be reviewed to determine if these actions will give the results anticipated or if something has been missed.

The tree diagram encourages team members to think creatively, makes large projects manageable, and generates a problem-solving atmosphere.

MATRIX DIAGRAM

The matrix diagram allows individuals or teams to identify, analyze, and rate the relationship among two or more variables. Data are presented in table form and can be objective

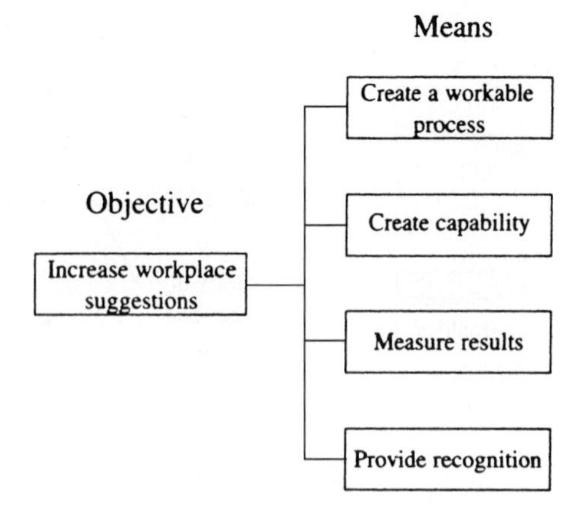

(a) Objective and means

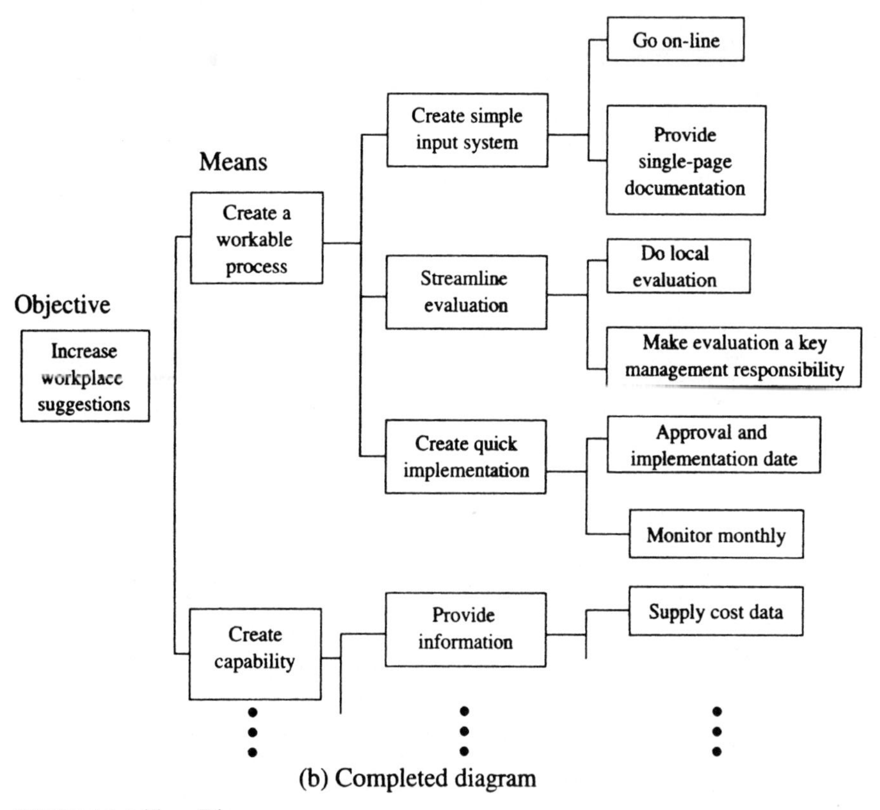

(b) Completed diagram

FIGURE 12-4 Tree Diagram

or subjective, which can be given symbols with or without numerical values. Quality function deployment (QFD), which was briefly discussed in Chapter 1, is an outstanding example of the use of the matrix diagram and is shown in Figure 12-5. There are at least five standard formats: L-shaped (2 variables), T-shaped (3 variables), Y-shaped (3 variables), C-shaped (3 variables), and X-shaped (4 variables). Our discussion will be limited to the L-shaped format, which is the most common.[4]

The procedure for the diagram is for the team to first select the factors affecting a successful plan, which for this situation are creativity, analysis, consensus, and action. Next, select the appropriate format, which in this case is the L-shaped diagram. That step is followed by determining the relationship symbols. Any symbols can be adopted, provided the diagram contains a legend as shown in the figure. Numerical values are sometimes associated with the symbol, as is done with QFD. The last step is to complete the matrix by analyzing each cell and inserting the appropriate symbol.

The matrix diagram clearly shows the relationship of the two variables. It encourages the team to think in terms of relationships, their strength, and any patterns.

[4] Detailed information on the other formats is available from Michael Brassard. *The Memory Jogger Plus+* (Salem, NH: GOAL/QPC, 1996).

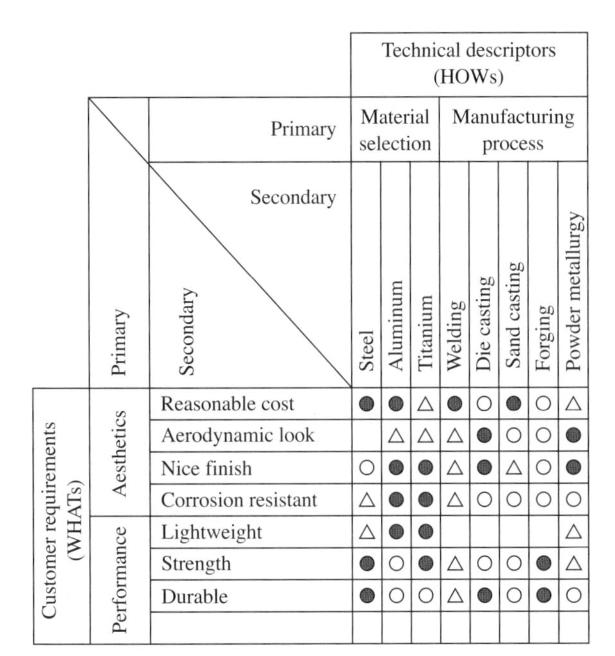

Relationship between customer requirements and technical descriptors WHATs vs. HOWs

+9 ● Strong +3 ○ Medium +1 △ Weak

FIGURE 12-5 Matrix Diagram for a QFD Process

PRIORITIZATION MATRICES

Prioritization matrices tools prioritize issues, tasks, characteristics, and so forth, based on weighted criteria using a combination of tree and matrix diagram techniques. Given prioritization, effective decisions can be made. Prioritization matrices are designed to reduce the team's options rationally before detailed implementation planning occurs. They utilize a combination of tree and matrix diagrams as shown in Figure 12-6. There are 15 implementation options; however, only the first three, beginning at "Train Supervisors," and the last one, "Purchase Fork Trucks," are shown in the tree diagram. There are four implementation criteria, however, as shown at the top of the matrix. These four are: quick to implement, accepted by users, available technology, and low cost. Prioritization matrices are the most difficult of the tools in this chapter, therefore, we will list the steps for creating one.

1. Construct an L-shaped matrix combining the options, which are the lowest level of detail of the tree diagram with the criteria. This information is given in the first column of Table 12-1.

2. Determine the implementation criteria using the nominal group technique (NGT) or any other technique that will satisfactorily weight the criteria. Using NGT, each team member submits the most important criteria on a piece of paper. They are listed on a flip chart, and the team members submit another piece of paper rank-ordering those listed on the flip chart. Those criteria with the greatest value are the most important. The team decides how many of the criteria to use. In this situation, the team decides to use the four criteria shown at the top of the matrix.

3. Prioritize the criteria using the NGT. Each team member weights the criteria so the total weight equals 1.00, and the results are totaled for the entire team as shown in Table 12-2.

4. Using NDT, rank-order the options in terms of importance by each criterion, average the results, and round to the nearest whole number. Thus, this ranking should be from 1 to the number of options for each criterion. For example, "train operators" is ranked 13 for quick to implement.

5. Compute the option importance score under each criterion by multiplying the rank by the criteria weight as shown in Table 12-1. The options with the highest total are those that should be implemented first.

There are two other techniques that are more complicated, and the reader is referred to *The Memory Jogger Plus+* for information.

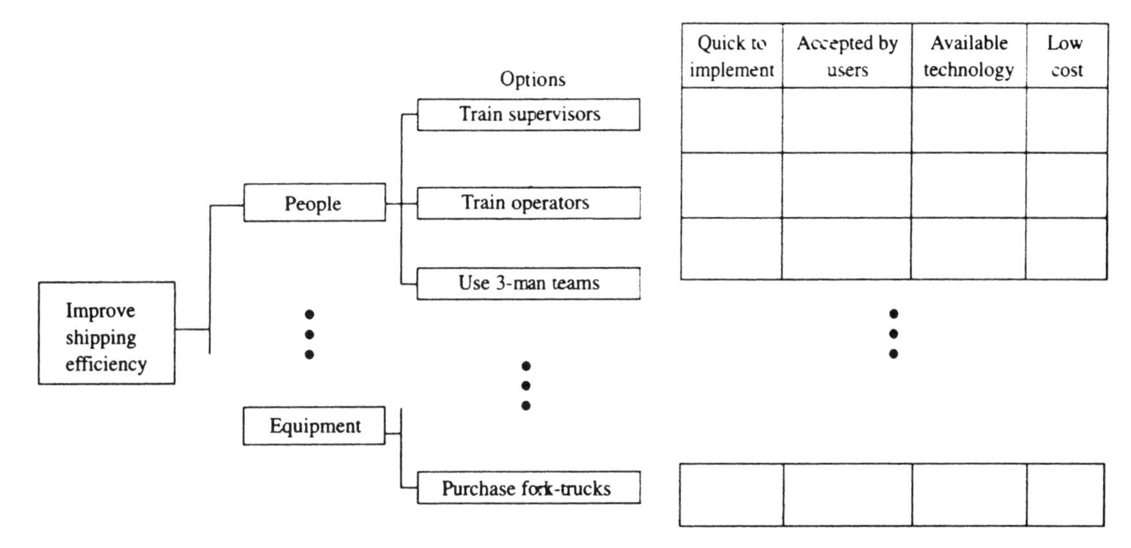

FIGURE 12-6 Prioritization Matrix for Improving Shipping Efficiency

TABLE 12-1 Improve Shipping Efficiency Using the Consensus Criteria Method

	Criteria				
Options	Quick to Implement	Accepted by Users	Available Technology	Low Cost	Total
Train operators	13(2.10) = 27.3	15(1.50) = 22.5	11(0.45) = 5.0	13(0.35) = 4.6	59.4
Train supervisors	12(2.10) = 25.2	11(1.50) = 16.5	12(0.45) = 5.4	8(0.35) = 2.8	49.9
Use 3-person teams	8(2.10) = 16.8	3(1.50) = 4.5	13(0.45) = 5.9	14(0.35) = 4.9	32.1
•	•	•	•	•	•
•	•	•	•	•	•
•	•	•	•	•	•
Purchase fork trucks	6(2.10) = 12.5	12(1.50) = 18	10(0.45) = 4.5	1(0.35) = 0.4	35.5

TABLE 12-2 Team Member Evaluation

Criteria	Member #1	Member #2		Total
Accepted by users	0.30	0.25		1.50
Low cost	0.15	0.20	• • •	0.35
Quick to implement	0.40	0.30		2.10
Available technology	0.15	0.25		0.45
	1.00	1.00		

PROCESS DECISION PROGRAM CHART

Programs to achieve particular objectives do not always go according to plan, and unexpected developments may have serious consequences. The process decision program chart (PDPC) avoids surprises and identifies possible countermeasures. Figure 12-7 illustrates the PDPC.

The procedure starts with the team stating the objective, which is to plan a successful conference. That activity is

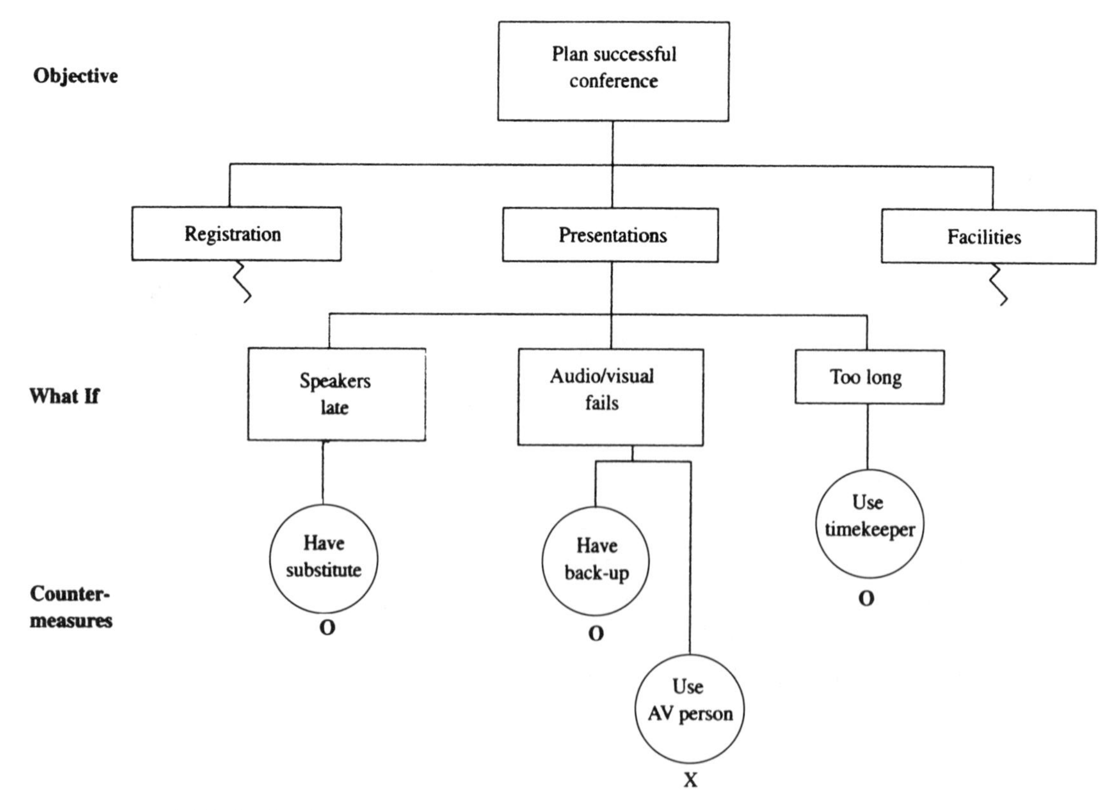

FIGURE 12-7 PDPC for Conference Presentation

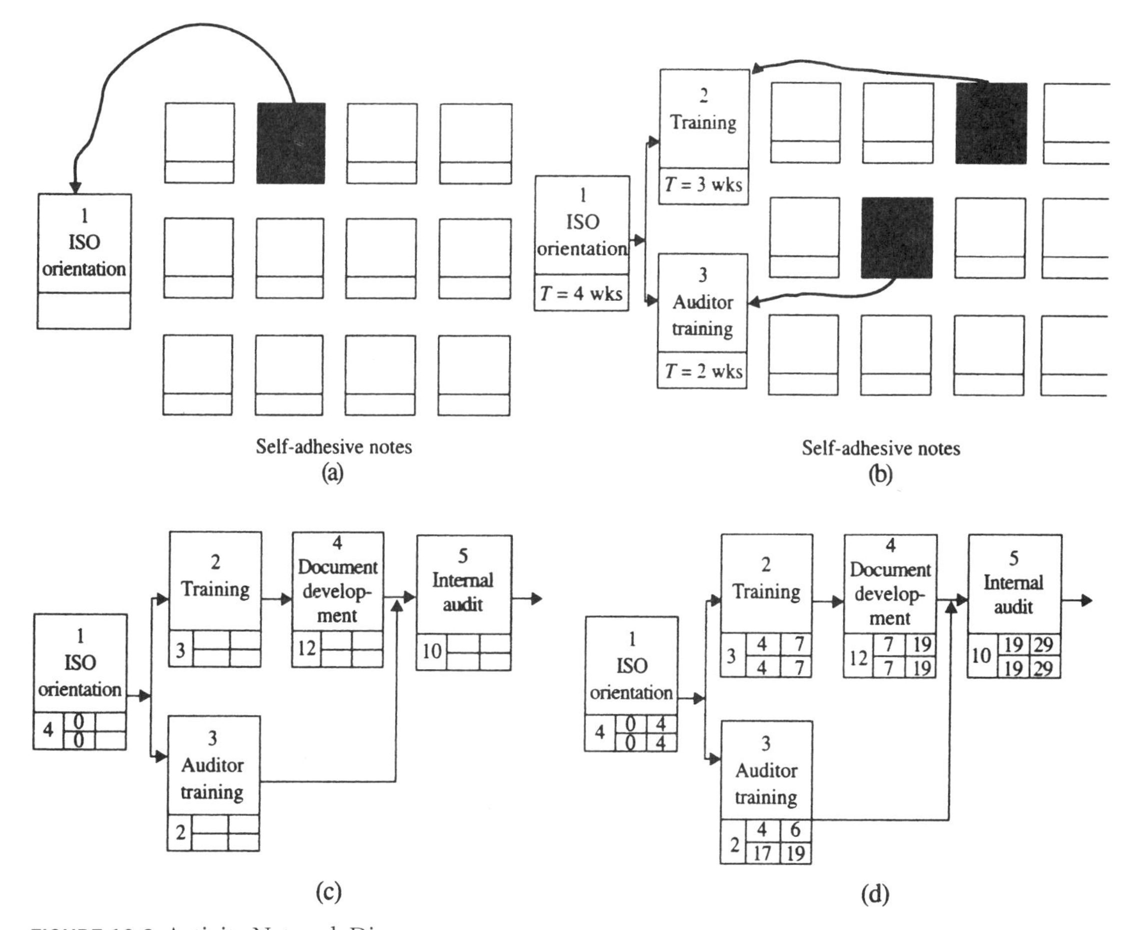

FIGURE 12-8 Activity Network Diagram

followed by the first level, which is the conference activities of registration, presentations, and facilities. Only the presentation activity is illustrated. In some cases a second level of detailed activities may be used. Next, the team brainstorms to determine what could go wrong with the conference, and these are shown as the "what if" level. Countermeasures are brainstormed and placed in a balloon in the last level. The last step is to evaluate the countermeasures and select the optimal ones by placing an *O* underneath. Place an *X* under those that are rejected.

The example has used a graphical format. PDPC can also use an outline format with the activities listed. The probability, in percent, that a "what if" will occur can be included in the box. Countermeasures should be plausible. PDPC should be used when the task is new or unique, complex, or potential failure has great risks. This tool encourages team members to think about what can happen to a process and how countermeasures can be taken. It provides the mechanism to effectively minimize uncertainty in an implementation plan.

ACTIVITY NETWORK DIAGRAM

The activity network diagram goes by a number of different names and deviations, such as program evaluation and review technique (PERT), critical path method (CPM), arrow diagram, and activity on node (AON). It allows the team to schedule a project efficiently. The diagram shows completion times, simultaneous tasks, and critical activity path. Given below is the procedure to follow.

1. The team brainstorms or documents all the tasks to complete a project. These tasks are recorded on self-adhesive notes so all members can see them.
2. The first task is located and placed on the extreme left of a large view work surface, as shown in Figure 12-8(a).
3. Any tasks that can be done simultaneously are placed below, as shown in Figure 12-8(b).
4. Repeat steps 2 and 3 until all tasks are placed in their correct sequence, as illustrated in Figure 12-8(c). *Note:* Because of space limitations, not all of the tasks are shown.
5. Number each task and draw connecting arrows. Determine the task completion time and post it in the lower left box. Completion times are recorded in hours, days, or weeks.
6. Determine the critical path by completing the four remaining boxes in each task. As shown below, these boxes are used for the earliest start time (ES), earliest finish (EF), latest start (LS), and latest finish (LF).

Activity time [T]	Earliest Start [ES]	Earliest Finish [EF]
	Latest Start [LS]	Latest Finish [LF]

The ES for Task 1 is 0, and the EF is 4 weeks later using the equation $EF = ES \div T$; the ES for Task 2 is 4 weeks, which is the same as the EF of Task 1, and the EF of Task 2 is $4 + 3 = 7$. This process is repeated for Tasks 4 and 5, which gives a total time of 29 weeks through the completion of the internal audit. If the project is to stay on schedule, the LS and LF for each of these tasks must equal the ES and EF, respectively. These values can be calculated by working backwards—subtracting the task time. They are shown in Figure 12-8(d).

Task 3, auditor training, does not have to be in sequence with the other tasks. It does have to be completed during the 19th week, because the ES for Task 5 is 19. Therefore, the LF for Task 3 is also 19 and the LS is 17. Auditor training could start after Task 1, which would give an ES of 4 and an EF of 6. The slack for Task 3 equals LS − ES [$17 - 4 = 13$]. The critical path is the longest cumulative time of connecting activities and occurs when the slack of each task is zero; thus, it is 1, 2, 4, and 5.

The benefits of an activity network diagram are (1) a realistic timetable determined by the users, (2) team members understand their role in the overall plan, (3) bottlenecks can be discovered and corrective action taken, and (4) members focus on the critical tasks. For this tool to work, the task times must be correct or reasonably close.

SUMMARY

The first three tools we discussed can be used in a wide variety of situations. They are simple to use by individuals and/or teams.

The last seven tools in the chapter are called the seven management and planning tools. Although these tools can be used individually, they are most effective when used as a system to implement an improvement plan. Figure 12-9 shows a suggested flow diagram for this integration.

The team may wish to follow this sequence or modify it to meet their needs.

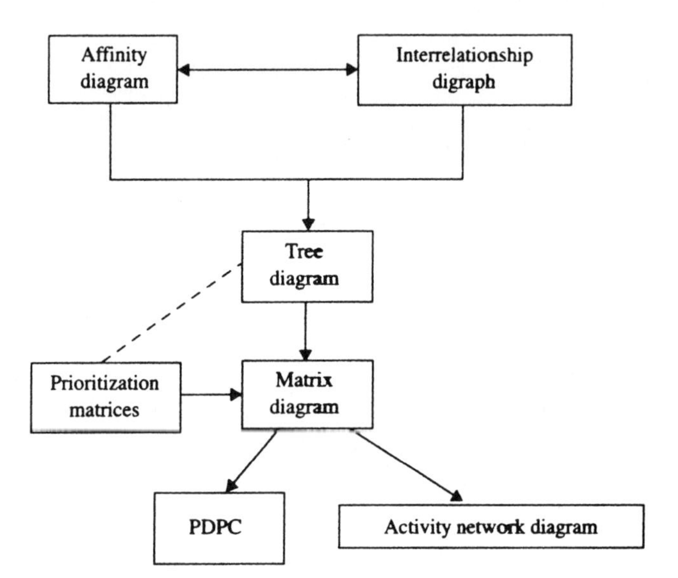

FIGURE 12-9 System Flow Diagram

EXERCISES

1. Determine why you did poorly on a recent examination by using the why, why tool.
2. Use the forced field analysis to
   a. lose weight;
   b. improve your GPA;
   c. increase your athletic ability in some sport.
3. Prepare an affinity diagram, using a team of three or more people, to plan
   a. an improvement in the cafeteria;
   b. a spring-break vacation;
   c. a field trip to a local organization.
4. Using a team of three or more people, prepare an interrelationship digraph for the
   a. computer networking of nine locations in the organization's facility;
   b. implementation of a recognition and reward system;
   c. performance improvement of the accounting department or any other work group.
5. Develop a tree diagram, using a team of three or more people, for
   a. the customer requirements for a product or service;
   b. planning a charity walkathon.
6. The church council is planning the activities for a successful carnival. Using a team of three or more people, design a tree diagram to determine detailed assignments.
7. Develop a matrix diagram to design an organization-wide training or employee involvement program. Use a team of three or more people.
8. Using a team of three or more people, construct a matrix diagram to
   a. determine customer requirements for a new product or service;
   b. allocate team assignments to implement a project such as new student week;
   c. compare teacher characteristics with potential student performance.
9. Develop a prioritization matrix, using the tree diagram developed in Exercise 6.
10. Construct a PDPC for
    a. a charity walkathon (see Exercise 5);
    b. the church carnival of Exercise 6;
    c. the matrix diagram developed in Exercise 7.
11. Using a team of three or more people, construct an activity network diagram for
    a. constructing a cardboard boat;
    b. an implementation schedule for a university event such as graduation;
    c. developing a new instructional laboratory.
12. With a team of three or more people, select a problem or situation and use the seven management and planning tools to implement an action plan. If one of the tools doesn't fit, justify its exclusion.

CHAPTER THIRTEEN

EXPERIMENTAL DESIGN

OBJECTIVES

Upon completion of this chapter, the reader is expected to

- know the applicable terminology;
- understand the concept of hypothesis testing;
- be able to determine significant factors using the *t* test;
- know the concept of point and interval estimate and be able to calculate them;
- be able to determine significant factors and their levels using the *F* test;
- understand the concept of fraction factorials.

INTRODUCTION

Organizations have become increasingly aware of the importance of quality, which is being used as a business strategy to increase market share. They are achieving world-class quality by using designed experiments. Experimental design is one of the most powerful techniques for improving quality and increasing productivity. Through experimentation, changes are intentionally introduced into the process or system in order to observe their effect on the performance characteristic or response of the system or process. A statistical approach is the most efficient method for optimizing these changes. A practitioner can do investigation without statistics using ad hoc and heuristic approaches. However, he/she can be much more efficient in his/her investigation if armed with statistical tools.

Any experiment that has the flexibility to make desired changes in the input variables of a process to observe the output response is known as experimental design. Experimental design is a systematic manipulation of a set of variables in which the effect of these manipulations is determined, conclusions are made, and results are implemented. The primary goals of a designed experiment are to

determine the variable(s) and their magnitude that influences the response;

determine the levels for these variables;

determine how to manipulate these variables to control the response.

A good experiment must be efficient. It is not an isolated test but a well-planned investigation that points the way toward understanding the process. Knowledge of the process is essential to obtain the required information and achieve the objective. Resources in the form of money, people, equipment, materials, and, most important, time, must be allocated. Efficiency does not mean producing only conforming units. Knowledge is also gained from nonconforming units.

Statistical process control (SPC) methods and experimental design techniques are powerful tools for the improvement and optimization of a process, system, design, and so forth. SPC assumes that the right variable is being controlled, the right target is known, and that the tolerance is correct. In SPC, the process gives information that leads to a useful change in the process—hence, the term *passive* statistical method. On the other hand, experimental design is known as an *active* statistical method. Information is extracted for process improvement based on tests done on the process, changes made in the input, and observations of the output. Consequently, experimental design should precede SPC, except when specifications are given by the customer.

Statistically designed experiments provide a structured plan of attack. They are more efficient than one-variable-at-a-time experiments, complement SPC, and force the experimenter to organize thoughts in a logical sequence. Experimental design can be used to

improve a process by increasing its performance and eliminate troubles;

establish statistical control of a process variable; that is, identify the variables to control the process;

improve an existing product or develop a new product.

Throughout this chapter and the next, the following terminology will be used:

Factor: A variable such as time, temperature, operator, and so forth, that is changed and results observed.

Level: A value that is assigned to change the factor. For example, two levels for the factor temperature could be 110° C and 150° C.

Treatment condition: The set of conditions (factors and their levels) for a test in an experiment.

Replicate: A repeat of a treatment condition. It requires a change in the setup.

Repetition: Multiple results of a treatment condition.

Randomization: Treatment conditions are run in a chance order to prevent any buildup in the results.

Orthogonal array: A simplified method of putting together the treatment conditions so that the design is balanced and factors can be analyzed singly or in combination.

Interaction: Two or more factors that, together, produce a result different than their separate effects.

BASIC STATISTICS

For any group of data, two parameters are of greatest interest, the mean and variance. For a group of data, $X_1, X_2, \ldots, X_n$, where n is the number of observations in the group, the mean or average is a measure of the central tendency of the group of data; that is,

$$X = \frac{1}{n}\sum_{i=1}^{n} X_i$$

The variance is a measure of the dispersion about the mean of the group of data; that is,

$$s_X^2 = \frac{\sum_{i=1}^{n}(X_i - \overline{X})^2}{n-1} = \frac{\sum_{i=1}^{n} X_i^2 - \frac{1}{n}\left(\sum_{i=1}^{n} X_i\right)^2}{n-1}$$

The standard deviation is often stated as the square root of the variance. The variance is also referred to as the mean square, MS, which is the sum of the squares, SS, divided by the number of degrees of freedom, ν; that is,

$$s_X^2 = MS = \frac{SS}{\nu}$$

The variance consists of n quantities $(X_i - \overline{X})$ in the numerator; however, there are only $n - 1$ independent quantities, because a sample statistic is used rather than the population parameter. In this case, $\overline{X}$ is used rather than μ. Thus, the number of degrees of freedom, ν, is given by

$$\nu = n - 1$$

As the sample size of a population increases, the variance of the sample approaches the variance of the population; that is,

$$MS = \lim_{n \to \infty} \frac{\sum_{i=1}^{n}(X_i - \overline{X})^2}{n-1} = \frac{\sum_{i=1}^{n}(X_i - \mu)^2}{n}$$

Example Problem 13-1

The ages of four adults are 31, 33, 28, and 36. Determine the mean, sum of squares, degrees of freedom, mean square (variance), and standard deviation of the data.

$$\overline{X} = \frac{1}{n}\sum_{i=1}^{n} X_i = \frac{1}{4}(31 + 33 + 28 + 36) = 32$$

$$SS = \sum_{i=1}^{n}(X_i - \overline{X})^2 = (-1)^2 + (+1)^2 + (-4)^2 + (+4)^2 = 34$$

$$v = n - 1 = 4 - 1 = 3$$

$$MS = s_X^2 = \frac{SS}{v} = \frac{34}{3} = 11.33$$

$$s_X = \sqrt{MS} = \sqrt{11.33} = 3.37$$

HYPOTHESES

Hypotheses testing is a statistical decision-making process in which inferences are made about the population from a sample. A probability statement is made regarding the population. Thus, hypotheses testing is subject to error. No difference, change, guilt, and so forth, is assumed between the samples until shown otherwise.

Primarily, hypotheses testing is concerned only with whether or not two samples from identical populations differ—that is, whether or not their respective means differ. Statistical hypotheses are stated in null form as

$$H_0: \mu_1 = \mu_2 \text{ or } \mu_1 - \mu_2 = 0$$

The probability of error is shown in Table 13-1.

A Type I error occurs if the null hypothesis is rejected when, in reality, it is true. Conversely, a Type II error occurs if the null hypothesis is accepted when, in reality, it is false. If either a Type I or II error occurs, alternative hypotheses are stated as

$$H_0: \mu_1 \neq \mu_2 \text{ (nondirectional, both tails)}$$
$$H_0: \mu_1 > \mu_2 \text{ (directional, right tail)}$$
$$H_0: \mu_1 < \mu_2 \text{ (directional, left tail)}$$

Alternative hypotheses are quantified by assigning a risk to the relative degree of Type I or II errors. The degree of risk in making a decision can be alternatively stated as the confidence in a decision. Types of decisions, their risk and/or

TABLE 13-1 Probability of Error for Hypothesis Testing

	SAMPLE SAYS	
Decision	Parts Are Different (Reject H_0)	Parts Are the Same (Accept H_0)
Parts are really different (reject H_0)	No error (OK)	Type II (α) error (10%, or 0.10)
Parts are really the same (accept H_0)	Type I (α) error (5%, or 0.05)	No error (OK)

TABLE 13-2 Level of Confidence and Consequences of a Wrong Decision

Designation	Risk α	Confidence $1 - \alpha$	Description
Supercritical	0.001 (0.1%)	0.999 (99.9%)	More than $200 million (large loss of life, e.g., nuclear disaster)
Critical	0.01 (1%)	0.99 (99%)	Less than $200 million (a few lives lost)
Important	0.05 (5%)	0.95 (95%)	Less than $200 thousand (no lives lost, injuries occur)
Moderate	0.10 (10%)	0.90 (90%)	Less than $100 (no injuries occur)

confidence, and the consequences of the decision are generalized in Table 13-2.

When decisions are needed on process or system improvement, product improvement, and new products, the consequences of the decision need to be evaluated and assigned an appropriate risk and/or confidence.

t TEST

The *t* test utilizes the *t* distribution to test the hypotheses of a sample from a population when the sample size is small. It compares two averages by separating difference, if there is one, from variance within the groups. Also, it can compare an average for one sample to a population mean or reference value. The *t* test assumes that the population is normally distributed. Furthermore, the *t* test can be used only when one or two samples are available for testing. For more samples, the *F* test, which is discussed in the next section, is used.

The *t* Distribution

Consider a normal random variable with mean, μ, and standard deviation, σ. If a small sample of a response variable $(y_1, y_2, \ldots, y_n)$ is taken from this normal distribution, the average, $\bar{y}$, and standard deviation, s, of the sample might not be closely related to μ or σ, respectively. The symbol X is defined as a control variable, and y is defined as a response variable with the relationship

$$y = b_1X_1 + b_2X_2 + \cdots + b_kX_k$$

Thus, when the sample size is small, the random variable

$$t = \frac{\bar{y} - \mu}{s/\sqrt{n}}$$

is governed by a t probability distribution with ν degrees of freedom, where $\nu = n - 1$. When n is small, the random variable t is not governed by a standard normal distribution; however, as n approaches infinity, the t distribution approaches

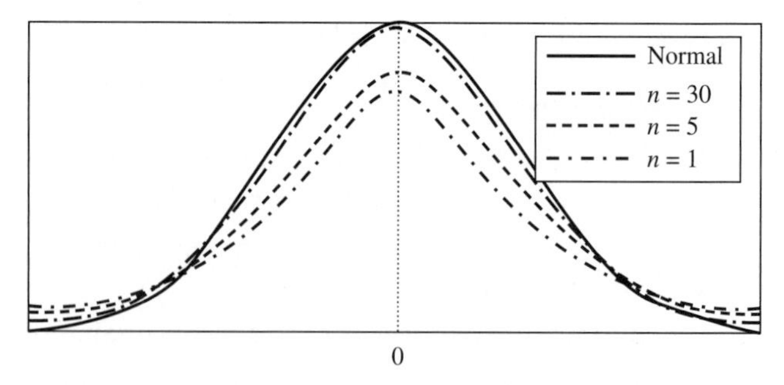

FIGURE 13-1 Normal and *t* Probability Distribution Functions

a standard normal distribution, as depicted in Figure 13-1. Consider a t distribution for a random sample with ν degrees of freedom, as shown in Figure 13-2. A value t on the abscissa represents the 100(1−) percentile of the t distribution, where the critical value α is the area under the curve to the right of t.

The value t is often designated by $t_{\alpha,\nu}$ to indicate the critical value and number of degrees freedom associated with the test. This is analogous to the terminology used in the Z test for a normal random variable when the sample size is large.

One-Sample *t* Test

The null statistical hypothesis for the one-sample t test is stated as

$$H_0: \mu = \mu_0 \text{ or } \mu - \mu_0 = 0$$

where the test statistic is

$$t = \frac{\bar{y} - \mu_0}{s/\sqrt{n}}$$

If the null statistical hypothesis is false, alternative statistical hypotheses are stated as

$$H_0: \mu \neq \mu_0 \text{ (nondirectional, both tails)}$$
$$H_0: \mu > \mu_0 \text{ (directional, right tail)}$$
$$H_0: \mu < \mu_0 \text{ (directional, left tail)}$$

and the corresponding critical regions are given by

$$t \geq t_{\alpha/2,\nu} \text{ or } t \leq -t_{\alpha/2,\nu}$$
$$t \geq t_{\alpha,\nu}$$
$$t \leq -t_{\alpha,\nu}$$

Values for $t_{\alpha,\nu}$ have been tabulated for various ranges of α and ν in Table F of the Appendix.

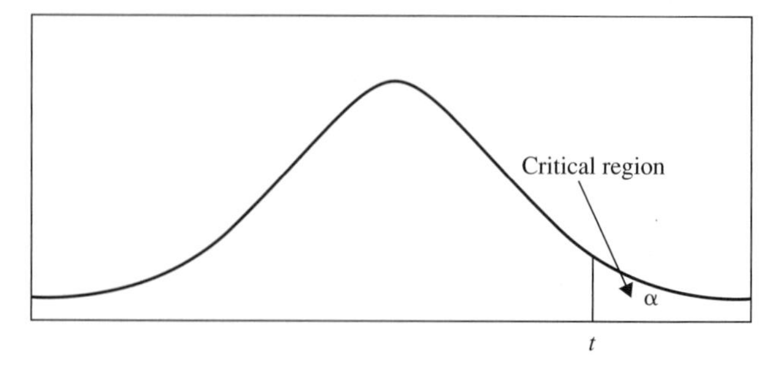

FIGURE 13-2 Critical Region of *t* Distribution

Example Problem 13-2

A lawn mower-manufacturing company would like to advertise that their top-of-the-line, self-propelled mower can go 250 hours before experiencing starting problems (more than two pulls) and subsequent service. During year-round field tests of 10 identical lawn mowers at a proving ground in a warm climate, the 10 lawn mowers required service after 205, 430, 301, 199, 222, 124, 89, 289, 260, and 214 hours. Based on this data, can the manufacturer make this advertising claim?

$$n = 10$$

$$v = n - 1 = 9$$

$$\mu_0 = 250$$

$$\bar{y} = \frac{1}{n}\sum_{i=1}^{n} y_i = \frac{1}{10}(2333) = 233.3$$

$$s = \sqrt{\frac{\sum_{i=1}^{n} y_i^2 - \frac{1}{n}\left(\sum_{i=1}^{n} y_i\right)^2}{n-1}}$$

$$= \sqrt{\frac{626{,}625 - \frac{1}{10}(2333)^2}{9}} = 95.64$$

$$t = \frac{\bar{y} - \mu_0}{s/\sqrt{n}} = \frac{233.3 - 250}{95.64/\sqrt{10}} = -0.552$$

At levels $\alpha = 0.01, 0.05$, and 0.10, $t_{\alpha,\nu}$ equals 2.821, 1.833, and 1.383, respectively, from Table F of the Appendix. The manufacturer would like to claim $\mu > 250$, so the alternative hypothesis is $\mu < 250$, and the critical (rejection) region is $t \le -t_{\alpha,\nu}$. Because $t \ge -t_{\alpha,}$ for all levels, the alternative hypothesis cannot be rejected, and hence the manufacturer cannot make the claim. After looking at the actual data, the primary reason the manufacturer cannot make the claim is the large spread in the data.

Two-Sample *t* Test

Consider two normal random variables with means μ_1 and μ_2 and respective standard deviations σ_1 and σ_2. If a small sample of n_1 random variables $(y_{11}, y_{12}, \ldots, y_{1n1})$ is taken from the first normal distribution, the mean, $\bar{y}_1$, and standard deviation, s_1, of the sample might not be close to μ_1 or σ_1, respectively. Similarly, if a small sample of n_2 random variables $(y_{21}, y_{22}, \ldots, y_{2n2})$ is taken from the second normal distribution, the mean, $\bar{y}_2$, and standard deviation, s_2, of the sample might not be close to μ_2 or σ_2, respectively.

The null statistical hypothesis for the two-sample t test is stated as

$$H_0: \mu_1 - \mu_2 = \mu_0$$

The test statistic is

$$t = \frac{\bar{y}_1 - \bar{y}_2}{s_p\sqrt{1/n_1 + 1/n_2}}$$

where the "pooled" estimator is given by

$$s_p = \sqrt{\frac{\nu_1 s_1^2 + \nu_2 s_2^2}{\nu_1 + \nu_2}}$$

and the degrees of freedom for each sample are

$$\nu_1 = n_1 - 1$$

$$\nu_2 = n_2 - 1$$

If the null statistical hypothesis is false, alternative statistical hypotheses are stated as

$$H_0: \mu_1 - \mu_2 \neq \mu_0 \text{ (nondirectional, both tails)}$$

$$H_0: \mu_1 - \mu_2 > \mu_0 \text{ (directional, right tail)}$$

$$H_0: \mu_1 - \mu_2 < \mu_0 \text{ (directional, left tail)}$$

and the corresponding critical (rejection) regions are given by

$$t \ge t_{\alpha/2,\nu} \text{ or } t \le -t_{\alpha/2,\nu}$$

$$t \ge t_{\alpha,\nu}$$

$$t \le -t_{\alpha,\nu}$$

where the total degrees of freedom are $\nu = \nu_1 + \nu_2$.

Example Problem 13-3

A study was done to determine the satisfaction with the implementation of a telephone-based class registration system at a small liberal arts college. Twenty undergraduate students were interviewed and asked to rate the current system (sample 1) of seeing an advisor and registering in person. Complaints for the current system included long lines, closed classes, and set hours. The new telephone-based system (sample 2), which is in its infancy, was rated by eight seniors. Complaints for the new system included loss of personal contact, unfamiliarity with the system, and computer downtime. The ratings were done on a scale of 1 to 100, with the following results:

Sample	Data									
1	65	70	92	54	88	83	81	75	40	95
	99	100	64	77	79	81	50	60	95	75
2	55	71	95	88	66	79	83	91		

Based on this data, use $\alpha = 0.05$ to determine whether the college should proceed with implementing the new system.

$$n_1 = 20$$

$$n_2 = 8$$

$$v_1 = n_1 - 1 = 19$$

$$v_2 = n_2 - 1 = 7$$

$$v = v_1 + v_2 = 26$$

$$\bar{y}_1 = \frac{1}{n_1}\sum_{i=1}^{n_1} y_{1i} = \frac{1}{20}(1523) = 76.15$$

$$\bar{y}_2 = \frac{1}{n_2}\sum_{i=1}^{n_2} y_{2i} = \frac{1}{8}(628) = 78.50$$

$$s_1 = \sqrt{\frac{\sum_{i=1}^{n_1} y_{1i}^2 - \frac{1}{n_1}\left(\sum_{i=1}^{n_1} y_{1i}\right)^2}{n_1 - 1}}$$

$$= \sqrt{\frac{121{,}327 - \frac{1}{20}(1523)^2}{19}} = 16.78$$

$$s_2 = \sqrt{\frac{\sum_{i=1}^{n_2} y_{2i}^2 - \frac{1}{n_2}\left(\sum_{i=1}^{n_2} y_{2i}\right)^2}{n_2 - 1}}$$

$$= \sqrt{\frac{50{,}602 - \frac{1}{8}(628)^2}{7}} = 13.65$$

$$s_p = \sqrt{\frac{\nu_1 s_1^2 + \nu_2 s_2^2}{\nu_1 + \nu_2}} = \sqrt{\frac{19(16.78)^2 + 7(13.65)^2}{19 + 7}}$$

$$= 16.00$$

$$t = \frac{\bar{y}_1 - \bar{y}_2}{s_p\sqrt{1/n_1 + 1/n_2}} = \frac{76.15 - 78.50}{16.00\sqrt{1/20 + 1/8}}$$

$$= -0.351$$

At $\alpha = 0.05$, $t_{\alpha,\nu}$ equals 1.706 from Table F of the Appendix. The college would like to claim $\mu_2 > \mu_1$ or $\mu_1 - \mu_2 < 0$. The critical (rejection) region is $t \leq -t_{\alpha,\nu}$. Because $t \geq -t_{\alpha,\nu}$, for an $\alpha = 0.05$ test, the school should not implement the new system.

F TEST

Completely randomized designs are a basic type of experimental design. Completely randomized design involve a single-treatment variable (factor) applied at two or more levels. A random group of subjects are assigned (selected) to different treatment levels. Subjects must be selected from the same population, that is, same process using different levels of one variable of the process.

For a test involving three factor levels, three two-sample *t* tests for factor levels 1 and 2, 2 and 3, and 1 and 3 would need to be performed. Therefore, for three *t* tests with each $\alpha = 0.05$, the probability of a correct decision is given by

$$(1 - \alpha)(1 - \alpha)(1 - \alpha) = (1 - 0.05)(1 - 0.05) \times (1 - 0.05) = 0.86$$

when three *t* tests are considered. Similarly, for a test involving four factor levels, six *t* tests would need to be performed, and the probability of a correct decision is 0.817. Thus, for more than two factors or levels, a more efficient technique is used to compare the means of more than two samples. Comparison between the means of more than two groups is accomplished by the use of the *F* distribution and *an*alysis *of* *va*riance (ANOVA).

The *F* Distribution

Consider an *F* distribution, as shown in Figure 13-3. Similar to the *t* distribution, a value *F* on the abscissa represents the

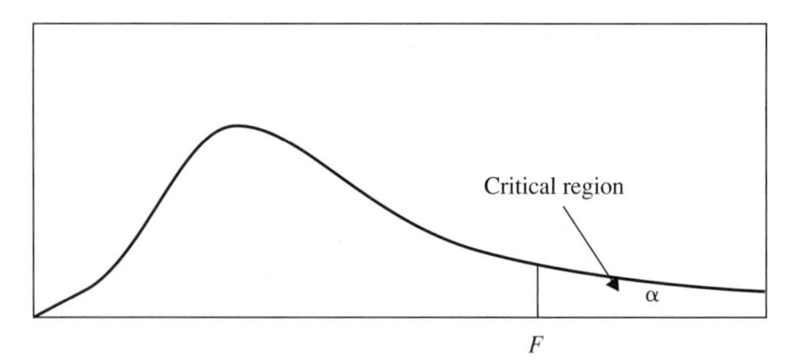

FIGURE 13-3 Critical Region of *F* Distribution

100(1 − α) percentile of the *F* distribution where the critical value α is the area under the curve to the right of *F*. In a terminology similar to the *t* test, the critical value for *F* is designated by $F_{\alpha,\nu 1,\nu 2}$ where α, ν_1, and ν_2 are the risk, degrees of freedom for the numerator, and degrees of freedom for the denominator, respectively. For simplicity, the mathematical expression for the *F* distribution function is not stated here.

Analysis of Variance

The observations of an experiment with more than two factor levels are tabulated as shown in Table 13-3. For completeness in the formulation that follows, the number of observations are different at each factor level.

In general, the samples selected are known as the levels, the specimens in each level are known as observations, or replicates, and the number of variables under study at a time is the number of factors. For the data in Table 13-3, there is only one level for each factor—thus, the term *factor level* in the table. The observations of an experiment can be written in a mathematical model as

$$y_{ij} = \mu + f_i + \varepsilon_{ij}$$

where
y_{ij} = *j*th observation at the *i*th factor level ($i = 1, \ldots, a, j = 1, \ldots, n_i$)
μ = mean value for all of the observations
f_i = difference due to treatment (factor effect) at each factor level ($i = 1, \ldots, a$)
ε_{ij} = error or noise within factor levels ($i = 1, \ldots, a, j = 1, \ldots, n_i$)

TABLE 13-3 Data for Factor Level Experimental Design

Factor Level	Data			
1	y_{11}	y_{12}	...	y_{1n1}
2	y_{21}	y_{22}	...	y_{2n2}
•	•	•	•	•
•	•	•	•	•
•	•	•	•	•
a	y_{a1}	y_{a2}	...	y_{ana}

a = number of factor levels
n_i = number of observations at each factor level ($i = 1, \ldots, a$)

In general, the observations of the experiments are assumed to be normally and independently distributed with the same variance in each factor level. Thus, the sum of the factor effects f_i can be defined as

$$\sum_{i=1}^{a} f_i = 0$$

From the preceding equation, the null statistical hypothesis for the F test is stated as

$$H_0: f_1 = f_2 = \cdots = f_a = 0$$

If this hypothesis is false, an alternative hypothesis is stated as

$$H_0: f_i \neq 0 \text{ for at least one } i$$

Testing for the validity of the hypotheses is accomplished by analysis of variance (ANOVA). The total variability in the data is given as

$$\sum_{i=1}^{a}\sum_{j=1}^{n_i}\left(y_{ij} - \frac{T}{N}\right)^2 = \sum_{i=1}^{a}\sum_{j=1}^{n_i}\left[\left(y_{ij} - \frac{A_i}{n_i}\right) + \left(\frac{A_i}{n_i} - \frac{T}{N}\right)\right]^2$$

where the total number of observations, total of the observations under the ith factor level, and total of the observations under all factor levels are given by

$$N = \sum_{i=1}^{a} n_i$$

$$A_i = \sum_{j=1}^{n_i} y_{ij}$$

$$T = \sum_{i=1}^{a} A_i$$

respectively. After expanding the square in the relationship for the total variability and simplifying the equation, the total variability in the data can be expressed as

$$SS_T = SS_F + SS_E$$

where the total sum of squares, sum of squares due to the factor, and sum of squares due to the error are given by

$$SS_T = \sum_{i=1}^{a}\sum_{j=1}^{n_i} y_{ij}^2 - \frac{T^2}{N}$$

$$SS_F = \sum_{i=1}^{a}\frac{A_i^2}{n_i} - \frac{T^2}{N}$$

$$SS_E = SS_T - SS_F$$

respectively. If there is a large difference among the means at different factor levels, SS_F will be large. The variability due to changing factor levels and error can be obtained by comparing the magnitude of SS_F and SS_E. To facilitate this, the mean squares of the factor and error are calculated by normalizing the sum of squares by their respective number of degrees of freedom; that is,

TABLE 13-4 Results for One-Factor Experimental Design

Source	Sum of Squares	Degrees of Freedom	Mean Square	F	F_{α,ν_1,ν_2}	Significant (yes/no)
Factor	SS_F	ν_F	MS_F	F	$F_{\alpha,\nu F,\nu E}$	
Error	SS_E	ν_E	MS_E			
Total	SS_T					

$$MS_F = \frac{SS_F}{\nu_F}$$

$$MS_E = \frac{SS_E}{\nu_E}$$

where the degrees of freedom for the factor and error are given by

$$\nu_F = a - 1$$

$$\nu_E = N - a$$

respectively. Analysis of variance is a technique to determine significant factors based on the F test that is used to compare ratio of factor variance with error variance; that is,

$$F = \frac{MS_F}{MS_E}$$

If $F > F_{\alpha,\nu 1,\nu 2}$, where $\nu_1 = \nu_F$ and $\nu_2 = \nu_E$, it can be said that the factor-level means are different for a level α test. Results of the F test for a one-factor experiment can be nicely depicted using a format similar to Table 13-4.

Example Problem 13-4

Consider the following data for four levels of a factor:

Level	Data					Total A	$\bar{y}$	s^2
1	26	25	32	29	30	142	28.4	8.30
2	34	31	26			91	30.3	16.33
3	15	16	25	18	13	87	17.4	21.30
4	12	15	13	11		51	12.8	2.92

Based on these data, use $\alpha = 0.05$ to determine if there is any significant difference in the data.

$$a = 4$$

$$n_1 = 5,\ n_2 = 3,\ n_3 = 5,\ n_4 = 4$$

$$N = \sum_{i=1}^{a} n_i = n_1 + n_2 + n_3 + n_4 = 17$$

$$\nu_F = a - 1 = 3$$

$$\nu_E = N - a = 13$$

$$A_i = \sum_{j=1}^{n_i} y_{ij};\ A_1 = 142,\ A_2 = 91,\ A_3 = 87,\ A_4 = 51$$

$$T = \sum_{i=1}^{a} A_i = 371$$

$$SS_T = \sum_{i=1}^{a}\sum_{j=1}^{n_i} y_{ij}^2 - \frac{T^2}{N} = 9117 - \frac{371^2}{17} = 1020.50$$

$$SS_F = \sum_{i=1}^{a}\frac{A_i^2}{n_i} - \frac{T^2}{N} = \frac{142^2}{5} + \frac{91^2}{3} + \frac{87^2}{5} + \frac{51^2}{4} - \frac{371^2}{17} = 860.65$$

$$SS_E = SS_T - SS_F = 159.85$$

$$MS_F = \frac{SS_F}{v_F} = \frac{860.65}{3} = 286.88$$

$$MS_E = \frac{SS_E}{v_E} = \frac{159.85}{13} = 12.30$$

$$F = \frac{MS_F}{MS_E} = 23.33$$

At $\alpha = 0.05$, with $\nu_1 = \nu_F$ and $\nu_2 = \nu_E$, F_{α,ν_1,ν_2} equals 3.41 from Table G-2 of the Appendix.

Source	Sum of Squares	Degrees of Freedom	Mean Square	F	F_{α,ν_1,ν_2}	Significant (yes/no)
Factor	860.65	3	286.88	23.33	3.41	Yes
Error	159.85	13	12.30			
Total	1020.50					

Because $F > F_{\alpha,\nu_1,\nu_2}$ it can be stated that the factor-level means are different.

ONE FACTOR AT A TIME

Consider an experimental design with N_F factors with two levels for each factor, as shown in Table 13-5. At treatment condition 1, all the factors are run at level 1 to obtain the benchmark or reference value. The next treatment condition is to run factor A at level 2 while the other factors are kept at level 1. For the third treatment condition, factor B is run at level 2 while the other factors are kept at level 1. This sequence is continued until all factors are run at level 2 while holding the other factors at level 1.

TABLE 13-5 Effects and Data for a One-Factor-at-a-Time Experimental Design

Treatment Condition	FACTORS/LEVELS A	B	C	D	...	N_F	Response (Results)
1	1	1	1	1	...	1	y_0
2	2	1	1	1	...	1	y_A
3	1	2	1	1	...	1	y_B
4	1	1	2	1	...	1	y_C
5	1	1	1	2	...	1	y_D
•	•	•	•	•	•	•	•
•	•	•	•	•	•	•	•
•	•	•	•	•	•	•	•
n_{tc}	1	1	1	1	...	2	y_{NF}

The change in response due to a change in the level of the factor is known as the *effect* of a factor. The main effect of a factor is defined as the difference between the average response at the two levels. In order to determine the effect of each factor, the responses at level 2 are compared to the reference value (level 1); that is,

$$e_A = y_A - y_0$$
$$e_B = y_B - y_0$$
$$e_C = y_C - y_0$$

and so on, for each factor. In this manner, the factors that have a strong effect on the response can be easily determined. In addition, the sign of the effect indicates whether an increase in a factor level increases the response, or vice versa.

As can be seen, this type of experimental design varies one factor at a time and evaluates its effect. To understand this concept, it is necessary to look at an example of a one-factor-at-a-time experiment that is neither balanced nor efficient.

Example Problem 13-5

An experiment involves maximizing a response variable. Three factors (speed, pressure, and time) are compared at two levels. The factors and levels are given in the following table:

Factors	Level 1	Level 2
Speed (m/s)	25	35
Pressure (Pa)	50	75
Time (min)	5	10

At treatment condition 1, all the factors are run at level 1 to obtain the benchmark value. The table below shows the results for a one-factor-at-a-time experimental design. The next treatment condition is to run speed at level 2 while pressure and time are kept at level 1. For the third treatment condition, pressure is run at level 2 while speed and time are kept at level 1. Similarly, treatment condition 4 is run with time at level 2, while speed and pressure are run at level 1.

Treatment Condition	FACTORS/LEVELS Speed	Pressure	Time	Response (Results)
1	1 (25 m/s)	1 (50 Pa)	1 (5 min)	2.8
2	2 (35 m/s)	1 (50 Pa)	1 (5 min)	3.4
3	1 (25 m/s)	2 (75 Pa)	1 (5 min)	4.6
4	1 (25 m/s)	1 (50 Pa)	2 (10 min)	3.7

The effects of speed, pressure, and time are:

$$e_S = y_S - y_0 = 3.4 - 2.8 = 0.6$$
$$e_P = y_P - y_0 = 4.6 - 2.8 = 1.8$$
$$e_T = y_T - y_0 = 3.7 - 2.8 = 0.9$$

It appears that pressure, time, and perhaps speed have a strong positive effect on the process. In other words, increasing any of the factors increases the response.

ORTHOGONAL DESIGN

Orthogonality means that the experimental design is balanced. A balanced experiment leads to a more efficient one. Table 13-6 shows an orthogonal experimental design for three factors with two levels each.

Although there are more treatment conditions, the design is balanced and efficient. Treatment conditions 1, 2, 3, and 5 are the same as treatment conditions 1, 2, 3, and 4 for the one-factor-at-a-time experiment in the previous section. A review of this particular design shows that there are four level 1s and four level 2s in each column. Although different designs have a different number of columns, rows, and levels, there will be an equal number of occurrences for each level. The idea of balance produces statistically independent results. Looking at the relationship between one column and another, we find that for each level within one column, each level in any other column will occur an equal number of times. Therefore, the four level 1s for factor *A* have two level 1s and two level 2s for factor *B*, and two level 1s and two level 2s for factor *C*. The effect of factor *A* at each level is statistically independent, because the effects of factor *B* and factor *C* for each level are averaged into both levels of the factor *A*. Statistical independence is also true for factor *B* and factor *C*.

In order to obtain the main effects of each factor/level, the results must be averaged. The average of the response for factor *A* due to level i $(i = 1, 2)$ is given by

$$\overline{A}_i = \frac{1}{(2)^{N_F-1}} \sum y_{ijk\cdots m}$$

Similarly, the average of the response for factor *B* due to level j $(j = 1, 2)$ is

$$\overline{B}_j = \frac{1}{(2)^{N_F-1}} \sum y_{ijk\cdots m}$$

and so on, for all the factors. While the number of treatment conditions increased by a factor of two, the number of values that were used to calculate the factor/level average increased by a factor of four. This improvement in the number of values used to determine the average makes the orthogonal design very efficient in addition to being balanced.

TABLE 13-6 Effects and Data for an Orthogonal Experimental Design

Treatment Condition	FACTORS/LEVELS A	B	C	Response (Results)
1	1	1	1	y_{111}
2	2	1	1	y_{211}
3	1	2	1	y_{121}
4	2	2	1	y_{221}
5	1	1	2	y_{112}
6	2	1	2	y_{212}
7	1	2	2	y_{122}
8	2	2	2	y_{222}

The main effects of factors *A* and *B* are then given by

$$e_A = \overline{A}_2 - \overline{A}_1$$
$$e_B = \overline{B}_2 - \overline{B}_1$$

and so on, for all the factors.

Example Problem 13-6

The following table shows an orthogonal experimental design for the same experiment given in the previous section for a one-factor-at-a-time experimental design:

Treatment Condition	FACTORS/LEVELS Speed	Pressure	Time	Response (Results)
1	1 (25 m/s)	1 (50 Pa)	1 (5 min)	2.8
2	2 (35 m/s)	1 (50 Pa)	1 (5 min)	3.4
3	1 (25 m/s)	2 (75 Pa)	1 (5 min)	4.6
4	2 (35 m/s)	2 (75 Pa)	1 (5 min)	3.8
5	1 (25 m/s)	1 (50 Pa)	2 (10 min)	3.7
6	2 (35 m/s)	1 (50 Pa)	2 (10 min)	2.7
7	1 (25 m/s)	2 (75 Pa)	2 (10 min)	3.1
8	2 (35 m/s)	2 (75 Pa)	2 (10 min)	4.4

The average of the response for the factors due to each level *i* is:

$$\overline{S}_1 = \frac{1}{4}(2.8 + 4.6 + 3.7 + 3.1) = 3.550$$

$$\overline{S}_2 = \frac{1}{4}(3.4 + 3.8 + 2.7 + 4.4) = 3.575$$

$$\overline{P}_1 = \frac{1}{4}(2.8 + 3.4 + 3.7 + 2.7) = 3.150$$

$$\overline{P}_2 = \frac{1}{4}(4.6 + 3.8 + 3.1 + 4.4) = 3.975$$

$$\overline{T}_1 = \frac{1}{4}(2.8 + 3.4 + 4.6 + 3.8) = 3.650$$

$$\overline{T}_2 = \frac{1}{4}(3.7 + 2.7 + 3.1 + 4.4) = 3.475$$

The main effects of speed, pressure, and time are then given by

$$e_S = \overline{S}_2 - \overline{S}_1 = 3.575 - 3.550 = 0.025$$
$$e_P = \overline{P}_2 - \overline{P}_1 = 3.975 - 3.150 = 0.825$$
$$e_T = \overline{T}_2 - \overline{T}_1 = 3.475 - 3.650 = -0.175$$

Based on these calculations, an increase in pressure (level 1 to level 2) has a strong effect on the process, whereas the small effects of speed and time are most likely due to the natural variation in the process. Clearly, this information is better than that obtained from the one-factor-at-a-time approach.

POINT AND INTERVAL ESTIMATE

In any statistical study, one of the objectives is to determine the value of a parameter or factor. Analysis and calculations yield a single point or value, which is called a *point estimate.* The true value is usually unknown; therefore, an *interval estimate* is needed to convey the degree of accuracy. It is the range or band within which the parameter is presumed to fall and is

$$CI = \bar{y} \pm t_{\alpha/2}\sqrt{s^2/n}$$

where CI = confidence interval.

Example Problem 13-7

Data for 24 observations has an average of 25.2 mm and a variance of 2.6 mm. Determine the interval that the point estimate lies within with 95% confidence. From the t table for 23 degrees of freedom and $\alpha/2 = 0.025$, the $t_{\alpha/2}$ value is 2.069. The confidence interval is

$$CI = \bar{y} \pm t_{\alpha/2}\sqrt{s^2/n} = 25.2 \pm 2.069\sqrt{2.6/24}$$
$$= 24.5 \text{ to } 25.9 \text{ mm}$$

t Reference Distribution

The t reference distribution is a technique to determine which factor levels are significant. In the previous example problem, it was found, from the F test, that the levels were significant, but it was not known which ones were significant. The t reference distribution uses the pooled variance of the data to find an interval estimate of the data. If two averages are outside this distribution, then they are different; if they are inside, there is no difference. The equation is similar to the confidence interval one. Both can be derived from the control limit equation for the $\bar{X}$ chart. The equation is

$$Ref.\ Dist. = \pm t_{\alpha/2}\sqrt{s_p^2/\bar{n}}$$

Example Problem 13-8

Using the t reference distribution, determine which pairs of levels are significant in Example Problem 13-4.

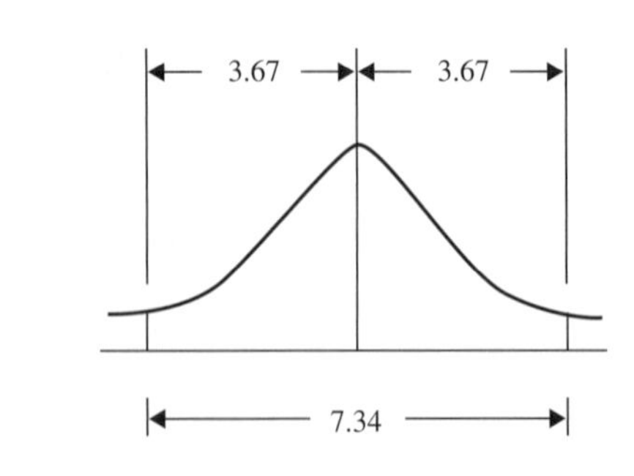

$$\bar{n} = \frac{1}{4}(5 + 3 + 5 + 4) = 4.25$$

$$s_p^2 = \frac{4(8.30) + 2(16.33) + 4(21.30) + 3(2.92)}{4 + 2 + 4 + 3}$$
$$= 12.29$$

$$Ref.\ Dist. = \pm 2.160\sqrt{12.29/4.25} = \pm 3.67$$

From the graph it is evident that there are no differences between A_1, A_2 or A_3, A_4, because the reference distribution of 7.34 easily covers these values. However, there is a difference between A_1, A_3; A_1, A_4; A_2, A_3; and A_2, A_4, because it does not cover these values.

There are a number of different ways to compare the factor levels from a significant F test. The t reference distribution is just one of them.

TWO FACTORS

Consider an experiment run N times with two factors A and B, with a and b levels, respectively. The observations are tabulated in Table 13-7. For simplicity, the number of observations are the same at each factor and level.

These observations can be written in a mathematical model as

$$y_{ijk} = \mu + (f_A)_i + (f_B)_j + (f_{AB})_{ij} + \varepsilon_{ijk}$$

where y_{ijk} = kth observation at the ith and jth levels of factors A and B, respectively ($i = 1, \ldots, a$, $j = 1, \ldots, b$, $k = 1, \ldots, n$)

μ = mean value for all of the observations

$(f_A)_i$ = difference due to treatment (factor effect A) at each factor A level ($i = 1, \ldots, a$)

$(f_B)_j$ = difference due to treatment (factor effect B) at each factor B level ($j = 1, \ldots, b$)

$(f_{AB})_{ij}$ = interaction between factors ($i = 1, \ldots, a, j = 1, \ldots, b$)

ε_{ijk} = error or noise within factors and levels ($i = 1, \ldots, a, j = 1, \ldots, b, k = 1, \ldots, n$)

a = number of levels for factor A

b = number of levels for factor B

n = number of observations for each run (replicate)

The total number of observations, total of the observation under the ith and jth levels for factors A and B, total of the observations under the ith level for factor A, total of the observations under the jth level for factor B, and total of the observations under all factor levels are given by

$$N = abn$$

$$A_{ij} = \sum_{k=1}^{n} y_{ijk}$$

$$A_{Ai} = \sum_{j=1}^{b} A_{ij}, \quad A_{Bj} = \sum_{i=1}^{a} A_{ij}$$

$$T = \sum_{i=1}^{a} A_{Ai} = \sum_{j=1}^{b} A_{Bj}$$

TABLE 13-7 Data for Two-Factor Experimental Design

Factor A Levels	FACTOR *B* LEVELS 1	2 Data	...	*b*
1	$y_{111}, \ldots, y_{11n}$	$y_{121}, \ldots, y_{12n}$	...	$y_{1b1}, \ldots, y_{1bn}$
2	$y_{211}, \ldots, y_{11n}$	$y_{221}, \ldots, y_{22n}$	...	$y_{2b1}, \ldots, y_{2bn}$
•	•	•	•	•
•	•	•	•	•
•	•	•	•	•
a	$y_{a11}, \ldots, y_{a1n}$	$y_{a21}, \ldots, y_{a2n}$	...	$y_{ab1}, \ldots, y_{abn}$

respectively. After expanding the square in the relationship for the total variability in the data and simplifying the equation, the total variability in the data can be expressed as

$$SS_T = SS_A + SS_B + SS_{AB} + SS_E$$

where the total sum of squares, sum of squares due to factor A, sum of squares due to factor B, sum of squares due to the interaction between factors A and B, and sum of square due to the error are given by

$$SS_T = \sum_{i=1}^{a}\sum_{j=1}^{b}\sum_{k=1}^{n} y_{ijk}^2 - \frac{T^2}{N}$$

$$SS_A = \sum_{i=1}^{a} \frac{A_{Ai}^2}{bn} - \frac{T^2}{N}, \quad SS_B = \sum_{j=1}^{b} \frac{A_{Bj}^2}{an} - \frac{T^2}{N}$$

$$SS_{AB} = \sum_{i=1}^{a}\sum_{j=1}^{b} \frac{A_{ij}^2}{n} - \frac{T^2}{N} - SS_A - SS_B$$

$$SS_E = SS_T - SS_A - SS_B - SS_{AB}$$

respectively. The mean squares are calculated by normalizing the sum of squares by the respective number of degrees of freedom. The mean squares for factors A and B, interaction between factors, and error are given by

$$MS_A = \frac{SS_A}{\nu_A}, \quad MS_B = \frac{SS_B}{\nu_B}$$

$$MS_{AB} = \frac{SS_{AB}}{\nu_{AB}}$$

$$MS_E = \frac{SS_E}{\nu_E}$$

where the degrees of freedom for the factor and error are given by

$$\nu_A = a - 1, \quad \nu_B = b - 1$$

$$\nu_{AB} = (a - 1)(b - 1)$$

$$\nu_E = N - ab$$

respectively. The ratio of the variance in factor A, factor B, and the interaction between factors A and B to the error variance is given by

$$F_A = \frac{MS_A}{MS_E}, \quad F_B = \frac{MS_B}{MS_E}$$

$$F_{AB} = \frac{MS_{AB}}{MS_E}$$

respectively. If $F_A > F_{\alpha,\nu1,\nu2}$, where $\nu_1 = \nu_A$ and $\nu_2 = \nu_E$, it can be said that factor A is significant for a particular value of α. Similarly, if $F_B > F_{\alpha,\nu1,\nu2}$, where $\nu_1 = \nu_B$ and $\nu_2 = \nu_E$, it can be said that factor B is significant for a particular value of α. Furthermore, if $F_{AB} > F_{\alpha,\nu1,\nu2}$, where $\nu_1 = \nu_{AB}$ and $\nu_2 = \nu_E$, it can be said that the interaction between factors A and B is significant for a particular value of α.

Results for a two-factor experimental design shown in Table 13-8 are depicted using a format similar to Table 13-4.

Example Problem 13-9

Consider an experiment on the volatility of a fluid with two factors, concentration (factor *A*) and temperature (factor *B*). Determine the significance of each factor

TABLE 13-8 Results for a Two-Factor Experimental Design

Source	Sum of Squares	Degrees of Freedom	Mean Square	F	F_{α,ν_1,ν_2}	Significant (yes/no)
Factor *A*	SS_A	ν_A	MS_A	F_A	$F_{\alpha,\nu A,\nu E}$	
Factor *B*	SS_B	ν_B	MS_B	F_B	$F_{\alpha,\nu B,\nu E}$	
Factor *AB*	SS_{AB}	ν_{AB}	MS_{AB}	F_{AB}	$F_{\alpha,\nu AB,\nu E}$	
Error	SS_E	ν_E	MS_E			
Total	SS_T					

for a level $\alpha = 0.05$. The results (response is volatility) of the experiment are given in the following table:

	FACTOR *B* LEVELS		
Factor A Levels	**1**	**2**	**3**
1	41, 38	43, 49	46, 47
2	42, 43	46, 45	48, 49

In order to obtain the significance of each factor and their interaction, the sum of the squares, degrees of freedom, mean square, and F are calculated first. These results are summarized in the following table:

Source	Sum of Squares	Degrees of Freedom	Mean Square	F	F_{α,ν_1,ν_2}	Significant (yes/no)
A	6.75	1	6.7500	1.6535	5.99	No
B	90.50	2	45.2500	11.0816	5.14	Yes
AB	6.50	2	3.2500	0.7959	5.14	No
Error	24.50	6	4.0833			
Total	128.25	11				

Based on these calculations, the effect of concentration on the volatility of the fluid is significant, whereas the effects of temperature and the interaction between concentration and temperature are not significant at a level $\alpha = 0.05$.

FULL FACTORIALS

Factorial experiments are suitable when there are many factors of interest in an experiment. Consider an experiment consisting of three factors, A, B, and C, with a and b levels for each factor, respectively. In a factorial experiment, all possible combinations of the factors are investigated. For example, when there are only two levels for each factor, all of the possible combinations (treatment conditions) can be represented using a sign table, as shown in Table 13-9.

In Table 13-9, the plus and minus signs represent the high and low levels for each factor, respectively. The method of determining the sign for the factors is to first make treatment condition (TC) 1 all negative. Then, use the equation 2^{c-1} to determine the number of consecutive like signs in a column, where c is the column number. Thus for the A column, with $c = 1, 2^{c-1} = 2^0 = 1$; for the B column, with $c = 2, 2^{c-1} = 2^1 = 2$; and for the C column, with $c = 3, 2^{c-1} = 2^2 = 4$. Once the columns for the individual factors have been determined, the columns for the interactions are the multiplication of the respective columns. Thus, the sign for the AB interaction for TC 1 is $(-)(-) = +$; TC 2 is $(+)(-) = -$; TC 7 is $(-)(+) = -$; and TC 8 is $(+)(+) = +$. For the ABC interaction, multiply the signs of the three individual factors for each TC. Thus, the sign for the ABC interaction for TC 3 is $(-)(+)(-) = +$. Similar tables can be developed for more than three factors and more than two levels; however, if each factor had three levels, a different terminology would need to be used.

The procedure for determining the sum of squares, mean square, and so forth, is similar to the procedure presented in the previous section for two factors. However, because of the increase in the number of factors, a simplified terminology is used. The sum of all responses at each treatment condition and total of all responses are given by

$$A_i = \sum_{j=1}^{n} y_{ij}$$

$$T = \sum_{i=1}^{8} A_i$$

respectively, where n is the number of runs (repetitions) per treatment condition. The sum of the responses due to each factor and level is then given by

$$A_+ = \sum_{+A's} A_i = A_2 + A_4 + A_6 + A_8,$$

$$A_- = \sum_{-A's} A_i = A_1 + A_3 + A_5 + A_7$$

$$B_+ = \sum_{+B's} A_i = A_3 + A_4 + A_7 + A_8,$$

TABLE 13-9 Signs for Effects and Data for a Three-Factor Experimental Design

	FACTORS							
Treatment Condition	***A***	***B***	***C***	***AB***	***AC***	***BC***	***ABC***	**Response (Results)**
1	−	−	−	+	+	+	−	$y_{11}, y_{12}, \ldots, y_{1n}$
2	+	−	−	−	−	+	+	$y_{21}, y_{22}, \ldots, y_{2n}$
3	−	+	−	−	+	−	+	$y_{31}, y_{32}, \ldots, y_{3n}$
4	+	+	−	+	−	−	−	$y_{41}, y_{42}, \ldots, y_{4n}$
5	−	−	+	+	−	−	+	$y_{51}, y_{52}, \ldots, y_{5n}$
6	+	−	+	−	+	−	−	$y_{61}, y_{62}, \ldots, y_{6n}$
7	−	+	+	−	−	+	−	$y_{71}, y_{72}, \ldots, y_{7n}$
8	+	+	+	+	+	+	+	$y_{81}, y_{82}, \ldots, y_{8n}$

$$B_- = \sum_{-B's} A_i = A_1 + A_2 + A_5 + A_6,$$
$$C_+ = \sum_{+C's} A_i = A_5 + A_6 + A_7 + A_8,$$
$$C_- = \sum_{-C's} A_i = A_1 + A_2 + A_3 + A_4$$
$$AB_+ = \sum_{+AB's} A_i = A_1 + A_4 + A_5 + A_8,$$
$$AB_- = \sum_{-AB's} A_i = A_2 + A_3 + A_6 + A_7$$
$$AC_+ = \sum_{+AC's} A_i = A_1 + A_3 + A_6 + A_8,$$
$$AC_- = \sum_{-AC's} A_i = A_2 + A_4 + A_5 + A_7$$
$$BC_+ = \sum_{+BC's} A_i = A_1 + A_2 + A_7 + A_8,$$
$$BC_- = \sum_{-BC's} A_i = A_3 + A_4 + A_5 + A_6$$
$$ABC_+ = \sum_{+ABC's} A_i = A_2 + A_3 + A_5 + A_8,$$
$$ABC_- = \sum_{-ABC's} A_i = A_1 + A_4 + A_6 + A_7$$

The total sum of squares, sum of squares for each factor, sum of squares due to interaction between factors, and sum of squares due to the error are given by

$$SS_T = \sum_{i=1}^{abc}\sum_{j=1}^{n} y_{ij}^2 - \frac{T^2}{N}$$
$$SS_A = \frac{A_+^2 + A_-^2}{n(2)^{k-1}} - \frac{T^2}{N}, \quad SS_B = \frac{B_+^2 + B_-^2}{n(2)^{k-1}} - \frac{T^2}{N},$$
$$SS_C = \frac{C_+^2 + C_-^2}{n(2)^{k-1}} - \frac{T^2}{N}$$
$$SS_{AB} = \frac{AB_+^2 + AB_-^2}{n(2)^{k-1}} - \frac{T^2}{N}, \quad SS_{AC} = \frac{AC_+^2 + AC_-^2}{n(2)^{k-1}} - \frac{T^2}{N}$$
$$SS_{BC} = \frac{BC_+^2 + BC_-^2}{n(2)^{k-1}} - \frac{T^2}{N},$$
$$SS_{ABC} = \frac{ABC_+^2 + ABC_-^2}{n(2)^{k-1}} - \frac{T^2}{N}$$
$$SS_E = SS_T - SS_A - SS_B - SS_C - SS_{AB} - SS_{BC} - SS_{AC} - SS_{ABC}$$

respectively, where k is the number of factors ($k = 3$ for this development). The mean squares are calculated by normalizing the sum of squares by the respective number of degrees of freedom. The mean squares for each factor, interaction between factors, and error are given by

$$MS_A = \frac{SS_A}{\nu_A}, \quad MS_B = \frac{SS_B}{\nu_B}, \quad MS_C = \frac{SS_C}{\nu_C}$$
$$MS_{AB} = \frac{SS_{AB}}{\nu_{AB}}, \quad MS_{AC} = \frac{SS_{AC}}{\nu_{AC}}, \quad MS_{BC} = \frac{SS_{BC}}{\nu_{BC}},$$
$$MS_{ABC} = \frac{SS_{ABC}}{\nu_{ABC}}$$
$$MS_E = \frac{SS_E}{\nu_E}$$

where the degrees of freedom for each factor, interaction between factors, and the error are given by

$$\nu_A = a - 1, \quad \nu_B = b - 1, \quad \nu_C = c - 1$$
$$\nu_{AB} = (a-1)(b-1), \quad \nu_{AC} = (a-1)(c-1),$$
$$\nu_{BC} = (b-1)(c-1)$$
$$\nu_{ABC} = (a-1)(b-1)(c-1)$$
$$\nu_E = N - abc$$

respectively. The ratio of the variance in each factor and the interaction between factor to the error variance is given by

$$F_A = \frac{MS_A}{MS_E}, \quad F_B = \frac{MS_B}{MS_E}, \quad F_C = \frac{MS_C}{MS_E}$$
$$F_{AB} = \frac{MS_{AB}}{MS_E}, \quad F_{AC} = \frac{MS_{AC}}{MS_E},$$
$$F_{BC} = \frac{MS_{BC}}{MS_E}, \quad F_{ABC} = \frac{MS_{ABC}}{MS_E}$$

respectively. If $F_A > F_{\alpha,\nu 1,\nu 2}$, where $\nu_1 = \nu_A$ and $\nu_2 = \nu_E$, it can be said that factor A is significant for a particular value of α. If $F_{AB} > F_{\alpha,\nu 1,\nu 2}$, where $\nu_1 = \nu_{AB}$ and $\nu_2 = \nu_E$, it can be said that the interaction between factors A and B is significant for a particular value of α. Similar comparisons can be made for factors B and C, and the interactions between factors A and C; B and C; and A, B, and C.

As previously stated, the change in response due to a change in the level of the factor is known as the factor effect. The effect of each factor is obtained by averaging the response from each level and then evaluating the difference between each average; that is,

$$e_A = \frac{A_+ - A_-}{n(2)^{k-1}}, \quad e_B = \frac{B_+ - B_-}{n(2)^{k-1}}, \quad e_C = \frac{C_+ - C_-}{n(2)^{k-1}}$$
$$e_{AB} = \frac{AB_+ - AB_-}{n(2)^{k-1}}, \quad e_{AC} = \frac{AC_+ - AC_-}{n(2)^{k-1}},$$
$$e_{BC} = \frac{BC_+ - BC_-}{n(2)^{k-1}}$$
$$e_{ABC} = \frac{ABC_+ - ABC_-}{n(2)^{k-1}}$$

Interaction between the factors is said to exist when the difference in response between the levels of one factor is different at all levels of the other factors. For a large interaction, the corresponding main effects are small. In other words, prominent interaction effect may overshadow the significance of the main effect. Hence, knowledge of interaction is often more useful than the main effect.

Results for a full factorial experimental design including factor effect and significance are depicted in Table 13-10.

Example Problem 13-10

Consider an experiment with three factors, each at two levels. Determine the significance of each factor for a level $\alpha = 0.05$ and the factor effects. The results of the experiment are given in the following table:

TABLE 13-10 Results for Three-Factor Experimental Design

Source (Factor)	Effect	Sum of Squares	ν	Mean ISquare	F	F_{α,ν_1,ν_2}	Significant (yes/no)
A	e_A	SS_A	ν_A	MS_A	F_A	$F_{,\alpha\nu A,\nu E}$	
B	e_B	SS_B	ν_B	MS_B	F_B	$F_{,\alpha\nu B,\nu E}$	
C	e_C	SS_C	ν_C	MS_C	F_C	$F_{,\alpha\nu C,\nu E}$	
AB	e_{AB}	SS_{AB}	ν_{AB}	MS_{AB}	F_{AB}	$F_{,\alpha\nu AB,\nu E}$	
AC	e_{AC}	SS_{AC}	ν_{AC}	MS_{AC}	F_{AC}	$F_{,\alpha\nu AC,\nu E}$	
BC	e_{BC}	SS_{BC}	ν_{BC}	MS_{BC}	F_{BC}	$F_{,\alpha\nu BC,\nu E}$	
ABC	e_{ABC}	SS_{ABC}	ν_{ABC}	MS_{ABC}	F_{ABC}	$F_{,\alpha\nu ABC,\nu E}$	
Error		SS_E	ν_E	MS_E			
Total		SS_T					

Treatment Condition	FACTORS							Response (Results)
	A	B	C	AB	AC	BC	ABC	
1	−	−	−	+	+	+	−	30, 28
2	+	−	−	−	−	+	+	28, 31
3	−	+	−	−	+	−	+	25, 37
4	+	+	−	+	−	−	−	36, 33
5	−	−	+	+	−	−	+	50, 45
6	+	−	+	−	+	−	−	45, 48
7	−	+	+	−	−	+	−	38, 41
8	+	+	|	|	+	+	+	44, 37

In order to obtain the significance and effect of each factor and their interaction, the sum of the squares, degrees of freedom, mean square, and F are calculated first. These results are summarized in the following table:

Source (Factor)	Effect	Sum of Squares	ν	Mean Square	F	F_{α,ν_1,ν_2}	Significant (yes/no)
A	1.00	4.00	1	4.00	0.248	5.532	No
B	1.75	12.25	1	12.25	0.760	5.532	No
C	12.50	625.00	1	625.00	38.760	5.532	Yes
AB	1.25	6.25	1	6.25	0.388	5.532	No
AC	1.00	4.00	1	4.00	0.248	5.532	No
BC	−2.75	110.25	1	110.25	6.837	5.532	Yes
ABC	−0.25	0.25	1	0.25	0.016	5.532	No
Error		129.00	8	16.125			
Total		891.00	15				

Based on these calculations, factor C has a strong effect on the experiment; for example, increasing factor C from level 1 to level 2 increases the response by 12.50. The interaction between factors B and C is also significant.

FRACTIONAL FACTORIALS

The perfect experimental design is a full factorial, with replications, that is conducted in a random manner. Unfortunately, this type of experimental design may make the number of experimental runs prohibitive, especially if the experiment is conducted on production equipment with lengthy setup times. The number of treatment conditions is determined by

$$\text{TC} = l^f$$

where TC = number of treatment conditions
l = number of levels
f = number of factors

Thus, for a two-level design, $2^2 = 4; 2^3 = 8; 2^4 = 16; 2^5 = 32; 2^6 = 64, \ldots$, and for a three-level design $3^2 = 9; 3^3 = 27; 3^4 = 81; \ldots$. If each treatment condition is replicated only once, the number of experimental runs is doubled. Thus, for a three-level design with five factors and one replicate, the number of runs is 486.

Table 13-11 shows a three-factor full factorial design. The design space is composed of the seven columns with + or −, and the design matrix is composed of the three individual factor columns A, B, and C. The design matrix tells us how to run the TCs, whereas the design space is used to make calculations to determine significance.

TABLE 13-11 Signs for Effects for a Three-Factor Experimental Design

Treatment Condition	FACTORS A	B	C	AB	AC	BC	ABC	Response (Results)
1	−	−	−	+	+	+	−	y_1
2	+	−	−	−	−	+	+	y_2
3	−	+	−	−	+	−	+	y_3
4	+	+	−	+	−	−	−	y_4
5	−	−	+	+	−	−	+	y_5
6	+	−	+	−	+	−	−	y_6
7	−	+	+	−	−	+	−	y_7
8	+	+	+	+	+	+	+	y_8

TABLE 13-12 Signs for Effects for a Fractional Factorial

Treatment Condition	FACTORS A	B	C = AB	Response (Results)
5	−	−	+	y_5
2	+	−	−	y_2
3	−	+	−	y_3
8	+	+	+	y_8

Three-factor interactions with a significant effect on the process are rare, and some two-factor interactions will not occur or can be eliminated by using engineering experience and judgment. If our engineering judgment showed that there was no three-factor interaction (*ABC*), we could place a Factor *D* in that column and make it part of the design matrix. Of course, we would need to have a high degree of confidence that factor *D* does not interact with the other columns, *A*, *B*, *C*, *AB*, *AC*, or *BC*. Similarly, we could place a Factor *E* in the column headed *BC* if we thought there was no *BC* interaction. This approach keeps the number of runs the same and adds factors.

Another approach is to reduce the number of runs while maintaining the number of factors. If we assume no two-factor or three-factor interactions, we can set $C = AB$. This approach is accomplished by selecting the rows that have *C* and *AB* of the same sign, as shown in Table 13-12.

The error term for a fractionalized experiment can be obtained by replication of the treatment conditions or by pooling the nonsignificant effects (see Chapter 14).

EXAMPLES

There are numerous examples of organizations employing design of experiments in new product development, improvement of existing products, process improvement, and improvements in service organizations. The following examples give a wide range of organizations that were capable of improving some aspect of their company or product by using design of experiments:

1. Evans Clay of McIntyre, Georgia, won the 1996 RIT/*USA Today* Quality Cup for small business by using design of experiments to increase production of kaolin clay by more than 10%, which helped the company post a profit for the first time in five years[1]. Evans Clay produces processed clay used to strengthen fiberglass used in boats and auto parts and used by the paper manufacturers. Evans Clay had not kept pace with changing times and was in serious financial danger having not posted a profit for several years. The problem the company faced was that one of their two mills consistently produced less processed clay than the other mill. To address the problem, Evans Clay conducted a robust design of experiment with numerous factors and levels: hammer configurations, air pressure, air flow, nozzles, dust collector bags, and the "horn" (used to produce sonic blasts to dislodge dust and increase air flow). The result of the design of experiments was to change the frequency of the horn blasts and to replace the dust collector bags made of polydacron with ones made of gortex that allow more air flow. As a result of their work, production increased by 10% and a profit was posted for the first time in five years.

2. John Deere Engine Works of Waterloo, Iowa, used design of experiments to eliminate an expensive chromate-conversion procedure used in the adhesion of their green paint to aluminum.[2] This change resulted in an annual savings of $500,000.

3. Nabisco Corporation had a problem with excessive scrap cookie dough and out-of-roundness conditions. They performed a factorial design to identify the important factors and the insignificant ones.[3] As a result, they saved 16

[1] 1995 RIT/*USA Today* Quality Cup for Small Business.

[2] Rich Burnham, "How to Select Design of Experiments Software," *Quality Digest* (November 1998): 32–36.

[3] Ibid.

pounds of cookie dough per shift and eliminated the out-of-roundness condition.

4. A team of employees at Wilkes-Barre General Hospital in Northeastern Pennsylvania won the 2001 RIT/*USA Today* Quality Cup for health care using a simple experimental design technique.[4] Prior to the team's work, 3 out of every 100 open-heart surgery patients developed infections that could cost $100,000 per patient, involving weeks on a ventilator, more surgery, and increased death rate. The team from Wilkes-Barre General Hospital found that a $12.47 prescription antibiotic ointment reduced the rate of infection from 3 out of 100 patients to less than 1 out of 100 patients. After studying nearly 2000 patients over three years, statistics showed that application of the ointment reduced the infection rate from 2.7% to 0.9%.

5. Eastman Kodak of Rochester, New York, used design of experiments to determine that they only needed to retool an existing machine instead of making a large capital investment in a new machine.[5] As a result, machine setup times were reduced from 8 hours to 20 minutes, scrap reduced by a factor of 10, repeatability increased to 100%, and $200,000 was saved on the capital investment in a new machine.

6. Two 9th grade students from Farmington High School in Farmington, Michigan, used design of experiments to test the AA battery life for their remote-control model cars.[6] The students used 2^3 factorial design with three control factors and two levels for each factor: AA battery (high cost and low cost), connector design type (gold-plated contacts and standard contacts), and battery temperature (cold and ambient). A battery box with contacts and a test circuit constituted the experimental setup. The students also performed an ANOVA, resulting in the battery cost being a significant factor and the other two factors not significant and possibly indistinguishable from experimental error. Their recommendations included increasing the sample size and studying the effects of noise levels.

7. K2 Corporation of Vashon, Washington, used design of experiments to reduce the high scrap rate in their new ski's complex design.[7] This resulted in a press downtime from 250 hours per week to only 2.5 hours per week.

8. Hercules Corporation had a problem with tin and lead leakage into circuit boards during plating causing costly rework or scrapping.[8] They performed a six-factor, sixteen-experiment design of experiments to find a correlation between antitarnish level and delamination. As a result, a slightly acidic rinse water was used, which eliminated the seepage problem.

As the previous examples show, design of experiments is used by a vast array of organizations, from industries to companies in the service sector, and even by young adults for a science project.

CONCLUSION

In order for experimental design to be a powerful and useful technique, careful attention must be given to the following considerations:[9]

Set good objectives.
- Selection of factors and number of factors
- Selection of the value for the levels of each factor and number of levels
- Selection of response variable

Measure response variables quantitatively.

Replicate conditions to dampen noise.

Run the experiments in a random order to eliminate any biasing.

Block out known (or unwanted) sources of variation.

Know (beforehand) which effects will be aliased from interactions.

Use results from one design of experiments to the next experiment.

Confirm results with a second experiment.

When performing the actual experiment, scientific procedures should be followed to ensure the validity of the results. Once the experiment has been performed, the statistical methods presented in this chapter should be implemented to arrive at conclusions that are objective rather than subjective.

COMPUTER PROGRAM

Microsoft's EXCEL has the capability of performing calculations under the Formulas/More Functions/Statistical Tabs. These formulas are t distribution, F distribution, and confidence. Sum of squares is performed under the Formulas/Math & Trig tabs.

[4]George Hager, "Low-Cost Antibiotic Saves Patients Pain, Thousands of Dollars," *USA Today* (May 9, 2001): 3B.

[5]Rich Burnham, "How to Select Design of Experiments Software," *Quality Digest* (November 1998): 32–36.

[6]Eric Wasiloff and Curtis Hargitt, "Using DOE to Determine AA Battery Life," *Quality Progress* (March 1999): 67–72.

[7]Rich Burnham, 32–36.

[8]Ibid.

[9]Mark J. Anderson and Shari L. Kraber, "Eight Keys to Successful DOE," *Quality Digest* (July 1999): 39–43.

EXERCISES

1. Using the judicial system as an example, explain Type I and II errors in hypothesis testing.
2. Identify the appropriate level of risk for the following items and justify your answer.
   a. Space shuttle
   b. X-ray machine
   c. Camera
   d. Canned soup
   e. Pencil
   f. Computer keyboard
   g. Chair
   h. Running shoe
   i. Automobile
   j. Fresh vegetables
   k. Child's toy
   l. Golf club
   m. Baby food
   n. Restaurant food
   o. Haircut
   p. Breast implants
   q. Child's soccer game
   r. Driving to work
   s. Taking an airplane flight
3. A law enforcement organization claims that their radar gun is accurate to within 0.5 miles per hour for a court hearing. An independent organization hired to test the accuracy of the radar gun conducted 12 tests on a projectile moving at 60 miles per hour. Given the data in the following table, determine if the radar gun can be used in court. State the hypothesis and base your comments on an appropriate α level.

Data											
63	61.5	59	63.5	57.5	61.5	61	63	60	60.5	64	62

4. An organization developing a new drug to combat depression is performing testing on 10 subjects. After giving 10 mg of the drug to each subject for 30 days, psychological tests are given to evaluate the drug's performance. The results of the test on a 100 scale are given in the following table. If the mean and variance of the psychological test prior to taking the drug was 61 and 89, respectively, comment on the performance of the drug using the two-sample t test with $\alpha = 0.01$.

Data									
88	73	66	95	69	73	81	48	59	72

5. Given the data in the following table for wave-soldering nonconformities, determine if there are any differences between the two processes. Based on various α levels, would you recommend a change to the proposed process?

Process	Data											
Current	8	9	6	8	7	6	7	8	9	8	7	6
Proposed	10	9	8	8	9							

6. An archeologist has recently uncovered an ancient find. Based upon carbon dating of bones and tools found at the site, the following data has been acquired. Use an α level of 0.10 to determine whether the bones and tools are from the same time period. Also determine whether the bones or tools are from the early Bronze Age, which dates back to before 3000 BC.

Artifact	Data (BC)				
Bones	2850	3125	3200	3050	2900
Tools	2700	2875	3250	3325	2995

7. Using the information in the following table for three levels of a factor, determine if there is any significant difference in the data for $\alpha = 0.05$. If significance occurs, determine the appropriate levels using the t reference distribution.

Level	Data							
1	12	14	13	14	15	13		
2	17	16	16	15	14	15	15	
3	11	9	10	12	11	7	11	9

8. Because of an influx of nonconforming bolts, an aircraft manufacturer decides to check the ultimate strength of four different size bolts in their last order using a tensile testing machine. Based on the following data, use an appropriate α level to determine if there is significance. Furthermore, if the ultimate strength should be greater than 120 kip/in^2, identify the bolt size(s) that you feel may be nonconforming.

Bolt Diameter (in)	Strength (kip/in^2)			
0.250	120	123	122	119
0.375	131	128	126	140
0.500	115	124	123	120
0.625	118	119	121	120

9. An engineer at a casting company performed an experiment to determine which factor (quenching temperature, alloy content, or machine tool speed) had the largest effect on the surface roughness. The following table gives the levels for each factor in the experiment.

Using a one-factor-at-a-time experiment, determine which factor has the most effect on surface roughness if the responses are $y_0 = 1.10\ \mu$ in, $y_A = 1.20\ \mu$ in, $y_B = 0.80\ \mu$ in, and $y_c = 1.60\ \mu$ in.

Factor	Description	Level 1	Level 2
A	Quenching temperature (°F)	800	400
B	Machine tool speed (in/s)	0.005	0.01
C	Nickel Alloy content (%)	1	4

10. Repeat Exercise 9 using an orthogonal experiment if the responses are $y_{111} = 1.10\ \mu$ in, $y_{211} = 0.80\ \mu$ in, $y_{121} = 0.90\ \mu$ in, $y_{221} = 0.70$ in, $y_{112} = 1.05\ \mu$ in, $y_{212} = 1.35\ \mu$ in, $y_{122} = 1.10\ \mu$ in, and $y_{222} = 1.40\ \mu$ in. Compare your results to Exercise 9 and comment on the effect of each factor on the machining process.
11. A university administrator is interested in determining the effect of SAT score and high school GPA on a university student's GPA. The university GPAs for 12 students are given in the following table. For an α level of 0.10, comment on the significance of SAT score, high school GPA, and their interaction on university GPA.

	SAT SCORE	
High School GPA	Greater than 1050	Less than 1050
Greater than 3.0/4.0	3.57, 3.75, 3.42	2.98, 2.59, 3.17
Less than 3.0/4.0	3.10, 3.22, 3.08	3.24, 2.55, 2.76

12. An insurance company is analyzing the cost of a femur fracture and tibia fracture at two hospitals. Given the following costs in dollars, determine if there is a significant difference in the cost of surgery at each hospital using an appropriate α level and comment on the results.

	TYPE OF SURGERY	
Hospital	Femur Fracture	Tibia Fracture
A	1325, 1250	900, 850
B	1125, 1075	1050, 1025

13. Develop a table similar to Table 13-9 for an experiment with four factors.
14. Formulate a full factorial experiment to determine what factors affect the retention rate in college. Identify four factors each at two levels that you feel influence retention rate. For example, one factor could be external work with two levels, 20 and 40 hours per week.
15. Three factors, each at two levels, are studied using a full factorial design. Factors and levels are given in the following table:

Factor	Description	Level 1	Level 2
A	Formulation	I	II
B	Cycle Time (s)	10	20
C	Pressure (lb/in^2)	300	400

Responses in the following table are for each treatment condition as depicted in Table 13-9. Set up a full factorial design and evaluate the significance and factor effects of this study for α levels of 0.10 and 0.05. Based on your results, comment on reducing this study to a fractional factorial experiment.

	TREATMENT CONDITION							
Replicate	1	2	3	4	5	6	7	8
1	1	2	3	4	6	6	7	9
2	2	4	5	6	8	8	9	11

16. Reduce the full factorial experiment in Exercise 14 to a three-factor fractional factorial experiment by using your experience and personal judgment. For instance, eliminate the four-factor interaction and some of the two- and three-factor interactions. Construct a table similar to Table 13-12.
17. Working individually or in a team, design a full factorial experiment and determine responses for one or more of the following items and one of your own choice. Your experiments should have at least three factors at two levels each and two replicates.
    a. Accuracy of weather forecaster (examples of factors are time of day, location, channel, and so forth)
    b. Quality of pizza (examples of factors are type of establishment, price, temperature, and so forth)
    c. Computer mouse performance (examples of factors are ergonomics, sensitivity, price, and so forth)
    d. Television remote control channel switching (examples of factors are time of day, gender, age, and so forth)

TAGUCHI'S QUALITY ENGINEERING

OBJECTIVES

Upon completion of this chapter, the reader is expected to

- know the loss function and be able to calculate the four types;
- understand the concept of orthogonal arrays with degrees of freedom, linear graphs, and interactions;
- know the signal-to-noise ratio and be able to calculate the three types;
- understand the concept of parameter design and be able to determine the strong factors and their levels;
- understand the concept of tolerance design and be able to determine percent contribution and low cost tolerance design.

INTRODUCTION

Most of the body of knowledge associated with the quality sciences was developed in the United Kingdom as design of experiments and in the United States as statistical quality control. More recently, Dr. Genichi Taguchi, a mechanical engineer who has won four Deming Awards, has added to this body of knowledge. In particular, he introduced the loss function concept, which combines cost, target, and variation into one metric with specifications being of secondary importance. Furthermore, he developed the concept of robustness, which means that noise factors are taken into account to ensure that the system functions correctly. Noise factors are uncontrollable variables that can cause significant variability in the process or the product.

LOSS FUNCTION

Taguchi has defined quality as the loss imparted to society from the time a product is shipped. Societal losses include failure to meet customer requirements, failure to meet ideal performance, and harmful side effects. Many practitioners have included the losses due to production, such as raw material, energy, and labor consumed on unusable products or toxic by-products.

The loss-to-society concept can be illustrated by an example associated with the production of large vinyl covers to protect materials from the elements. Figure 14-1 shows three stages in the evolution of vinyl thickness. At (1), the process is just capable of meeting the specifications (USL and LSL); however, it is on the target *tau*, τ.[1] After considerable effort, the production process was improved by reducing the variability about the target, as shown at (2). In an effort to reduce its production costs, the organization decided to shift the target closer to the LSL, as shown at (3). This action resulted in a substantial improvement by lowering the cost to the organization; however, the vinyl covers were not as strong as before. When farmers used the covers to protect wheat from the elements, they tore and a substantial loss occurred to the farmers. In addition, the cost of wheat increased as a result of supply-and-demand factors, thereby causing an increase in wheat prices and a further loss to society. The company's reputation suffered, which created a loss of market share with its unfavorable loss aspects.

Assuming the target is correct, losses of concern are those caused by a product's critical performance characteristics deviating from the target. The importance of concentrating on "hitting the target" is documented by Sony. In spite of the fact that the design and specifications were identical, U.S. customers preferred the color density of shipped TV sets produced by Sony–Japan over those produced by Sony–USA. Investigation of this situation revealed that the frequency distributions were markedly different, as shown in Figure 14-2. Even though Sony–Japan had 0.3% outside the specifications, the distribution was normal and centered on the target. The distribution of Sony–USA was uniform between the specifications with no values outside specifications. It was clear that customers perceived quality as meeting the target (Japan) rather than just meeting the specifications (USA). Ford Motor Company had a similar experience with transmissions.

[1] Taguchi uses the symbol *m* for the target.

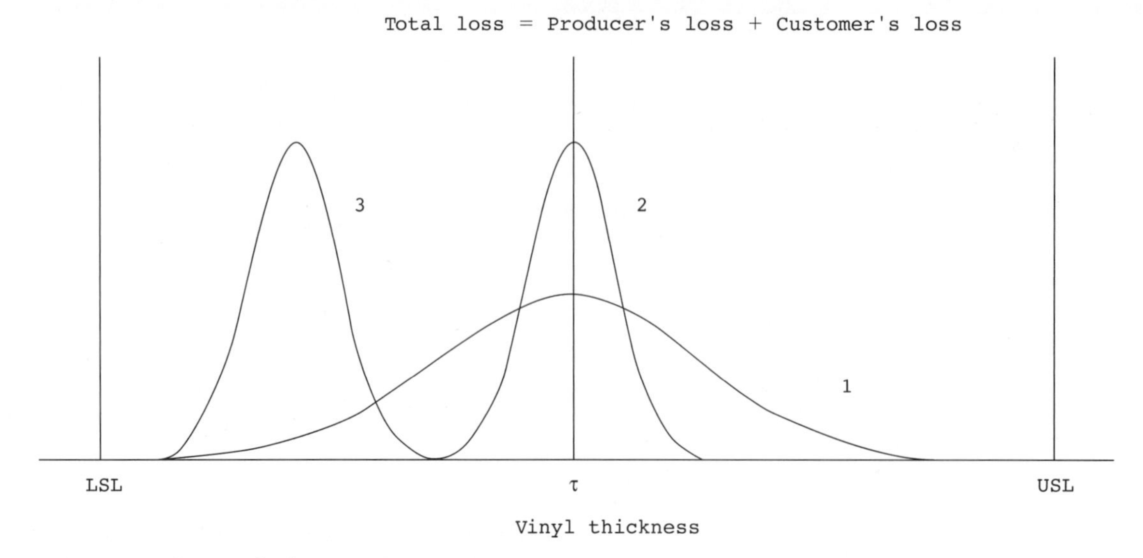

FIGURE 14-1 Loss to Society

Reproduced, with permission, from *Taguchi Methods: Introduction to Quality Engineering* (Bingham Farms, Mich.: American Supplier Institute, Inc., 1991).

Out of specification is the common measure of quality loss. Although this concept may be appropriate for accounting, it is a poor concept for all other areas. It implies that all products that meet specifications are good, whereas those that do not are bad. From the customer's point of view, the product that barely meets specification is as good (or bad) as the product that is barely out of specification. It appears the wrong measuring system is being used. The loss function corrects for the deficiency described above by combining cost, target, and variation into one metric.

Nominal-the-Best

Although Taguchi developed more than 68 loss functions, many situations are approximated by the quadratic function that is called the nominal-the-best type. Figure 14-3(a) shows the step function that describes the Sony–USA situation. When the value for the performance characteristic, y, is within specifications the loss is \$0, and when it is outside the specifications the loss is \$$A$. The quadratic function is shown at 14-3(b) and describes the Sony–Japan situation. In this situation loss occurs as soon as the performance characteristic, y, departs from the target, τ.

The quadratic loss function is described by the equation

$$L = k(y - \tau)^2$$

where L = cost incurred as quality deviates from the target
y = performance characteristic
τ = target
k = quality loss coefficient

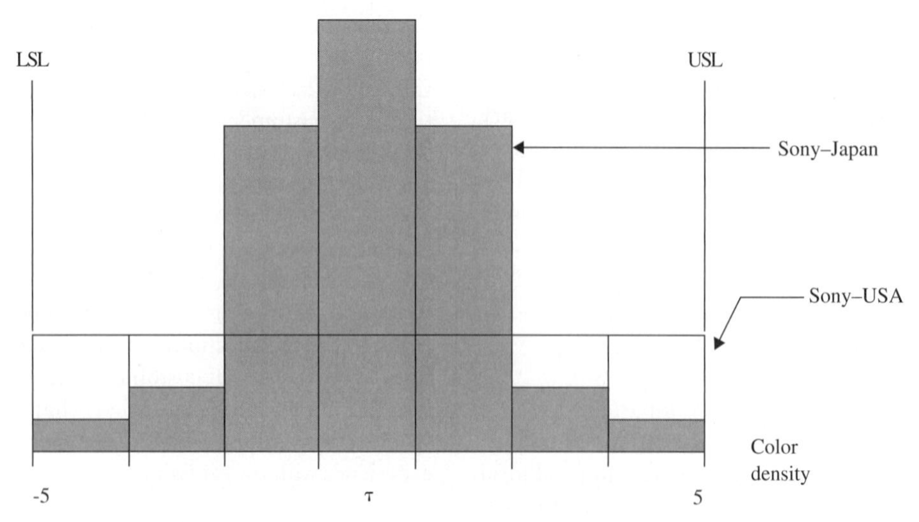

FIGURE 14-2 Distribution of Color Density for Sony–USA and Sony–Japan

Source: *The Asahi*, April 17, 1979.

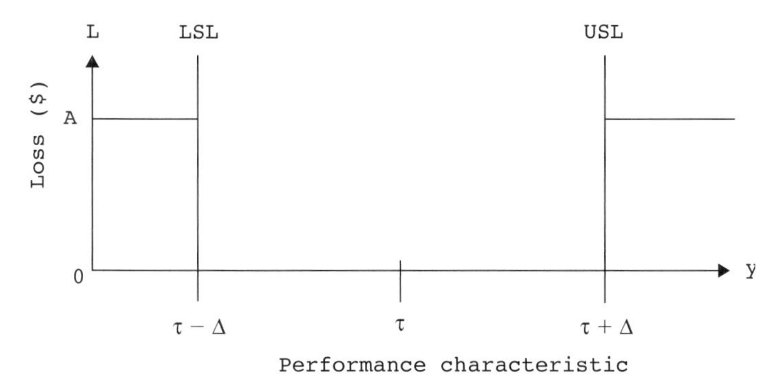

(a) Step function (Sony – USA)

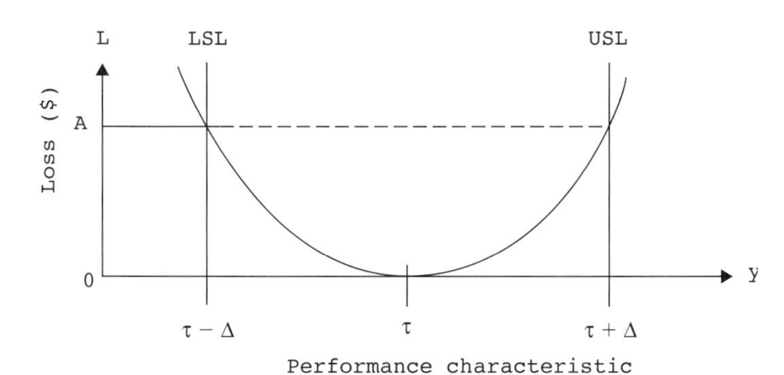

(b) Quadratic function (Sony – Japan)

FIGURE 14-3 Step and Quadratic Loss Functions

The loss coefficient is determined by setting $\Delta = (y - \tau)$, the deviation from the target. When Δ is at the USL or LSL, the loss to the customer of repairing or discarding the product is \$$A$. Thus,

$$k = A/(y - \tau)^2 = A/\Delta^2$$

Example Problem 14-1

If the specifications are 10 ± 3 for a particular quality characteristic and the average repair cost is \$230, determine the loss function. Determine the loss at $y = 12$.

$$k = A/\Delta^2 = 230/3^2 = 25.6$$

Thus, $L = 25.6\,(y - 10)^2$ and at $y = 12$,

$$\begin{aligned} L &= 25.6(y - 10)^2 \\ &= 25.6(12 - 10)^2 \\ &= \$102.40 \end{aligned}$$

Average Loss

The loss described here assumes that the quality characteristic is static. In reality, one cannot always hit the target, τ. It is varying due to noise, and the loss function must reflect the variation of many pieces rather than just one piece. Noise factors are classified as external and internal, with internal being further classified as unit-to-unit and deterioration.

A refrigerator temperature control will serve as an example to help clarify the noise concept. External noise is due to the actions of the user, such as the number of times the door is opened and closed, amount of food inside, the initial temperature, and so forth. Unit-to-unit internal noise is due to variation during production such as seal tightness, control sensor variations and so forth. Although this type of noise is inevitable, every effort should be made to keep it to a minimum. Noise due to deterioration is caused by leakage of refrigerant, mechanical wear of compressor parts, and so forth. This type of noise is primarily a function of the design. Noise factors cause deviation from the target, which causes a loss to society.

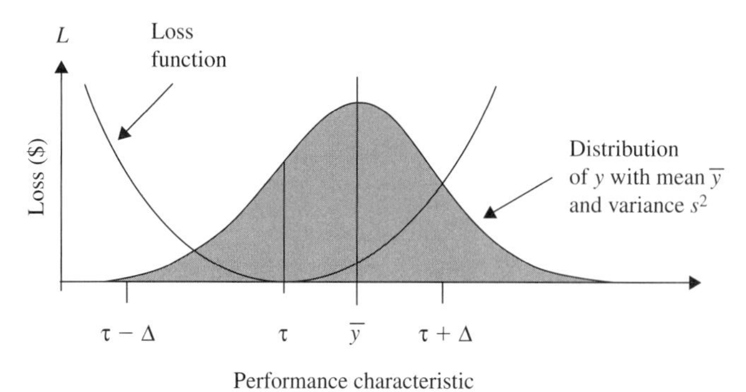

FIGURE 14-4 Average or Expected Loss

Reproduced, with permission, from Madhav S. Phadke, *Quality Engineering Using Robust Design* (Englewood Cliffs, N.J.: Prentice Hall, 1989).

Figure 14-4 shows the nominal-the-best loss function with the distribution of the noise factors. An equation can be derived by summing the individual loss values and dividing by their number to give

$$\bar{L} = k[\sigma^2 + (\bar{y} - \tau)^2]$$

where $\bar{L}$ = the average or expected loss.

Because the population standard deviation, σ, will rarely be known, the sample standard deviation, s, will need to be substituted. This action will make the value somewhat larger; however, the average loss is a very conservative value.

The loss can be lowered by first reducing the variation, σ, and then adjusting the average, $\bar{y}$, to bring it on target, τ. The loss function "speaks the language of things," which is engineering's measure, and money, which is management's measure. Examples where the nominal-the-best loss function would be applicable are the performance characteristics of color density, voltage, dimensions, and so forth.

Example Problem 14-2

Compute the average loss for a process that produces steel shafts. The target value is 6.40 mm and the coefficient is 9500. Eight samples give 6.36, 6.40, 6.38, 6.39, 6.43, 6.39, 6.46, and 6.42.

$$s = 0.0315945 \quad \bar{y} = 6.40375$$

$$\begin{aligned} \bar{L} &= k[s^2 + (\bar{y} - \tau)^2] \\ &= 9500[0.0315945^2 + (6.40375 - 6.40)^2] \\ &= \$9.62 \end{aligned}$$

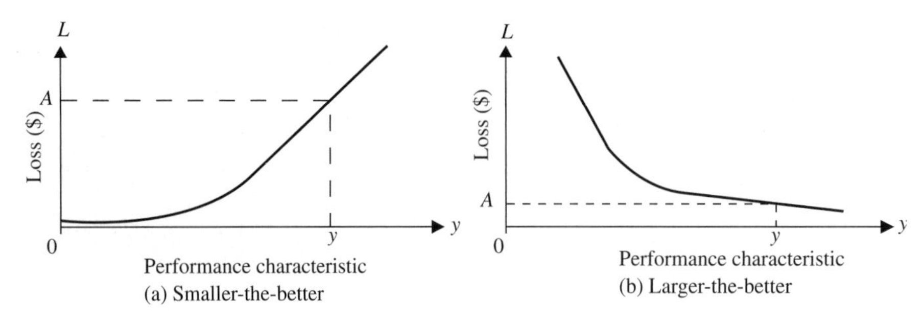

FIGURE 14-5 Smaller-the-Better and Larger-the-Better Loss Functions

Other Loss Functions

There are two other loss functions that are quite common, smaller-the-better and larger-the-better. Figure 14-5 illustrates the concepts.

As shown in the figure, the target value for smaller-the-better is 0, and there are no negative values for the performance characteristic. Examples of performance characteristics are radiation leakage from a microwave appliance, response time for a computer, pollution from an automobile, out of round for a hole, etc.

In the larger-the-better situation, shown in Figure 14-5(b), the target value is ∞, which gives a zero loss. There are no negative values and the worst case is at $y = 0$. Actually, larger-the-better is the reciprocal of smaller-the-better, and because of the difficulty of working with ∞, some practitioners prefer to work with the reciprocal. Thus, a larger-the-better performance characteristic of meters/second becomes a smaller-the-better performance characteristic of seconds/meter. Examples of performance characteristics are bond strength of adhesives, welding strength, automobile gasoline consumption, and so forth.

Summary of the Equations

Table 14-1 gives a summary of the equations for the three common loss functions. It also shows the relationship of the loss function to the mean squared deviation (MSD).

These three common loss functions will cover most situations. After selecting one of the loss functions, one point on the curve needs to be determined in order to obtain the coefficient. It is helpful to work with accounting to obtain this one point. Knowing the coefficient, the equation is complete and can be used to justify the use of resources and as a benchmark to measure improvement. It is much easier to use the loss function to obtain cost information than to develop an elaborate quality cost system. Cost data are usually quite conservative; therefore, it is not necessary for the loss function to be perfect for it to be effective.

Sometimes the loss function curves are modified for particular situations. For example, larger-the-better can be represented by one-half the nominal-the-best curve. Another situation occurs where the performance characteristic is weld strength. In such a case the larger-the-better curve can terminate at the strength of the parent metal rather than ∞. If the three common loss functions do not seem to be representative of a particular situation, then individual points can be plotted.

TABLE 14-1 Summary of the Equations for the Three Common Loss Functions

Nominal-the-best	$L = k(y - \tau)^2$ where $k = A/\Delta^2$
	$\bar{L} = k$ (MSD) where MSD $= [\Sigma(y - \tau)^2]/n$
	$\bar{L} = k[\sigma^2 + (\bar{y} - \tau)^2]$
Smaller-the-better	$L = ky^2$ where $k = A/y^2$
	$\bar{L} = k$ (MSD) where MSD $= [\Sigma y^2]/n$
	$\bar{L} = k[\bar{y}^2 + \sigma^2]$
Larger-the-better	$L = k(1/y^2)$ where $k = Ay^2$
	$\bar{L} = k$ (MSD) where MSD $= [\Sigma(1/y^2)]/n$
	$\bar{L} = k[\Sigma(1/y^2)]/n$

ORTHOGONAL ARRAYS[2]

Orthogonal arrays (OA) are a simplified method of putting together an experiment. The original development of the concept was by Sir R. A. Fischer of England in the 1930s. Taguchi added three OAs to the list in 1956, and the National Institute of Science and Technology (NIST) of the United States added three.

An orthogonal array is shown in Table 14-2. The 8 in the designation OA8 represents the number of rows, which is also the number of treatment conditions (TC) and the degrees of freedom. Across the top of the orthogonal array is the maximum number of factors that can be used, which in this case is seven. The levels are designated by 1 and 2. If more levels occur in the array, then 3, 4, 5, and so forth, are used. Other schemes such as −, 0, and + can be used.

The orthogonal property of the OA is not compromised by changing the rows or the columns. Taguchi changed the rows from a traditional design so that TC 1 was composed of all level 1s and, if the team desired, could thereby represent the existing conditions. Also, the columns were switched so that the least amount of change occurs in the columns on the left. This arrangement can provide the team with the capability

[2]Orthogonal arrays, interaction tables, and linear graphs in this chapter are reproduced, with permission, from *Taguchi Methods: Introduction to Quality Engineering* (Bingham Farms, Mich.: American Supplier Institute, Inc., 1991).

TABLE 14-2 Orthogonal Array (OA8)*

TC	1	2	3	4	5	6	7
1	1	1	1	1	1	1	1
2	1	1	1	2	2	2	2
3	1	2	2	1	1	2	2
4	1	2	2	2	2	1	1
5	2	1	2	1	2	1	2
6	2	1	2	2	1	2	1
7	2	2	1	1	2	2	1
8	2	2	1	2	1	1	2

*Taguchi uses a more elaborate system of identification for the orthogonal arrays. It is the authors' opinion that a simple system using OA is more than satisfactory.

to assign factors with long setup times to those columns. Orthogonal arrays can handle dummy factors and can be modified. Refer to the bibliography for these techniques.

To determine the appropriate orthogonal array, use the following procedure:

1. Define the number of factors and their levels.
2. Determine the degrees of freedom.
3. Select an orthogonal array.
4. Consider any interactions.

The project team completes the first step.

Degrees of Freedom

The number of degrees of freedom is a very important value because it determines the minimum number of treatment conditions. It is equal to the sum of

(Number of levels - 1) for each factor.

(Number of levels - 1)(number of levels - 1) for each interaction.

One for the average.

An example problem will illustrate the concept.

Example Problem 14-3

Given four two-level factors, *A, B, C, D,* and two suspected interactions, *BC* and *CD*, determine the degrees of freedom, df. What is the answer if the factors are three-level?

$$df = 4(2 - 1) + 2(2 - 1)(2 - 1) + 1 = 7$$
$$df = 4(3 - 1) + 2(3 - 1)(3 - 1) + 1 = 17$$

At least 7 treatment conditions are needed for the two-level, and 17 conditions are needed for the three-level. As can be seen by the example, the number of levels has considerable influence on the number of treatment conditions. Although a three-level design provides a great deal more information about the process, it can be costly in terms of the number of treatment conditions.

TABLE 14-3 Maximum Degrees of Freedom for a Four-Factor, Two-Level Experimental Design

Design Space				df
A	*B*	*C*	*D*	4
AB	*AC*	*AD*	*BC*	6
BD	*CD*			
ABC	*ABD*	*ACD*	*BCD*	4
ABCD				1
Average				1
			Sum	16

The maximum degrees of freedom is equal to

$$df = v = l^f$$

where l = number of levels
f = number of factors

For the example problem with two levels, df $= 2^4 = 16$. Table 14-3 shows the maximum degrees of freedom.

In the example problem, it was assumed that four of the two-factor interactions (*AB, AC, AD,* and *BD*), four of the three-factor interactions (*ABC, ABD, ACD,* and *BCD*), and the four-factor interaction (*ABCD*) would not occur. Interactions are discussed later in the chapter.

Selecting the Orthogonal Array

Once the degrees of freedom are known, the next step, selecting the orthogonal array (OA), is easy. The number of treatment conditions is equal to the number of rows in the OA and must be equal to or greater than the degrees of freedom. Table 14-4 shows the orthogonal arrays that are available, up to OA36. Thus, if the number of degrees of freedom is 13, then the next available OA is OA16. The second column of the table has the number of rows and is redundant with the designation in the first column. The third column gives the maximum number of factors that can be used, and the last four columns give the maximum number of columns available at each level.

Analysis of the table shows that there is a geometric progression for the two-level arrays of OA4, OA8, OA16, OA32, ..., which is $2^2, 2^3, 2^4, 2^5, \ldots$, and for the three-level arrays of OA9, OA27, OA81, ..., which is $3^2, 3^3, 3^4, \ldots$. Orthogonal arrays can be modified. Refer to the references for more information.

Interaction Table

Confounding is the inability to distinguish among the effects of one factor from another factor and/or interaction. In order to prevent confounding, one must know which columns to use for the factors. This knowledge is provided by an interaction table, which is shown in Table 14-6. The orthogonal array (OA8) is repeated in Table 14-5 for the convenience of the reader.

TABLE 14-4 Orthogonal Array Information

OA	Number of Rows	Maximum Number of Factors	MAXIMUM NUMBER OF COLUMNS 2-Level	3-Level	4-Level	5-Level
OA4	4	3	3	—	—	—
OA8	8	7	7	—	—	—
OA9	9	4	—	4	—	—
OA12	12	11	11	—	—	—
OA16	16	15	15	—	—	—
OA16[1]	16	5	—	—	5	—
OA18	18	8	1	7	—	—
OA25	25	6	—	—	—	6
OA27	27	13	—	13	—	—
OA32	32	31	31	—	—	—
OA32[1]	32	10	1	—	9	—
OA36	36	23	11	12	—	—
OA36[1]	36	16	3	13	—	—
•	•	•	•	•	•	•
•	•	•	•	•	•	•
•	•	•	•	•	•	•

Adapted, with permission, from Madhav S. Phadke, *Quality Engineering Using Robust Design* (Englewood Cliffs, NJ: Prentice Hall, 1989).

TABLE 14-5 Orthogonal Array OA8

TC	1	2	3	4	5	6	7
1	1	1	1	1	1	1	1
2	1	1	1	2	2	2	2
3	1	2	2	1	1	2	2
4	1	2	2	2	2	1	1
5	2	1	2	1	2	1	2
6	2	1	2	2	1	2	1
7	2	2	1	1	2	2	1
8	2	2	1	2	1	1	2

Let's assume that factor *A* is assigned to column 1 and factor *B* to column 2. If there is an interaction between factors *A* and *B*, then column 3 is used for the interaction, *AB*. Another factor, say, *C*, would need to be assigned to column 4. If there is an interaction between factor *A* (column 1) and factor *C* (column 4), then interaction *AC* will occur in column 5. The columns that are reserved for interactions are used so that calculations can be made to determine whether there is a strong interaction. If there are no interactions, then all the columns can be used for factors. The actual experiment is conducted using the columns designated for the factors, and these columns are referred to as the design matrix. All the columns are referred to as the design space.

TABLE 14-6 Interaction Table for OA8

Column	1	2	3	4	5	6	7
1	(1)	3	2	5	4	7	6
2		(2)	1	6	7	4	5
3			(3)	7	6	5	4
4				(4)	1	2	3
5					(5)	3	2
6						(6)	1
7							(7)

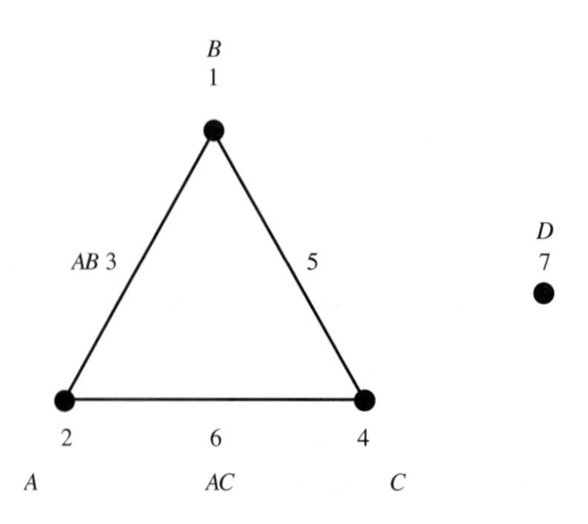

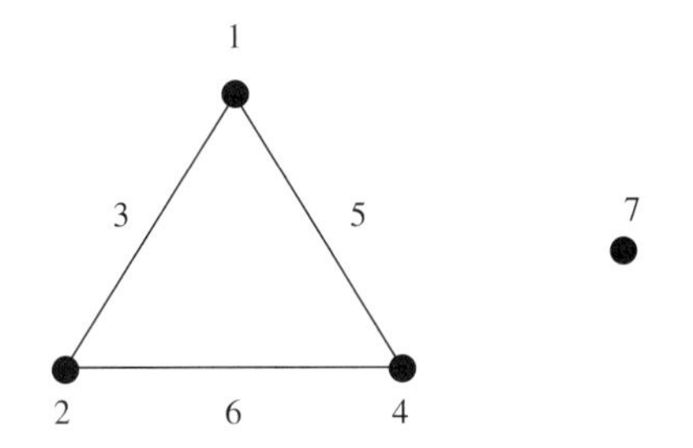

FIGURE 14-6 Linear Graphs for OA8

Linear Graphs

Taguchi developed a simpler method of working with interactions using linear graphs. Two are shown in Figure 14-6 for OA8. They make it easier to assign factors and interactions to the various columns of an array. Factors are assigned to the points. If there is an interaction between two factors, then it is assigned to the line segment between the two points. For example, using the linear graph on the left in the figure, if factor *B* is assigned to column 2 and factor *C* is assigned to column 4, then interaction *BC* is assigned to column 6. If there is no interaction, then column 6 can be used for a factor.

The linear graph on the right would be used when one factor has three two-level interactions. Three-level orthogonal arrays must use two columns for interactions, because one column is for the linear interaction and one for the quadratic interaction. The linear graphs—and, for that matter, the interaction tables—are not designed for three or more factor interactions, which rarely occur. Linear graphs can be modified; refer to the references for modification techniques. Use of the linear graphs requires some trial-and-error activity, and a number of solutions may be possible, as shown by the example problem.

Example Problem 14-4

An experimental design has four two-level factors (*A, B, C, D*) where only main effects are possible for factor *D* and there is no *BC* interaction. Thus, only interactions *AB* and *AC* are possible, and they can be assigned the line segments 3 and 5, 3 and 6, or 5 and 6, with their apex for factor *A*. Factors *B* and *C* are then assigned to the adjacent points. Column 7 or a line segment that does not have an interaction is used for factor *D*. A number of solutions are possible; one is shown here. The one chosen might well be a function of the setup time when the experiment is run. Column 5 is not used, so it is given the symbol *UX* for unexplained, and calculations for this column should show no effect (very small variation).

Orthogonal Array OA8

	B	*A*	*AB*	*C*	*UX*	*AC*	*D*
TC	**1**	**2**	**3**	**4**	**5**	**6**	**7**
1	1	1	1	1	1	1	1
2	1	1	1	2	2	2	2
3	1	2	2	1	1	2	2
4	1	2	2	2	2	1	1
5	2	1	2	1	2	1	2
6	2	1	2	2	1	2	1
7	2	2	1	1	2	2	1
8	2	2	1	2	1	1	2

Interactions

The fourth step in the procedure is to consider interactions. Figure 14-7 shows the graphical relationship between two factors. At (a) there is no interaction because the lines are parallel; at (b) there is some interaction; and at (c) there is a strong interaction. The graph is constructed by plotting the points A_1B_1, A_1B_2, A_2B_1, and A_2B_2 drawing lines B_1 and B_2. Taguchi's approach to interactions is given in the following list:

1. Interactions use degrees of freedom; therefore, more treatment conditions are needed or fewer factors can be used.

2. Orthogonal arrays are used in parameter design to obtain optimal factor/levels for robustness and cost in order to improve product and process performance. On

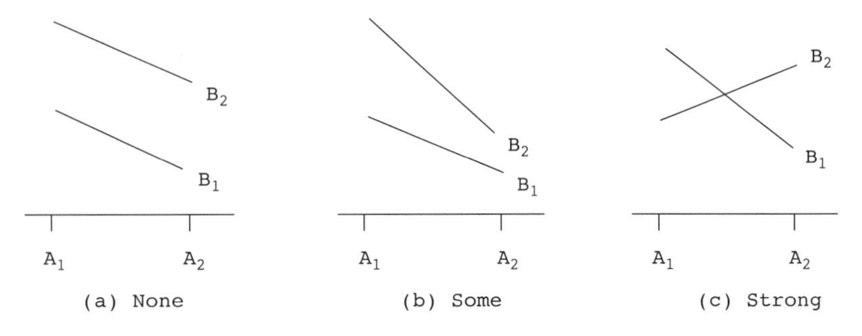

FIGURE 14-7 Interaction Between Two Factors

the other hand, statistics are applied in pure and applied research to find relationships and a mathematical model. The emphasis is different—one engineering and the other mathematical.

3. Interactions are primarily between control factors and noise factors. Control and noise factors are discussed later.

4. As long as interactions are relatively mild, main effect analysis will give the optimal results and good reproducibility.

5. OA12 (two-level) and OA18 (three-level) are recommended so that if interactions are present, they are dispersed among all the factors.

6. Engineers should strive to develop a design that uses main effects only.

7. Control factors that will not interact should be selected. For example, the dimensions length and width will frequently interact, whereas the area may provide the same information and save two degrees of freedom.

8. Energy-related outputs, such as braking distance, should be selected whenever possible. This concept is discussed in the next section.

9. An unsuccessful confirmation run may indicate an interaction.

Table H in the Appendix gives the common orthogonal arrays with their interaction tables and linear graphs. OA12 and OA18 do not have linear graphs, because the effect of any interactions is dispersed within the array.

SIGNAL-TO-NOISE (S/N) RATIO

Another of Taguchi's contributions is the signal-to-noise (S/N) ratio. It was developed as a proactive equivalent to the reactive loss function. Figure 14-8 illustrates the concept of the S/N ratio. When a person puts his/her foot on the brake pedal of a car, energy is transformed with the intent to slow the car, which is the signal. However, some of the energy is wasted by squeal, pad wear, heat, and so forth. The figure emphasizes that energy is neither created nor destroyed. At the bottom of the figure the concept is written in the form of a ratio.

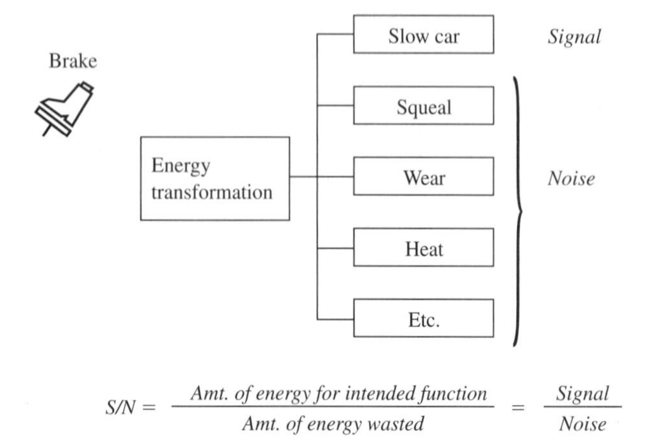

FIGURE 14-8 Concept of Signal-to-Noise (S/N) Ratio

Signal factors ($\bar{y}$) are set by the designer or operator to obtain the intended value of the response variable. Noise factors (s^2) are not controlled or are very expensive or difficult to control. Both the average, $\bar{y}$, and the variance, s^2, need to be controlled with a single figure of merit. In elementary form, S/N is $\bar{y}/s$, which is the inverse of the coefficient of variation and a unitless value. Squaring and taking the log transformation gives

$$S/N_N = \log_{10}(\bar{y}^2/s^2)$$

Adjusting for small sample sizes and changing from Bels to decibels by multiplying by ten yields

$$S/N_N = 10 \quad \log_{10}[(\bar{y}^2/s^2) - (1/n)],$$

which is the nominal-the-best S/N_N ratio. The average and sample standard deviation are squared to eliminate any negative averages and to use the variance, which is an unbiased measure of the dispersion. By taking the log transformation, the calculated value becomes a relative one.

The S/N units are decibels (dB), which are tenths of a Bel and are a very common unit in electrical engineering. Why use decibels? If someone says that the voltage is 6 volts too low, does it indicate a problem? Of course, it depends if you are describing a 745,000-volt power line or the battery in your car. It makes more sense to say that your car battery voltage is 50% low or is only half of the target value of 12 volts. A useful system for describing this condition is logarithms. Another advantage of using decibels as the unit of measure is that they are a relative measure. For example, the difference between 74 dB and 71 dB is the same as the difference between 7 dB and 10 dB. Both are twice as good or bad. Decibels also are not affected by different units. Temperature in Celsius or in Kelvin will give different answers but the same amount of change. Table 14-7 shows some linear and percent changes for the decibel change. The percent change is found by subtracting the linear change from 1.000, dividing by 1.000, and multiplying by 100. Thus, a 2.0-dB change is

$$\frac{(1.59 - 1.000)}{1.000}100 = 59\%$$

TABLE 14-7 Decibel, Linear, and Percent Change

dB Change (log)	Linear Change (nonlog) 10^8	Percent Change
0.001	1.00	0
0.5	1.12	12
1.0	1.26	26
1.5	1.41	41
2.0	1.59	59
3.0	2.00	100
6.0	4.00	300
10.0	10.00	900

There are many different S/N ratios. Six basic ones are

1. Nominal-the-best
2. Target-the-best
3. Smaller-the-better
4. Larger-the-better
5. Classified attribute
6. Dynamic

We will discuss those ratios that parallel the loss function.

Nominal-the-Best

The equation for nominal-the-best was given in the initial discussion. It is used wherever there is a nominal or target value and a variation about that value, such as dimensions, voltage, weight, and so forth. The target (τ) is finite but not zero. For robust (optimal) design, the S/N ratio should be maximized. The nominal-the-best S/N value is a maximum when the average is large and the variance is small. When the average is off target on the high side, the S/N_N value can give more favorable information; when off target on the low side, the value can give less favorable information. Taguchi's approach is to reduce variation and then bring the average on target. Another S/N_T ratio, called target-the-best, eliminates these problems provided the target is known.[3]

Example Problem 14-5

Determine the S/N ratio for a process that has a temperature average of 21°C and a sample standard deviation of 2°C for four observations.

$$\begin{aligned} S/N_N &= 10 \log_{10}[(\bar{y}^2/s^2) - (1/n)] \\ &= 10 \log_{10}[(21^2/2^2) - (1/4)] \\ &= 20.41 \text{ dB} \end{aligned}$$

The adjustment for the small sample size in the example problem has little effect on the answer. If it had not been used, the answer would have been 20.42 dB.

Smaller-the-Better

The S/N_S ratio for smaller-the-better is used for situations where the target value (τ) is zero, such as computer response time, automotive emissions, or corrosion. The equation is

$$S/N_N = -10 \log_{10}[\text{MSD}] = -10 \log_{10}[(\Sigma y^2)/n]$$

The negative sign is used to ensure that the largest value gives the optimum value for the response variable and, therefore, robust design. Mean standard deviation (MSD) is given to show the relationship to the loss function.

Example Problem 14-6

A bread-stuffing producer is comparing the calorie content of the original process with a new process. Which has the lower content and what is the difference? Results are

Original	130	135	128	127
Light	115	112	120	113

$$\begin{aligned} S/N_S &= -10 \log_{10}[(\Sigma y^2)/n] \\ &= -10 \log_{10}[(130^2 + 135^2 + 128^2 + 127^2)/4] \\ &= -42.28 \text{ dB} \end{aligned}$$

$$\begin{aligned} S/N_S &= -10 \log_{10}[(\Sigma y^2)/n] \\ &= -10 \log_{10}[(115^2 + 112^2 + 120^2 + 113^2)/4] \\ &= -41.22 \text{ dB} \end{aligned}$$

$$\Delta = [-41.22 - (-42.28)] = 1.06 \text{ dB}$$

Light is lower in calories by 26%.

Larger-the-Better

The third S/N ratio is larger-the-better. It is used where the largest value is desired, such as weld strength, gasoline mileage, or yield. From a mathematical viewpoint, the target value is infinity. Like the loss function, it is the reciprocal of smaller-the-better. The equation is

$$S/N_L = -10 \log_{10}[\text{MSD}] = -10 \log_{10}[\Sigma(1/y^2)/n]$$

Example Problem 14-7

Using the existing design, the lives of three AA batteries are 20, 22, and 21 hours. An experimental design produces batteries with values of 17, 21, and 25 hours. Which is the better design and by how much? What is your next step?

$$\begin{aligned} S/N_L &= -10 \log_{10}[\Sigma(1/y^2)/n] \\ &= -10 \log_{10}[(20^2 + 22^2 + 21^2)/3] \\ &= 26.42 \text{ dB} \end{aligned}$$

$$\begin{aligned} S/N_L &= -10 \log_{10}[\Sigma(1/y^2)/n] \\ &= -10 \log_{10}[(17^2 + 21^2 + 25^2)/3] \\ &= 26.12 \text{ dB} \end{aligned}$$

$$\Delta = [26.42 - 26.12] = 0.3 \text{ dB}$$

The new design is 7% better. It is suggested that more data be collected.

Although signal-to-noise ratios have achieved good results, they have not been accepted by many in the statistical community. The controversy has focused more attention on variation or noise, whereas in the past the entire focus was on the average or signal. It is our opinion that with computer programs, it is quite easy to use three metrics—average, variance, and signal-to-noise. Also, note that the advantages of the log transformation can be used with the average and the variance.

PARAMETER DESIGN

Introduction

There are three product-development stages: product design, process design, and production. These stages are shown in Table 14-8, along with the three previously discussed sources of variation or noise: environmental variables, product

[3]Thomas B. Barker, *Engineering Quality by Design* (New York: Marcel Dekker, 1990).

TABLE 14-8 Product Development Stages

Product Development Stages	SOURCES OF VARIATION (NOISE) Environmental Variables	Product Deterioration	Production Variations
Product design	0	0	0
Process design	X	X	0
Production	X	X	0

0—Countermeasures possible; X—Countermeasures impossible

Adapted, with permission, from Raghu N. Kackar, "Taguchi's Quality Philosophy: Analysis and Commentary," *Quality Progress* (December 1986): 21–29.

deterioration, and production variations. Only at the product design stage are countermeasures possible against all the sources of variation.

The cornerstone of Taguchi's philosophy is robust design. Figure 14-9 illustrates the three design components: system design, parameter design, and tolerance design, with robust design encompassing the latter two. System design is composed of traditional research and development. Until recently, Japan spent approximately 40%, 40%, and 20% and the United States spent 70%, 2%, and 28%, respectively, of their time and money on the three design components.

System design is the development of the prototype. It uses engineering and scientific knowledge to define initial setting of product and process parameters. Knowledge of customer requirements is important at this stage.

Parameter design is the selection of the optimal conditions (parameters), so that the product is least sensitive to noise variables. The idea is to start with inferior-grade, low-cost components and raw materials. Nothing is more foolish than research using high-priced materials and components. Variability is reduced by identifying control and noise factors and treating them separately in order to innovate and develop a product that is high in quality and low in cost. The concept uses OAs, response tables, and the metrics of S/N ratios, variances, and averages to obtain the appropriate parameters.

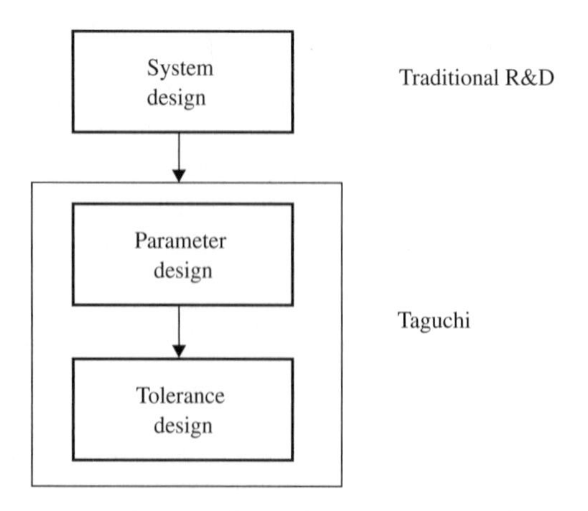

FIGURE 14-9 Robust Design is the Cornerstone of Taguchi's Philosophy

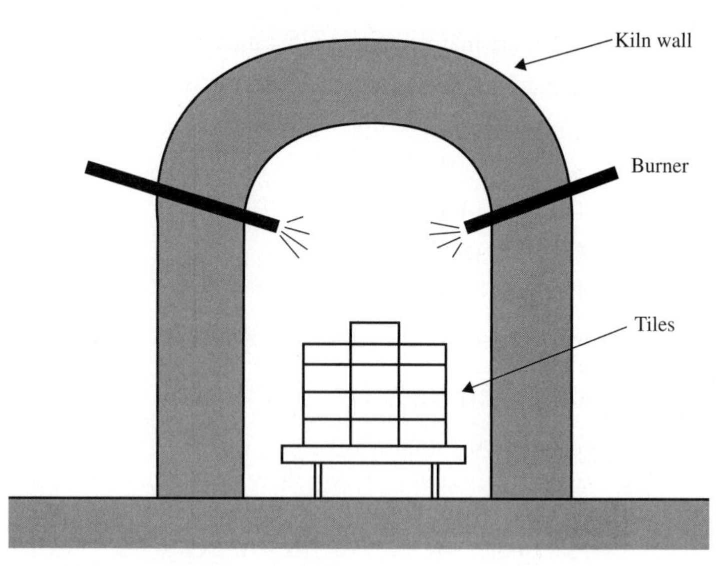

FIGURE 14-10 Ina Tile Company

Reproduced, with permission, from Madhav S. Phadke, *Quality Engineering Using Robust Design* (Englewood Cliffs, NJ: Prentice Hall, 1989).

An excellent example of robust design is provided by the Ina Tile Company. Figure 14-10 shows a cross section of an 80-meter-long kiln for baking tiles. The kiln, purchased from West Germany, had quality problems because the tiles at the center were baked at a lower temperature than the outer tiles, which resulted in nonuniform dimensions. One hundred percent inspection was used to screen for improper dimensions; however, this activity was very expensive. The kiln could be redesigned to give uniform dimensions, but the redesign would be very expensive and there was no guarantee that it would correct the problem. It was decided to run an experiment using an OA8 design with seven variables at two levels. The results showed that by increasing the lime content of the clay from 1% to 5%, the excessive dimensional variation was eliminated. Lime is an inexpensive material in the content of the clay. It was also found that an expensive material in the process could be reduced.

Parameter Design Example

An example of an improved product at the design stage is illustrated by the development of a paper feeder for a copy machine. It is important that the prototype, as illustrated by the schematic in Figure 14-11, be constructed so that experiments can be run on the factors. The first step for the project team is to determine the performance characteristic from the customer's viewpoint, which most likely will be "no

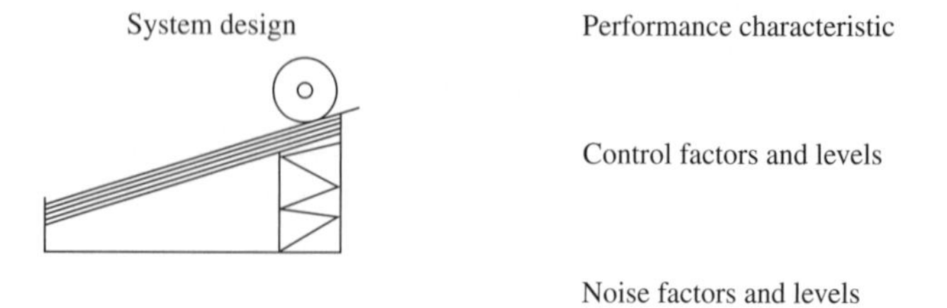

FIGURE 14-11 Parameter Design Concept Using a Paper Feed Device

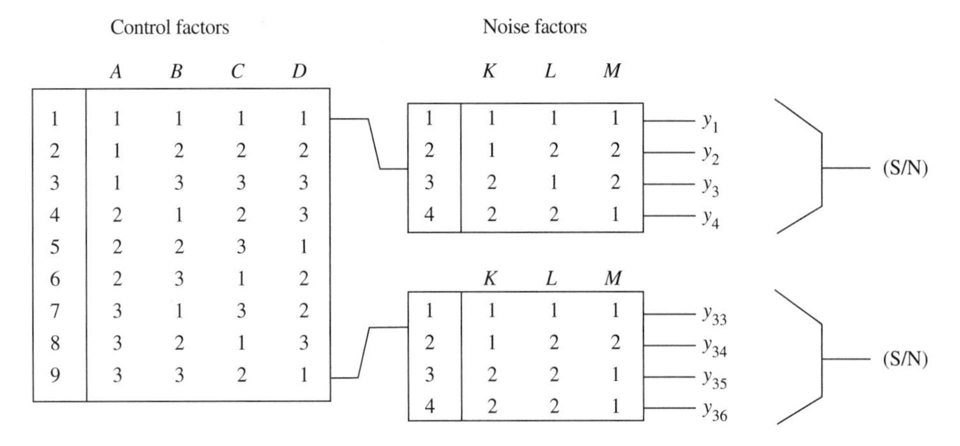

FIGURE 14-12 Parameter Design Concept

Adapted, with permission, from Raghu N. Kackar, "Taguchi's Quality Philosophy: Analysis and Commentary," *Quality Progress* (December 1986): 21–29.

multisheets and no misses." Next, the factors and their levels for the control factors are determined. The team decides to use a three-level design with four control factors. They are factor *A*—spring tension with the levels low, medium, and high; factor *B*—roller diameter with the levels small, medium, and large; factor *C*—speed with the levels slow, medium, and fast; and factor *D*—roller durometer with the levels soft, medium, and hard. The third step is for the team to determine the noise or uncontrollable factors that can, however, be controlled in an experiment. Note that uncontrollable factors can also be very expensive to control. The team identifies three noise factors and decides to experiment using two levels. They are factor *K*—paper weight with the levels 12 and 20 lb; factor *L*—moisture with the levels 30% RH and 80% RH; and factor *M*—number of sheets with the levels 1 and 100. Not all sources of noise can be included because of lack of knowledge. Additional information on the inclusion of noise factors is given later.

The experimental design that results from the preceding example is shown in Figure 14-12. Basically there are two arrays: the control factor array, which is also called the inner, or controllable, array, and the noise array, which is also called the outer, or uncontrollable (except in the experiment), array. An OA9 with its nine treatment conditions is used for the inner array, and an OA4 with its four treatment conditions is used for the outer array. Treatment condition 1 is set up with all level 1s, and four runs are made—one for each of the noise treatment conditions. These four runs produce four results y_1, y_2, y_3, and y_4, which are combined to produce the S/N_1 ratio for TC 1 of the control factors. This process is repeated for the other nine treatment conditions of the control factors. The results are then used in a response table to determine the strong factors and their levels.

A number of case studies will be used to illustrate the approach. Each case builds on the preceding ones.

Case I: Iron Casting

This case illustrates the basic technique using a two-level, smaller-the-better performance characteristic with the maximum number of factors. Wirco Castings, Inc., designed an experiment to evaluate the percent of casting that required finish grinding, with the objective of reducing this labor-intensive operation. It was decided there were seven factors that influenced the grinding operation, and they are shown in Table 14-9, along with their levels. Noise factors were not considered to be important, so they were not included in this experiment.

TABLE 14-9 Factors and Levels for Iron Casting Design

Factors	Level 1	Level 2
A. Sand compact	$A_1 = 55$ mm	$A_2 = 49$ mm
B. Iron temperature	$B_1 =$ FT	$B_2 =$ Chill
C. Clay addition	$C_1 = 6.5$ lb	$C_2 = 16$ lb
D. Mold hardness	$D_1 = 1000$ lb/in.^{2}	$D_2 = 750$ lb/in.^{2}
E. Mulling time	$E_1 = 4$ min	$E_2 = 1.7$ min
F. Seacoal addition	$F_1 = 6.7$ lb	$F_2 = 15$ lb
G. Sand addition	$G_1 = 0$	$G_2 = 150$ lb

Design and Results

An OA8 was used for the design, as shown in Table 14-10 with the treatment condition results. Each treatment condition was run and produced 16 molds with 4 cavities per mold, for a total of 64 castings per TC.

The effect of each factor and its levels are calculated here:

$$A_1 = (89 + 55 + 38 + 44)/4 = 56.5$$
$$A_2 = (83 + 16 + 66 + 55)/4 = 55.0$$
$$B_1 = (89 + 55 + 83 + 16)/4 = 60.8$$
$$B_2 = (38 + 44 + 66 + 55)/4 = 50.8$$
$$C_1 = (89 + 55 + 66 + 55)/4 = 66.3$$
$$C_2 = (38 + 44 + 83 + 16)/4 = 45.3$$
$$D_1 = (89 + 38 + 83 + 66)/4 = 69.0$$
$$D_2 = (55 + 44 + 16 + 66)/4 = 42.5$$

TABLE 14-10 Orthogonal Array and Results for Iron Castings

	A	B	C	D	E	F	G	Results
TC	1	2	3	4	5	6	7	%
1	1	1	1	1	1	1	1	89
2	1	1	1	2	2	2	2	55
3	1	2	2	1	1	2	2	38
4	1	2	2	2	2	1	1	44
5	2	1	2	1	2	1	2	83
6	2	1	2	2	1	2	1	16
7	2	2	1	1	2	2	1	66
8	2	2	1	2	1	1	2	55

$$E_1 = (89 + 38 + 16 + 55)/4 = 49.5$$
$$E_2 = (55 + 44 + 83 + 66)/4 = 62.0$$
$$F_1 = (89 + 44 + 83 + 55)/4 = 67.8$$
$$F_2 = (55 + 38 + 16 + 66)/4 = 43.8$$
$$G_1 = (89 + 44 + 16 + 66)/4 = 53.8$$
$$G_2 = (55 + 38 + 83 + 55)/4 = 56.8$$

Calculations for level 2 can be simplified by subtracting level 1 from the total.

Response Table and Graph

Values from the preceding calculations are placed in a *response table*, as shown in Table 14-11. The absolute difference, Δ, between level 1 and level 2 is calculated and placed in the table. A *response graph*, as shown by Figure 14-13, can also be constructed to aid in visualizing the strong effects. This graph is Paretoized, with the largest difference on the left and the smallest difference on the right.

A simple rule is used as a guideline to analyze which of the factors has a strong effect on the process and which is merely a natural variation. Take the largest difference, which in this case is 26.5, and divide in half, to give 13.3. All differences equal to or above 13.3 are considered to be strong effects. Because this experiment is a smaller-the-better performance, the strong factors and their levels are D_2, F_2, and C_2.

Confirmation Run

The next step is to predict the outcome of the confirmation run using the equation

TABLE 14-11 Response Table for Iron Casting

	A	B	C	D	E	F	G
Level 1	56.5	60.8	66.3	69.0	49.5	67.8	53.8
Level 2	55.0	50.8	45.3	42.5	62.0	43.8	57.8
Δ	1.5	10.0	21.0	26.5	12.5	24.0	4.0

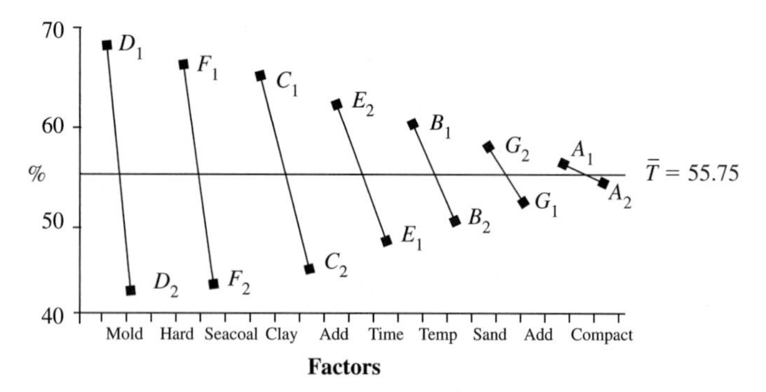

FIGURE 14-13 Response Graph for Iron Casting

$$\hat{\mu} = \overline{T} - (\overline{T} - D_2) - (\overline{T} - F_2) - (\overline{T} - C_2)$$
$$= D_2 + F_2 + C_2 - (N - 1)\overline{T}$$

where $\hat{\mu}$ = estimate of the response for y

$\overline{T}$ = overall average of the response data

Calculations are

$$\overline{T} = \sum y/n = (89 + 55 + 38 + 44 + 83 + 16 + 66 + 55)/8 = 55.8$$
$$\hat{\mu} = D_2 + F_2 + C_2 - (N - 1)\overline{T}$$
$$= 42.5 + 43.8 + 45.3 - (3 - 1)55.8$$
$$= 20.0\%$$

A fundamental part of the Taguchi approach is the confirmation run. For this experiment the factors and their levels are D_2, F_2, C_2, E_1, B_1, G_2, and A_1, and the result from the confirmation run was 19.2%, which is very close to the predicted value. TC 6 also had good results at 16%, using most of the same factors and their levels. Thus, the experiment is a success, and the new parameters are the strong effects of D_2, F_2, and C_2. The levels of the other four factors might not make any difference. If the confirmation run does not show good producibility, like this case, then (1) the factor levels may have been set too close, (2) an important factor was missing, or (3) interactions were present due to the improper selection of control factors.

As a result of this experiment, less finish grinding was needed, thereby reducing the workforce from six to five people. In addition, the parameter changes resulted in a reduction in scrap from 8% to 4% and in smoother castings.

Although the S/N ratio and variance have good additivity, the percent nonconforming should not be outside the range of 20% to 80%. Fortunately, in this experiment, the data were within that range. If such had not been the case, an omega transformation would have been made on each of the eight results. For information on the omega transformation, see the references.

Case II: Grille[4]

This case is a two-level design with a smaller-the-better performance characteristic and interactions. Automotive grille

[4]Adapted, with permission, from P. I. Hsieh and D. E. Goodwin, "Sheet Molded Compound Process Improvement," *Fourth Symposium on Taguchi Methods* (Bingham Farms, Mich.: American Supplier Institute, Inc., 1986).

TABLE 14-12 Factors and Levels for Grille

	Variable	Level 1	Level 2
A	Mold pressure	Low	High
B	Mold temp.	Low	High
C	Mold cycle	Low	High
D	Cut pattern	Method I	Method II
E	Priming	Method I	Method II
F	Viscosity	Low	High
G	Weight	Low	High
H	Mat'l thickness	Process I	Process II
I	Glass type	Type I	Type II

opening panels were shipped to Chrysler by a supplier and assembled in the car prior to painting. Surface imperfections in the finish (pops) caused a first run capability of 77% conforming, and the condition was not detected prior to painting. A joint supplier/customer team, using the problem-solving techniques of flow process and cause-and-effect diagrams, decided on a two-level design with nine control variables and five potential interactions, as shown in Table 14-12.

Five potential interactions were possible: *AB, AC, AD, BF*, and *FH*.

Because there are 15 degrees of freedom, an OA16 was needed. A modified linear graph was used to determine the experimental design, which is shown in Table 14-13. No factor or interaction was assigned to column 15, so it is labeled unexplained, *UX*.

The experiment was run with two repetitions per TC and a response table calculated as shown in Table 14-14.

TABLE 14-14 Response Table for Grille

Factor	Level 1	Level 2	Difference
E	14.31	3.19	11.12
A	13.75	3.75	10.00
C	12.06	5.44	6.62
D	7.19	10.31	3.12
B	7.56	9.94	2.38
I	7.50	10.00	2.50
H	7.62	9.88	2.26
F	8.38	9.12	0.74
G	8.62	8.88	0.26
FH	14.12	3.38	10.74
AC	13.19	4.31	8.88
BF	13.06	4.44	8.62
AD	6.81	10.69	3.88
AB	8.06	9.44	1.38
UX	13.25	4.25	9.00

Using the one-half guideline, the factors *E, A*, and *C* and the interactions *FH, AC*, and *BF*, are the strong effects. It is also noted that *UX* has a strong effect.

The interactions are plotted to determine the level for those factors involved in the interactions. Using the results in Table 14-13, the calculations are

$$A_1C_1 = (56 + 10 + 2 + 1 + 3 + 1 + 50 + 49)/8 = 21.5$$

$$A_1C_2 = (17 + 2 + 4 + 3 + 4 + 13 + 2 + 3)/8 = 6.0$$

$$A_2C_1 = (1 + 3 + 3 + 2 + 3 + 4 + 0 + 8)/8 = 3.0$$

TABLE 14-13 Experimental Design Using an OA16 and Results for Grille

1	2	3	4	5	6	7	8	9	10	11	12	13	14	15		
A	*B*	*AB*	*F*	*G*	*BF*	*E*	*C*	*AC*	*H*	*I*	*D*	*AD*	*FH*	*UX*	*R1*	*R2*
1	1	1	1	1	1	1	1	1	1	1	1	1	1	1	56	10
1	1	1	1	1	1	1	2	2	2	2	2	2	2	2	17	2
1	1	1	2	2	2	2	1	1	1	1	2	2	2	2	2	1
1	1	1	2	2	2	2	2	2	2	2	1	1	1	1	4	3
1	2	2	1	1	2	2	1	1	2	2	1	1	2	2	3	1
1	2	2	1	1	2	2	2	2	1	1	2	2	1	1	4	13
1	2	2	2	2	1	1	1	1	2	2	2	2	1	1	50	49
1	2	2	2	2	1	1	2	2	1	1	1	1	2	2	2	3
2	1	2	1	2	1	2	1	2	1	2	1	2	1	2	1	3
2	1	2	1	2	1	2	2	1	2	1	2	1	2	1	0	3
2	1	2	2	1	2	1	1	2	1	2	2	1	2	1	3	2
2	1	2	2	1	2	1	2	1	2	1	1	2	1	2	12	2
2	2	1	1	2	2	1	1	2	2	1	1	2	2	1	3	4
2	2	1	1	2	2	1	2	1	1	2	2	1	1	2	4	10
2	2	1	2	1	1	2	1	2	2	1	2	1	1	2	0	5
2	2	1	2	1	1	2	2	1	1	2	1	2	2	1	0	8

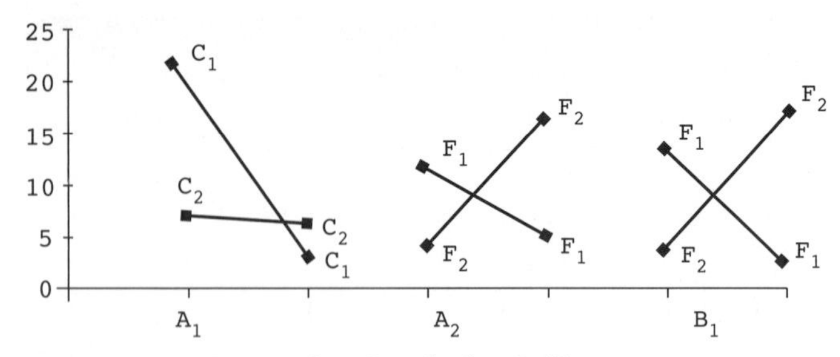

FIGURE 14-14 Interaction Graphs for Grille

$$A_2C_2 = (0 + 3 + 12 + 2 + 4 + 10 + 0 + 8)/8 = 4.9$$
$$B_1F_1 = (56 + 10 + 17 + 2 + 1 + 3 + 0 + 3)/8 = 11.5$$
$$B_1F_2 = (2 + 1 + 4 + 3 + 3 + 2 + 12 + 2)/8 = 3.6$$
$$B_2F_1 = (3 + 1 + 4 + 13 + 3 + 4 + 4 + 10)/8 = 5.3$$
$$B_2F_2 = (50 + 49 + 2 + 3 + 0 + 5 + 0 + 8)/8 = 14.6$$
$$F_1H_1 = (56 + 10 + 4 + 13 + 1 + 3 + 4 + 10)/8 = 12.6$$
$$F_1H_2 = (17 + 2 + 3 + 1 + 0 + 3 + 3 + 4)/8 = 4.1$$
$$F_2H_1 = (2 + 1 + 2 + 3 + 3 + 2 + 0 + 8)/8 = 2.6$$
$$F_2H_2 = (4 + 3 + 50 + 49 + 12 + 2 + 0 + 5)/8 = 15.6$$

These values are plotted in Figure 14-14. Analysis of the *AC* interaction shows that A_2 and C_1 give the best interaction results; however, C_2 gives the best results from a factor viewpoint, as seen in Table 14-14. If we use C_1, the gain will be 1.7 (4.6 − 2.9) from the interaction, but there will be a loss of 6.7 (12.1 − 5.4) from the factor. Thus, A_2 and C_2 will give the optimum results. For the *BF* interaction, the best values are at B_1 and F_2, although B_2 and F_1 are close. The decision may be based on some criterion other than "pops." The level chosen for *F* influences the choice of the level for *H*. Thus, if F_2 is chosen, then H_1 would be used; if F_1 is chosen, then H_2 would be used. The preceding analysis could just as well have started with the *FH* interaction rather than the *BF* interaction.

The confirmation run gave a first-time capability of 96%, for a savings of $900,000 per year. Future experiments are planned to find the reason for *UX* being too high.

Case III: Tube[5]

This case is a three-level control array and a two-level noise array, with a larger-the-better performance characteristic using the average and a signal-to-noise ratio. The experiment involves the joining of a small flexible tube to a rigid plastic connector, with the objectives of minimum assembly effort and maximum pull-off force in pounds; the focus is on the latter. A project team determines the control factors and their levels, and they are listed in Table 14-15.

TABLE 14-15 Factors and Levels for Tube

Control Factors	Levels		
A. Interference	Low	Med.	High
B. Wall thickness	Thin	Med.	Thick
C. Insertion depth	Shal.	Med.	High
D. % adhesive	Low	Med.	High

Noise Factors	Levels	
E. Time	24 h	120 h
F. Temp.	72°F	150°F
G. RH	25%	75%

An OA9 is used for the four control factors, and an OA4 is used for the three noise factors. The layout, along with the results of the experiment, is shown in Table 14-16. For TC 1, there are four observations for the three noise factors. These observations are 15.6, 19.9, 19.6, and 20.0; their average is 18.8.lb, and their signal-to-noise ratio for larger-the-better is 25.3 lb. The process is repeated for the other eight treatment conditions.

TABLE 14-16 OA9 and OA4 Layout with Experimental Results for Tube

					E_1	E_1	E_2	E_2		
					F_1	F_2	F_1	F_2		
TC	*A*	*B*	*C*	*D*	G_1	G_2	G_2	G_1	S/N_L	$\bar{y}$
1	1	1	1	1	15.6	19.9	19.6	20.0	25.3	18.8
2	1	2	2	2	15.0	19.6	19.8	24.2	25.5	19.7
3	1	3	3	3	16.3	15.6	18.2	23.3	25.0	18.4
4	2	1	2	3	18.3	18.6	18.9	23.2	25.8	19.8
5	2	2	3	1	19.7	25.1	21.4	27.5	27.2	23.4
6	2	3	1	2	16.2	19.8	19.6	22.5	25.6	19.5
7	3	1	3	2	16.4	23.6	18.6	24.3	26.0	20.7
8	3	2	1	3	14.2	16.8	19.6	23.3	24.9	18.5
9	3	3	2	1	16.1	17.3	22.7	22.6	25.6	19.7

The response table is shown in Table 14-17 for S/N and $\bar{y}$, along with the maximum difference between the levels. A

TABLE 14-17 S/N and $\bar{y}$ Response Table for Tube

	S/N_L		Δ
$A_1 = 25.3$	$A_2 = 26.2$	$A_3 = 25.5$	0.9
$B_1 = 25.7$	$B_2 = 25.9$	$B_3 = 25.4$	0.5
$C_1 = 25.3$	$C_2 = 25.6$	$C_3 = 26.1$	0.8
$D_1 = 26.0$	$D_2 = 25.7$	$D_3 = 25.2$	0.8
	$\bar{y}$		Δ
$A_1 = 18.9$	$A_2 = 20.9$	$A_3 = 19.6$	2.0
$B_1 = 19.8$	$B_2 = 20.5$	$B_3 = 19.2$	1.3
$C_1 = 18.9$	$C_2 = 19.7$	$C_3 = 20.8$	1.9
$D_1 = 20.6$	$D_2 = 20.0$	$D_3 = 18.9$	1.7

[5]Adapted, with permission, from Diane M. Byrne and Shin Taguchi, "The Taguchi Approach to Parameter Design," *Quality Progress* (December 1987): 19–26.

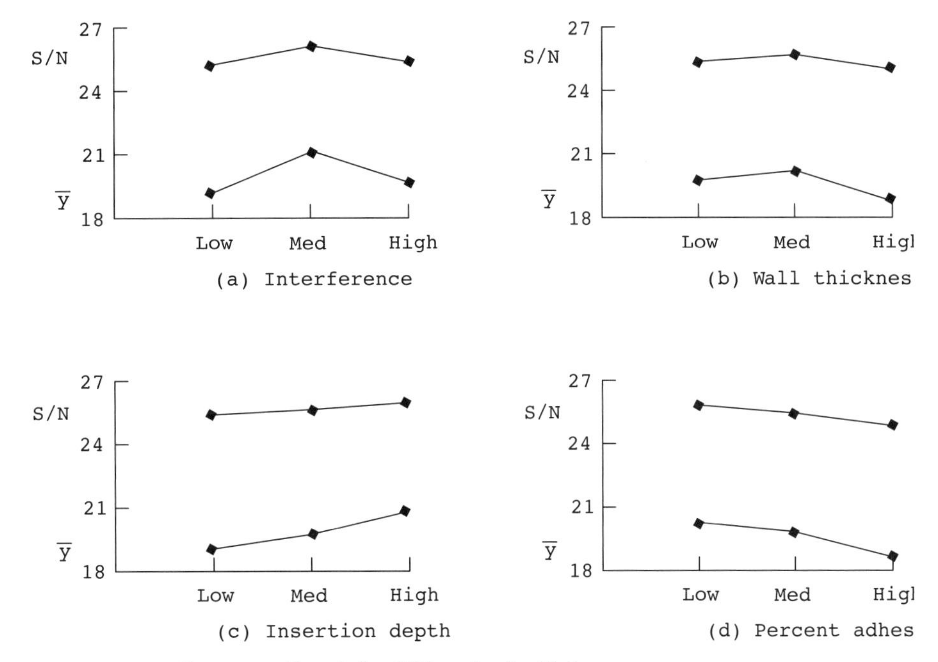

FIGURE 14-15 Response Graph for S/N and y for Tube

better evaluation of the factors and their levels can be seen in the response graph given in Figure 14-15. For factor A, interference, level 2 (medium) is obviously the best one for both S/N and $\bar{y}$; for factor B, wall thickness, level 2 (medium) is best for $\bar{y}$, but there does not appear to be a clear choice for the S/N for factor C, insertion depth, level 3 (deep) is best for both S/N and $\bar{y}$; and for factor D, percent adhesive, level 1 (low) is best for both S/N and $\bar{y}$.

A summary of results is shown in Table 14-18; it includes cost and ease of assembly, as well as information on the noise factors. For factor B, wall thickness, level 1 (thin wall thickness) was selected for its lower cost and ease of assembly as well as for the fact that the S/N did not show any difference. It should be noted that these factors and their levels are least sensitive to the three noise factors; therefore, it is a robust design. The predicted equation was

TABLE 14-18 Summary of Results

Factors	Levels	Assembly Effort	Pull-off Force	Cost Rating	Overall Rating
Interference	1. Low	8.1	18.9	Least	—
(*A*)	2. Medium	8.3	20.9	—	X
	3. High	8.7	19.6	Most	—
Wall	1. Thin	7.8	19.8	Least	X
Thickness (*B*)	2. Medium	8.3	20.5	—	—
	3. Thick	8.4	19.2	Most	—
Insertion	1. Shallow	7.7	18.9	Least	—
Depth (*C*)	2. Medium	8.3	19.7	—	X
	3. Deep	9.1	20.8	Most	—
Percent	1. Low	8.3	20.6	Least	X
Adhesive (*D*)	2. Medium	8.4	20.0	—	—
	3. High	8.4	18.9	Most	—
Conditioning	1. 24 h		18.0		
Time (*E*)	2. 120 h		21.6		
Conditioning	1. 75°F		18.1		
Temp. (*F*)	2. 150°F		21.5		
Conditioning	1. 25%		19.9		
R.H. (*G*)	2. 75%		19.7		

$$\overline{T} = \sum y/n = (18.8 + 19.7 + \cdots + 19.7)/8 = 19.7$$

$$\begin{aligned}\hat{\mu} &= \overline{T} + (A_2 - \overline{T}) + (B_1 - \overline{T}) + (C_3 - \overline{T}) + (D_1 - \overline{T})\\ &= A_2 + B_1 + C_3 + D_1 - (N-1)\overline{T}\\ &= 20.9 + 19.8 + 20.8 + 20.6 - (4-1)19.7\\ &= 23.0\end{aligned}$$

The S/N ratio could also have been used for the prediction. Results of the confirmation run were very close to the predicted value and also close to TC 5, which is the same combination except for the difference in the *B* level. The actual combination was not run during the experiment; this tendency is often the case with highly fractionalized experiments. In addition, the operators were quite pleased with the ease of assembly.

Treating Noise

Before continuing to the next case, let's discuss ways to treat noise. There are three techniques:

1. **Repetition.** When the process is very noisy, it is necessary to run only a few repetitions.

2. **Strongest.** When there is one strong noise factor, then two levels for that factor will be sufficient. For example: If temperature is a strong noise factor, it would be set at, say, 20° and 40°, and there would be two runs for each TC.

3. **Compounded.** When neither repetition nor strongest is applicable, then compounded is used. It requires an initial experiment for noise using a prototype or some units from production. The objective is to determine the extremes. Table 14-19 shows an OA4 with results for three noise factors and the response table next to it. The two extreme noise situations are

$$N_1 = U_1, V_2, \text{and } W_2$$
$$N_2 = U_2, V_1, \text{and } W_1$$

Each of these techniques can be used to minimize the number of runs while maintaining the concept of the noise array.

Case IV: Metal Stamping[6]

This case is a two-level design with a noise array and a nominal-the-best performance characteristic that is the distance from the center of the hole to the edge of a metal stamping. The target value is 0.40 in. Three control factors and their levels and three noise factors and their levels are determined by the team and are shown in Table 14-20. The experiment is run using an OA4 for the control array and an OA4 for the noise array. This layout, along with the results, is shown in Table 14-21.

The nominal-the-best strategy is to identify two types of control factors:

1. Those that affect variation and that are to be minimized. They are determined first using the S/N ratio.
2. Those that affect the average and are called adjustment or signal factors. They are used to adjust the average, y, to the target, τ.

Figure 14-16 shows the response table and response graph for the S/N ratio. A strong effect is given by factor *B*, with level 1 the appropriate level. Thus variation is minimized by using either level of factors *A* and *C* and factor B_1. Figure 14-17 shows the response table and response graph for the average, $\bar{y}$. A strong effect is given by factor *A*, which is the roller height. Thus the adjustment factor becomes the roller height, and it is adjusted to obtain the target value of 0.400. This adjustment or signal factor can make the design very robust by providing this additional capability for the process or the product. An excellent example of this concept for a product is the zero-adjustment feature of a bathroom scale.

TABLE 14-19 Compounded Noise Example for Three Noise Factors

TC	U	V	W	y
1	1	1	1	50
2	1	2	2	10
3	2	1	2	45
4	2	2	1	40

Level	U	V	W
1	30.0	47.5	45.0
2	42.5	25.0	27.5
Δ	12.5	22.5	17.5

TABLE 14-20 Control and Noise Factors with Their Levels for Metal Stamping

Control Factors	Level 1	Level 2
A. Roller Height	Sm	Lg
B. Material Supplier	SAE	SQC
C. Feed Adjustment	I	II
Noise Factors	**Level 1**	**Level 2**
U. Amount of Oil	Sm	Lg
V. Material Thickness	Low	High
W. Material Hardness	Low	High

TABLE 14-21 Experimental Design With Results for Metal Stamping

TC	A 1	B 2	C 3	U_1 V_1 W_1	U_2 V_2 W_1	U_2 V_1 W_2	U_1 V_2 W_2	$\bar{y}$	S/N_N
1	1	1	1	0.37	0.38	0.36	0.37	0.370	33.12
2	1	2	2	0.35	0.39	0.40	0.33	0.368	20.92
3	2	1	2	0.45	0.44	0.44	0.46	0.448	33.39
4	2	2	1	0.41	0.52	0.46	0.42	0.443	19.13

[6]Adapted, with permission, from *Taguchi Methods: Introduction to Quality Engineering* (Bingham Farms, Mich.: American Supplier Institute, Inc., 1991).

Level	A	B	C
1	27.02	33.26	26.12
2	26.27	20.03	27.15
Δ	0.75	13.23	1.03

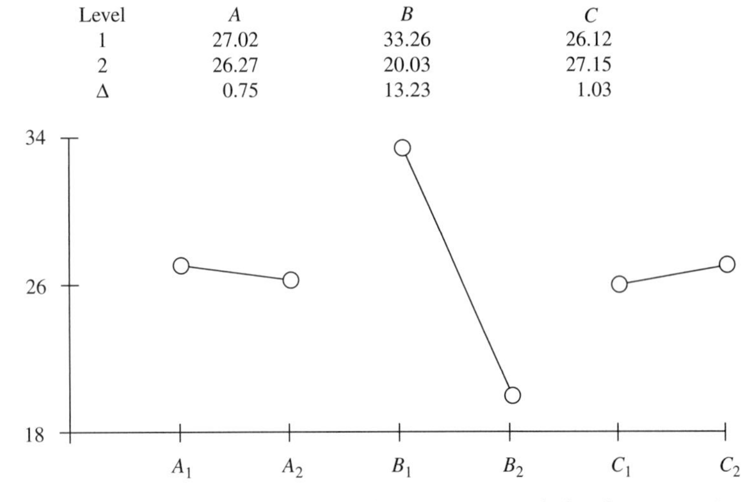

FIGURE 14-16 Response Table and Response Graph for the S/N Ratio

Regular analysis for the average, as used in this case, is appropriate when there is a small change in the average. If you like the log transform concept, then 10 log $\bar{y}$ could be used. Sensitivity analysis, *Sm*, is used when there is a large change in the average, which is true in research. The equation is

$$Sm\,(\text{dB}) = 10 \log_{10} (T^2/n)$$

The expression T^2/n is the sum of squares for the average, SS_{avg}, which is a common measure used in traditional design of experiments.

Regular analysis for the S/N ratio, such as used in this case, is appropriate when variability in relation to the average, $\bar{y}$, is measured in terms of a plus or minus percentage of $\bar{y}$. However, when variability in relation to the average is measured in terms of plus or minus absolute units and negative values are possible, then use the equation

$$\text{S/N}_N = -10 \log_{10} s^2$$

This situation occurs in few cases.

TOLERANCE DESIGN

Tolerance design is the process of determining the statistical tolerance around the target. During the parameter-design stage, low-cost tolerancing should be used. Only when the values are beyond the low-cost tolerancing limits is this concept implemented. Tolerance design is the selective tightening of tolerances and/or upgrading to eliminate excessive variation. It uses analysis of variance (ANOVA) to determine which factors contribute to the total variability and the loss function to obtain a trade-off between quality and cost.

Percent Contribution

In order to determine the percent contribution of the factors to the total variability, the iron casting case is used as the learning vehicle. The first step is to calculate the sum of squares for each of the factors and the total:

Level	A	B	C
1	0.369	0.409	0.412
2	0.446	0.411	0.408
Δ	0.077	0.002	0.004

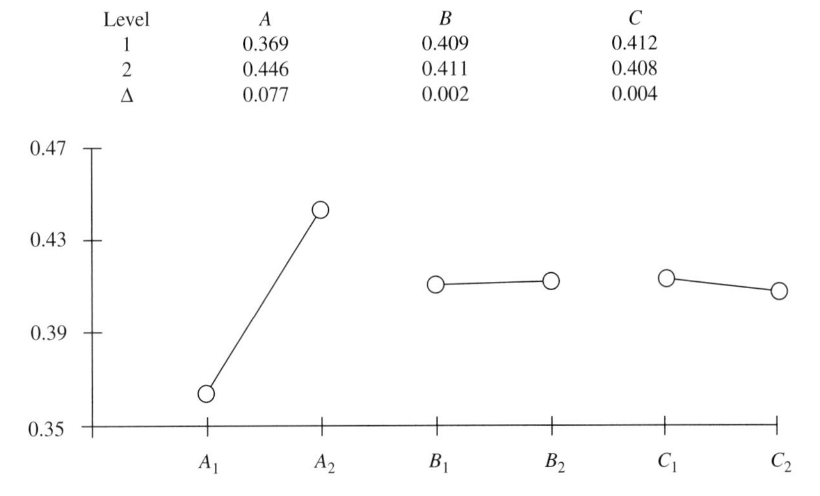

FIGURE 14-17 Response Table and Response Graph for the Average

$$SS_A = \Sigma(A_2/n) - T^2/n = \left(\frac{226^2}{4} + \frac{220^2}{4}\right) - \frac{446^2}{8}$$
$$= 4.5$$
$$SS_B = \Sigma(B_2/n) - T^2/n = \left(\frac{243^2}{4} + \frac{203^2}{4}\right) - \frac{446^2}{8}$$
$$= 200$$
$$SS_C = \Sigma(C_2/n) - T^2/n = \left(\frac{265^2}{4} + \frac{181^2}{4}\right) - \frac{446^2}{8}$$
$$= 882$$
$$SS_D = \Sigma(D_2/n) - T^2/n = \left(\frac{276^2}{4} + \frac{370^2}{4}\right) - \frac{446^2}{8}$$
$$= 28,405$$
$$SS_E = \Sigma(E_2/n) - T^2/n = \left(\frac{198^2}{4} + \frac{248^2}{4}\right) - \frac{446^2}{8}$$
$$= 313$$
$$SS_F = \Sigma(F_2/n) - T^2/n = \left(\frac{271^2}{4} + \frac{175^2}{4}\right) - \frac{446^2}{8}$$
$$= 1152$$
$$SS_G = \Sigma(G_2/n) - T^2/n = \left(\frac{215^2}{4} + \frac{231^2}{4}\right) - \frac{446^2}{8}$$
$$= 32$$
$$SS_{Total} = \Sigma y^2 - T^2/n = (89^2 + 55^2 + \cdots + 55^2) - 446^2/8 = 3987.5$$

These values are placed in an ANOVA table, as shown in Table 14-22, along with their degrees of freedom, which for a two-level is 1 (2 − 1). This information is given in the first three columns. Note that the smaller the difference in the response table, the smaller the *SS*. The fourth column is the mean square (*MS*) column for each factor, and its equation is

$$MS = \frac{SS}{\text{df}}$$

Because the number of degrees of freedom is 1, the *MS* value is the same as the *SS* value. If this were a three-level design, the number of degrees of freedom would have been 2: one for the linear component and one for the quadratic component, as shown in Table 14-24.

TABLE 14-22 ANOVA for the Iron Casting Case

Source	*df*	*SS*	*MS*	*F*
A	1	4.5	4.5	
B	1	200.0	200.0	
C	1	882.0	882.0	
D	1	1404.5	1404.5	
E	1	312.5	312.5	
F	1	1152.0	1152.0	
G	1	32.0	32.0	
Total	7	3987.5	3987.5	

The last column in the table, labeled *F*, has not been completed. This value is determined by the equation

$$F = \frac{MS_{\text{factor}}}{MS_{\text{error}}}$$

We do not have a value for MS_{error}. If the treatment conditions had been repeated, a value for the error could have been obtained. Regardless, a pooling-up technique can be used to obtain an estimate.

The pooling-up technique maximizes the number of significant factors. In other words, a factor can be classified as significant when, in truth, it is not. This action is referred to as the alpha error, or risk. Opposite to the alpha error is the beta error, or risk, which can classify a factor as being nonsignificant when, in truth, it is significant. The *F*-test, developed by Fischer, is used to determine significance at a designated alpha risk, which is usually 0.05.

The pooling-up procedure is to *F*-test the factor with the smallest *SS* against the next largest. If not significant, the *SS* is pooled and tested against the next largest. The process is continued until a factor is significant or one-half the total number of degrees of freedom is used.

Calculations for the first step using factors *A* and *G* and the iron casting case are

$$F = \frac{MS_G}{MS_e} = \frac{SS_G/\text{df}}{SS_A/\text{df}} = \frac{32/1}{4.5/1} = 7.11, \text{ thus, n.s.}$$

An *F* table for an alpha value of 0.05 and 1 degree of freedom for both the numerator and the denominator gives a critical value of 161. Because the calculated value is less than the critical value, the factor is not significant (n.s.). The next step is to pool the *SS* for the factors *G* and *A* and test against the next factor. Calculations are

$$F = \frac{MS_B}{MS_e} = \frac{SS_B/\text{df}}{SS_{A,G}/\text{df}} = \frac{200/1}{36.5/1} = 10.93 \text{ n.s.}$$

From an *F* table, the critical value for 1 and 2 degrees of freedom is 18.5, which is greater than the calculated value, so factor *B* is also not significant. Factor *B* is now pooled with the other two factors to obtain *SS* of 236.5 and *MS* of 78.8. However, because one-half the total degrees of freedom was obtained, the pooling process is complete.

The *F* column for the ANOVA is completed and shown in Table 14-23. *F* values in the column are compared with the critical *F*, which is obtained from the *F* table and shown at the bottom of the table. Three of the factors are significant at an alpha of 0.05 and one factor is nonsignificant. Two additional columns, labeled *SS*′, which is the pure *SS* after the error is subtracted, and %, which is the percent contribution, are shown in the table. Calculations for *SS*′ are

$$SS'_C = SS_C - (MS_e * \text{df}_C) = 882.0 - 78.8(1) = 803.2$$
$$SS'_D = SS_D - (MS_e * \text{df}_D) = 1404.5 - 78.8(1) = 1325.7$$
$$SS'_E = SS_E - (MS_e * \text{df}_E) = 312.5 - 78.8(1) = 233.7$$
$$SS'_F = SS_F - (MS_e * \text{df}_F) = 1152.0 - 78.8(1) = 1076.2$$
$$SS'_e = SS_e - (MS_e * \text{df}_{C,D,E,F}) = 236.5 + 78.8(4) = 551.7$$

TABLE 14-23 Percent Contribution for the Iron Casting Case

Source	df	SS	MS	F	SS′	%
A	[1]	[4.5]				
B	[1]	[200.0]				
C	1	882.0	882.0	11.2*	803.2	20.1
D	1	1404.5	1404.5	17.8*	1325.7	33.2
E	1	312.5	312.5	4.0 n.s.	233.7	5.9
F	1	1152.0	1152.0	14.6*	1073.2	26.9
G	[1]	[32.0]				
Pooled e	3	236.5	78.8		551.7	13.8
Total	7	3987.5			3987.5	99.9

*Significance at 95% confidence, $F(0.05; 1, 3) = 10.1$.

Values for the percent contribution for each factor are obtained by dividing by the total and are shown in the last column. The percent contribution for the error is 13.8, which is satisfactory. If the percent contribution for the error is a high value, say, 40% or more, then some important factors were omitted, conditions were not precisely controlled, or measurement error was excessive.

Tables 14-24 and 14-25 provide further illustration of the percent contribution technique for two of the cases discussed in this chapter.

Case I: TV Power Circuit[7]

A TV power circuit specification is 115 ± 15 volts. If the circuit goes out of this range, the customer will need to correct the problem at an average cost for all TV sets of $45.00. The problem can be corrected by recalibration at the end of the production line for a cost of $1.60. Using the loss function, the customer's loss is

$$L = k(y - \tau)^2$$
$$A_0 = k(\Delta_0)^2$$
$$k = A_0/\Delta_0^2$$

The manufacturer's loss function is

$$L = k(y - \tau)^2$$

Substituting gives

$$A = \frac{A_0}{\Delta_0^2}(\Delta)^2$$

Thus,

$$\Delta = \Delta_0\sqrt{\frac{A}{A_0}}$$
$$= 15\sqrt{\frac{1.60}{45.00}}$$
$$= 2.6 \text{ or about } 3\text{ V}$$

The tolerance is ±3 V, and the specifications are 112 V to 118 V. If the TV circuit is outside the specifications, it is recalibrated at a cost of $1.60.

Case II: Butterfly[8]

The plastic butterfly for the carburetor of a small engine has experienced a rash of complaints due to breakage. A larger-the-better loss function showed a loss of $39.00 at an average strength of 105 lb/in.^{2} and a standard deviation of 30 lb/in.^{2} for the defective items. Four factors of the plastic-molding process are identified by the project team, along with the experiment goal of $\bar{y} = 160\text{ lb/in.}^2$ and $s = 16\text{ lb/in.}^2$.

TABLE 14-24 Percent Contribution for the Tube Case

Source		SS	df	MS	F	SS′	%
Intr.	(L)	[0.73500]	[1]				
	(Q)	5.28125	1	5.28125	15.2149*	4.93419	25.9
Tube	(L)	[0.48167]	[1]				
	(Q)	2.20500	1	2.20500	6.3525	1.85789	9.8
Ins. dp.	(L)	5.46260	1	5.46260	15.7374	5.11549	26.9
	(Q)	[0.07031]	[1]				
% Adh.	(L)	4.68167	1	4.68167	13.4876	4.33456	22.8
	(Q)	[0.10125]	[1]				
e		0.0	0				
Pooled e		1.38843	4	0.34711		2.77687	14.6
Total		19.01875	8			19.01875	100.0

*$F_{1,4}(\alpha = 0.05) = 7.7088$.

[7]Adapted, with permission, from *Quality Engineering: Dynamic Characteristics and Measurement Engineering* (Bingham Farms, Mich.: American Supplier Institute, 1990).

[8]Adapted, with permission, from Thomas B. Baker, "Quality Engineering by Design: Taguchi's Philosophy," *Quality Progress* (December 1986): 32–42.

TABLE 14-25 Percent Contribution for the Grille Case

Source	*df*	*SS*	*MS*	*F*	*SS'*	%
A	1	800.000	800.000	11.728**	731.788	11.04
B	[1]	[45.125]				
AB	[1]	[15.125]				
F	[1]	[4.500]				
G	[1]	[0.500]				
BF	1	595.125	595.1215	8.725**	526.913	7.95
E	1	990.125	990.125	14.515**	921.913	13.91
C	1	351.125	351.125	5.148**	282.913	4.27
AC	1	630.125	630.125	9.128**	561.913	8.48
H	[1]	[40.500]				
I	[1]	[50.000]				
D	[1]	[78.125]				
AD	1	120.125	120.125	1.761	51.913	0.78
FH	1	924.500	924.500	13.553**	856.288	12.92
UX	1	648.000	648.000	9.500**	579.788	8.75
e	16	1335.000	83.438			
Pooled *e*	23	1568.873	68.212		2114.571	31.9
Total	31	6628.000			6628.000	100.0

*Significance at 95% confidence, $F(0.05; 1, 23) = 4.28$.

**Significance at 99% confidence, $F(0.01; 1, 23) = 7.88$.

Table 14-26 shows the factors, range of interest, and the low-cost tolerancing. A new temperature-control system was recently installed on the machine, and its tolerance is ±0.1%.

The parameter design resulted in a feed rate (*FR*) of 1200, a first rpm (*1R*) of 480, a second rpm (*2R*) of 950, and a temperature (*T*) of 360. A confirmation run was made using an OA9 with the low-cost tolerances as the outer levels and the parameter as the nominal. In other words, the three levels were 960, 1200, and 1440 for *FR*; 432, 480, and 528 for *IR*; 855, 950, and 1045 for *2R*; and 360, 360, and 360 for *T*. The results were $\bar{y} = 168.5$ lb/in.^{2} and $s = 28.4$ lb/in.^{2}, which was a substantial improvement, but the goal of $s = 16$ was not met. From the confirmation run, the percent contribution was calculated,

Feed rate	17.9%
First rpm	65.5%
Second rpm	9.4%
Residual (error)	7.3%
Total	100.1%

TABLE 14-26 Factors, Range of Interest, and Low-cost Tolerancing for Butterfly

Factor	Range of Interest	Low-cost Tolerance
A. Feed rate (*FR*)	1000–1400 g/min	±20%
B. First rpm (*1R*)	400–480 rev/min	±10%
C. Second rpm (*2R*)	850–950 rev/min	±10%
D. Temp. (*T*)	320–400°F	±0.1%

Temperature did not contribute to the percent contribution, so it is not included.

The next step is the rational (selective) reduction in tolerances by the project team. Variances are additive and standard deviations are not; therefore, calculations are made using variance. We have a variance of 806.6 (28.4^2) and need 256.0 (16^2). Thus, the variance needs to be reduced by 256.0/806.6, or 0.317 of its current value. Using this value, the factors, and the percent contribution, we can write the equation

$$0.317 = [(FR)^2 0.179 + (IR)^2 0.655 + (2R)^2 0.094 + e^2(0.073)]$$

where $e^2 = 1$ and is the residual or error.

The solution to the equation is by trial and error. After deliberation, the project team decides that the feed rate can be reduced by 33% of its original value; the calculations are

$$0.317 = (0.33)^2 0.179 + \cdots$$
$$(0.33)^2 0.179 = 0.020$$
$$0.317 - 0.020 = 0.297 \text{ left}$$

The project team decides that the first rpm can be reduced by 50% of its original value; the calculations are

$$0.297 = \cdots + (0.50)^2 0.655 + \cdots$$
$$(0.50)^2 0.655 = 0.164$$
$$0.297 - 0.164 = 0.133 \text{ left}$$

TABLE 14-27 New Tolerances for Butterfly

Factor	Old		New
FR	±20%	20(0.33)	±6.6%
1R	±10%	10(0.50)	±5.0%
2R	±10%	10(0.80)	±8.0%

The remaining factor is the second rpm. Removing the residual of 7.3%, the result is 0.060(0.133 − 0.073). The calculations are

$$(2R)^2\,0.094 = 0.060$$
$$2R = 0.80$$

New tolerances are shown in Table 14-27. If these need modification, the preceding process can be repeated with different values.

Levels for the final run are

FR 6.6% of 1200 = 80 (1120, 1200, 1280)

1R 5.05% of 480 = 24 (456, 480, 504)

2R 8.0% of 950 = 76 (874, 950, 1026)

Using an OA9 and these levels, the results were $\bar{y} = 179.9$, $s = 12.1$, and S/N = 45.0, which is almost 200% better than the original. The appropriate parameter settings and realistic tolerance settings are now known. From the loss function graph, the average loss is $2.00 per engine, for a savings of $370,000 for 10,000 engines. Statistical process control (SPC) is used to maintain the target and variation.

Noise can be based on (1) percent of the parameter, as in this case, (2) experience, or (3) standard deviation (Taguchi recommends 1σ for a two-level and 1.22σ for a three-level). A two-level design would occur when the tolerance is unilateral or one-directional or only the upper and lower specifications are used without the nominal or target. It should be noted that the rational reduction concept was not developed by Taguchi.

Case III: Control Circuit[9]

This case concerns the tolerance design of a control circuit after the parameter design has determined the factors and their target or nominal values. The design team decides to establish the tolerance levels at $\pm 1\sigma$. Table 14-28 shows the 12 factors with their nominal and tolerance specifications.

The design team established a target for the performance characteristic of 570 cycles and functional limits of ±150 cycles, with the loss at the limit of $100.00. Using these values, the average loss equation is

$$\bar{L} = 0.004444[\sigma^2 + (y - 570)^2]$$

The experiment was run using an OA16, which is shown in Table 14-29 along with the results. Note that three of the columns are unexplained or classified as error. From the results, the percent contribution is determined using the ANOVA and pooling-up technique, as shown in Table 14-30. Note the error includes the last three columns plus the pooling of factors *E*, *H*, and *J*. The error percent contribution is very low at 1.3%.

The mean squared deviation for the total is calculated from the percent contribution table using the equation

$$MSD_T = \frac{SS_T}{df_T} = \frac{105{,}361.5}{15} = 7024.1$$

TABLE 14-28 Factors and Levels for the Tolerance Design of the Control Circuit

Factors	Units	Level 1 $-\sigma$	Target or Nominal	Level 2 $+\sigma$
Resistor *A*	kilohms	2.09	2.20	2.31
Resistor *B*	ohms	446.5	4.700	493.5
Capacitor *C*	microfarads	0.65	0.68	0.71
Resistor *D*	kilohms	95.0	100.0	105.0
Capacitor *E*	microfarads	8.0	10.0	12.0
Transistor *F*	hfe	90.0	180.0	270.0
Resistor *G*	kilohms	9.5	10.0	10.5
Resistor *H*	kilohms	1.43	1.50	1.57
Resistor *I*	kilohms	9.5	10.0	10.5
Resistor *J*	kilohms	9.5	10.0	10.5
Transistor *K*	hfe	90.0	180.0	270.0
Voltage *L*	volts	6.2	6.5	6.8

TABLE 14-29 OA16 with Results for Control Circuit

TC	*A*	*B*	*C*	*D*	*E*	*F*	*G*	*H*	*I*	*J*	*K*	*L*	*e*	*e*	*e*	Data
1	1	1	1	1	1	1	1	1	1	1	1	1	1	1	1	523
2	1	1	1	1	1	1	1	2	2	2	2	2	2	2	2	430
3	1	1	1	2	2	2	2	1	1	1	1	2	2	2	2	674
4	1	1	1	2	2	2	2	2	2	2	2	1	1	1	1	572
5	1	2	2	1	1	2	2	1	1	2	2	1	1	2	2	609
6	1	2	2	1	1	2	2	2	2	1	1	2	2	1	1	534
7	1	2	2	2	2	1	1	1	1	2	2	2	2	1	1	578
8	1	2	2	2	2	1	1	2	2	1	1	1	1	2	2	527
9	2	1	2	1	2	1	2	1	2	1	2	1	2	1	2	605
10	2	1	2	1	2	1	2	2	1	2	1	2	1	2	1	707
11	2	1	2	2	1	2	1	1	2	1	2	2	1	2	1	541
12	2	1	2	2	1	2	1	2	1	2	1	1	2	1	2	669
13	2	2	1	1	2	2	1	1	2	2	1	1	1	2	1	430
14	2	2	1	1	2	2	1	2	1	1	2	2	2	1	2	480
15	2	2	1	2	1	1	2	1	2	2	1	2	2	1	2	578
16	2	2	1	2	1	1	2	2	1	1	2	1	1	2	1	668

[9] *Source:* Robert Moesta, American Supplier, Inc.

TABLE 14-30 Percent Contribution for Control Circuit

Source	*df*	*SS*	*MS*	*SS'*	%
A	1	3,335.1	3,335.1	3,247.1	3.1
B	1	6,280.6	6,280.6	6,192.6	5.9
C	1	10,764.1	10,764.1	10,676.1	10.1
D	1	14,945.1	14,945.1	14,857.1	14.1
E	[1]	27.6			
F	1	715.6	715.6	627.6	0.6
G	1	36,960.1	36,960.1	36,875.1	35.0
H	[1]	150.0			
I	1	29,842.6	29,842.6	29,754.6	28.2
J	[1]	27.6			
K	1	1,580.1	1,580.1	1,492.1	1.4
L	1	410.1	410.1	322.1	0.3
e	[3]	322.8			
e (pooled)	6	528.1	88.0	1320.3	1.3
Total	15	105,361.5		105,361.5	100.0

TABLE 14-31 Quality Loss by Factor

Factor	Type	%	*MSD*	Loss—$
A	Resistor	3.1	217.7	0.97
B	Resistor	5.9	414.4	1.84
C	Capacitor	10.1	709.4	3.15
D	Resistor	14.1	990.4	4.40
E	Capacitor	0.0		
F	Transistor	0.6	42.1	0.19
G	Resistor	35.0	2458.4	10.93
H	Resistor	0.0		
I	Resistor	28.2	1980.8	8.80
J	Resistor	0.0		
K	Transistor	1.4	98.3	0.44
L	Voltage	0.3	21.1	0.09
e	Error	1.3	91.3	0.41
Total		100.0		$31.22

This amount is then apportioned to each of the factors based on its percent contribution and placed in the *MSD* column of Table 14-31. Using the loss function equation, $L = k(MSD)$, the quality loss for each factor is determined and placed in the loss column of the table. Calculations for factor *A* are

$$MSD_A = MSD_T(\%_A/100) = 7024.1(3.1/100) = 217.7$$
$$L_A = k(MSD_A) = 0.004444(217.7) = \$0.97$$

Calculations are repeated for the other factors and placed in the table. The total quality loss is $31.22, shown at the bottom of the table.

The next step in the process is to evaluate the upgrade performance of the factors. With electronic components, a higher quality component is obtained by reducing the tolerance. For example, factor *A*, a resistor, changes from a 5% tolerance to a 1% tolerance, and Δ in the loss function equation is $5/1 = 5$. Thus, Δ^2 is 25. This information is shown in Table 14-32. Also shown are the upgraded *MSD*, new loss, and quality gain; for factor *A* the calculations are

$$MSD'_A = MSD_A/\Delta^2 = 217.7/25 = 8.7$$
$$L'_A = k(MSD'_A) = 0.004444(8.7) = \$0.04$$
$$\text{Gain}_A = \$0.97 - \$0.04 = \$0.93$$

The final step in the process is to make the upgrade decision by comparing the quality gain to the upgrade cost

TABLE 14-32 Upgrade Performance

Factor	Type	Loss—$	Upgrade Effect	Upgrade Factor	Upgraded *MSD'*	New Loss—$	Quality Gain—$
A	Resistor	0.97	5%-1%	25	8.7	0.04	0.93
B	Resistor	1.84	5%-1%	25	16.6	0.07	1.77
C	Capacitor	3.15	5%-1%	25	28.4	0.13	3.02
D	Resistor	4.40	5%-1%	25	39.6	0.18	4.22
E	Capacitor						
F	Transistor	0.19	12%-3%	16	2.6	0.01	0.18
G	Resistor	10.93	5%-1%	25	98.3	0.44	10.49
H	Resistor						
I	Resistor	8.80	5%-1%	25	72.2	0.35	8.45
J	Resistor						
K	Transistor	0.44	12%-1%	16	6.1	0.03	0.41
L	Voltage	0.09					
e	Error	0.41					
Total		31.22					

TABLE 14-33 Upgrade Decision for Control Circuit

Factor	Type	Quality Gain—$	Upgrade Cost—$	Net Gain—$	Make Upgrade?
A	Resistor	0.93	0.06	0.87	Yes
B	Resistor	11.77	0.06	1.71	Yes
C	Capacitor	3.02	1.00	2.02	Yes
D	Resistor	4.22	0.06	4.16	Yes
E	Capacitor				No
F	Transistor	0.18	2.00	(1.82)	No
G	Resistor	10.49	0.06	10.41	Yes
H	Resistor				No
I	Resistor	8.45	0.06	8.40	Yes
J	Resistor				No
K	Transistor	0.41	1.00	(0.59)	No
L	Voltage				No
e	Error				
Total			1.30	27.57	

to obtain the net gain. As shown in Table 14-33, a resistor upgrade is inexpensive and a transistor upgrade is expensive; this information affects the final decision. For factor *A* the net gain calculations are

$$\begin{aligned} \text{Net gain} &= \text{quality gain} - \text{upgrade cost} \\ &= \$0.93 - \$0.06 \\ &= \$0.87 \end{aligned}$$

Of course, because $0.87 is saved per unit, the upgrade decision is "yes."

As a result of this tolerance design, there was a substantial improvement in cost and performance. Upgrade costs were $1.30 and the net gain was $27.57 per unit. The C_{pk} went from 0.6 to 2.2.

CONCLUSION

The Taguchi approach has built on traditional design of experimental methods to improve the design of products and processes. These unique and relatively simple concepts result in substantial improvements in quality at lower costs.

EXERCISES

1. The specifications of a steel shaft are 6.40 ± 0.10 mm. The device sometimes fails when the shaft exceeds the specification. When failure occurs, repair or replacement is necessary at an average cost of $95.00.
   a. What is the loss coefficient k?
   b. What is the loss function equation?
   c. What is the loss at 6.55 mm?
2. The specifications for an electronic device are 24 ± 0.4 Amps and the average repair cost is $32.00.
   a. Determine the loss function.
   b. Determine the loss at 24.6 Amps.
3. Determine the average loss for the information in Exercise 2 if 13 samples give 24.2, 24.0, 24.1, 23.8, 24.3, 24.2, 23.9, 23.8, 24.0, 23.6, 23.8, 23.9, and 23.7.
4. For an out-of-round condition (smaller-the-better) of a steel shaft, the true indicator readings (TIR) for eight shafts are 0.05, 0.04, 0.04, 0.03, 0.04, 0.02, 0.04, and 0.03 mm.
   a. If the average loss at 0.03 is $15.00, what is the loss function?
   b. What is the loss at 0.05?
   c. What is the average loss?
5. When the tensile strength of a plastic part is 120 lb/in.^{2}, there is an average loss of $34.00 due to breakage. Determine the average loss for sample tests of 130, 132, 130, 132, and 131.
   Hint: Use larger-the-better formula.
6. A new process is proposed for the manufacture of steel shafts, as given in the example problem on page 566. Data are 6.38, 6.40, 6.41, 6.38, 6.39, 6.36, and 6.37.
   a. What is the expected loss?
   b. Is the new process better than the process of the example?
   c. What future improvements might be tried?
   Hint: Compare average with target and the standard deviations of both processes.
7. Given three two-level factors and three suspected two-factor interactions, determine the degrees of freedom and the OA.
8. If a three-factor interaction was also suspected in Exercise 7, what are the degrees of freedom and the OA? What type of OA is this design?
9. What are the degrees of freedom and OA if the factors in Exercise 7 are three-level? Why does a three-level design require so much more design space?
10. An experimental design has five two-level factors (A, B, C, D, E), where only main effects are possible for factor C and there are no suspected AB and three-factor or higher interactions. Using a linear graph, assign the factors and their interactions to the columns of the OA.
11. Using a linear graph, assign the factors and their interactions to the columns of the OA determined in Exercise 7.
12. Using a linear graph, assign the factors and their interactions to the columns of the OA determined in Exercise 9.
13. A new process has been developed and the temperature results are 21°C for the average and 0.8°C for the sample standard deviation ($n = 5$).
    a. What is the S/N ratio for nominal-the-best?
    b. How much improvement has occurred? Compare to the nominal-the-best Example Problem 14-5 answer of 20.41 dB.
    c. If you change the units of the example problem and this exercise, will you get the same results? Prove your conclusion.
14. Suppose the results of the new process for the bread stuffing example problem are 125, 132, 138, 137, 128, and 131. What conclusions can be drawn?
15. The yield on a new chemical process for five days is 61, 63, 58, 57, and 60 and the old process had recent yields of 54, 56, 52, 56, 53, 51, 54, 53, and 52. Is the new process better? If so, how much better?
16. The results for a larger-the-better experimental design that was run in random order with seven factors are as follows:

	A	*B*	*C*	*D*	*E*	*F*	*G*		
TC	*1*	*2*	*3*	*4*	*5*	*6*	*7*	*R1*	*R2*
1	1	1	1	1	1	1	1	19	25
2	1	1	1	2	2	2	2	20	24
3	1	2	2	1	1	2	2	24	22
4	1	2	2	2	2	1	1	22	25
5	2	1	2	1	2	1	2	26	20
6	2	1	2	2	1	2	1	25	26
7	2	2	1	1	2	2	1	25	20
8	2	2	1	2	1	1	2	25	21

    a. Determine the response table, response graph, strong effects, and prediction for the average and the S/N ratio.
    b. If the confirmation run is 27.82, what can you say about the experiment? If the confirmation run is 27.05, what can you say about the experiment?
17. The results of a nominal-the-best experimental design are as follows:

	A	*B*	*C*				
TC	*1*	*2*	*3*	*N1*	*N2*	*y*	*S/N*
1	1	1	1	1.75	1.84	1.80	29.01
2	1	2	2	1.34	2.13	1.74	9.84
3	2	1	2	2.67	2.43	2.55	23.54
4	2	2	1	2.23	2.73	2.48	16.92

a. Determine the response table, response graph, and strong effects.

b. Analyze your results in terms of adjustment factors and variation factors.

18. The results for a smaller-the-better saturated experimental design using an OA16 with 15 factors where the factors *A*, *B*, . . . , *O* are located in columns 1, 2, . . . , 15, respectively, are as follows:

R1	*R2*	*R3*	*R4*
0.49	0.54	0.46	0.45
0.55	0.60	0.57	0.58
0.07	0.09	0.11	0.08
0.16	0.16	0.19	0.19
0.13	0.22	0.20	0.23
0.16	0.17	0.13	0.12
0.24	0.22	0.19	0.25
0.13	0.19	0.19	0.19
0.08	0.10	0.14	0.18
0.07	0.04	0.19	0.18
0.48	0.49	0.44	0.41
0.54	0.53	0.53	0.54
0.13	0.17	0.21	0.17
0.28	0.26	0.26	0.30
0.34	0.32	0.30	0.41
0.58	0.62	0.59	0.54

a. Determine the response table, response graph, strong effects, and prediction for the average and the S/N ratio.

b. If the results of the confirmation run are 0.13, 0.07, 0.06, and 0.08, what can you say about the experiment?

19. The results of a larger-the-better experimental design with an outer array for noise are as follows:

TC	A	B	C	D	N1	N2
1	1	1	1	1	7.9	11.9
2	1	2	2	2	7.3	12.1
3	1	3	3	3	8.6	10.5
4	2	1	2	3	10.6	11.2
5	2	2	3	1	12.0	13.7
6	2	3	1	2	8.5	11.9
7	3	1	3	2	8.7	10.9
8	3	2	1	3	6.5	11.9
9	3	3	2	1	8.4	15.0

a. Determine the response table, response graph, strong effects, and prediction for the S/N ratio.

b. What value for the confirmation run would you consider satisfactory?

20. The results of a smaller-the-better experimental design are as follows:

	B	*A*	*AB*	*C*	*UX*	*AC*	*D*	
TC	1	2	3	4	5	6	7	S/N
1	1	1	1	1	1	1	1	32.1
2	1	1	1	2	2	2	2	33.6
3	1	2	2	1	1	2	2	32.8
4	1	2	2	2	2	1	1	31.7
5	2	1	2	1	2	1	2	31.2
6	2	1	2	2	1	2	1	33.7
7	2	2	1	1	2	2	1	32.3
8	2	2	1	2	1	1	2	33.6

a. Determine the response table, response graph, and strong effects.

b. Explain the results.

21. Determine the percent contributions of Exercise 16.

22. Determine the percent contributions of Exercise 20.

23. The confirmation run for the experimental design of an electronic device gave the following percent contributions for unpooled factors from an OA12 design. Also given is the upgrade effect and the upgrade cost.

Factor	Type	*df*	%	Ungrade Effect	Upgrade Cost—$
A	Capacitor	1	41.0	5%-1%	1.81
B	Resistor	1	12.4	5%-1%	0.15
C	Transistor	1	32.1	12%-3%	3.92
D	Resistor	1	20.9	5%-1%	0.15
e	Error	7	5.6		
		11			

If the total *SS* is 1301.2 and $k = 0.05$, determine the net gain per unit from upgrading.

24. A four-factor experiment gives the following percent contributions for the confirmation run: *A* (43%), *B*(9%), *C*(28%), *D*(13%), and residual (7%). If the variance is currently 225 and the desired value is 100, determine two possible reduction schemes.

25. Design and conduct a Taguchi experiment for

a. growth of a house plant;

b. flight of a paper airplane;

c. baking brownies, chocolate-chip cookies, and so forth;

d. making coffee, popcorn, etc.;

e. any organization listed in Chapter 1, Exercise 5.

APPENDIX

TABLE A Areas Under the Normal Curve[a]

$\frac{X_i - \mu}{\sigma}$	0.09	0.08	0.07	0.06	0.05	0.04	0.03	0.02	0.01	0.00
−3.5	0.00017	0.00017	0.00018	0.00019	0.00019	0.00020	0.00021	0.00022	0.00022	0.00023
−3.4	0.00024	0.00025	0.00026	0.00027	0.00028	0.00029	0.00030	0.00031	0.00033	0.00034
−3.3	0.00035	0.00036	0.00038	0.00039	0.00040	0.00042	0.00043	0.00045	0.00047	0.00048
−3.2	0.00050	0.00052	0.00054	0.00056	0.00058	0.00060	0.00062	0.00064	0.00066	0.00069
−3.1	0.00071	0.00074	0.00076	0.00079	0.00082	0.00085	0.00087	0.00090	0.00094	0.00097
−3.0	0.00100	0.00104	0.00107	0.00111	0.00114	0.00118	0.00122	0.00126	0.00131	0.00135
−2.9	0.0014	0.0014	0.0015	0.0015	0.0016	0.0016	0.0017	0.0017	0.0018	0.0019
−2.8	0.0019	0.0020	0.0021	0.0021	0.0022	0.0023	0.0023	0.0024	0.0025	0.0026
−2.7	0.0026	0.0027	0.0028	0.0029	0.0030	0.0031	0.0032	0.0033	0.0034	0.0035
−2.6	0.0036	0.0037	0.0038	0.0039	0.0040	0.0041	0.0043	0.0044	0.0045	0.0047
−2.5	0.0048	0.0049	0.0051	0.0052	0.0054	0.0055	0.0057	0.0059	0.0060	0.0062
−2.4	0.0064	0.0066	0.0068	0.0069	0.0071	0.0073	0.0075	0.0078	0.0080	0.0082
−2.3	0.0084	0.0087	0.0089	0.0091	0.0094	0.0096	0.0099	0.0102	0.0104	0.0107
−2.2	0.0110	0.0113	0.0116	0.0119	0.0122	0.0125	0.0129	0.0132	0.0136	0.0139
−2.1	0.0143	0.0146	0.0150	0.0154	0.0158	0.0162	0.0166	0.0170	0.0174	0.0179
−2.0	0.0183	0.0188	0.0192	0.0197	0.0202	0.0207	0.0212	0.0217	0.0222	0.0228
−1.9	0.0233	0.0239	0.0244	0.0250	0.0256	0.0262	0.0268	0.0274	0.0281	0.0287
−1.8	0.0294	0.0301	0.0307	0.0314	0.0322	0.0329	0.0336	0.0344	0.0351	0.0359
−1.7	0.0367	0.0375	0.0384	0.0392	0.0401	0.0409	0.0418	0.0427	0.0436	0.0446
−1.6	0.0455	0.0465	0.0475	0.0485	0.0495	0.0505	0.0516	0.0526	0.0537	0.0548
−1.5	0.0559	0.0571	0.0582	0.0594	0.0606	0.0618	0.0630	0.0643	0.0655	0.0668
−1.4	0.0681	0.0694	0.0708	0.0721	0.0735	0.0749	0.0764	0.0778	0.0793	0.0808
−1.3	0.0823	0.0838	0.0853	0.0869	0.0885	0.0901	0.0918	0.0934	0.0951	0.0968
−1.2	0.0895	0.1003	0.1020	0.1038	0.1057	0.1075	0.1093	0.1112	0.1131	0.1151
−1.1	0.1170	0.1190	0.1210	0.1230	0.1251	0.1271	0.1292	0.1314	0.1335	0.1357
−1.0	0.1379	0.1401	0.1423	0.1446	0.1469	0.1492	0.1515	0.1539	0.1562	0.1587
−0.9	0.1611	0.1635	0.1660	0.1685	0.1711	0.1736	0.1762	0.1788	0.1814	0.1841
−0.8	0.1867	0.1894	0.1922	0.1949	0.1977	0.2005	0.2033	0.2061	0.2090	0.2119
−0.7	0.2148	0.2177	0.2207	0.2236	0.2266	0.2297	0.2327	0.2358	0.2389	0.2420
−0.6	0.2451	0.2483	0.2514	0.2546	0.2578	0.2611	0.2643	0.2676	0.2709	0.2743
−0.5	0.2776	0.2810	0.2843	0.2877	0.2912	0.2946	0.2981	0.3015	0.3050	0.3085

TABLE A (*Continued*)

$\frac{X_i - \mu}{\sigma}$	0.09	0.08	0.07	0.06	0.05	0.04	0.03	0.02	0.01	0.00
−0.4	0.3121	0.3156	0.3192	0.3228	0.3264	0.3300	0.3336	0.3372	0.3409	0.3446
−0.3	0.3483	0.3520	0.3557	0.3594	0.3632	0.3669	0.3707	0.3745	0.3783	0.3821
−0.2	0.3859	0.3897	0.3936	0.3974	0.4013	0.4052	0.4090	0.4129	0.4168	0.4207
−0.1	0.4247	0.4286	0.4325	0.4364	0.4404	0.4443	0.4483	0.4522	0.4562	0.4602
−0.0	0.4641	0.4681	0.4721	0.4761	0.4801	0.4840	0.4880	0.4920	0.4960	0.5000

$\frac{X_i - \mu}{\sigma}$	0.00	0.01	0.02	0.03	0.04	0.05	0.06	0.07	0.08	0.09
+0.0	0.5000	0.5040	0.5080	0.5120	0.5160	0.5199	0.5239	0.5279	0.5319	0.5359
+0.1	0.5398	0.5438	0.5478	0.5517	0.5557	0.5596	0.5636	0.5675	0.5714	0.5753
+0.2	0.5793	0.5832	0.5871	0.5910	0.5948	0.5987	0.6026	0.6064	0.6103	0.6141
+0.3	0.6179	0.6217	0.6255	0.6293	0.6331	0.6368	0.6406	0.6443	0.6480	0.6517
+0.4	0.6554	0.6591	0.6628	0.6664	0.6700	0.6736	0.6772	0.6808	0.6844	0.6879
+0.5	0.6915	0.6950	0.6985	0.7019	0.7054	0.7088	0.7123	0.7157	0.7190	0.7224
+0.6	0.7257	0.7291	0.7324	0.7357	0.7389	0.7422	0.7454	0.7486	0.7517	0.7549
+0.7	0.7580	0.7611	0.7642	0.7673	0.7704	0.7734	0.7764	0.7794	0.7823	0.7852
+0.8	0.7881	0.7910	0.7939	0.7967	0.7995	0.8023	0.8051	0.8079	0.8106	0.8133
+0.9	0.8159	0.8186	0.8212	0.8238	0.8264	0.8289	0.8315	0.8340	0.8365	0.8389
+1.0	0.8413	0.8438	0.8461	0.8485	0.8508	0.8531	0.8554	0.8577	0.8599	0.8621
+1.1	0.8643	0.8665	0.8686	0.8708	0.8729	0.8749	0.8770	0.8790	0.8810	0.8830
+1.2	0.8849	0.8869	0.8888	0.8907	0.8925	0.8944	0.8962	0.8980	0.8997	0.9015
+1.3	0.9032	0.9049	0.9066	0.9082	0.9099	0.9115	0.9131	0.9147	0.9162	0.9177
+1.4	0.9192	0.9207	0.9222	0.9236	0.9251	0.9265	0.9279	0.9292	0.9306	0.9319
+1.5	0.9332	0.9345	0.9357	0.9370	0.9382	0.9394	0.9406	0.9418	0.9429	0.9441
+1.6	0.9452	0.9463	0.9474	0.9484	0.9495	0.9505	0.9515	0.9525	0.9535	0.9545
+1.7	0.9554	0.9564	0.9573	0.9582	0.9591	0.9599	0.9608	0.9616	0.9625	0.9633
+1.8	0.9641	0.9649	0.9656	0.9664	0.9671	0.9678	0.9686	0.9693	0.9699	0.9706
+1.9	0.9713	0.9719	0.9726	0.9732	0.9738	0.9744	0.9750	0.9756	0.9761	0.9767
+2.0	0.9773	0.9778	0.9783	0.9788	0.9793	0.9798	0.9803	0.9808	0.9812	0.9817
+2.1	0.9821	0.9826	0.9830	0.9834	0.9838	0.9842	0.9846	0.9850	0.9854	0.9857
+2.2	0.9861	0.9864	0.9868	0.9871	0.9875	0.9878	0.9881	0.9884	0.9887	0.9890
+2.3	0.9893	0.9896	0.9898	0.9901	0.9904	0.9906	0.9909	0.9911	0.9913	0.9916
+2.4	0.9918	0.9920	0.9922	0.9925	0.9927	0.9929	0.9931	0.9932	0.9934	0.9936
+2.5	0.9938	0.9940	0.9941	0.9943	0.9945	0.9946	0.9948	0.9949	0.9951	0.9952
+2.6	0.9953	0.9955	0.9956	0.9957	0.9959	0.9960	0.9961	0.9962	0.9963	0.9964
+2.7	0.9965	0.9966	0.9967	0.9968	0.9969	0.9970	0.9971	0.9972	0.9973	0.9974
+2.8	0.9974	0.9975	0.9976	0.9977	0.9977	0.9978	0.9979	0.9979	0.9980	0.9981
+2.9	0.9981	0.9982	0.9983	0.9983	0.9984	0.9984	0.9985	0.9985	0.9986	0.9986
+3.0	0.99865	0.99869	0.99874	0.99878	0.99882	0.99886	0.99889	0.99893	0.99896	0.99900
+3.1	0.99903	0.99906	0.99910	0.99913	0.99915	0.99918	0.99921	0.99924	0.99926	0.99929
+3.2	0.99931	0.99934	0.99936	0.99938	0.99940	0.99942	0.99944	0.99946	0.99948	0.99950
+3.3	0.99952	0.99953	0.99955	0.99957	0.99958	0.99960	0.99961	0.99962	0.99964	0.99965
+3.4	0.99966	0.99967	0.99969	0.99970	0.99971	0.99972	0.99973	0.99974	0.99975	0.99976
+3.5	0.99977	0.99978	0.99978	0.99979	0.99980	0.99981	0.99981	0.99982	0.99983	0.99983

[a] Proportion of total area under the curve that is under the proportion of the curve from $-\infty$ to $(X_i - \mu)/\sigma$ (X_i represents any desired value of the variable X).

TABLE B Factors for Computing Central Lines and 3σ Control Limits for $\bar{X}$, s, and R Charts

Observations in Sample, n	CHART FOR AVERAGES			CHART FOR STANDARD DEVIATIONS					CHART FOR RANGES					
	Factors for Control Limits			Factor for Central Line	Factors for Control Limits				Factor for Central Line	Factors for Control Limits				
	A	A_2	A_3	c_4	B_3	B_4	B_5	B_6	d_2	d_3	D_1	D_2	D_3	D_4
2	2.121	1.880	2.659	0.7979	0	3.267	0	2.606	1.128	0.853	0	3.686	0	3.267
3	1.732	1.023	1.954	0.8862	0	2.568	0	2.276	1.693	0.888	0	4.358	0	2.574
4	1.500	0.729	1.628	0.9213	0	2.266	0	2.088	2.059	0.880	0	4.698	0	2.282
5	1.342	0.577	1.427	0.9400	0	2.089	0	1.964	2.326	0.864	0	4.918	0	2.114
6	1.225	0.483	1.287	0.9515	0.030	1.970	0.029	1.874	2.534	0.848	0	5.078	0	2.004
7	1.134	0.419	1.182	0.9594	0.118	1.882	0.113	1.806	2.704	0.833	0.204	5.204	0.076	1.924
8	1.061	0.373	1.099	0.9650	0.185	1.815	0.179	1.751	2.847	0.820	0.388	5.306	0.136	1.864
9	1.000	0.337	1.032	0.9693	0.239	1.761	0.232	1.707	2.970	0.808	0.547	5.393	0.184	1.816
10	0.949	0.308	0.975	0.9727	0.284	1.716	0.276	1.669	3.078	0.797	0.687	5.469	0.223	1.777
11	0.905	0.285	0.927	0.9754	0.321	1.679	0.313	1.637	3.173	0.787	0.811	5.535	0.256	1.744
12	0.866	0.266	0.886	0.9776	0.354	1.646	0.346	1.610	3.258	0.778	0.922	5.594	0.283	1.717
13	0.832	0.249	0.850	0.9794	0.382	1.618	0.374	1.585	3.336	0.770	1.025	5.647	0.307	1.693
14	0.802	0.235	0.817	0.9810	0.406	1.594	0.399	1.563	3.407	0.763	1.118	5.696	0.328	1.672
15	0.775	0.223	0.789	0.9823	0.428	1.572	0.421	1.544	3.472	0.756	1.203	5.741	0.347	1.653
16	0.750	0.212	0.763	0.9835	0.448	1.552	0.440	1.526	3.532	0.750	1.282	5.782	0.363	1.637
17	0.728	0.203	0.739	0.9845	0.466	1.534	0.458	1.511	3.588	0.744	1.356	5.820	0.378	1.622
18	0.707	0.194	0.718	0.9854	0.482	1.518	0.475	1.496	3.640	0.739	1.424	5.856	0.391	1.608
19	0.688	0.187	0.698	0.9862	0.497	1.503	0.490	1.483	3.689	0.734	1.487	5.891	0.403	1.597
20	0.671	0.180	0.680	0.9869	0.510	1.490	0.504	1.470	3.735	0.729	1.549	5.921	0.415	1.585

TABLE C The Poisson Distribution $P(c) = \frac{(np_0)^c}{c!} e^{-np_0}$ (Cumulative Values Are in Parentheses)

c np_0	0.1	0.2	0.3	0.4	0.5
0	0.905 (0.905)	0.819 (0.819)	0.741 (0.741)	0.670 (0.670)	0.607 (0.607)
1	0.091 (0.996)	0.164 (0.983)	0.222 (0.963)	0.268 (0.938)	0.303 (0.910)
2	0.004 (1.000)	0.016 (0.999)	0.033 (0.996)	0.054 (0.992)	0.076 (0.986)
3		0.010 (1.000)	0.004 (1.000)	0.007 (0.999)	0.013 (0.999)
4		0.001 (1.000)	0.001 (1.000)		

c np_0	0.6	0.7	0.8	0.9	1.0
0	0.549 (0.549)	0.497 (0.497)	0.449 (0.449)	0.406 (0.406)	0.368 (0.368)
1	0.329 (0.878)	0.349 (0.845)	0.359 (0.808)	0.366 (0.772)	0.368 (0.736)
2	0.099 (0.977)	0.122 (0.967)	0.144 (0.952)	0.166 (0.938)	0.184 (0.920)
3	0.020 (0.997)	0.028 (0.995)	0.039 (0.991)	0.049 (0.987)	0.061 (0.981)
4	0.003 (1.000)	0.005 (1.000)	0.008 (0.999)	0.011 (0.998)	0.016 (0.997)
5			0.001 (1.000)	0.002 (1.000)	0.003 (1.000)

c np_0	1.1	1.2	1.3	1.4	1.5
0	0.333 (0.333)	0.301 (0.301)	0.273 (0.273)	0.247 (0.247)	0.223 (0.223)
1	0.366 (0.699)	0.361 (0.662)	0.354 (0.627)	0.345 (0.592)	0.335 (0.558)
2	0.201 (0.900)	0.217 (0.879)	0.230 (0.857)	0.242 (0.834)	0.251 (0.809)
3	0.074 (0.974)	0.087 (0.966)	0.100 (0.957)	0.113 (0.947)	0.126 (0.935)
4	0.021 (0.995)	0.026 (0.992)	0.032 (0.989)	0.039 (0.986)	0.047 (0.982)
5	0.004 (0.999)	0.007 (0.999)	0.009 (0.998)	0.011 (0.997)	0.014 (0.996)
6	0.001 (1.000)	0.001 (1.000)	0.002 (1.000)	0.003 (1.000)	0.004 (1.000)

c np_0	1.6	1.7	1.8	1.9	2.0
0	0.202 (0.202)	0.183 (0.183)	0.165 (0.165)	0.150 (0.150)	0.135 (0.135)
1	0.323 (0.525)	0.311 (0.494)	0.298 (0.463)	0.284 (0.434)	0.271 (0.406)
2	0.258 (0.783)	0.264 (0.758)	0.268 (0.731)	0.270 (0.704)	0.271 (0.677)
3	0.138 (0.921)	0.149 (0.907)	0.161 (0.892)	0.171 (0.875)	0.180 (0.857)
4	0.055 (0.976)	0.064 (0.971)	0.072 (0.964)	0.081 (0.956)	0.090 (0.947)
5	0.018 (0.994)	0.022 (0.993)	0.026 (0.990)	0.031 (0.987)	0.036 (0.983)
6	0.005 (0.999)	0.006 (0.999)	0.008 (0.998)	0.010 (0.997)	0.012 (0.995)
7	0.001 (1.000)	0.001 (1.000)	0.002 (1.000)	0.003 (1.000)	0.004 (0.999)
8					0.001 (1.000)

c np_0	2.1	2.2	2.3	2.4	2.5
0	0.123 (0.123)	0.111 (0.111)	0.100 (0.100)	0.091 (0.091)	0.082 (0.082)
1	0.257 (0.380)	0.244 (0.355)	0.231 (0.331)	0.218 (0.309)	0.205 (0.287)
2	0.270 (0.650)	0.268 (0.623)	0.265 (0.596)	0.261 (0.570)	0.256 (0.543)
3	0.189 (0.839)	0.197 (0.820)	0.203 (0.799)	0.209 (0.779)	0.214 (0.757)
4	0.099 (0.938)	0.108 (0.928)	0.117 (0.916)	0.125 (0.904)	0.134 (0.891)
5	0.042 (0.980)	0.048 (0.976)	0.054 (0.970)	0.060 (0.964)	0.067 (0.958)
6	0.015 (0.995)	0.017 (0.993)	0.021 (0.991)	0.024 (0.988)	0.028 (0.986)
7	0.004 (0.999)	0.005 (0.998)	0.007 (0.998)	0.008 (0.996)	0.010 (0.996)
8	0.001 (1.000)	0.002 (1.000)	0.002 (1.000)	0.003 (0.999)	0.003 (0.999)
9				0.001 (1.000)	0.001 (1.000)

TABLE C (*Continued*)

c np_0	2.6	2.7	2.8	2.9	3.0
0	0.074 (0.074)	0.067 (0.067)	0.061 (0.061)	0.055 (0.055)	0.050 (0.050)
1	0.193 (0.267)	0.182 (0.249)	0.170 (0.231)	0.160 (0.215)	0.149 (0.199)
2	0.251 (0.518)	0.245 (0.494)	0.238 (0.469)	0.231 (0.446)	0.224 (0.423)
3	0.218 (0.736)	0.221 (0.715)	0.223 (0.692)	0.224 (0.670)	0.224 (0.647)
4	0.141 (0.877)	0.149 (0.864)	0.156 (0.848)	0.162 (0.832)	0.168 (0.815)
5	0.074 (0.951)	0.080 (0.944)	0.087 (0.935)	0.094 (0.926)	0.101 (0.916)
6	0.032 (0.983)	0.036 (0.980)	0.041 (0.976)	0.045 (0.971)	0.050 (0.966)
7	0.012 (0.995)	0.014 (0.994)	0.016 (0.992)	0.019 (0.990)	0.022 (0.988)
8	0.004 (0.999)	0.005 (0.999)	0.006 (0.998)	0.007 (0.997)	0.008 (0.996)
9	0.001 (1.000)	0.001 (1.000)	0.002 (1.000)	0.002 (0.999)	0.003 (0.999)
10				0.001 (1.000)	0.001 (1.000)

c np_0	3.1	3.2	3.3	3.4	3.5
0	0.045 (0.045)	0.041 (0.041)	0.037 (0.037)	0.033 (0.033)	0.030 (0.030)
1	0.140 (0.185)	0.130 (0.171)	0.122 (0.159)	0.113 (0.146)	0.106 (0.136)
2	0.216 (0.401)	0.209 (0.380)	0.201 (0.360)	0.193 (0.339)	0.185 (0.321)
3	0.224 (0.625)	0.223 (0.603)	0.222 (0.582)	0.219 (0.558)	0.216 (0.537)
4	0.173 (0.798)	0.178 (0.781)	0.182 (0.764)	0.186 (0.744)	0.189 (0.726)
5	0.107 (0.905)	0.114 (0.895)	0.120 (0.884)	0.126 (0.870)	0.132 (0.858)
6	0.056 (0.961)	0.061 (0.956)	0.066 (0.950)	0.071 (0.941)	0.077 (0.935)
7	0.025 (0.986)	0.028 (0.984)	0.031 (0.981)	0.035 (0.976)	0.038 (0.973)
8	0.010 (0.996)	0.011 (0.995)	0.012 (0.993)	0.015 (0.991)	0.017 (0.990)
9	0.003 (0.999)	0.004 (0.999)	0.005 (0.998)	0.006 (0.997)	0.007 (0.997)
10	0.001 (1.000)	0.001 (1.000)	0.002 (1.000)	0.002 (0.999)	0.002 (0.999)
11				0.001 (1.000)	0.001 (1.000)

c np_0	3.6	3.7	3.8	3.9	4.0
0	0.027 (0.027)	0.025 (0.025)	0.022 (0.022)	0.020 (0.020)	0.018 (0.018)
1	0.098 (0.125)	0.091 (0.116)	0.085 (0.107)	0.079 (0.099)	0.073 (0.091)
2	0.177 (0.302)	0.169 (0.285)	0.161 (0.268)	0.154 (0.253)	0.147 (0.238)
3	0.213 (0.515)	0.209 (0.494)	0.205 (0.473)	0.200 (0.453)	0.195 (0.433)
4	0.191 (0.706)	0.193 (0.687)	0.194 (0.667)	0.195 (0.648)	0.195 (0.628)
5	0.138 (0.844)	0.143 (0.830)	0.148 (0.815)	0.152 (0.800)	0.157 (0.785)
6	0.083 (0.927)	0.088 (0.918)	0.094 (0.909)	0.099 (0.899)	0.104 (0.889)
7	0.042 (0.969)	0.047 (0.965)	0.051 (0.960)	0.055 (0.954)	0.060 (0.949)
8	0.019 (0.988)	0.022 (0.987)	0.024 (0.984)	0.027 (0.981)	0.030 (0.979)
9	0.008 (0.996)	0.009 (0.996)	0.010 (0.994)	0.012 (0.993)	0.013 (0.992)
10	0.003 (0.999)	0.003 (0.999)	0.004 (0.998)	0.004 (0.997)	0.005 (0.997)
11	0.001 (1.000)	0.001 (1.000)	0.001 (0.999)	0.002 (0.999)	0.002 (0.999)
12			0.001 (1.000)	0.001 (1.000)	0.001 (1.000)

c np_0	4.1	4.2	4.3	4.4	4.5
0	0.017 (0.017)	0.015 (0.015)	0.014 (0.014)	0.012 (0.012)	0.011 (0.011)
1	0.068 (0.085)	0.063 (0.078)	0.058 (0.072)	0.054 (0.066)	0.050 (0.061)
2	0.139 (0.224)	0.132 (0.210)	0.126 (0.198)	0.119 (0.185)	0.113 (0.174)
3	0.190 (0.414)	0.185 (0.395)	0.180 (0.378)	0.174 (0.359)	0.169 (0.343)
4	0.195 (0.609)	0.195 (0.590)	0.193 (0.571)	0.192 (0.551)	0.190 (0.533)

(*Continued*)

TABLE C (*Continued*)

c np_0	4.1	4.2	4.3	4.4	4.5
5	0.160 (0.769)	0.163 (0.753)	0.166 (0.737)	0.169 (0.720)	0.171 (0.704)
6	0.110 (0.879)	0.114 (0.867)	0.119 (0.856)	0.124 (0.844)	0.128 (0.832)
7	0.064 (0.943)	0.069 (0.936)	0.073 (0.929)	0.078 (0.922)	0.082 (0.914)
8	0.033 (0.976)	0.036 (0.972)	0.040 (0.969)	0.043 (0.965)	0.046 (0.960)
9	0.015 (0.991)	0.017 (0.989)	0.019 (0.988)	0.021 (0.986)	0.023 (0.983)
10	0.006 (0.997)	0.007 (0.996)	0.008 (0.996)	0.009 (0.995)	0.011 (0.994)
11	0.002 (0.999)	0.003 (0.999)	0.003 (0.999)	0.004 (0.999)	0.004 (0.998)
12	0.001 (1.000)	0.001 (1.000)	0.001 (1.000)	0.001 (1.000)	0.001 (0.999)
13					0.001 (1.000)

c np_0	4.6	4.7	4.8	4.9	5.0
0	0.010 (0.010)	0.009 (0.009)	0.008 (0.008)	0.008 (0.008)	0.007 (0.007)
1	0.046 (0.056)	0.043 (0.052)	0.039 (0.047)	0.037 (0.045)	0.034 (0.041)
2	0.106 (0.162)	0.101 (0.153)	0.095 (0.142)	0.090 (0.135)	0.084 (0.125)
3	0.163 (0.325)	0.157 (0.310)	0.152 (0.294)	0.146 (0.281)	0.140 (0.265)
4	0.188 (0.513)	0.185 (0.495)	0.182 (0.476)	0.179 (0.460)	0.176 (0.441)
5	0.172 (0.685)	0.174 (0.669)	0.175 (0.651)	0.175 (0.635)	0.176 (0.617)
6	0.132 (0.817)	0.136 (0.805)	0.140 (0.791)	0.143 (0.778)	0.146 (0.763)
7	0.087 (0.904)	0.091 (0.896)	0.096 (0.887)	0.100 (0.878)	0.105 (0.868)
8	0.050 (0.954)	0.054 (0.950)	0.058 (0.945)	0.061 (0.939)	0.065 (0.933)
9	0.026 (0.980)	0.028 (0.978)	0.031 (0.976)	0.034 (0.973)	0.036 (0.969)
10	0.012 (0.992)	0.013 (0.991)	0.015 (0.991)	0.016 (0.989)	0.018 (0.987)
11	0.005 (0.997)	0.006 (0.997)	0.006 (0.997)	0.007 (0.996)	0.008 (0.995)
12	0.002 (0.999)	0.002 (0.999)	0.002 (0.999)	0.003 (0.999)	0.003 (0.998)
13	0.001 (1.000)	0.001 (1.000)	0.001 (1.000)	0.001 (1.000)	0.001 (0.999)
14					0.001 (1.000)

c np_0	6.0	7.0	8.0	9.0	10.0
0	0.002 (0.002)	0.001 (0.001)	0.000 (0.000)	0.000 (0.000)	0.000 (0.000)
1	0.015 (0.017)	0.006 (0.007)	0.003 (0.003)	0.001 (0.001)	0.000 (0.000)
2	0.045 (0.062)	0.022 (0.029)	0.011 (0.014)	0.005 (0.006)	0.002 (0.002)
3	0.089 (0.151)	0.052 (0.081)	0.029 (0.043)	0.015 (0.021)	0.007 (0.009)
4	0.134 (0.285)	0.091 (0.172)	0.057 (0.100)	0.034 (0.055)	0.019 (0.028)
5	0.161 (0.446)	0.128 (0.300)	0.092 (0.192)	0.061 (0.116)	0.038 (0.066)
6	0.161 (0.607)	0.149 (0.449)	0.122 (0.314)	0.091 (0.091)	0.063 (0.129)
7	0.138 (0.745)	0.149 (0.598)	0.140 (0.454)	0.117 (0.324)	0.090 (0.219)
8	0.103 (0.848)	0.131 (0.729)	0.140 (0.594)	0.132 (0.456)	0.113 (0.332)
9	0.069 (0.917)	0.102 (0.831)	0.124 (0.718)	0.132 (0.588)	0.124 (0.457)
10	0.041 (0.958)	0.071 (0.902)	0.099 (0.817)	0.119 (0.707)	0.125 (0.582)
11	0.023 (0.981)	0.045 (0.947)	0.072 (0.889)	0.097 (0.804)	0.114 (0.696)
12	0.011 (0.992)	0.026 (0.973)	0.048 (0.937)	0.073 (0.877)	0.095 (0.791)
13	0.005 (0.997)	0.014 (0.987)	0.030 (0.967)	0.050 (0.927)	0.073 (0.864)
14	0.002 (0.999)	0.007 (0.994)	0.017 (0.984)	0.032 (0.959)	0.052 (0.916)
15	0.001 (1.000)	0.003 (0.997)	0.009 (0.993)	0.019 (0.978)	0.035 (0.951)
16		0.002 (0.999)	0.004 (0.997)	0.011 (0.989)	0.022 (0.973)
17		0.001 (1.000)	0.002 (0.999)	0.006 (0.995)	0.013 (0.986)

TABLE C (*Continued*)

c np_0	6.0	7.0	8.0	9.0	10.0
18			0.001 (1.000)	0.003 (0.998)	0.007 (0.993)
19				0.001 (0.999)	0.004 (0.997)
20				0.001 (1.000)	0.002 (0.999)
21					0.001 (1.000)

c np_0	11.0	12.0	13.0	14.0	15.0
0	0.000 (0.000)	0.000 (0.000)	0.000 (0.000)	0.000 (0.000)	0.000 (0.000)
1	0.000 (0.000)	0.000 (0.000)	0.000 (0.000)	0.000 (0.000)	0.000 (0.000)
2	0.001 (0.001)	0.000 (0.000)	0.000 (0.000)	0.000 (0.000)	0.000 (0.000)
3	0.004 (0.005)	0.002 (0.002)	0.001 (0.001)	0.000 (0.000)	0.000 (0.000)
4	0.010 (0.015)	0.005 (0.007)	0.003 (0.004)	0.001 (0.001)	0.001 (0.001)
5	0.022 (0.037)	0.013 (0.020)	0.007 (0.011)	0.004 (0.005)	0.002 (0.003)
6	0.041 (0.078)	0.025 (0.045)	0.015 (0.026)	0.009 (0.014)	0.005 (0.008)
7	0.065 (0.143)	0.044 (0.089)	0.028 (0.054)	0.017 (0.031)	0.010 (0.018)
8	0.089 (0.232)	0.066 (0.155)	0.046 (0.100)	0.031 (0.062)	0.019 (0.037)
9	0.109 (0.341)	0.087 (0.242)	0.066 (0.166)	0.047 (0.109)	0.032 (0.069)
10	0.119 (0.460)	0.105 (0.347)	0.086 (0.252)	0.066 (0.175)	0.049 (0.118)
11	0.119 (0.579)	0.114 (0.461)	0.101 (0.353)	0.084 (0.259)	0.066 (0.184)
12	0.109 (0.688)	0.114 (0.575)	0.110 (0.463)	0.099 (0.358)	0.083 (0.267)
13	0.093 (0.781)	0.106 (0.681)	0.110 (0.573)	0.106 (0.464)	0.096 (0.363)
14	0.073 (0.854)	0.091 (0.772)	0.102 (0.675)	0.106 (0.570)	0.102 (0.465)
15	0.053 (0.907)	0.072 (0.844)	0.088 (0.763)	0.099 (0.669)	0.102 (0.567)
16	0.037 (0.944)	0.054 (0.898)	0.072 (0.835)	0.087 (0.756)	0.096 (0.663)
17	0.024 (0.968)	0.038 (0.936)	0.055 (0.890)	0.071 (0.827)	0.085 (0.748)
18	0.015 (0.983)	0.026 (0.962)	0.040 (0.930)	0.056 (0.883)	0.071 (0.819)
19	0.008 (0.991)	0.016 (0.978)	0.027 (0.957)	0.041 (0.924)	0.056 (0.875)
20	0.005 (0.996)	0.010 (0.988)	0.018 (0.975)	0.029 (0.953)	0.042 (0.917)
21	0.002 (0.998)	0.006 (0.994)	0.011 (0.986)	0.019 (0.972)	0.030 (0.947)
22	0.001 (0.999)	0.003 (0.997)	0.006 (0.992)	0.012 (0.984)	0.020 (0.967)
23	0.001 (1.000)	0.002 (0.999)	0.004 (0.996)	0.007 (0.991)	0.013 (0.980)
24		0.001 (1.000)	0.002 (0.998)	0.004 (0.995)	0.008 (0.988)
25			0.001 (0.999)	0.003 (0.998)	0.005 (0.993)
26			0.001 (1.000)	0.001 (0.999)	0.003 (0.996)
27				0.001 (1.000)	0.002 (0.998)
28					0.001 (0.999)
29					0.001 (1.000)

TABLE D Random Numbers

63271	59986	71744	51102	15141	80714	58683	93108
88547	09896	95436	79115	08303	01041	20030	63754
55957	57243	83865	09911	19761	66535	40102	26646
46276	87453	44790	67122	45573	84358	21625	16999
55363	07449	34835	15290	76616	67191	12777	21861
69393	92785	49902	58447	42048	30378	87618	26933
13186	29431	88190	04588	38733	81290	89541	70290
17726	28652	56836	78351	47327	18518	92222	55201
36520	64465	05550	30157	82242	29520	69753	72602
81628	36100	39254	56835	37636	02421	98063	89641
84649	48968	75215	75498	49539	74240	03466	49292
63291	11618	12613	75055	43915	26488	41116	64531
70502	53225	03655	05915	37140	57051	48393	91322
06426	24771	59935	49801	11081	66762	94477	02494
20711	55609	29430	70165	45406	78484	31699	52009
41990	70538	77191	25860	55204	73417	83920	69468
72452	36618	76298	26678	89334	33938	95567	29380
37042	40318	57099	10528	09925	89773	41335	96244
53766	52875	15987	46962	67342	77592	57651	95508
90585	58955	53122	16025	84299	53310	67380	84249
32001	96293	37203	64516	51530	37069	40261	61374
62606	64324	46354	72157	67248	20135	49804	09226
10078	28073	85389	50324	14500	15562	64165	06125
91561	46145	24177	15294	10061	98124	75732	08815
13091	98112	53959	79607	52244	63303	10413	63839
73864	83014	72457	26682	03033	61714	88173	90835
66668	25467	48894	51043	02365	91726	09365	63167
84745	41042	29493	01836	09044	51926	43630	63470
48068	26805	94595	47907	13357	38412	33318	26098
54310	96175	97594	88616	42035	38093	36745	56702
14877	33095	10924	58013	61439	21882	42059	24177
78295	23179	02771	43464	59061	71411	05697	67194
67524	02865	39593	54278	04237	92441	26602	63835
58268	57219	68124	73455	83236	08710	04284	55005
97158	28672	50685	01181	24262	19427	52106	34308
04230	16831	69085	30802	65559	09205	71829	06489
94879	56606	30401	02602	57658	70091	54986	41394
71446	15232	66715	26385	91518	70566	02888	79941
32886	05644	79316	09819	00813	88407	17461	73925
62048	33711	25290	21526	02223	75947	66466	06232

TABLE E Commonly Used Conversion Factors

Quantity	Conversion	Multiply By
Length	in. to–m	2.54[a] E − 02
Area	in.^{2} to–m^2	6.451 600 E − 04
Volume	in.^{3} to–m^3	1.638 706 E − 05
	U.S. gallon to–m^3	3.785 412 E − 03
Mass	oz (avoir) to–kg	2.834 952 E − 02
Acceleration	ft/s^2 to–m/s^2	3.048[a] E − 01
Force	poundal to–N	1.382 550 E − 01
Pressure, stress	poundal/ft^2 to–Pa	1.488 164 E + 00
	lb$_f$/in.^{2} to–Pa	6.894 757 E + 03
Energy, work	(ft) (lb$_f$) to–J	1.355 818 E + 00
Power	hp (550 ft) (lb$_f$/s) to–W	7.456 999 E + 02

[a]Relationship is exact and needs no additional decimal points.

TABLE F Critical Values, $t_{\alpha,m}$, of t Distribution

	α							
v	0.25	0.10	0.05	0.025	0.01	0.005	0.001	0.0005
1	1.000	3.078	6.314	12.706	21.821	63.657	318.31	636.62
2	0.816	1.886	2.920	4.303	6.965	9.925	22.326	31.598
3	0.765	1.638	2.353	3.182	4.541	5.841	10.213	12.924
4	0.741	1.533	2.132	2.776	3.747	4.604	7.173	8.610
5	0.727	1.476	2.015	2.571	3.365	4.032	5.893	6.869
6	0.718	1.440	1.943	2.447	3.143	3.707	5.208	5.959
7	0.711	1.415	1.895	2.365	2.998	3.499	4.785	5.408
8	0.706	1.397	1.860	2.306	2.896	3.355	4.501	5.041
9	0.703	1.383	1.833	2.262	2.821	3.250	4.297	4.781
10	0.700	1.372	1.812	2.228	2.764	3.169	4.144	4.587
11	0.697	1.363	1.796	2.201	2.718	3.106	4.025	4.437
12	0.695	1.356	1.782	2.179	2.681	3.055	3.930	4.318
13	0.694	1.350	1.771	2.160	2.650	3.012	3.852	4.221
14	0.692	1.345	1.761	2.145	2.624	2.977	3.787	4.140
15	0.691	1.341	1.753	2.131	2.602	2.947	3.733	4.073
16	0.690	1.337	1.746	2.120	2.583	2.921	3.686	4.015
17	0.689	1.333	1.740	2.110	2.567	2.898	3.646	3.965
18	0.688	1.330	1.734	2.101	2.552	2.878	3.610	3.922
19	0.688	1.328	1.729	2.093	2.539	2.861	3.579	3.883
20	0.687	1.325	1.725	2.086	2.528	2.845	3.552	3.850
21	0.686	1.323	1.721	2.080	2.518	2.831	3.527	3.819
22	0.686	1.321	1.717	2.074	2.508	2.819	3.505	3.792
23	0.685	1.319	1.714	2.069	2.500	2.807	3.485	3.767
24	0.685	1.318	1.711	2.064	2.492	2.797	3.467	3.745
25	0.684	1.316	1.708	2.060	2.485	2.787	3.450	3.725
26	0.684	1.315	1.706	2.056	2.479	2.779	3.435	3.707
27	0.684	1.314	1.703	2.052	2.473	2.771	3.421	3.690

(*Continued*)

TABLE F (*Continued*)

	α							
v	0.25	0.10	0.05	0.025	0.01	0.005	0.001	0.0005
28	0.683	1.313	1.701	2.048	2.467	2.763	3.408	3.674
29	0.683	1.311	1.699	2.045	2.462	2.756	3.396	3.659
30	0.683	1.310	1.697	2.042	2.457	2.750	3.385	3.646
40	0.681	1.303	1.684	2.021	2.423	2.704	3.307	3.551
60	0.679	1.296	1.671	2.000	2.390	2.660	3.232	3.460
∞	0.674	1.282	1.645	1.960	2.326	2.576	3.090	3.291

TABLE G–1 Critical Values, F_{α,ν_1,ν_2}, of F Distribution ($\alpha = 0.1$)

	ν_1 (NUMERATOR)												
ν_2	1	2	3	4	6	8	10	12	15	20	50	100	∞
1	4052	4999	5403	5625	5859	5981	6056	6106	6157	6209	6300	6330	6366
2	98.5	99.0	99.2	99.2	99.3	99.4	99.4	99.4	99.4	99.4	99.5	99.5	99.5
3	34.1	30.8	29.5	28.7	27.9	27.5	27.2	27.1	26.9	26.7	26.4	26.2	26.1
4	21.2	18.0	16.7	16.0	15.2	14.8	14.5	14.4	14.2	14.0	13.7	13.6	13.5
5	16.3	13.3	12.1	11.4	10.7	10.3	10.1	9.89	9.72	9.55	9.24	9.13	9.02
6	13.7	10.9	9.78	9.15	8.47	8.10	7.87	7.72	7.56	7.40	7.09	6.99	6.88
7	12.2	9.55	8.45	7.85	7.19	6.84	6.62	6.47	6.31	6.16	5.86	5.75	5.65
8	11.3	8.65	7.59	7.01	6.37	6.03	5.81	5.67	5.52	5.36	5.07	4.96	4.86
9	10.6	8.02	6.99	6.42	5.80	5.47	5.26	5.11	4.96	4.81	4.52	4.42	4.31
10	10.0	7.56	6.55	5.99	5.39	5.06	4.85	4.71	4.56	4.41	4.12	4.01	3.91
11	9.65	7.21	6.22	5.67	5.07	4.74	4.54	4.40	4.25	4.10	3.81	3.71	3.60
12	9.33	6.93	5.95	5.41	4.82	4.50	4.30	4.16	4.01	3.86	3.57	3.47	3.36
13	9.07	6.70	5.74	5.21	4.62	4.30	4.10	3.96	3.82	3.66	3.38	3.27	3.17
14	8.86	6.51	5.56	5.04	4.46	4.14	3.94	3.80	3.66	3.51	3.22	3.11	3.00
15	8.68	6.36	5.42	4.89	4.32	4.00	3.80	3.67	3.52	3.37	3.08	2.98	2.87
16	8.53	6.23	5.29	4.77	4.20	3.89	3.69	3.55	3.41	3.26	2.97	2.86	2.75
17	8.40	6.11	5.18	4.67	4.10	3.79	3.59	3.46	3.31	3.16	2.87	2.76	2.65
18	8.29	6.01	5.09	4.58	4.01	3.71	3.51	3.37	3.23	3.08	2.78	2.68	2.57
19	8.18	5.93	5.01	4.50	3.94	3.63	3.43	3.30	3.15	3.00	2.71	2.60	2.49
20	8.10	5.85	4.94	4.43	3.87	3.56	3.37	3.23	3.09	2.94	2.64	2.54	2.42
22	7.95	5.72	4.82	4.31	3.76	3.45	3.26	3.12	2.98	2.83	2.53	2.42	2.31
24	7.82	5.61	4.72	4.22	3.67	3.36	3.17	3.03	2.89	2.74	2.44	2.33	2.21
26	7.72	5.53	4.64	4.14	3.59	3.29	3.09	2.96	2.81	2.66	2.36	2.25	2.13
28	7.64	5.45	4.57	4.07	3.53	3.23	3.03	2.90	2.75	2.60	2.30	2.19	2.06
30	7.56	5.39	4.51	4.02	3.47	3.17	2.98	2.84	2.70	2.55	2.25	2.13	2.01
40	7.31	5.18	4.31	3.83	3.29	2.99	2.80	2.66	2.52	2.37	2.06	1.94	1.80
60	7.08	4.98	4.13	3.65	3.12	2.82	2.63	2.50	2.35	2.20	1.88	1.75	1.60
120	6.85	4.79	3.95	3.48	2.96	2.66	2.47	2.34	2.19	2.03	1.70	1.56	1.38
200	6.76	4.71	3.88	3.41	2.89	2.60	2.41	2.27	2.13	1.97	1.63	1.48	1.28
∞	6.63	4.61	3.78	3.32	2.80	2.51	2.32	2.18	2.04	1.88	1.52	1.36	1.00

TABLE G-2 Critical Values, F_{α,ν_1,ν_2}, of F Distribution ($\alpha = 0.05$)

	ν_1 (NUMERATOR)												
ν_2	1	2	3	4	6	8	10	12	15	20	50	100	∞
1	161	200	216	225	234	239	242	244	246	248	252	253	254
2	18.5	19.0	19.2	19.2	19.3	19.4	19.4	19.4	19.4	19.4	19.5	19.5	19.5
3	10.1	9.55	9.28	9.12	8.94	8.85	8.79	8.74	8.70	8.66	8.58	8.55	8.53
4	7.71	6.94	6.59	6.39	6.16	6.04	5.96	5.91	5.86	5.80	5.70	5.66	5.63
5	6.61	5.79	5.41	5.19	4.95	4.82	4.74	4.68	4.62	4.56	4.44	4.41	4.36
6	5.99	5.14	4.76	4.53	4.28	4.15	4.06	4.00	3.94	3.87	3.75	3.71	3.67
7	5.59	4.74	4.35	4.12	3.87	3.73	3.64	3.57	3.51	3.44	3.32	3.27	3.23
8	5.32	4.46	4.07	3.84	3.58	3.44	3.35	3.28	3.22	3.15	3.02	2.97	2.93
9	5.12	4.26	3.86	3.63	3.37	3.23	3.14	3.07	3.01	2.94	2.80	2.76	2.71
10	4.96	4.10	3.71	3.48	3.22	3.07	2.98	2.91	2.85	2.77	2.64	2.59	2.54
11	4.84	3.98	3.59	3.36	3.09	2.95	2.85	2.79	2.72	2.65	2.51	2.46	2.40
12	4.75	3.89	3.49	3.26	3.00	2.85	2.75	2.69	2.62	2.54	2.40	2.35	2.30
13	4.67	3.81	3.41	3.18	2.92	2.77	2.67	2.60	2.53	2.46	2.31	2.26	2.21
14	4.60	3.74	3.34	3.11	2.85	2.70	2.60	2.53	2.46	2.39	2.24	2.19	2.13
15	4.54	3.68	3.29	3.06	2.79	2.64	2.54	2.48	2.40	2.33	2.18	2.12	2.07
16	4.49	3.63	3.24	3.01	2.74	2.59	2.49	2.42	2.35	2.28	2.12	2.07	2.01
17	4.45	3.59	3.20	2.96	2.70	2.55	2.45	2.38	2.31	2.23	2.08	2.02	1.96
18	4.41	3.55	3.16	2.93	2.66	2.51	2.41	2.34	2.27	2.19	2.04	1.98	1.92
19	4.38	3.52	3.13	2.90	2.63	2.48	2.38	2.31	2.23	2.16	2.00	1.94	1.88
20	4.35	3.49	3.10	2.87	2.60	2.45	2.35	2.28	2.20	2.12	1.97	1.91	1.84
22	4.30	3.44	3.05	2.82	2.55	2.40	2.30	2.23	2.15	2.07	1.91	1.85	1.78
24	4.26	3.40	3.01	2.78	2.51	2.36	2.25	2.18	2.11	2.03	1.86	1.80	1.73
26	4.23	3.37	2.98	2.74	2.47	2.32	2.22	2.15	2.07	1.99	1.82	1.76	1.69
28	4.20	3.34	2.95	2.71	2.45	2.29	2.19	2.12	2.04	1.96	1.79	1.73	1.65
30	4.17	3.32	2.92	2.69	2.42	2.27	2.16	2.09	2.01	1.93	1.76	1.70	1.62
40	4.08	3.23	2.84	2.61	2.34	2.18	2.08	2.00	1.92	1.84	1.66	1.59	1.51
60	4.00	3.15	2.76	2.53	2.25	2.10	1.99	1.92	1.84	1.75	1.56	1.48	1.39
120	3.92	3.07	2.68	2.45	2.17	2.02	1.91	1.83	1.75	1.66	1.46	1.37	1.25
200	3.89	3.04	2.65	2.42	2.14	1.98	1.88	1.80	1.72	1.62	1.41	1.32	1.19
∞	3.84	3.00	2.60	2.37	2.10	1.94	1.83	1.75	1.67	1.57	1.35	1.24	1.00

TABLE G–3 Critical Values, F_{α,ν_1,ν_2}, of F Distribution ($\alpha = 0.01$)

	ν_1 (NUMERATOR)												
ν_2	1	2	3	4	6	8	10	12	15	20	50	100	∞
1	39.9	49.5	53.6	55.8	58.2	59.4	60.2	60.7	61.2	61.7	62.7	63.0	63.3
2	8.53	9.00	9.16	9.24	9.33	9.37	9.39	9.41	9.42	9.44	9.47	9.48	9.49
3	5.54	5.46	5.39	5.34	5.28	5.25	5.23	5.22	5.20	5.18	5.15	5.14	5.13
4	4.54	4.32	4.19	4.11	4.01	3.95	3.92	3.90	3.87	3.84	3.80	3.78	3.76
5	4.06	3.78	3.62	3.52	3.40	3.34	3.30	3.27	3.24	3.21	3.15	3.13	3.10
6	3.78	3.46	3.29	3.18	3.05	2.98	2.94	2.90	2.87	2.84	2.77	2.75	2.72
7	3.59	3.26	3.07	2.96	2.83	2.75	2.70	2.67	2.63	2.59	2.52	2.50	2.47
8	3.46	3.11	2.92	2.81	2.67	2.59	2.54	2.50	2.46	2.42	2.35	2.32	2.29
9	3.36	3.01	2.81	2.69	2.55	2.47	2.42	2.38	2.34	2.30	2.22	2.19	2.16
10	3.28	2.92	2.73	2.61	2.46	2.38	2.32	2.28	2.24	2.20	2.12	2.09	2.06
11	3.23	2.86	2.66	2.54	2.39	2.30	2.25	2.21	2.17	2.12	2.04	2.00	1.97
12	3.18	2.81	2.61	2.48	2.33	2.24	2.19	2.15	2.10	2.06	1.97	1.94	1.90
13	3.14	2.76	2.56	2.43	2.28	2.20	2.14	2.10	2.05	2.01	1.92	1.88	1.85
14	3.10	2.73	2.52	2.39	2.24	2.15	2.10	2.05	2.01	1.96	1.87	1.83	1.80
15	3.07	2.70	2.49	2.36	2.21	2.12	2.06	2.02	1.97	1.92	1.83	1.79	1.76
16	3.05	2.67	2.46	2.33	2.18	2.09	2.03	1.99	1.94	1.89	1.79	1.76	1.72
17	3.03	2.64	2.44	2.31	2.15	2.06	2.00	1.96	1.91	1.86	1.76	1.73	1.69
18	3.01	2.62	2.42	2.29	2.13	2.04	1.98	1.93	1.89	1.84	1.74	1.70	1.66
19	2.99	2.61	2.40	2.27	2.11	2.02	1.96	1.91	1.86	1.81	1.71	1.67	1.63
20	2.97	2.59	2.38	2.25	2.09	2.00	1.94	1.89	1.84	1.79	1.69	1.65	1.61
22	2.95	2.56	2.35	2.22	2.06	1.97	1.90	1.86	1.81	1.76	1.65	1.61	1.57
24	2.93	2.54	2.33	2.19	2.04	1.94	1.88	1.83	1.78	1.73	1.62	1.58	1.53
26	2.91	2.52	2.31	2.17	2.01	1.92	1.86	1.81	1.76	1.71	1.59	1.55	1.50
28	2.89	2.50	2.29	2.16	2.00	1.90	1.84	1.79	1.74	1.69	1.37	1.53	1.48
30	2.88	2.49	2.28	2.14	1.98	1.88	1.82	1.77	1.72	1.67	1.55	1.51	1.46
40	2.84	2.44	2.23	2.09	1.93	1.83	1.76	1.71	1.66	1.61	1.48	1.43	1.38
60	2.79	2.39	2.18	2.04	1.87	1.77	1.71	1.66	1.60	1.54	1.41	1.36	1.29
120	2.75	2.35	2.13	1.99	1.82	1.72	1.65	1.60	1.55	1.48	1.34	1.27	1.19
200	2.73	2.33	2.11	1.97	1.80	1.70	1.63	1.57	1.52	1.46	1.31	1.24	1.14
∞	2.71	2.30	2.08	1.94	1.77	1.67	1.60	1.55	1.49	1.42	1.26	1.18	1.00

TABLE H Orthogonal Arrays, Interaction Tables, and Linear Graphs

Orthogonal Array (OA4)

	COLUMN		
TC	1	2	3
1	1	1	1
2	1	2	2
3	2	1	2
4	2	2	1

Orthogonal Array (OA8)

	COLUMN						
TC	1	2	3	4	5	6	7
1	1	1	1	1	1	1	1
2	1	1	1	2	2	2	2
3	1	2	2	1	1	2	2
4	1	2	2	2	2	1	1
5	2	1	2	1	2	1	2
6	2	1	2	2	1	2	1
7	2	2	1	1	2	2	1
8	2	2	1	2	1	1	2

Interaction Table for OA8

	COLUMN						
Column	1	2	3	4	5	6	7
1	(1)	3	2	5	4	7	6
2		(2)	1	6	7	4	5
3			(3)	7	6	5	4
4				(4)	1	2	3
5					(5)	3	2
6						(6)	1
7							(7)

Orthogonal Array (OA9)

	COLUMN			
TC	1	2	3	4
1	1	1	1	1
2	1	2	2	2
3	1	3	3	3
4	2	1	2	3
5	2	2	3	1
6	2	3	1	2
7	3	1	3	2
8	3	2	1	3
9	3	3	2	1

(*Continued*)

TABLE H (*Continued*)

Orthogonal Array (OA12)

TC	COLUMN										
	1	2	3	4	5	6	7	8	9	10	11
1	1	1	1	1	1	1	1	1	1	1	1
2	1	1	1	1	1	2	2	2	2	2	2
3	1	1	2	2	2	1	1	1	2	2	2
4	1	2	1	2	2	1	2	2	1	1	2
5	1	2	2	1	2	2	1	2	1	2	1
6	1	2	2	2	1	2	2	1	2	1	1
7	2	1	2	2	1	1	2	2	1	2	1
8	2	1	2	1	2	2	2	1	1	1	2
9	2	1	1	2	2	2	1	2	2	1	1
10	2	2	2	1	1	1	1	2	2	1	2
11	2	2	1	2	1	2	1	1	1	2	2
12	2	2	1	1	2	1	2	1	2	2	1

Orthogonal Array (OA16)

TC	COLUMN														
	1	2	3	4	5	6	7	8	9	10	11	12	13	14	15
1	1	1	1	1	1	1	1	1	1	1	1	1	1	1	1
2	1	1	1	1	1	1	1	2	2	2	2	2	2	2	2
3	1	1	1	2	2	2	2	1	1	1	1	2	2	2	2
4	1	1	1	2	2	2	2	2	2	2	2	1	1	1	1
5	1	2	2	1	1	2	2	1	1	2	2	1	1	2	2
6	1	2	2	1	1	2	2	2	2	1	1	2	2	1	1
7	1	2	2	2	2	1	1	1	1	2	2	2	2	1	1
8	1	2	2	2	2	1	1	2	2	1	1	1	1	2	2
9	2	1	2	1	2	1	2	1	2	1	2	1	2	1	2
10	2	1	2	1	2	1	2	2	1	2	1	2	1	2	1
11	2	1	2	2	1	2	1	1	2	1	2	2	1	2	1
12	2	1	2	2	1	2	1	2	1	2	1	1	2	1	2
13	2	2	1	1	2	2	1	1	2	2	1	1	2	2	1
14	2	2	1	1	2	2	1	2	1	1	2	2	1	1	2
15	2	2	1	2	1	1	2	1	2	2	1	2	1	1	2
16	2	2	1	2	1	1	2	2	1	1	2	1	2	2	1

TABLE H (*Continued*)

Interaction Table for OA16

Column	COLUMN 1	2	3	4	5	6	7	8	9	10	11	12	13	14	15
1	(1)	3	2	5	4	7	6	9	8	11	10	13	12	13	14
2		(2)	1	6	7	4	5	10	11	8	9	14	15	12	13
3			(3)	7	6	5	4	11	10	9	8	15	14	13	12
4				(4)	1	2	3	12	13	14	15	8	9	10	11
5					(5)	3	2	13	12	15	14	9	8	11	10
6						(6)	1	14	15	12	13	10	11	8	9
7							(7)	15	14	13	12	11	10	9	8
8								(8)	1	2	3	4	5	6	7
9									(9)	3	2	5	4	7	6
10										(10)	1	6	7	4	5
11											(11)	7	6	5	4
12												(12)	1	2	3
13													(13)	3	2
14														(14)	1
15															(15)

Orthogonal Array (OA18)

TC	COLUMN 1	2	3	4	5	6	7	8
1	1	1	1	1	1	1	1	1
2	1	1	2	2	2	2	2	2
3	1	1	3	3	3	3	3	3
4	1	2	1	1	2	2	3	3
5	1	2	2	2	3	3	1	1
6	1	2	3	3	1	1	2	2
7	1	3	1	2	1	3	2	3
8	1	3	2	3	2	1	3	1
9	1	3	3	1	3	2	1	2
10	2	1	1	3	3	2	2	1
11	2	1	2	1	1	3	3	2
12	2	1	3	2	2	1	1	3
13	2	2	1	2	3	1	3	2
14	2	2	2	3	1	2	1	3
15	2	2	3	1	2	3	2	1
16	2	3	1	3	2	3	1	2
17	2	3	2	1	3	1	2	3
18	2	3	3	2	1	2	3	1

(*Continued*)

TABLE H (*Continued*)

Orthogonal Array (OA27)

	COLUMN												
TC	**1**	**2**	**3**	**4**	**5**	**6**	**7**	**8**	**9**	**10**	**11**	**12**	**13**
1	1	1	1	1	1	1	1	1	1	1	1	1	1
2	1	1	1	1	2	2	2	2	2	2	2	2	2
3	1	1	1	1	3	3	3	3	3	3	3	3	3
4	1	2	2	2	1	1	1	2	2	2	3	3	3
5	1	2	2	2	2	2	2	3	3	3	1	1	1
6	1	2	2	2	3	3	3	1	1	1	2	2	2
7	1	3	3	3	1	1	1	3	3	3	2	2	2
8	1	3	3	3	2	2	2	1	1	1	3	3	3
9	1	3	3	3	3	3	3	2	2	2	1	1	1
10	2	1	2	3	1	2	3	1	2	3	1	2	3
11	2	1	2	3	2	3	1	2	3	1	2	3	1
12	2	1	2	3	3	1	2	3	1	2	3	1	2
13	2	2	3	1	1	2	3	2	3	1	3	1	2
14	2	2	3	1	2	3	1	3	1	2	1	2	3
15	2	2	3	1	3	1	2	1	2	3	2	3	1
16	2	3	1	2	1	2	3	3	1	2	2	3	1
17	2	3	1	2	2	3	1	1	2	3	3	1	2
18	2	3	1	2	3	1	2	2	3	1	1	2	3
19	3	1	3	2	1	3	2	1	3	2	1	3	2
20	3	1	3	2	2	1	3	2	1	3	2	1	3
21	3	1	3	2	3	2	1	3	2	1	3	2	1
22	3	2	1	3	1	3	2	2	1	3	3	2	1
23	3	2	1	3	2	1	3	3	2	1	1	3	2
24	3	2	1	3	3	2	1	1	3	2	2	1	3
25	3	3	2	1	1	3	2	3	2	1	2	1	3
26	3	3	2	1	2	1	3	1	3	2	3	2	1
27	3	3	2	1	3	2	1	2	1	3	1	3	2

TABLE H (*Continued*)

Interaction Table for OA27

	Column											
Column	**2**	**3**	**4**	**5**	**6**	**7**	**8**	**9**	**10**	**11**	**12**	**13**
1	3	2	2	6	5	5	9	8	8	12	11	11
	4	4	3	7	7	6	10	10	9	13	13	12
2		1	1	8	9	10	5	6	7	5	6	7
		4	3	11	12	13	11	12	13	8	9	10
3			1	9	10	8	7	8	6	6	7	5
			2	13	11	12	12	13	11	10	8	9
4				10	8	9	6	7	5	7	5	6
				12	13	11	13	11	12	9	10	8
5					1	1	2	3	4	2	4	3
					7	6	11	13	12	8	10	9
6						1	4	2	3	3	2	4
						5	3	12	11	10	9	8
7							3	4	2	4	3	2
							12	11	13	9	8	10
8								1	1	2	3	4
								10	9	5	7	6
9									1	4	2	3
									8	7	6	5
10										3	4	2
										6	7	7
11											1	1
											13	12
12												1
												11

SELECTED BIBLIOGRAPHY

ALUKAL, GEORGE AND ANTHONY MANOS, *Lean Kaizen: A Simplified Approach to Process Improvement,* Milwaukee, WI: ASQ Press, 2006.

ANSI/ASQ B1–B3—1996, *Quality Control Chart Methodologies.* Milwaukee, WI: American Society for Quality, 1996.

ANSI/ASQ SI—1996, *An Attribute Skip-Lot Sampling Program.* Milwaukee, WI: American Society for Quality, 1996.

ANSI/ASQ S2—1995, *Introduction to Attribute Sampling.* Milwaukee, WI: American Society for Quality, 1995.

ANSI/ASQ Standard Q3—1988, *Sampling Procedures and Tables for Inspection by Isolated Lots by Attributes.* Milwaukee, WI: American Society for Quality, 1988.

ANSI/ASQ Z1.4—2003, *Sampling Procedures and Tables for Inspection by Attributes.* Milwaukee, WI: American Society for Quality, 2008.

ANSI/ASQ Z1.9—2003, *Sampling Procedures and Tables for Inspection by Variables for Percent Nonconforming.* Milwaukee, WI: American Society for Quality, 2008.

ANSI/ISO/ASQ A3534-1—2003, *Statistics—Vocabulary and Symbols—Probability and General Statistical Terms.* Milwaukee, WI: American Society for Quality, 2003.

ANSI/ISO/ASQ A3534-2—2003, *Statistics—Vocabulary and Symbols—Statistical Quality.* Milwaukee, WI: American Society for Quality, 2003.

ASQ/AIAG TASK FORCE, *Fundamental Statistical Process Control.* Troy, MI: Automobile Industry Action Group, 1991.

ASQ STATISTICS DIVISION, *Glossary and Tables for Statistical Quality Control.* Milwaukee, WI: American Society for Quality, 1983.

BESTERFIELD, DALE, CAROL BESTERFIELD–MICHNA, GLEN BESTERFIELD, AND MARY BESTERFIELD-SACRE, *Total Quality Management,* 3rd ed. Upper Saddle River, N.J.: Prentice Hall, 2003.

BOSSERT, JAMES L., *Quality Function Deployment: A Practitioner's Approach.* Milwaukee, WI: ASQ Quality Press, 1991.

BRASSARD, MICHAEL, *The Memory Jogger Plus.* Methuen, MA: GOAL/QPC, 1989.

BRASSARD, MICHAEL, *The Memory Jogger Plus +. Featuring the Seven Management and Planning Tools.* Methuen, MA: GOAL/QPC, 1996.

CAMP, ROBERT C., *Benchmarking: The Search for Industry Best Practices That Lead to Superior Practice.* Milwaukee, WI: ASQ Quality Press, 1989.

DENNIS, PASCAL, *Lean Production Simplified, 2nd Edition.* New York: Productivity Press, 2007.

DUNCAN, ACHESON J., *Quality Control and Industrial Statistics,* 5th ed. Homewood, IL: Irwin, 1986.

FELLERS, GARY, *SPC for Practitioners: Special Cases and Continuous Processes.* Milwaukee, WI: ASQ Quality Press, 1991.

GEORGE, MICHAEL L., MARK PRICE, JOHN MAXEY, AND DAVID ROWLANDS, *The Lean Six Sigma Pocket Toolbook.* New York: McGraw-Hill Companies, 2010.

GROSH, DORIS L., *A Primer of Reliability Theory.* New York: John Wiley & Sons, 1989.

HENLEY, ERNEST J., AND HIROMITSU KUMAMOTO, *Reliability Engineering and Risk Assessment.* Englewood Cliffs, NJ: Prentice Hall, 1981.

HICKS, CHARLES R., *Fundamental Concepts in the Design of Experiments.* New York: Holt, Rinehart and Winston, 1973.

JURAN, JOSEPH M. (ed.), *Quality Control Handbook,* 4th ed. New York: McGraw-Hill Book Company, 1988.

JURAN, JOSEPH M., AND FRANK M. GRYNA, JR., *Quality Planning and Analysis,* 2nd ed. New York: McGraw-Hill Book Company, 1980.

MONTGOMERY, DOUGLAS C., *Introduction to Statistical Quality Control,* 5th ed. Indianapolis, IN: Wiley Publishing, 2004.

PEACE, STUART GLEN, *Taguchi Methods: A Hands-On Approach.* New York: Addison-Wesley Publishing Company, Inc., 1992.

PEACH, ROBERT W. (ed.), *The ISO 9000 Handbook.* Fairfax, VA: CEEM Information Services, 1992.

PYZDEK, THOMAS AND PAUL KELLER, *The Six Sigma Handbook, 3rd Edition,* New York: McGraw-Hill Professionals, 2009.

SHAPIRO, SAMUEL S., *The ASQ Basic References in Quality Control: Statistical Techniques,* Edward J. Dudewicz, Ph.D., Editor, Volume 3: *How to Test Normality and Other Distributional Assumptions.* Milwaukee, WI: American Society for Quality, 1980.

TAGUCHI, G., *Introduction to Quality Engineering.* Tokyo: Asian Productivity Organization, 1986.

WHEELER, DONALD J., *Short Run SPC.* Knoxville, TN: SPC Press, 1991.

WHEELER, DONALD J., *Understanding Industrial Experimentation.* Knoxville, TN: Statistical Process Controls, Inc., 1988

GLOSSARY

Acceptance Quality Limit The worst process average that can be considered satisfactory for the purpose of acceptance sampling.

Acceptance Sampling A system by which a lot is accepted or rejected based on the results of inspecting samples in that lot.

Activity Network Diagram A group of diagrams that facilitates the efficient scheduling of a project.

Affinity Diagram A diagram that allows the team to creatively generate a large number of issues/ideas and group them logically.

Analysis of Variance A statistical technique to evaluate the relationship among two or more factors and their interactions.

Assignable Cause A cause of variation that is large in magnitude and easily identified; also called special cause.

Attribute A quality characteristic classified as either conforming or not conforming to specifications.

Availability A time-related factor that measures the ability of a product or service to perform its designated function.

Average The sum of all observations divided by the total number of observations.

Average Outgoing Quality (AOQ) Curve A curve that shows the average quality level for lots leaving the acceptance sampling system for different percent nonconforming values.

Average Run Length The average number of points plotted on a control chart before one is out of control by chance.

Average Sample Number (ASN) Curve A curve that shows the average amount inspected per lot by the consumer for different percent nonconforming values.

Average Total Inspection (ATI) Curve A curve that shows the amount inspected by both the consumer and the producer for different percent nonconforming values.

Cause-and-Effect Diagram A picture composed of lines and symbols designed to represent a meaningful relationship between an effect and its causes.

Cell A grouping of observed values within specified upper and lower boundaries.

Chance Cause A cause of variation that is small in magnitude and difficult to identify; also called random or common cause.

Check Sheets A device used to carefully and accurately record data.

Combination A counting technique that requires the unordered arrangement of a set of objects.

Consumer's Risk The probability that an unacceptable lot will be accepted.

Control Chart A graphical record of the variation in quality of a particular characteristic or characteristics during a specified time period.

Control Limits The limits on a control chart used to evaluate the variations in quality from subgroup to subgroup—not to be confused with specification limits.

DMAIC Acronym for the define, measure, analyze, improve, and control Six Sigma improvement methodology.

Factor A variable that is changed and results observed.

Failure Rate The probability that a unit of product will fail in a specific unit of time or cycles.

Five S's The organization of the workplace using the concepts of sort, straighten, shine, standardize and sustain.

Forced Field Analysis A technique to identify the forces and factors that may influence the problem or goal.

Frequency Distribution The arrangement of data to show the repetition of values in a category.

Histogram A graphical display, in rectangle form, of a frequency distribution.

Hypothesis A statistical decision-making process to make inferences about the population from a sample.

Interaction The relationship of two or more factors that may affect the intended response.

Interrelationship Diagram A diagram that clarifies the interrelationship of many factors of a complex situation.

Interval Estimate A range to convey the true accuracy of a point estimate.

Kaizen The process of continuous improvement in small increments without expensive equipment.

Kurtosis A value used to describe the peakedness of a distribution.

Level A value that is assigned to change a factor.

Limiting Quality (LQ) The percent nonconforming in a lot or batch that, for acceptance sampling purposes, the consumer wishes the probability of acceptance to be low.

Loss function A value that represents the loss to society from the time it was shipped.

Maintainability The ease with which preventative and corrective maintenance on a product or service can be achieved.

Matrix A diagram that provides for the identity, analysis, and evaluation of the interrelationship among variables.

Mean The average of a population.

Median The value that divides a series of ordered observations so that the number of items above it is equal to the number of items below it.

Mode The value that occurs with the greatest frequency in a set of numbers.

Multi-Vari Chart A graphical record of variation due to within parts, between parts, and time-to-time variation.

Nominal Group Technique A technique that provides issues/ideas from everyone on the team and for effective decisions.

Nonconforming Unit A product or service that contains at least one nonconformity.

Nonconformity The departure of a quality characteristic from its intended level with a severity sufficient to cause a product or service not to meet specifications.

Operating Characteristic (OC) Curve A curve that shows the probability that a lot with a certain percent nonconforming will be accepted.

Orthogonal Design An experiment that has a balanced array, which produces independent results.

Pareto Diagram A method used to identify and communicate the vital few causes and the useful many causes of a situation.

Percent Tolerance Percent Chart A chart that evaluates the percent deviation from target for precontrol data.

Permutation A counting technique that requires the ordered arrangement of a set of objects.

Point Estimate A single point or value, which is an indication of the true value.

Population The total set of observations being considered in a statistical procedure.

Precontrol A technique to compare samples of two with specifications.

Prioritization Matrices A technique that prioritizes issues, tasks, characteristics, and so forth, based on weighted criteria.

Probability The mathematical calculation of the likelihood that an event will occur.

Process Capability The spread of the process. It is equal to six standard deviations when the process is in a state of statistical control.

Process Decision Program Chart A chart that avoids surprises and identifies possible countermeasures.

Process Flow Diagram A diagram that shows the movement of a product or service as it travels through various processing stations.

Producer's Risk The probability that an acceptable lot is not accepted.

Quality Meeting or exceeding customer's expectations; customer satisfaction.

Range The difference between the largest and smallest observed values.

Reliability The probability that a product will perform its intended function for a prescribed life under certain stated conditions.

Repetition Multiple results of a treatment condition.

Replication A repeat run of a treatment condition with a new setup.

Sample A small portion of a population used to represent the entire population.

Signal-to-Noise A ratio of the signal or target to the noise or deviation.

Six Sigma A term used to indicate both a statistical term and an improvement methodology.

Skewness The departure of data from symmetrical.

Specification Limits The limits that define the boundaries of an acceptable product or service.

Standard Deviation A measure of the dispersion about the mean of a population or average of a sample.

Target The intended value of a quality characteristic; also called the nominal value.

Tolerance The permissible variation in the size of a quality characteristic.

Treatment Condition A set of factors and levels that are assigned for an experimental test.

Tree Diagram A tool used to reduce a broad objective into increasing levels of detail.

Value Stream Map A graphical description of the sequence and movement of the activities associated with value added to the process.

Variable A quality characteristic that is measurable, such as weight, length, and so forth.

Waste Any effort that does not add value to the product or service for the customer.

ANSWERS TO SELECTED EXERCISES

CHAPTER 1

1. 27.2 ppm

CHAPTER 4

1. 30.1%, 19.0%, 17.4%, 12.7%, 8.8%, 3.5%, 3.0%, 5.5%
3. 30.1%, 28.1%, 8.1%, 3.6%, 3.3%, 3.2%, 23.6%
5. 30.9%, 23.1%, 12.1%, 11.3%, 6.7%, 5.8%, 2.7%
7. 39.1%, 21.7%, 13.1%, 10.9%, 8.7%, 4.3%, 2.2%

CHAPTER 5

1. (a) 0.86, (b) 0.63, (c) 0.15, (d) 0.48
3. (a) 0.0006, (b) 0.001, (c) 0.002, (d) 0.3
5. (a) 66.4, (b) 379.1, (c) 5, (d) 4.652, (e) 6.2×10^2
7. Frequencies starting at 5.94 are 1, 2, 4, 8, 16, 24, 20, 17, 13, 3, 1, 1
9. Frequencies starting at 0.3 are 3, 15, 34, 29, 30, 22, 15, 2
11. (a) Relative frequencies starting at 5.94 (in %) are 0.9, 1.8, 3.6, 7.3, 14.5, 21.8, 18.2, 15.4, 11.8, 2.7, 0.9, 0.9
    (b) Cumulative frequencies starting at 5.945 are 1, 3, 7, 15, 31, 55, 75, 92, 105, 108, 109, 110
    (c) Relative cumulative frequencies starting at 5.945 (in %) are 0.9, 2.7, 6.4, 13.6, 28.2, 50.0, 68.2, 83.6, 95.4, 98.2, 99.1, 100.0
13. (a) Relative frequencies starting at 0.3 are 0.020, 0.100, 0.227, 0.193, 0.200, 0.147, 0.100, 0.013
    (b) Cumulative frequencies starting at 0.3 are 3, 18, 52, 81, 111, 133, 148, 150
    (c) Relative cumulative frequencies starting at 0.3 are 0.020, 0.120, 0.347, 0.540, 0.740, 0.888, 0.987, 1.000
17. 116
19. 95
21. 3264
23. (a) 15; (b) 35.5
25. (a) 55, (b) none, (c) 14, 17
27. (a) 11, (b) 6, (c) 14, (d) 0.11
29. 0.004
31. (a) 0.8 (b) 20
35. (b) Frequencies beginning at 0.5 are 1, 17, 29, 39, 51, 69, 85, 88
39. (a) Relative frequencies beginning at 0.5 (in %) are 1.1, 18.2, 13.6, 11.4, 13.6, 20.5, 18.2, 3.4
    (b) Cumulative relative frequencies beginning at 0.5 (in %) are 1.1, 19.3, 33.0, 44.3, 58.0, 78.4, 96.6, 100.0
41. (b) −0.14, 3.11
43. Process is not capable—5 out of 65 above specification and 6 out of 65 below specification.
45. (a) 0.0268, (b) 0.0099, (c) 0.9914 (based on rounded Z values)
47. 0.606
49. (a) Not normal
    (b) Normal
53. −0.92, $y = 1.624 + (-0.43)\,x$, 0.46
55. 0.89

CHAPTER 6

1. $\overline{\overline{X}} = 72$; CLs $= 0.83, 0.61$ $\overline{R} = 0.148$; CLs $= 0.34, 0$
3. $\overline{X}_0 = 482$; CLs $= 500, 464$; $R_0 = 25$; CLs $= 57, 0$
5. $\overline{X}_0 = 2.08$; CLs $= 2.42, 1.74$; $R_0 = 0.47$; CLs $= 1.08, 0$
7. $\overline{X}_0 = 81.9$; CLs $= 82.8, 81.0$; $s_0 = 0.7$; CLs $= 1.4, 0.0$
11. 0.47% scrap, 2.27% rework, $\overline{X}_0 = 305.32$ mm, 6.43% rework
13. 0.13
15. $6\sigma = 160$
17. (a) $6\sigma = 0.80$, (b) 1.38
19. 0.82, change specifications or reduce σ
21. $C_{pk} = 0.82; 0.41; 0; -0.41$
23. $\overline{\overline{X}} = 4.56$; CLs $= 4.76, 4.36$; $\overline{R} = 0.20$; CLs $= 0.52, 0$
25. $\text{Md}_{\text{Md}} = 6.3$; CLs $= 7.9, 4.7$; $R_{\text{Md}} = 1.25$; CLs $= 3.4, 0$
27. $\overline{X} = 7.59$; CLs $= 8.47, 6.71$; MR $= 0.33$; CLs $= 1.08, 0$
29. $\overline{X}_0 = 20.40$; RLs $= 20.46, 20.34$
31. Histogram is symmetrical, while run chart slopes downward.

CHAPTER 7

1. 200, $r = 4$
3. At 1400 h, time-to-time variation occurred and within-piece variation increased at 2100 h.
5. $\overline{X}_0 = 25.0$; CLs $= 25.15, 24.85$; $R_0 = 0.11$; CLs $= 47, 0$
7. Yes, ratio is 1.17
9. $\overline{Z}_0 = 0$; $CLs = +1.023, -1.023$; $W_0 = 1.00$; CLs $= 2.574, 0$; $\overline{Z}$ plotted points $= -0.2, 1.6, 0.4$; W plotted points $= 0.8, 0.6, 1.2$
11. −3.00, 4.67
13. PC $= 31.5, 32.5$

15. 73.5%, 24.5%

17. (a) −10, 80, (b) −80, 0, (c) −10, 64, (d) −20, −30

CHAPTER 8

1. 1.000, 0
3. 0.833
5. 0.50, 0.81
7. 0.40
9. 0.57
11. 0.018
13. 0.989
15. 520
17. 3.13×10^{15}
19. 161,700
21. 25,827,165
23. $1.50696145 \times 10^{-16}$
25. $C_r^n = C_{n-r}^n$
27. If $n = r$, then $C_r^n = 1$
29. 0.255, 0.509, 0.218, 0.018, P(4) is impossible.
31. 0.087, 0.997
33. 0.0317
35. 0.246
37. 0.075
39. 0.475
41. 0.435
43. 0.084

CHAPTER 9

1. (a) Plotted points = 0.010, 0.018, 0.023, 0.015, 0.025
   (b) $\bar{p} = 0.0297$; CLs = 0.055, 0.004
   (c) $p_0 = 0.0242$; CLs = 0.047, 0.001
3. $100p_0 = 1.54$; CLs = 0.0367, 0; $q_0 = 0.9846$; CLs = 1.0000, 0.9633; $100q_0 = 98.46$; CLs = 100.00, 96.33
5. CLs = 0.188, 0; in control
7. $p_0 = 0.011$
9. $p_0 = 0.144$, CLs = 0.227; 0.061; Nov. 15
11. (a) $100p_0 = 1.54$; CLs = 3.67, 0
    (b) $100p_0 = 2.62$; CLs = 3.76, 1.48
13. $np_0 = 45.85$; CLs = 66, 26; np
15. (a) $q_0 = 0.9846$; CLs = 1.00, 0.9633
    $100q_0 = 98.46$; CLs = 100; 96.33
    $nq_0 = 295$; CLs = 300, 289
    (b) $q_0 = 0.9738$; CLs = 0.9850, 0.9624
    $100q_0 = 97.38$; CLs = 98.50, 96.24
    $nq_0 = 1704$; CLs = 1724, 1684
17. $n = 10$
19. $c_0 = 6.86$; CLs = 14.72, 0
21. Process in control; $\bar{c} = 10.5$; CLs = 20.2, 0.78
23. $u_0 = 3.34$; CLs = 5.08, 161
25. $u_0 = 1.092$; CLs = 1.554, 0.630

CHAPTER 10

1. (p, P_a) pairs are (0.01, 0.974), (0.02, 0.820), (0.04, 0.359), (0.05, 0.208), (0.06, 0.109), (0.08, 0.025), (0.10, 0.005)
3. $(P_a)_{\mathrm{I}} = P$ (2 or less)
   $(P_a)_{\mathrm{II}} = P(3)_{\mathrm{I}}\, P(3 \text{ or less})_{\mathrm{II}} + P(4)_{\mathrm{I}}\, P(2 \text{ or less})_{\mathrm{II}} + P(5)_{\mathrm{I}}$ $P(1 \text{ or less})_{\mathrm{II}}\, (P_a)_{\text{both}} = (P_a)_{\mathrm{I}} + (P_a)_{\mathrm{II}}$
5. (100p, AOQ) pairs are (1, 0.974), (2, 1.640), (4, 1.436), (5, 1.040), (6, 0.654), (8, 0.200); AOQL ≅ 1.7%
7. AQL = 0.025%; AOQL = 0.19%
9. (p, ASN) pairs are (0, 125), (0.01, 140), (0.02, 169), (0.03, 174), (0.04, 165), (0.05, 150), (0.06, 139)
11. (p, ATI) pairs are (0, 80), (0.00125, 120), (0.01, 311), (0.02, 415), (0.03, 462), (0.04, 483)
13. (100p, AOQ) pairs are (0.5, 0.493), (1.0, 0.848), (1.5, 0.885), (2.0, 0.694), (2.5, 0.430)
15. 3,55; 6, 155; 12, 407
17. 2, 82; 6, 162; 14, 310
19. 1, 332; 3, 502; 5, 655
21. 3, 195
23. 4, 266
25. 5, 175
27. 0.69

CHAPTER 11

1. 0.78
3. 0.980
5. 0.87
7. 0.99920
9. 6.5×10^{-3}
11. 0.0002
13. 0.0027; 370 h
15. Plotted points are (35, 0.0051), (105, 0.0016), (175, 0.0005), (245, 0.0004), (315, 0.0003), (385, 0.0004), (455, 0.0007), (525, 0.0010), (595, 0.0011), (665, 0.0015)
17. 0.527; 0.368; 0.278
19. 0.257
21. Plotted points are (1300, 0.993), (1000, 0.972), (500, 0.576), (400, 0.332), (300, 0.093), (250, 0.027), (750, 0.891), (600, 0.750)
23. $n = 30, r = 6, T = 70$ h
25. $n = 9, r = 3, T = 82$ h
27. $n = 60, r = 30, T = 396$ cycles

CHAPTER 13

1. Innocent until proven guilty. Type I error would occur 5% of the time when a guilty verdict would occur, but the party was actually innocent. Type II error would occur 10% of the time when an innocent verdict would occur, but the party was actually guilty.
3. t = 1.578, α for 0.05 = 1.796
5. t = 2.5389, α for 0.05 = 1.753
7. F = 35.127, Ref. Dist. = 1.012, Difference A1-A3 & A2-A3
9. Nickel shows most improvement. Speed has a negative effect.

11. SAT score is significant with an F value of 11.025.
13. 16 treatment conditions are needed.
15. F = 9.966 for factor B, F = 47.207 for factor C.

CHAPTER 14

1. (a) 9500, (c) $23.75
3. $9.45
5. $28.53
7. 7, OA8
9. 19, OA27
11. 1-A, 2-B, 3-AB, 4-C, 5-AC, 6-BC; Other solutions are possible
13. (a) 28.38 dB, (b) 7.97 dB (c) Yes
15. Average is better, but standard deviation is greater. S/N is not much better.
17. B1 and C1 are strong effects for S/N. A has too great a range—reduce it.
19. Strong effects-A2, C3, D1; Predicted S/N confirm run-22.1dB, which gives a difference of 2.4dB (75% improvement) as compared to $\overline{T}$ (S/N average) of 19.7dB.
21. C-32.1%, D-18.3%, E-17.5%, F-23.6%, Error-8.5%
23. Upgrade Factors A, B, D; Save $1.55 per unit

INDEX

W

Z